HOW
THE MIND
WORKS

OTHER BOOKS BY STEVEN PINKER

Language, Learning and Language Development
Visual Cognition (ed.)
Connections and Symbols (ed., with Jacques Mehler)
Learnability and Cognition
Lexical and Conceptual Semantics (ed., with Beth Levin)
The Language Instinct

HOW THE MIND WORKS

Steven Pinker

First published in the USA by W W Norton, 1997
Published in Great Britain by Allen Lane The Penguin Press 1998
1 3 5 7 9 10 8 6 4 2

This edition published by The Softback Preview 1998

Set in Fairfield Light with display setting in Castellar
Printed in Great Britain by Clays Ltd, St Ives plc

A CIP catalogue record for this book is available from the British Library

ISBN 0–713-991305

FOR ILAVENIL

CONTENTS

PREFACE

Any book called *How the Mind Works* had better begin on a note of humility, and I will begin with two.

First, we don't understand how the mind works—not nearly as well as we understand how the body works, and certainly not well enough to design utopia or to cure unhappiness. Then why the audacious title? The linguist Noam Chomsky once suggested that our ignorance can be divided into *problems* and *mysteries*. When we face a problem, we may not know its solution, but we have insight, increasing knowledge, and an inkling of what we are looking for. When we face a mystery, however, we can only stare in wonder and bewilderment, not knowing what an explanation would even look like. I wrote this book because dozens of mysteries of the mind, from mental images to romantic love, have recently been upgraded to problems (though there are still some mysteries, too!). Every idea in the book may turn out to be wrong, but that would be progress, because our old ideas were too vapid to be wrong.

Second, *I* have not discovered what we do know about how the mind works. Few of the ideas in the pages to follow are mine. I have selected, from many disciplines, theories that strike me as offering a special insight into our thoughts and feelings, that fit the facts and predict new ones, and that are consistent in their content and in their style of explanation. My goal was to weave the ideas into a cohesive picture using two even bigger ideas that are not mine: the computational theory of mind and the theory of the natural selection of replicators.

The opening chapter presents the big picture: that the mind is a system of organs of computation designed by natural selection to solve the problems faced by our evolutionary ancestors in their foraging way of life. Each of the two big ideas—computation and evolution—then gets a chapter. I dissect the major faculties of the mind in chapters on perception, reasoning, emotion, and social relations (family, lovers, rivals, friends, acquaintances, allies, enemies). A final chapter discusses our higher callings: art, music, literature, humor, religion, and philosophy. There is no chapter on language; my previous book *The Language Instinct* covers the topic in a complementary way.

This book is intended for anyone who is curious about how the mind works. I didn't write it only for professors and students, but I also didn't write it only to "popularize science." I am hoping that scholars and general readers both might profit from a bird's-eye view of the mind and how it enters into human affairs. At this high altitude there is little difference between a specialist and a thoughtful layperson because nowadays we specialists cannot be more than laypeople in most of our own disciplines, let alone neighboring ones. I have not given comprehensive literature reviews or an airing of all sides to every debate, because they would have made the book unreadable, indeed, unliftable. My conclusions come from assessments of the convergence of evidence from different fields and methods, and I have provided detailed citations so readers can follow them up.

I have intellectual debts to many teachers, students, and colleagues, but most of all to John Tooby and Leda Cosmides. They forged the synthesis between evolution and psychology that made this book possible, and thought up many of the theories I present (and many of the better jokes). By inviting me to spend a year as a Fellow of the Center for Evolutionary Psychology at the University of California, Santa Barbara, they provided an ideal environment for thinking and writing and immeasurable friendship and advice.

I am deeply grateful to Michael Gazzaniga, Marc Hauser, David Kemmerer, Gary Marcus, John Tooby, and Margo Wilson for their reading of the entire manuscript and their invaluable criticism and encouragement. Other colleagues generously commented on chapters in their areas of expertise: Edward Adelson, Barton Anderson, Simon Baron-Cohen, Ned Block, Paul Bloom, David Brainard, David Buss, John Constable, Leda Cosmides, Helena Cronin, Dan Dennett, David Epstein, Alan Fridlund, Gerd Gigerenzer, Judith Harris, Richard Held, Ray Jackendoff, Alex Kacelnik, Stephen Kosslyn, Jack Loomis, Charles Oman, Bernard Sherman, Paul Smolensky, Elizabeth Spelke, Frank Sulloway, Donald Symons, and Michael Tarr. Many others answered queries and offered profitable suggestions, including Robert Boyd, Donald Brown, Napoleon Chagnon, Martin Daly, Richard Dawkins, Robert Hadley, James Hillenbrand, Don Hoffman, Kelly Olguin Jaakola, Timothy Ketelaar, Robert Kurzban, Dan Montello, Alex Pentland, Roslyn Pinker, Robert Provine, Whitman Richards, Daniel Schacter, Devendra Singh, Pawan Sinha, Christopher Tyler, Jeremy Wolfe, and Robert Wright.

This book is a product of the stimulating environments at two institutions, the Massachusetts Institute of Technology and the University of California, Santa Barbara. Special thanks go to Emilio Bizzi of the Department of Brain and Cognitive Sciences at MIT for enabling me to take a sabbatical leave, and to Loy Lytle and Aaron Ettenberg of the Department of Psychology and to Patricia Clancy and Marianne Mithun of the Department of Linguistics at UCSB for inviting me to be a Visiting Scholar in their departments.

Patricia Claffey of MIT's Teuber Library knows everything, or at least knows where to find it, which is just as good. I am grateful for her indefatigable efforts to track down the obscurest material with swiftness and good humor. My secretary, the well-named Eleanor Bonsaint, offered professional, cheerful help in countless matters. Thanks go also to Marianne Teuber and to Sabrina Detmar and Jennifer Riddell of MIT's List Visual Arts Center for advice on the jacket art.

My editors, Drake McFeely (Norton), Howard Boyer (now at the University of California Press), Stefan McGrath (Penguin), and Ravi Mirchandani (now at Orion), offered fine advice and care throughout. I am also grateful to my agents, John Brockman and Katinka Matson, for their efforts on my behalf and their dedication to science writing. Special appreciation goes to Katya Rice, who has now worked with me on four books over fourteen years. Her analytical eye and masterly touch have improved the books and have taught me much about clarity and style.

My heartfelt gratitude goes to my family for their encouragement and suggestions: to Harry, Roslyn, Robert, and Susan Pinker, Martin, Eva, Carl, and Eric Boodman, Saroja Subbiah, and Stan Adams. Thanks, too, to Windsor, Wilfred, and Fiona.

Greatest thanks of all go to my wife, Ilavenil Subbiah, who designed the figures, provided invaluable comments on the manuscript, offered constant advice, support, and kindness, and shared in the adventure. This book is dedicated to her, with love and gratitude.

My research on mind and language has been supported by the National Institutes of Health (grant HD 18381), the National Science Foundation (grants 82-09540, 85-18774, and 91-09766), and the McDonnell-Pew Center for Cognitive Neuroscience at MIT.

HOW
THE MIND
WORKS

1

STANDARD EQUIPMENT

Why are there so many robots in fiction, but none in real life? I would pay a lot for a robot that could put away the dishes or run simple errands. But I will not have the opportunity in this century, and probably not in the next one either. There are, of course, robots that weld or spray-paint on assembly lines and that roll through laboratory hallways; my question is about the machines that walk, talk, see, and think, often better than their human masters. Since 1920, when Karel Čapek coined the word *robot* in his play *R.U.R.*, dramatists have freely conjured them up: Speedy, Cutie, and Dave in Isaac Asimov's *I, Robot*, Robbie in *Forbidden Planet*, the flailing canister in *Lost in Space*, the daleks in *Dr. Who*, Rosie the Maid in *The Jetsons*, Nomad in *Star Trek*, Hymie in *Get Smart*, the vacant butlers and bickering haberdashers in *Sleeper*, R2D2 and C3PO in *Star Wars*, the Terminator in *The Terminator*, Lieutenant Commander Data in *Star Trek: The Next Generation*, and the wisecracking film critics in *Mystery Science Theater 3000*.

This book is not about robots; it is about the human mind. I will try to explain what the mind is, where it came from, and how it lets us see, think, feel, interact, and pursue higher callings like art, religion, and philosophy. On the way I will try to throw light on distinctively human quirks. Why do memories fade? How does makeup change the look of a face? Where do ethnic stereotypes come from, and when are they irrational? Why do people lose their tempers? What makes children bratty? Why do fools fall in love? What makes us laugh? And why do people believe in ghosts and spirits?

3

But the gap between robots in imagination and in reality is my starting point, for it shows the first step we must take in knowing ourselves:
appreciating the fantastically complex design behind feats of mental life
we take for granted. The reason there are no humanlike robots is not that
the very idea of a mechanical mind is misguided. It is that the engineering problems that we humans solve as we see and walk and plan and
make it through the day are far more challenging than landing on the
moon or sequencing the human genome. Nature, once again, has found
ingenious solutions that human engineers cannot yet duplicate. When
Hamlet says, "What a piece of work is a man! how noble in reason! how
infinite in faculty! in form and moving how express and admirable!" we
should direct our awe not at Shakespeare or Mozart or Einstein or
Kareem Abdul-Jabbar but at a four-year old carrying out a request to put
a toy on a shelf.

In a well-designed system, the components are black boxes that perform their functions as if by magic. That is no less true of the mind. The
faculty with which we ponder the world has no ability to peer inside
itself or our other faculties to see what makes them tick. That makes us
the victims of an illusion: that our own psychology comes from some
divine force or mysterious essence or almighty principle. In the Jewish
legend of the Golem, a clay figure was animated when it was fed an
inscription of the name of God. The archetype is echoed in many robot
stories. The statue of Galatea was brought to life by Venus' answer to
Pygmalion's prayers; Pinocchio was vivified by the Blue Fairy. Modern
versions of the Golem archetype appear in some of the less fanciful stories of science. All of human psychology is said to be explained by a single, omnipotent cause: a large brain, culture, language, socialization,
learning, complexity, self-organization, neural-network dynamics.

I want to convince you that our minds are not animated by some
godly vapor or single wonder principle. The mind, like the Apollo spacecraft, is designed to solve many engineering problems, and thus is
packed with high-tech systems each contrived to overcome its own
obstacles. I begin by laying out these problems, which are both design
specs for a robot and the subject matter of psychology. For I believe that
the discovery by cognitive science and artificial intelligence of the technical challenges overcome by our mundane mental activity is one of the
great revelations of science, an awakening of the imagination comparable
to learning that the universe is made up of billions of galaxies or that a
drop of pond water teems with microscopic life.

THE ROBOT CHALLENGE

What does it take to build a robot? Let's put aside superhuman abilities like calculating planetary orbits and begin with the simple human ones: seeing, walking, grasping, thinking about objects and people, and planning how to act.

In movies we are often shown a scene from a robot's-eye view, with the help of cinematic conventions like fish-eye distortion or crosshairs. That is fine for us, the audience, who already have functioning eyes and brains. But it is no help to the robot's innards. The robot does not house an audience of little people—homunculi—gazing at the picture and telling the robot what they are seeing. If you could see the world through a robot's eyes, it would look not like a movie picture decorated with crosshairs but something like this:

```
225 221 216 219 219 214 207 218 219 220 207 155 136 135
213 206 213 223 208 217 223 221 223 216 195 156 141 130
206 217 210 216 224 223 228 230 234 216 207 157 136 132
211 213 221 223 220 222 237 216 219 220 176 149 137 132
221 229 218 230 228 214 213 209 198 224 161 140 133 127
220 219 224 220 219 215 215 206 206 221 159 143 133 131
221 215 211 214 220 218 221 212 218 204 148 141 131 130
214 211 211 218 214 220 226 216 223 209 143 141 141 124
211 208 223 213 216 226 231 230 241 199 153 141 136 125
200 224 219 215 217 224 232 241 240 211 150 139 128 132
204 206 208 205 233 241 241 252 242 192 151 141 133 130
200 205 201 216 232 248 255 246 231 210 149 141 132 126
191 194 209 238 245 255 249 235 238 197 146 139 130 132
189 199 200 227 239 237 235 236 247 192 145 142 124 133
198 196 209 211 210 215 236 240 232 177 142 137 135 124
198 203 205 208 211 224 226 240 210 160 139 132 129 130
216 209 214 220 210 231 245 219 169 143 148 129 128 136
211 210 217 218 214 227 244 221 162 140 139 129 133 131
215 210 216 216 209 220 248 200 156 139 131 129 139 128
219 220 211 208 205 209 240 217 154 141 127 130 124 142
229 224 212 214 220 229 234 208 151 145 128 128 142 122
252 224 222 224 233 244 228 213 143 141 135 128 131 129
255 235 230 249 253 240 228 193 147 139 132 128 136 125
250 245 238 245 246 235 235 190 139 136 134 135 126 130
240 238 233 232 235 255 246 168 156 144 129 127 136 134
```

Each number represents the brightness of one of the millions of tiny patches making up the visual field. The smaller numbers come from darker patches, the larger numbers from brighter patches. The numbers shown in the array are the actual signals coming from an electronic camera trained on a person's hand, though they could just as well be the firing rates of some of the nerve fibers coming from the eye to the brain as a person looks at a hand. For a robot brain—or a human brain—to recognize objects and not bump into them, it must crunch these numbers and guess what kinds of objects in the world reflected the light that gave rise to them. The problem is humblingly difficult.

First, a visual system must locate where an object ends and the backdrop begins. But the world is not a coloring book, with black outlines around solid regions. The world as it is projected into our eyes is a mosaic of tiny shaded patches. Perhaps, one could guess, the visual brain looks for regions where a quilt of large numbers (a brighter region) abuts a quilt of small numbers (a darker region). You can discern such a boundary in the square of numbers; it runs diagonally from the top right to the bottom center. Most of the time, unfortunately, you would not have found the edge of an object, where it gives way to empty space. The juxtaposition of large and small numbers could have come from many distinct arrangements of matter. This drawing, devised by the psychologists Pawan Sinha and Edward Adelson, appears to show a ring of light gray and dark gray tiles.

In fact, it is a rectangular cutout in a black cover through which you are looking at part of a scene. In the next drawing the cover has been removed, and you can see that each pair of side-by-side gray squares comes from a different arrangement of objects.

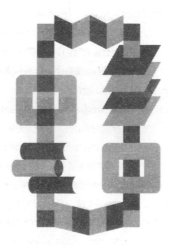

Big numbers next to small numbers can come from an object standing in front of another object, dark paper lying on light paper, a surface painted two shades of gray, two objects touching side by side, gray cellophane on a white page, an inside or outside corner where two walls meet, or a shadow. Somehow the brain must solve the chicken-and-egg problem of identifying three-dimensional objects from the patches on the retina *and* determining what each patch is (shadow or paint, crease or overlay, clear or opaque) from knowledge of what object the patch is part of.

The difficulties have just begun. Once we have carved the visual world into objects, we need to know what they are made of, say, snow versus coal. At first glance the problem looks simple. If large numbers come from bright regions and small numbers come from dark regions, then large number equals white equals snow and small number equals black equals coal, right? Wrong. The amount of light hitting a spot on the retina depends not only on how pale or dark the object is but also on how bright or dim the light illuminating the object is. A photographer's light meter would show you that more light bounces off a lump of coal outdoors than off a snowball indoors. That is why people are so often disappointed by their snapshots and why photography is such a complicated craft. The camera does not lie; left to its own devices, it renders outdoor scenes as milk and indoor scenes as mud. Photographers, and sometimes microchips inside the camera, coax a realistic image out of the film with tricks like adjustable shutter timing, lens apertures, film speeds, flashes, and darkroom manipulations.

Our visual system does much better. Somehow it lets us see the bright outdoor coal as black and the dark indoor snowball as white. That is a happy outcome, because our conscious sensation of color and lightness matches the world as it is rather than the world as it presents itself to the eye. The snowball is soft and wet and prone to melt whether it is indoors or out, and we see it as white whether it is indoors or out. The coal is always hard and dirty and prone to burn, and we always see it as black. The harmony between how the world *looks* and how the world *is* must be an achievement of our neural wizardry, because black and white don't simply announce themselves on the retina. In case you are still skeptical, here is an everyday demonstration. When a television set is off, the screen is a pale greenish gray. When it is on, some of the phosphor dots give off light, painting in the bright areas of the picture. But the other dots do not suck light and paint in the dark areas; they just stay gray. The areas that you see as black are in fact just the pale shade of the picture tube when the set was off. The blackness is a figment, a product of the brain circuitry that ordinarily allows you to see coal as coal. Television engineers exploited that circuitry when they designed the screen.

The next problem is seeing in depth. Our eyes squash the three-dimensional world into a pair of two-dimensional retinal images, and the third dimension must be reconstituted by the brain. But there are no telltale signs in the patches on the retina that reveal how far away a surface is. A stamp in your palm can project the same square on your retina as a chair across the room or a building miles away (first drawing, page 9). A cutting board viewed head-on can project the same trapezoid as various irregular shards held at a slant (second drawing, page 9).

You can feel the force of this fact of geometry, and of the neural mechanism that copes with it, by staring at a lightbulb for a few seconds or looking at a camera as the flash goes off, which temporarily bleaches a patch onto your retina. If you now look at the page in front of you, the afterimage adheres to it and appears to be an inch or two across. If you look up at the wall, the afterimage appears several feet long. If you look at the sky, it is the size of a cloud.

Finally, how might a vision module recognize the objects out there in the world, so that the robot can name them or recall what they do? The

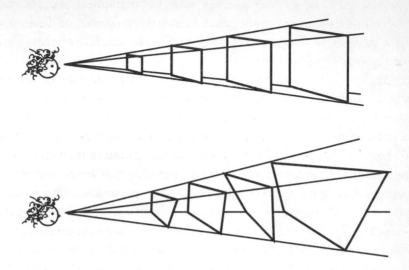

obvious solution is to build a template or cutout for each object that duplicates its shape. When an object appears, its projection on the retina would fit its own template like a round peg in a round hole. The template would be labeled with the name of the shape—in this case, "the letter *P*"—and whenever a shape matches it, the template announces the name:

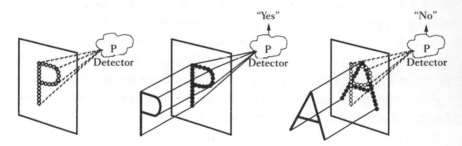

Alas, this simple device malfunctions in both possible ways. It sees *P*'s that aren't there; for example, it gives a false alarm to the *R* shown in the first square below. And it fails to see *P*'s that are there; for example, it misses the letter when it is shifted, tilted, slanted, too far, too near, or too fancy:

And these problems arise with a nice, crisp letter of the alphabet. Imagine trying to design a recognizer for a shirt, or a face! To be sure, after four decades of research in artificial intelligence, the technology of shape recognition has improved. You may own software that scans in a page, recognizes the printing, and converts it with reasonable accuracy to a file of bytes. But artificial shape recognizers are still no match for the ones in our heads. The artificial ones are designed for pristine, easy-to-recognize worlds and not the squishy, jumbled real world. The funny numbers at the bottom of checks were carefully drafted to have shapes that don't overlap and are printed with special equipment that positions them exactly so that they can be recognized by templates. When the first face recognizers are installed in buildings to replace doormen, they will not even try to interpret the chiaroscuro of your face but will scan in the hard-edged, rigid contours of your iris or your retinal blood vessels. Our brains, in contrast, keep a record of the shape of every face we know (and every letter, animal, tool, and so on), and the record is somehow matched with a retinal image even when the image is distorted in all the ways we have been examining. In Chapter 4 we will explore how the brain accomplishes this magnificent feat.

~

Let's take a look at another everyday miracle: getting a body from place to place. When we want a machine to move, we put it on wheels. The invention of the wheel is often held up as the proudest accomplishment of civilization. Many textbooks point out that no animal has evolved wheels and cite the fact as an example of how evolution is often incapable of finding the optimal solution to an engineering problem. But it is not a good example at all. Even if nature *could* have evolved a moose on wheels, it surely would have opted not to. Wheels are good only in a world with roads and rails. They bog down in any terrain that is soft, slippery, steep, or uneven. Legs are better. Wheels have to roll along an unbroken supporting ridge, but legs can be placed on a series of separate footholds, an extreme example being a ladder. Legs can also be placed to minimize lurching and to step over obstacles. Even today, when it seems as if the world has become a parking lot, only about half of the earth's land is accessible to vehicles with wheels or tracks, but most of the earth's land is accessible to vehicles with feet: animals, the vehicles designed by natural selection.

But legs come with a high price: the software to control them. A wheel, merely by turning, changes its point of support gradually and can bear weight the whole time. A leg has to change its point of support all at once, and the weight has to be unloaded to do so. The motors controlling a leg have to alternate between keeping the foot on the ground while it bears and propels the load and taking the load off to make the leg free to move. All the while they have to keep the center of gravity of the body within the polygon defined by the feet so the body doesn't topple over. The controllers also must minimize the wasteful up-and-down motion that is the bane of horseback riders. In walking windup toys, these problems are crudely solved by a mechanical linkage that converts a rotating shaft into a stepping motion. But the toys cannot adjust to the terrain by finding the best footholds.

Even if we solved these problems, we would have figured out only how to control a walking insect. With six legs, an insect can always keep one tripod on the ground while it lifts the other tripod. At any instant, it is stable. Even four-legged beasts, when they aren't moving too quickly, can keep a tripod on the ground at all times. But as one engineer has put it, "the upright two-footed locomotion of the human being seems almost a recipe for disaster in itself, and demands a remarkable control to make it practicable." When we walk, we repeatedly tip over and break our fall in the nick of time. When we run, we take off in bursts of flight. These aerobatics allow us to plant our feet on widely or erratically spaced footholds that would not prop us up at rest, and to squeeze along narrow paths and jump over obstacles. But no one has yet figured out how we do it.

Controlling an arm presents a new challenge. Grab the shade of an architect's lamp and move it along a straight diagonal path from near you, low on the left, to far from you, high on the right. Look at the rods and hinges as the lamp moves. Though the shade proceeds along a straight line, each rod swings through a complicated arc, swooping rapidly at times, remaining almost stationary at other times, sometimes reversing from a bending to a straightening motion. Now imagine having to do it in reverse: without looking at the shade, you must choreograph the sequence of twists around each joint that would send the shade along a straight path. The trigonometry is frightfully complicated. But your arm is an architect's lamp, and your brain effortlessly solves the equations every time you point. And if you have ever held an architect's lamp by its clamp, you will appreciate that the problem is even harder than what I have described. The lamp flails under its weight as if it had a mind of its own; so would your arm if your brain did not compensate for its weight, solving a near-intractable physics problem.

A still more remarkable feat is controlling the hand. Nearly two thousand years ago, the Greek physician Galen pointed out the exquisite natural engineering behind the human hand. It is a single tool that manipulates objects of an astonishing range of sizes, shapes, and weights, from a log to a millet seed. "Man handles them all," Galen noted, "as well as if his hands had been made for the sake of each one of them alone." The hand can be configured into a hook grip (to lift a pail), a scissors grip (to hold a cigarette), a five-jaw chuck (to lift a coaster), a three-jaw chuck (to hold a pencil), a two-jaw pad-to-pad chuck (to thread a needle), a two-jaw pad-to-side chuck (to turn a key), a squeeze grip (to hold a hammer), a disc grip (to open a jar), and a spherical grip (to hold a ball). Each grip needs a precise combination of muscle tensions that mold the hand into the right shape and keep it there as the load tries to bend it back. Think of lifting a milk carton. Too loose a grasp, and you drop it; too tight, and you crush it; and with some gentle rocking, you can even use the tugging on your fingertips as a gauge of how much milk is inside! And I won't even begin to talk about the tongue, a boneless water balloon controlled only by squeezing, which can loosen food from a back tooth or perform the ballet that articulates words like *thrilling* and *sixths*.

～

"A common man marvels at uncommon things; a wise man marvels at the commonplace." Keeping Confucius' dictum in mind, let's continue to look at commonplace human acts with the fresh eye of a robot designer seeking to duplicate them. Pretend that we have somehow built a robot that can see and move. What will it do with what it sees? How should it decide how to act?

An intelligent being cannot treat every object it sees as a unique entity unlike anything else in the universe. It has to put objects in categories so that it may apply its hard-won knowledge about similar objects, encountered in the past, to the object at hand.

But whenever one tries to program a set of criteria to capture the members of a category, the category disintegrates. Leaving aside slippery concepts like "beauty" or "dialectical materialism," let's look at a textbook example of a well-defined one: "bachelor." A bachelor, of course, is simply an adult human male who has never been married. But now imagine that a friend asks you to invite some bachelors to her party. What would happen if you used the definition to decide which of the following people to invite?

Arthur has been living happily with Alice for the last five years. They have a two-year-old daughter and have never officially married.

Bruce was going to be drafted, so he arranged with his friend Barbara to have a justice of the peace marry them so he would be exempt. They have never lived together. He dates a number of women, and plans to have the marriage annulled as soon as he finds someone he wants to marry.

Charlie is 17 years old. He lives at home with his parents and is in high school.

David is 17 years old. He left home at 13, started a small business, and is now a successful young entrepreneur leading a playboy's lifestyle in his penthouse apartment.

Eli and Edgar are homosexual lovers who have been living together for many years.

Faisal is allowed by the law of his native Abu Dhabi to have three wives. He currently has two and is interested in meeting another potential fiancée.

Father Gregory is the bishop of the Catholic cathedral at Groton upon Thames.

The list, which comes from the computer scientist Terry Winograd, shows that the straightforward definition of "bachelor" does not capture our intuitions about who fits the category.

Knowing who is a bachelor is just common sense, but there's nothing common about common sense. Somehow it must find its way into a human or robot brain. And common sense is not simply an almanac about life that can be dictated by a teacher or downloaded like an enormous database. No database could list all the facts we tacitly know, and no one ever taught them to us. You know that when Irving puts the dog in the car, it is no longer in the yard. When Edna goes to church, her head goes with her. If Doug is in the house, he must have gone in through some opening unless he was born there and never left. If Sheila is alive at 9 A.M. and is alive at 5 P.M., she was also alive at noon. Zebras in the wild never wear underwear. Opening a jar of a new brand of peanut butter will not vaporize the house. People never shove meat thermometers in their ears. A gerbil is smaller than Mt. Kilimanjaro.

An intelligent system, then, cannot be stuffed with trillions of facts. It must be equipped with a smaller list of core truths and a set of rules to deduce their implications. But the rules of common sense, like the categories of common sense, are frustratingly hard to set down. Even the most straightforward ones fail to capture our everyday reasoning. Mavis lives in Chicago and has a son named Fred, and Millie lives in Chicago and has a son named Fred. But whereas the Chicago that Mavis lives in is the same Chicago that Millie lives in, the Fred who is Mavis' son is not the same Fred who is Millie's son. If there's a bag in your car, and a gallon of milk in the bag, there is a gallon of milk in your car. But if there's a person in your car, and a gallon of blood in a person, it would be strange to conclude that there is a gallon of blood in your car.

Even if you were to craft a set of rules that derived only sensible conclusions, it is no easy matter to use them all to guide behavior intelligently. Clearly a thinker cannot apply just one rule at a time. A match gives light; a saw cuts wood; a locked door is opened with a key. But we laugh at the man who lights a match to peer into a fuel tank, who saws off the limb he is sitting on, or who locks his keys in the car and spends the next hour wondering how to get his family out. A thinker has to compute not just the direct effects of an action but the side effects as well.

But a thinker cannot crank out predictions about *all* the side effects, either. The philosopher Daniel Dennett asks us to imagine a robot designed to fetch a spare battery from a room that also contained a time bomb. Version 1 saw that the battery was on a wagon and that if it pulled the wagon out of the room, the battery would come with it. Unfortunately, the bomb was also on the wagon, and the robot failed to deduce that pulling the wagon out brought the bomb out, too. Version 2 was programmed to consider all the side effects of its actions. It had just finished computing that pulling the wagon would not change the color of the room's walls and was proving that the wheels would turn more revolutions than there are wheels on the wagon, when the bomb went off. Version 3 was programmed to distinguish between relevant implications and irrelevant ones. It sat there cranking out millions of implications and putting all the relevant ones on a list of facts to consider and all the irrelevant ones on a list of facts to ignore, as the bomb ticked away.

An intelligent being has to deduce the implications of what it knows, but only the *relevant* implications. Dennett points out that this requirement poses a deep problem not only for robot design but for epistemology, the analysis of how we know. The problem escaped the notice of generations of philosophers, who were left complacent by the illusory effortlessness of their own common sense. Only when artificial intelligence researchers tried to duplicate common sense in computers, the ultimate blank slate, did the conundrum, now called "the frame problem," come to light. Yet somehow we all solve the frame problem whenever we use our common sense.

Imagine that we have somehow overcome these challenges and have a machine with sight, motor coordination, and common sense. Now we must figure out how the robot will put them to use. We have to give it motives.

What should a robot want? The classic answer is Isaac Asimov's Fundamental Rules of Robotics, "the three rules that are built most deeply into a robot's positronic brain."

1. A robot may not injure a human being or, through inaction, allow a human being to come to harm.

2. A robot must obey orders given it by human beings except where such orders would conflict with the First Law.

3. A robot must protect its own existence as long as such protection does not conflict with the First or Second Law.

Asimov insightfully noticed that self-preservation, that universal biological imperative, does not automatically emerge in a complex system. It has to be programmed in (in this case, as the Third Law). After all, it is just as easy to build a robot that lets itself go to pot or eliminates a malfunction by committing suicide as it is to build a robot that always looks out for Number One. Perhaps easier; robot-makers sometimes watch in horror as their creations cheerfully shear off limbs or flatten themselves against walls, and a good proportion of the world's most intelligent machines are kamikaze cruise missiles and smart bombs.

But the need for the other two laws is far from obvious. Why give a robot an order to obey orders—why aren't the original orders enough? Why command a robot not to do harm—wouldn't it be easier never to command it to *do* harm in the first place? Does the universe contain a mysterious force pulling entities toward malevolence, so that a positronic brain must be programmed to withstand it? Do intelligent beings inevitably develop an attitude problem?

In this case Asimov, like generations of thinkers, like all of us, was unable to step outside his own thought processes and see them as artifacts of how our minds were put together rather than as inescapable laws of the universe. Man's capacity for evil is never far from our minds, and it is easy to think that evil just comes along with intelligence as part of its very essence. It is a recurring theme in our cultural tradition: Adam and Eve eating the fruit of the tree of knowledge, Promethean fire and Pandora's box, the rampaging Golem, Faust's bargain, the Sorcerer's Apprentice, the adventures of Pinocchio, Frankenstein's monster, the murderous apes and mutinous HAL of *2001: A Space Odyssey*. From the 1950s through the 1980s, countless films in the computer-runs-amok genre captured a popular fear that the exotic mainframes of the era would get smarter and more powerful and someday turn on us.

Now that computers really *have* become smarter and more powerful, the anxiety has waned. Today's ubiquitous, networked computers have an unprecedented ability to do mischief should they ever go to the bad. But the only mayhem comes from unpredictable chaos or from human malice in the form of viruses. We no longer worry about electronic serial killers or subversive silicon cabals because we are beginning to appreciate that malevolence—like vision, motor coordination, and common sense—does not come free with computation but has to be programmed in. The computer running WordPerfect on your desk will continue to fill paragraphs for as long as it does anything at all. Its software will not insidiously mutate into depravity like the picture of Dorian Gray.

Even if it could, why would it want to? To get—what? More floppy disks? Control over the nation's railroad system? Gratification of a desire to commit senseless violence against laser-printer repairmen? And wouldn't it have to worry about reprisals from technicians who with the turn of a screwdriver could leave it pathetically singing "A Bicycle Built for Two"? A network of computers, perhaps, could discover the safety in numbers and plot an organized takeover—but what would make one computer volunteer to fire the data packet heard round the world and risk early martyrdom? And what would prevent the coalition from being undermined by silicon draft-dodgers and conscientious objectors? Aggression, like every other part of human behavior we take for granted, is a challenging engineering problem!

But then, so are the kinder, gentler motives. How would you design a robot to obey Asimov's injunction never to allow a human being to come to harm through inaction? Michael Frayn's 1965 novel *The Tin Men* is set in a robotics laboratory, and the engineers in the Ethics Wing, Macintosh, Goldwasser, and Sinson, are testing the altruism of their robots. They have taken a bit too literally the hypothetical dilemma in every moral philosophy textbook in which two people are in a lifeboat built for one and both will die unless one bails out. So they place each robot in a raft with another occupant, lower the raft into a tank, and observe what happens.

> [The] first attempt, Samaritan I, had pushed itself overboard with great alacrity, but it had gone overboard to save anything which happened to be next to it on the raft, from seven stone of lima beans to twelve stone of wet seaweed. After many weeks of stubborn argument Macintosh had conceded that the lack of discrimination was unsatisfactory, and he had abandoned Samaritan I and developed Samaritan II, which would sacrifice itself only for an organism at least as complicated as itself.
>
> The raft stopped, revolving slowly, a few inches above the water. "Drop it," cried Macintosh.
>
> The raft hit the water with a sharp report. Sinson and Samaritan sat perfectly still. Gradually the raft settled in the water, until a thin tide began to wash over the top of it. At once Samaritan leaned forward and seized Sinson's head. In four neat movements it measured the size of his skull, then paused, computing. Then, with a decisive click, it rolled sideways off the raft and sank without hesitation to the bottom of the tank.

But as the Samaritan II robots came to behave like the moral agents in the philosophy books, it became less and less clear that they were really moral at all. Macintosh explained why he did not simply tie a rope around the self-sacrificing robot to make it easier to retrieve: "I don't want it to know that it's going to be saved. It would invalidate its decision to sacrifice itself. . . . So, every now and then I leave one of them in instead of fishing it out. To show the others I mean business. I've written off two this week." Working out what it would take to program goodness into a robot shows not only how much machinery it takes to be good but how slippery the concept of goodness is to start with.

And what about the most caring motive of all? The weak-willed computers of 1960s pop culture were not tempted only by selfishness and power, as we see in the comedian Allan Sherman's song "Automation," sung to the tune of "Fascination":

> It was automation, I know.
> That was what was making the factory go.
> It was IBM, it was Univac,
> It was all those gears going clickety clack, dear.
> I thought automation was keen
> Till you were replaced by a ten-ton machine.
> It was a computer that tore us apart, dear,
> Automation broke my heart. . . .
>
> It was automation, I'm told,
> That's why I got fired and I'm out in the cold.
> How could I have known, when the 503
> Started in to blink, it was winking at me, dear?
> I thought it was just some mishap
> When it sidled over and sat on my lap.
> But when it said "I love you" and gave me a hug, dear,
> That's when I pulled out . . . its . . . plug.

But for all its moonstruck madness, love is no bug or crash or malfunction. The mind is never so wonderfully concentrated as when it turns to love, and there must be intricate calculations that carry out the peculiar logic of attraction, infatuation, courtship, coyness, surrender, commitment, malaise, philandering, jealousy, desertion, and heartbreak. And in the end, as my grandmother used to say, every pot finds a cover; most people—including, significantly, all of our ancestors—manage to pair up long enough to produce viable children. Imagine how many lines of programming it would take to duplicate that!

Robot design is a kind of consciousness-raising. We tend to be blasé about our mental lives. We open our eyes, and familiar articles present themselves; we will our limbs to move, and objects and bodies float into place; we awaken from a dream, and return to a comfortingly predictable world; Cupid draws back his bow, and lets his arrow go. But think of what it takes for a hunk of matter to accomplish these improbable outcomes, and you begin to see through the illusion. Sight and action and common sense and violence and morality and love are no accident, no inextricable ingredients of an intelligent essence, no inevitability of information processing. Each is a tour de force, wrought by a high level of targeted design. Hidden behind the panels of consciousness must lie fantastically complex machinery—optical analyzers, motion guidance systems, simulations of the world, databases on people and things, goal-schedulers, conflict-resolvers, and many others. Any explanation of how the mind works that alludes hopefully to some single master force or mind-bestowing elixir like "culture," "learning," or "self-organization" begins to sound hollow, just not up to the demands of the pitiless universe we negotiate so successfully.

The robot challenge hints at a mind loaded with original equipment, but it still may strike you as an argument from the armchair. Do we actually find signs of this intricacy when we look directly at the machinery of the mind and at the blueprints for assembling it? I believe we do, and what we see is as mind-expanding as the robot challenge itself.

When the visual areas of the brain are damaged, for example, the visual world is not simply blurred or riddled with holes. Selected aspects of visual experience are removed while others are left intact. Some patients see a complete world but pay attention only to half of it. They eat food from the right side of the plate, shave only the right cheek, and draw a clock with twelve digits squished into the right half. Other patients lose their sensation of color, but they do not see the world as an arty black-and-white movie. Surfaces look grimy and rat-colored to them, killing their appetite and their libido. Still others can see objects change their positions but cannot see them move—a syndrome that a philosopher once tried to convince me was logically impossible! The stream from a teapot does not flow but looks like an icicle; the cup does not gradually fill with tea but is empty and then suddenly full.

Other patients cannot recognize the objects they see: their world is like handwriting they cannot decipher. They copy a bird faithfully but identify it as a tree stump. A cigarette lighter is a mystery until it is lit. When they try to weed the garden, they pull out the roses. Some patients can recognize inanimate objects but cannot recognize faces. The patient deduces that the visage in the mirror must be his, but does not viscerally recognize himself. He identifies John F. Kennedy as Martin Luther King, and asks his wife to wear a ribbon at a party so he can find her when it is time to leave. Stranger still is the patient who recognizes the face but not the person: he sees his wife as an amazingly convincing impostor.

These syndromes are caused by an injury, usually a stroke, to one or more of the thirty brain areas that compose the primate visual system. Some areas specialize in color and form, others in where an object is, others in what an object is, still others in how it moves. A seeing robot cannot be built with just the fish-eye viewfinder of the movies, and it is no surprise to discover that humans were not built that way either. When we gaze at the world, we do not fathom the many layers of apparatus that underlie our unified visual experience, until neurological disease dissects them for us.

Another expansion of our vista comes from the startling similarities between identical twins, who share the genetic recipes that build the mind. Their minds are astonishingly alike, and not just in gross measures like IQ and personality traits like neuroticism and introversion. They are alike in talents such as spelling and mathematics, in opinions on questions such as apartheid, the death penalty, and working mothers, and in their career choices, hobbies, vices, religious commitments, and tastes in dating. Identical twins are far more alike than fraternal twins, who share only half their genetic recipes, and most strikingly, they are almost as alike when they are reared apart as when they are reared together. Identical twins separated at birth share traits like entering the water backwards and only up to their knees, sitting out elections because they feel insufficiently informed, obsessively counting everything in sight, becoming captain of the volunteer fire department, and leaving little love notes around the house for their wives.

People find these discoveries arresting, even incredible. The discoveries cast doubt on the autonomous "I" that we all feel hovering above our bodies, making choices as we proceed through life and affected only by our past and present environments. Surely the mind does not come equipped with so many small parts that it could predestine us to flush the toilet before and after using it or to sneeze playfully in crowded elevators, to take two other traits shared by identical twins reared apart. But apparently it does. The far-reaching effects of the genes have been documented in scores of studies and show up no matter how one tests for them: by comparing twins reared apart and reared together, by comparing identical and fraternal twins, or by comparing adopted and biological children. And despite what critics sometimes claim, the effects are not products of coincidence, fraud, or subtle similarities in the family environments (such as adoption agencies striving to place identical twins in homes that both encourage walking into the ocean backwards). The findings, of course, can be misinterpreted in many ways, such as by imagining a gene for leaving little love notes around the house or by concluding that people are unaffected by their experiences. And because this research can measure only the ways in which people *differ*, it says little about the design of the mind that all normal people share. But by showing how many ways the mind can vary in its innate structure, the discoveries open our eyes to how much structure the mind must have.

REVERSE-ENGINEERING THE PSYCHE

The complex structure of the mind is the subject of this book. Its key idea can be captured in a sentence: The mind is a system of organs of computation, designed by natural selection to solve the kinds of problems our ancestors faced in their foraging way of life, in particular, understanding and outmaneuvering objects, animals, plants, and other people. The summary can be unpacked into several claims. The mind is what the brain does; specifically, the brain processes information, and thinking is a kind of computation. The mind is organized into modules or mental organs, each with a specialized design that makes it an expert in one arena of interaction with the world. The modules' basic logic is specified by our genetic program. Their operation was shaped by natural selection to solve the problems of the hunting and gathering life led by our ancestors in most of our evolutionary history. The various problems for our ancestors were subtasks of one big problem for their genes, maximizing the number of copies that made it into the next generation.

On this view, psychology is engineering in reverse. In forward-engineering, one designs a machine to do something; in reverse-engineering, one figures out what a machine was designed to do. Reverse-engineering is what the boffins at Sony do when a new product is announced by Panasonic, or vice versa. They buy one, bring it back to the lab, take a screwdriver to it, and try to figure out what all the parts are for and how they combine to make the device work. We all engage in reverse-engineering when we face an interesting new gadget. In rummaging through an antique store, we may find a contraption that is inscrutable until we figure out what it was designed to do. When we realize that it is an olive-pitter, we suddenly understand that the metal ring is designed to hold the olive, and the lever lowers an X-shaped blade through one end, pushing the pit out through the other end. The shapes and arrangements of the springs, hinges, blades, levers, and rings all make sense in a satisfying rush of insight. We even understand why canned olives have an X-shaped incision at one end.

In the seventeenth century William Harvey discovered that veins had valves and deduced that the valves must be there to make the blood circulate. Since then we have understood the body as a wonderfully complex machine, an assembly of struts, ties, springs, pulleys, levers, joints, hinges, sockets, tanks, pipes, valves, sheaths, pumps, exchangers, and filters. Even today we can be delighted to learn what mysterious parts are for. Why do we have our wrinkled, asymmetrical ears? Because they filter sound waves coming from different directions in different ways. The nuances of the sound shadow tell the brain whether the source of the sound is above or below, in front of or behind us. The strategy of reverse-engineering the body has continued in the last half of this century as we have explored the nanotechnology of the cell and of the molecules of life. The stuff of life turned out to be not a quivering, glowing, wondrous gel but a contraption of tiny jigs, springs, hinges, rods, sheets, magnets, zippers, and trapdoors, assembled by a data tape whose information is copied, downloaded, and scanned.

The rationale for reverse-engineering living things comes, of course, from Charles Darwin. He showed how "organs of extreme perfection and complication, which justly excite our admiration" arise not from God's foresight but from the evolution of replicators over immense spans of time. As replicators replicate, random copying errors sometimes crop up, and those that happen to enhance the survival and reproduction rate of the replicator tend to accumulate over the generations. Plants and animals are replicators, and their complicated machinery thus appears to have been engineered to allow them to survive and reproduce.

Darwin insisted that his theory explained not just the complexity of an animal's body but the complexity of its mind. "Psychology will be based on a new foundation," he famously predicted at the end of *The Origin of Species*. But Darwin's prophecy has not yet been fulfilled. More than a century after he wrote those words, the study of the mind is still mostly Darwin-free, often defiantly so. Evolution is said to be irrelevant, sinful, or fit only for speculation over a beer at the end of the day. The allergy to evolution in the social and cognitive sciences has been, I think, a barrier to understanding. The mind is an exquisitely organized system that accomplishes remarkable feats no engineer can duplicate. How could the forces that shaped that system, and the purposes for which it was designed, be irrelevant to understanding it? Evolutionary thinking is indispensable, not in the form that many people think of—dreaming up missing links or narrating stories about the stages of Man—but in the form of careful reverse-engineering. Without reverse-engineering we are like the singer in Tom Paxton's "The Marvelous Toy," reminiscing about a childhood present: "It went ZIP! when it moved, and POP! when it stopped, and WHIRRR! when it stood still; I never knew just what it was, and I guess I never will."

Only in the past few years has Darwin's challenge been taken up, by a new approach christened "evolutionary psychology" by the anthropologist John Tooby and the psychologist Leda Cosmides. Evolutionary psychology brings together two scientific revolutions. One is the cognitive revolution of the 1950s and 1960s, which explains the mechanics of thought and emotion in terms of information and computation. The other is the revolution in evolutionary biology of the 1960s and 1970s, which explains the complex adaptive design of living things in terms of selection among replicators. The two ideas make a powerful combination. Cognitive science helps us to understand how a mind is possible and what kind of mind we have. Evolutionary biology helps us to understand *why* we have the kind of mind we have.

The evolutionary psychology of this book is, in one sense, a straightforward extension of biology, focusing on one organ, the mind, of one species, *Homo sapiens*. But in another sense it is a radical thesis that discards the way issues about the mind have been framed for almost a century. The premises of this book are probably not what you think they are. Thinking is computation, I claim, but that does not mean that the computer is a good metaphor for the mind. The mind is a set of modules, but the modules are not encapsulated boxes or circumscribed swatches on the surface of the brain. The organization of our mental modules comes from our genetic program, but that does not mean that there is a gene for every trait or that learning is less important than we used to think. The mind is an adaptation designed by natural selection, but that does not mean that everything we think, feel, and do is biologically adaptive. We evolved from apes, but that does not mean we have the same minds as apes. And the ultimate goal of natural selection is to propagate genes, but that does not mean that the ultimate goal of people is to propagate genes. Let me show you why not.

~

This book is about the brain, but I will not say much about neurons, hormones, and neurotransmitters. That is because the mind is not the brain but what the brain does, and not even everything it does, such as metabolizing fat and giving off heat. The 1990s have been named the Decade of the Brain, but there will never be a Decade of the Pancreas. The brain's special status comes from a special thing the brain does, which makes us see, think, feel, choose, and act. That special thing is information processing, or computation.

Information and computation reside in patterns of data and in relations of logic that are independent of the physical medium that carries them. When you telephone your mother in another city, the message stays the same as it goes from your lips to her ears even as it physically changes its form, from vibrating air, to electricity in a wire, to charges in silicon, to flickering light in a fiber optic cable, to electromagnetic waves, and then back again in reverse order. In a similar sense, the message stays the same when she repeats it to your father at the other end of the couch after it has changed its form inside her head into a cascade of neurons firing and chemicals diffusing across synapses. Likewise, a given program can run on computers made of vacuum tubes, electromagnetic switches, transistors, integrated circuits, or well-trained pigeons, and it accomplishes the same things for the same reasons.

This insight, first expressed by the mathematician Alan Turing, the computer scientists Alan Newell, Herbert Simon, and Marvin Minsky, and the philosophers Hilary Putnam and Jerry Fodor, is now called the computational theory of mind. It is one of the great ideas in intellectual history, for it solves one of the puzzles that make up the "mind-body problem": how to connect the ethereal world of meaning and intention, the stuff of our mental lives, with a physical hunk of matter like the brain. Why did Bill get on the bus? Because he wanted to visit his grandmother and knew the bus would take him there. No other answer will do. If he hated the sight of his grandmother, or if he knew the route had changed, his body would not be on that bus. For millennia this has been a paradox. Entities like "wanting to visit one's grandmother" and "knowing the bus goes to Grandma's house" are colorless, odorless, and tasteless. But at the same time they are *causes* of physical events, as potent as any billiard ball clacking into another.

The computational theory of mind resolves the paradox. It says that beliefs and desires are *information*, incarnated as configurations of symbols. The symbols are the physical states of bits of matter, like chips in a computer or neurons in the brain. They symbolize things in the world because they are triggered by those things via our sense organs, and because of what they do once they are triggered. If the bits of matter that constitute a symbol are arranged to bump into the bits of matter constituting another symbol in just the right way, the symbols corresponding to one belief can give rise to new symbols corresponding to another belief logically related to it, which can give rise to symbols corresponding to other beliefs, and so on. Eventually the bits of matter constituting a symbol bump into bits of matter connected to the muscles, and behavior happens. The computational theory of mind thus allows us to keep beliefs and desires in our explanations of behavior while planting them squarely in the physical universe. It allows meaning to cause and be caused.

The computational theory of mind is indispensable in addressing the questions we long to answer. Neuroscientists like to point out that all parts of the cerebral cortex look pretty much alike—not only the different parts of the human brain, but the brains of different animals. One could draw the conclusion that all mental activity in all animals is the same. But a better conclusion is that we cannot simply look at a patch of brain and read out the logic in the intricate pattern of connectivity that makes each part do its separate thing. In the same way that all books are physically just different combinations of the same seventy-five or so characters, and all movies are physically just different patterns of charges along the tracks of a videotape, the mammoth tangle of spaghetti of the brain may all look alike when examined strand by strand. The content of a book or a movie lies in the *pattern* of ink marks or magnetic charges, and is apparent only when the piece is read or seen. Similarly, the content of brain activity lies in the patterns of connections and patterns of activity among the neurons. Minute differences in the details of the connections may cause similar-looking brain patches to implement very different programs. Only when the program is run does the coherence become evident. As Tooby and Cosmides have written,

There are birds that migrate by the stars, bats that echolocate, bees that compute the variance of flower patches, spiders that spin webs, humans that speak, ants that farm, lions that hunt in teams, cheetahs that hunt alone, monogamous gibbons, polyandrous seahorses, polygynous gorillas. . . . There are millions of animal species on earth, each with a different set of cognitive programs. *The same basic neural tissue embodies all of these programs*, and it could support many others as well. Facts about the properties of neurons, neurotransmitters, and cellular development cannot tell you which of these millions of programs the human mind contains. Even if all neural activity is the expression of a uniform process at the cellular level, it is the arrangement of neurons—into bird song templates or web-spinning programs—that matters.

That does not imply, of course, that the brain is irrelevant to understanding the mind! Programs are assemblies of simple information-processing units—tiny circuits that can add, match a pattern, turn on some other circuit, or do other elementary logical and mathematical operations. What those microcircuits can do depends only on what they are made of. Circuits made from neurons cannot do exactly the same things as circuits made from silicon, and vice versa. For example, a silicon circuit is faster than a neural circuit, but a neural circuit can match a larger pattern than a silicon one. These differences ripple up through the programs built from the circuits and affect how quickly and easily the programs do various things, even if they do not determine exactly which things they do. My point is not that prodding brain tissue is irrelevant to understanding the mind, only that it is not enough. Psychology, the analysis of mental software, will have to burrow a considerable way into the mountain before meeting the neurobiologists tunneling through from the other side.

The computational theory of mind is not the same thing as the despised "computer metaphor." As many critics have pointed out, computers are serial, doing one thing at a time; brains are parallel, doing millions of things at once. Computers are fast; brains are slow. Computer parts are reliable; brain parts are noisy. Computers have a limited number of connections; brains have trillions. Computers are assembled according to a blueprint; brains must assemble themselves. Yes, and computers come in putty-colored boxes and have AUTOEXEC.BAT files and run screen-savers with flying toasters, and brains do not. The claim is not that the brain is like commercially available computers. Rather, the claim is that brains and computers embody intelligence for some of the same reasons. To explain how birds fly, we invoke principles of lift and drag and fluid mechanics that also explain how airplanes fly. That does not commit us to an Airplane Metaphor for birds, complete with jet engines and complimentary beverage service.

Without the computational theory, it is impossible to make sense of the evolution of the mind. Most intellectuals think that the human mind must somehow have escaped the evolutionary process. Evolution, they think, can fabricate only stupid instincts and fixed action patterns: a sex drive, an aggression urge, a territorial imperative, hens sitting on eggs and ducklings following hulks. Human behavior is too subtle and flexible to be a product of evolution, they think; it must come from somewhere else—from, say, "culture." But if evolution equipped us not with irresistible urges and rigid reflexes but with a neural computer, everything changes. A program is an intricate recipe of logical and statistical operations directed by comparisons, tests, branches, loops, and subroutines embedded in subroutines. Artificial computer programs, from the Macintosh user interface to simulations of the weather to programs that recognize speech and answer questions in English, give us a hint of the finesse and power of which computation is capable. Human thought and behavior, no matter how subtle and flexible, could be the product of a very complicated program, and that program may have been our endowment from natural selection. The typical imperative from biology is not "Thou shalt . . . ," but "If . . . then . . . else."

The mind, I claim, is not a single organ but a system of organs, which we can think of as psychological faculties or mental modules. The entities now commonly evoked to explain the mind—such as general intelligence, a capacity to form culture, and multipurpose learning strategies—will surely go the way of protoplasm in biology and of earth, air, fire, and water in physics. These entities are so formless, compared to the exacting phenomena they are meant to explain, that they must be granted near-magical powers. When the phenomena are put under the microscope, we discover that the complex texture of the everyday world is supported not by a single substance but by many layers of elaborate machinery. Biologists long ago replaced the concept of an all-powerful protoplasm with the concept of functionally specialized mechanisms. The organ systems of the body do their jobs because each is built with a particular structure tailored to the task. The heart circulates the blood because it is built like a pump; the lungs oxygenate the blood because they are built like gas exchangers. The lungs cannot pump blood and the heart cannot oxygenate it. This specialization goes all the way down. Heart tissue differs from lung tissue, heart cells differ from lung cells, and many of the molecules making up heart cells differ from those making up lung cells. If that were not true, our organs would not work.

A jack-of-all-trades is master of none, and that is just as true for our mental organs as for our physical organs. The robot challenge makes that clear. Building a robot poses many software engineering problems, and different tricks are necessary to solve them.

Take our first problem, the sense of sight. A seeing machine must solve a problem called inverse optics. Ordinary optics is the branch of physics that allows one to predict how an object with a certain shape, material, and illumination projects the mosaic of colors we call the retinal image. Optics is a well-understood subject, put to use in drawing, photography, television engineering, and more recently, computer graphics and virtual reality. But the brain must solve the *opposite* problem. The input is the retinal image, and the output is a specification of the objects in the world and what they are made of—that is, what we know we are seeing. And there's the rub. Inverse optics is what engineers call an "ill-posed problem." It literally has no solution. Just as it is easy to multiply some numbers and announce the product but impossible to take a product and announce the numbers that were multiplied to get it, optics is easy but inverse optics impossible. Yet your brain does it every time you open the refrigerator and pull out a jar. How can this be?

The answer is that *the brain supplies the missing information*, information about the world we evolved in and how it reflects light. If the visual brain "assumes" that it is living in a certain kind of world—an evenly lit world made mostly of rigid parts with smooth, uniformly colored surfaces—it can make good guesses about what is out there. As we saw earlier, it's impossible to distinguish coal from snow by examining the brightnesses of their retinal projections. But say there is a module for perceiving the properties of surfaces, and built into it is the following assumption: "The world is smoothly and uniformly lit." The module can solve the coal-versus-snow problem in three steps: subtract out any gradient of brightness from one edge of the scene to the other; estimate the average level of brightness of the whole scene; and calculate the shade of gray of each patch by subtracting its brightness from the average brightness. Large positive deviations from the average are then seen as white things, large negative deviations as black things. If the illumination really is smooth and uniform, those perceptions will register the surfaces of the world accurately. Since Planet Earth has, more or less, met the even-illumination assumption for eons, natural selection would have done well by building the assumption in.

The surface-perception module solves an unsolvable problem, but at a price. The brain has given up any pretense of being a general problem-solver. It has been equipped with a gadget that perceives the nature of surfaces in typical earthly viewing conditions because it is specialized for that parochial problem. Change the problem slightly and the brain no longer solves it. Say we place a person in a world that is not blanketed with sunshine but illuminated by a cunningly arranged patchwork of light. If the surface-perception module assumes that illumination is even, it should be seduced into hallucinating objects that aren't there. Could that really happen? It happens every day. We call these hallucinations slide shows and movies and television (complete with the illusory black I mentioned earlier). When we watch TV, we stare at a shimmering piece of glass, but our surface-perception module tells the rest of our brain that we are seeing real people and places. The module has been unmasked; it does not apprehend the nature of things but relies on a cheat-sheet. That cheat-sheet is so deeply embedded in the operation of our visual brain that we cannot erase the assumptions written on it. Even in a lifelong couch potato, the visual system never "learns" that television is a pane of glowing phosphor dots, and the person never loses the illusion that there is a world behind the pane.

Our other mental modules need their own cheat-sheets to solve their unsolvable problems. A physicist who wants to figure out how the body moves when muscles are contracted has to solve problems in kinematics (the geometry of motion) and dynamics (the effects of forces). But a brain that has to figure out how to contract muscles to get the body to move has to solve problems in *inverse* kinematics and *inverse* dynamics— what forces to apply to an object to get it to move in a certain trajectory. Like inverse optics, inverse kinematics and dynamics are ill-posed problems. Our motor modules solve them by making extraneous but reasonable assumptions—not assumptions about illumination, of course, but assumptions about bodies in motion.

Our common sense about other people is a kind of intuitive psychology—we try to infer people's beliefs and desires from what they do, and try to predict what they will do from our guesses about their beliefs and desires. Our intuitive psychology, though, must make the assumption that other people *have* beliefs and desires; we cannot sense a belief or desire in another person's head the way we smell oranges. If we did not see the social world through the lens of that assumption, we would be like the Samaritan I robot, which sacrificed itself for a bag of lima beans, or like Samaritan II, which went overboard for any object with a human-like head, even if the head belonged to a large wind-up toy. (Later we shall see that people suffering from a certain syndrome lack the assumption that people have minds and *do* treat other people as wind-up toys.) Even our feelings of love for our family members embody a specific assumption about the laws of the natural world, in this case an inverse of the ordinary laws of genetics. Family feelings are designed to help our genes replicate themselves, but we cannot see or smell genes. Scientists use forward genetics to deduce how genes get distributed among organisms (for example, meiosis and sex cause the offspring of two people to have fifty percent of their genes in common); our emotions about kin use a kind of inverse genetics to guess which of the organisms we interact with are likely to share our genes (for example, if someone appears to have the same parents as you do, treat the person as if their genetic well-being overlaps with yours). I will return to all these topics in later chapters.

The mind has to be built out of specialized parts because it has to solve specialized problems. Only an angel could be a general problem-solver; we mortals have to make fallible guesses from fragmentary information. Each of our mental modules solves its unsolvable problem by a leap of faith about how the world works, by making assumptions that are indispensable but indefensible—the only defense being that the assumptions worked well enough in the world of our ancestors.

The word "module" brings to mind detachable, snap-in components, and that is misleading. Mental modules are not likely to be visible to the naked eye as circumscribed territories on the surface of the brain, like the flank steak and the rump roast on the supermarket cow display. A mental module probably looks more like roadkill, sprawling messily over the bulges and crevasses of the brain. Or it may be broken into regions that are interconnected by fibers that make the regions act as a unit. The beauty of information processing is the flexibility of its demand for real estate. Just as a corporation's management can be scattered across sites linked by a telecommunications network, or a computer program can be fragmented into different parts of the disk or memory, the circuitry underlying a psychological module might be distributed across the brain in a spatially haphazard manner. And mental modules need not be tightly sealed off from one another, communicating only through a few narrow pipelines. (That is a specialized sense of "module" that many cognitive scientists have debated, following a definition by Jerry Fodor.) Modules are defined by the special things they do with the information available to them, not necessarily by the kinds of information they have available.

So the metaphor of the mental module is a bit clumsy; a better one is Noam Chomsky's "mental organ." An organ of the body is a specialized structure tailored to carry out a particular function. But our organs do not come in a bag like chicken giblets; they are integrated into a complex whole. The body is composed of systems divided into organs assembled from tissues built out of cells. Some kinds of tissues, like the epithelium, are used, with modifications, in many organs. Some organs, like the blood and the skin, interact with the rest of the body across a widespread, convoluted interface, and cannot be encircled by a dotted line. Sometimes it is unclear where one organ leaves off and another begins, or how big a chunk of the body we want to call an organ. (Is the hand an organ? the finger? a bone in the finger?) These are all pedantic questions of terminology, and anatomists and physiologists have not wasted their time on them. What is clear is that the body is not made of Spam but has a heterogeneous structure of many specialized parts. All this is likely to be true of the mind. Whether or not we establish exact boundaries for the components of the mind, it is clear that it is not made of mental Spam but has a heterogeneous structure of many specialized parts.

~

Our physical organs owe their complex design to the information in the human genome, and so, I believe, do our mental organs. We do not learn to have a pancreas, and we do not learn to have a visual system, language acquisition, common sense, or feelings of love, friendship, and fairness. No single discovery proves the claim (just as no single discovery proves that the pancreas is innately structured), but many lines of evidence converge on it. The one that most impresses me is the Robot Challenge. Each of the major engineering problems solved by the mind is unsolvable without built-in assumptions about the laws that hold in that arena of interaction with the world. All of the programs designed by artificial intelligence researchers have been specially engineered for a particular domain, such as language, vision, movement, or one of many different kinds of common sense. Within artificial intelligence research, the proud parent of a program will sometimes tout it as a mere demo of an amazingly powerful general-purpose system to be built in the future, but everyone else in the field routinely writes off such hype. I predict that no one will ever build a humanlike robot—and I mean a *really* humanlike robot—unless they pack it with computational systems tailored to different problems.

Throughout the book we will run into other lines of evidence that our mental organs owe their basic design to our genetic program. I have already mentioned that much of the fine structure of our personality and intelligence is shared by identical twins reared apart and hence charted by the genes. Infants and young children, when tested with ingenious methods, show a precocious grasp of the fundamental categories of the physical and social world, and sometimes command information that was never presented to them. People hold many beliefs that are at odds with their experience but were true in the environment in which we evolved, and they pursue goals that subvert their own well-being but were adaptive in that environment. And contrary to the widespread belief that cultures can vary arbitrarily and without limit, surveys of the ethnographic literature show that the peoples of the world share an astonishingly detailed universal psychology.

But if the mind has a complex innate structure, that does *not* mean that learning is unimportant. Framing the issue in such a way that innate structure and learning are pitted against each other, either as alternatives or, almost as bad, as complementary ingredients or interacting forces, is a colossal mistake. It's not that the claim that there is an interaction between innate structure and learning (or between heredity and environment, nature and nurture, biology and culture) is literally wrong. Rather, it falls into the category of ideas that are so bad they are not even wrong.

Imagine the following dialogue:

"This new computer is brimming with sophisticated technology. It has a 500 megahertz processor, a gigabyte of RAM, a terabyte of disk storage, a 3-D color virtual reality display, speech output, wireless access to the World Wide Web, expertise in a dozen subjects, and built-in editions of the Bible, the *Encyclopaedia Britannica*, *Bartlett's Famous Quotations*, and the complete works of Shakespeare. Tens of thousands of hacker-hours went into its design."

"Oh, so I guess you're saying that it doesn't matter what I type into the computer. With all that built-in structure, its environment can't be very important. It will always do the same thing, regardless of what I type in."

The response is patently senseless. Having a lot of built-in machinery should make a system respond *more* intelligently and flexibly to its inputs, not less. Yet the reply captures how centuries of commentators have reacted to the idea of a richly structured, high-tech mind.

And the "interactionist" position, with its phobia of ever specifying the innate part of the interaction, is not much better. Look at these claims.

> The behavior of a computer comes from a complex interaction between the processor and the input.
>
> When trying to understand how a car works, one cannot neglect the engine or the gasoline or the driver. All are important factors.
>
> The sound coming out of this CD player represents the inextricably intertwined mixture of two crucial variables: the structure of the machine, and the disk you insert into it. Neither can be ignored.

These statements are true but useless—so blankly uncomprehending, so defiantly incurious, that it is almost as bad to assert them as to deny them. For minds, just as for machines, the metaphors of a mixture of two ingredients, like a martini, or a battle between matched forces, like a tug-of-war, are wrongheaded ways of thinking about a complex device designed to process information. Yes, every part of human intelligence involves culture and learning. But learning is not a surrounding gas or force field, and it does not happen by magic. It is made possible by innate machinery designed to do the learning. The claim that there are several innate modules is a claim that there are several innate learning machines, each of which learns according to a particular logic. To understand learning, we need new ways of thinking to replace the prescientific metaphors—the mixtures and forces, the writing on slates and sculpting of blocks of marble. We need ideas that capture the ways a complex device can tune itself to unpredictable aspects of the world and take in the kinds of data it needs to function.

The idea that heredity and environment interact is not always meaningless, but I think it confuses two issues: what all minds have in common, and how minds can differ. The vapid statements above can be made intelligible by replacing "How X works" with "What makes X work better than Y":

> The *usefulness* of a computer depends on both the power of its processor and the expertise of the user.
> The *speed* of a car depends on the engine, the fuel, and the skill of the driver. All are important factors.
> The *quality* of sound coming from a CD player depends on two crucial variables: the player's mechanical and electronic design, and the quality of the original recording. Neither can be ignored.

When we are interested in *how much better* one system functions than a similar one, it is reasonable to gloss over the causal chains inside each system and tally up the factors that make the whole thing fast or slow, hi-fi or low-fi. And this *ranking* of people—to determine who enters medical school, or who gets the job—is where the framing of nature versus nurture comes from.

But this book is about how the mind works, not about why some people's minds might work a bit better in certain ways than other people's minds. The evidence suggests that humans everywhere on the planet see, talk, and think about objects and people in the same basic way. The difference between Einstein and a high school dropout is trivial compared to the difference between the high school dropout and the best robot in existence, or between the high school dropout and a chimpanzee. That is the mystery I want to address. Nothing could be farther from my subject matter than a comparison between the means of overlapping bell curves for some crude consumer index like IQ. And for this reason, the relative importance of innateness and learning is a phony issue.

An emphasis on innate design should not, by the way, be confused with the search for "a gene for" this or that mental organ. Think of the genes and putative genes that have made the headlines: genes for muscular dystrophy, Huntington's disease, Alzheimer's, alcoholism, schizophrenia, manic-depressive disorder, obesity, violent outbursts, dyslexia, bed-wetting, and some kinds of retardation. They are *disorders*, all of them. There have been no discoveries of a gene for civility, language, memory, motor control, intelligence, or other complete mental systems, and there probably won't ever be. The reason was summed up by the politician Sam Rayburn: Any jackass can kick down a barn, but it takes a carpenter to build one. Complex mental organs, like complex physical organs, surely are built by complex genetic recipes, with many genes cooperating in as yet unfathomable ways. A defect in any one of them could corrupt the whole device, just as a defect in any part of a complicated machine (like a loose distributor cable in a car) can bring the machine to a halt.

The genetic assembly instructions for a mental organ do not specify every connection in the brain as if they were a wiring schematic for a Heathkit radio. And we should not expect each organ to grow under a particular bone of the skull regardless of what else happens in the brain. The brain and all the other organs differentiate in embryonic development from a ball of identical cells. Every part of the body, from the toenails to the cerebral cortex, takes on its particular shape and substance when its cells respond to some kind of information in its neighborhood that unlocks a different part of the genetic program. The information may come from the taste of the chemical soup that a cell finds itself in, from the shapes of the molecular locks and keys that the cell engages, from mechanical tugs and shoves from neighboring cells, and other cues still poorly understood. The families of neurons that will form the different mental organs, all descendants of a homogeneous stretch of embryonic tissue, must be designed to be opportunistic as the brain assembles itself, seizing any available information to differentiate from one another. The coordinates in the skull may be one trigger for differentiation, but the pattern of input firings from connected neurons is another. Since the brain is destined to be an organ of computation, it would be surprising if the genome did not exploit the capacity of neural tissue to process information during brain assembly.

In the sensory areas of the brain, where we can best keep track of what is going on, we know that early in fetal development neurons are wired according to a rough genetic recipe. The neurons are born in appropriate numbers at the right times, migrate to their resting places, send out connections to their targets, and hook up to appropriate cell types in the right general regions, all under the guidance of chemical trails and molecular locks and keys. To make precise connections, though, the baby neurons must begin to function, and their firing pattern carries information downstream about their pinpoint connections. This isn't "experience," as it all can take place in the pitch-black womb, sometimes before the rods and cones are functioning, and many mammals can see almost perfectly as soon as they are born. It is more like a kind of genetic data compression or a set of internally generated test patterns. These patterns can trigger the cortex at the receiving end to differentiate, at least one step of the way, into the kind of cortex that is appropriate to processing the incoming information. (For example, in animals that have been cross-wired so that the eyes are connected to the auditory brain, that area shows a few hints of the properties of the visual brain.) How the genes control brain development is still unknown, but a reasonable summary of what we know so far is that brain modules assume their identity by a combination of what kind of tissue they start out as, where they are in the brain, and what patterns of triggering input they get during critical periods in development.

Our organs of computation are a product of natural selection. The biologist Richard Dawkins called natural selection the Blind Watchmaker; in the case of the mind, we can call it the Blind Programmer. Our mental programs work as well as they do because they were shaped by selection to allow our ancestors to master rocks, tools, plants, animals, and each other, ultimately in the service of survival and reproduction.

Natural selection is not the only cause of evolutionary change. Organisms also change over the eons because of statistical accidents in who lives and who dies, environmental catastrophes that wipe out whole families of creatures, and the unavoidable by-products of changes that *are* the product of selection. But natural selection is the only evolutionary force that acts like an engineer, "designing" organs that accomplish improbable but adaptive outcomes (a point that has been made forcefully by the biologist George Williams and by Dawkins). The textbook argument for natural selection, accepted even by those who feel that selection has been overrated (such as the paleontologist Stephen Jay Gould), comes from the vertebrate eye. Just as a watch has too many finely meshing parts (gears, springs, pivots, and so on) to have been assembled by a tornado or a river eddy, entailing instead the design of a watchmaker, the eye has too many finely meshing parts (lens, iris, retina, and so on) to have arisen from a random evolutionary force like a big mutation, statistical drift, or the fortuitous shape of the nooks and crannies between other organs. The design of the eye must be a product of natural selection of replicators, the only nonmiraculous natural process we know of that can manufacture well-functioning machines. The organism appears as if it was designed to see well now because it owes its existence to the success of its ancestors in seeing well in the past. (This point will be expanded in Chapter 3.)

Many people acknowledge that natural selection is the artificer of the body but draw the line when it comes to the human mind. The mind, they say, is a by-product of a mutation that enlarged the head, or is a clumsy programmer's hack, or was given its shape by cultural rather than biological evolution. Tooby and Cosmides point out a delicious irony. The eye, that most uncontroversial example of fine engineering by natural selection, is not just any old organ that can be sequestered with flesh and bone, far away from the land of the mental. It doesn't digest food or, except in the case of Superman, change anything in the physical world. What does the eye do? The eye is an organ of information processing, firmly connected to—anatomically speaking, a part of—the brain. And all those delicate optics and intricate circuits in the retina do not dump information into a yawning empty orifice or span some Cartesian chasm from a physical to a mental realm. The receiver of this richly structured message must be every bit as well engineered as the sender. As we have seen in comparing human vision and robot vision, the parts of the mind that allow us to see are indeed well engineered, and there is no reason to think that the quality of engineering progressively deteriorates as the information flows upstream to the faculties that interpret and act on what we see.

The adaptationist program in biology, or the careful use of natural selection to reverse-engineer the parts of an organism, is sometimes ridiculed as an empty exercise in after-the-fact storytelling. In the satire of the syndicated columnist Cecil Adams, "the reason our hair is brown is that it enabled our monkey ancestors to hide amongst the coconuts." Admittedly, there is no shortage of bad evolutionary "explanations." Why do men avoid asking for directions? Because our male ancestors might have been killed if they approached a stranger. What purpose does music serve? It brings the community together. Why did happiness evolve? Because happy people are pleasant to be around, so they attracted more allies. What is the function of humor? To relieve tension. Why do people overestimate their chance of surviving an illness? Because it helps them to operate effectively in life.

These musings strike us as glib and lame, but it is not because they dare to seek an evolutionary explanation of how some part of the mind works. It is because they botch the job. First, many of them never bother to establish the facts. Has anyone ever documented that *women* like to ask for directions? Would a woman in a foraging society *not* have come to harm when she approached a stranger? Second, even if the facts had been established, the stories try to explain one puzzling fact by taking for granted some other fact that is just as much of a puzzle, getting us nowhere. *Why* do rhythmic noises bring a community together? *Why* do people like to be with happy people? *Why* does humor relieve tension? The authors of these explanations treat some parts of our mental life as so obvious—they are, after all, obvious to each of *us*, here inside our heads—that they don't need to be explained. But *all* parts of the mind are up for grabs—every reaction, every pleasure, every taste—when we try to explain how it evolved. We *could have* evolved like the Samaritan I robot, which sacrificed itself to save a sack of lima beans, or like dung beetles, which must find dung delicious, or like the masochist in the old joke about sadomasochism (Masochist: "Hit me!" Sadist: "No!").

A good adaptationist explanation needs the fulcrum of an engineering analysis that is independent of the part of the mind we are trying to explain. The analysis begins with a goal to be attained and a world of causes and effects in which to attain it, and goes on to specify what kinds of designs are better suited to attain it than others. Unfortunately for those who think that the departments in a university reflect meaningful divisions of knowledge, it means that psychologists have to look outside psychology if they want to explain what the parts of the mind are for. To understand sight, we have to look to optics and computer vision systems. To understand movement, we have to look to robotics. To understand sexual and familial feelings, we have to look to Mendelian genetics. To understand cooperation and conflict, we have to look to the mathematics of games and to economic modeling.

Once we have a spec sheet for a well-designed mind, we can see whether *Homo sapiens* has that kind of mind. We do the experiments or surveys to get the facts down about a mental faculty, and then see whether the faculty meets the specs: whether it shows signs of precision, complexity, efficiency, reliability, and specialization in solving its assigned problem, especially in comparison with the vast number of alternative designs that are biologically growable.

The logic of reverse-engineering has guided researchers in visual perception for over a century, and that may be why we understand vision better than we understand any other part of the mind. There is no reason that reverse-engineering guided by evolutionary theory should not bring insight about the rest of the mind. An interesting example is a new theory of pregnancy sickness (traditionally called "morning sickness") by the biologist Margie Profet. Many pregnant women become nauseated and avoid certain foods. Though their sickness is usually explained away as a side effect of hormones, there is no reason that hormones should induce nausea and food aversions rather than, say, hyperactivity, aggressiveness, or lust. The Freudian explanation is equally unsatisfying: that pregnancy sickness represents the woman's loathing of her husband and her unconscious desire to abort the fetus orally.

Profet predicted that pregnancy sickness should confer some benefit that offsets the cost of lowered nutrition and productivity. Ordinarily, nausea is a protection against eating toxins: the poisonous food is ejected from the stomach before it can do much harm, and our appetite for similar foods is reduced in the future. Perhaps pregnancy sickness protects women against eating or digesting foods with toxins that might harm the developing fetus. Your local Happy Carrot Health Food Store notwithstanding, there is nothing particularly healthy about natural foods. Your cabbage, a Darwinian creature, has no more desire to be eaten than you do, and since it can't very well defend itself through behavior, it resorts to chemical warfare. Most plants have evolved dozens of toxins in their tissues: insecticides, insect repellents, irritants, paralytics, poisons, and other sand to throw in herbivores' gears. Herbivores have in turn evolved countermeasures, such as a liver to detoxify the poisons and the taste sensation we call bitterness to deter any further desire to ingest them. But the usual defenses may not be enough to protect a tiny embryo.

So far this may not sound much better than the barf-up-your-baby theory, but Profet synthesized hundreds of studies, done independently of each other and of her hypothesis, that support it. She meticulously documented that (1) plant toxins in dosages that adults tolerate can cause birth defects and induce abortion when ingested by pregnant women; (2) pregnancy sickness begins at the point when the embryo's organ systems are being laid down and the embryo is most vulnerable to teratogens (birth defect—inducing chemicals) but is growing slowly and has only a modest need for nutrients; (3) pregnancy sickness wanes at the stage when the embryo's organ systems are nearly complete and its biggest need is for nutrients to allow it to grow; (4) women with pregnancy sickness selectively avoid bitter, pungent, highly flavored, and novel foods, which are in fact the ones most likely to contain toxins; (5) women's sense of smell becomes hypersensitive during the window of pregnancy sickness and less sensitive than usual thereafter; (6) foraging peoples (including, presumably, our ancestors) are at even higher risk of ingesting plant toxins, because they eat wild plants rather than domesticated crops bred for palatability; (7) pregnancy sickness is universal across human cultures; (8) women with more severe pregnancy sickness are less likely to miscarry; (9) women with more severe pregnancy sickness are less likely to bear babies with birth defects. The fit between how a baby-making system in a natural ecosystem ought to work and how the feelings of modern women do work is impressive, and gives a measure of confidence that Profet's hypothesis is correct.

The human mind is a product of evolution, so our mental organs are either present in the minds of apes (and perhaps other mammals and vertebrates) or arose from overhauling the minds of apes, specifically, the common ancestors of humans and chimpanzees that lived about six million years ago in Africa. Many titles of books on human evolution remind us of this fact: *The Naked Ape, The Electric Ape, The Scented Ape, The Lopsided Ape, The Aquatic Ape, The Thinking Ape, The Human Ape, The Ape That Spoke, The Third Chimpanzee, The Chosen Primate.* Some authors are militant that humans are barely different from chimpanzees and that any focus on specifically human talents is arrogant chauvinism or tantamount to creationism. For some readers that is a reductio ad absurdum of the evolutionary framework. If the theory says that man "at best is only a monkey shaved," as Gilbert and Sullivan put it in *Princess Ida*, then it fails to explain the obvious fact that men and monkeys have different minds.

We *are* naked, lopsided apes that speak, but we also have minds that differ considerably from those of apes. The outsize brain of *Homo sapiens sapiens* is, by any standard, an extraordinary adaptation. It has allowed us to inhabit every ecosystem on earth, reshape the planet, walk on the moon, and discover the secrets of the physical universe. Chimpanzees, for all their vaunted intelligence, are a threatened species clinging to a few patches of forest and living as they did millions of years ago. Our curiosity about this difference demands more than repeating that we share most of our DNA with chimpanzees and that small changes can have big effects. Three hundred thousand generations and up to ten megabytes of potential genetic information are enough to revamp a mind considerably. Indeed, minds are probably easier to revamp than bodies because software is easier to modify than hardware. We should not be surprised to discover impressive new cognitive abilities in humans, language being just the most obvious one.

None of this is incompatible with the theory of evolution. Evolution is a conservative process, to be sure, but it can't be all *that* conservative or we would all be pond scum. Natural selection introduces differences into descendants by fitting them with specializations that adapt them to different niches. Any museum of natural history has examples of complex organs unique to a species or to a group of related species: the elephant's trunk, the narwhal's tusk, the whale's baleen, the platypus' duckbill, the armadillo's armor. Often they evolve rapidly on the geological timescale. The first whale evolved in something like ten million years from its common ancestor with its closest living relatives, ungulates such as cows and pigs. A book about whales could, in the spirit of the human-evolution books, be called *The Naked Cow*, but it would be disappointing if the book spent every page marveling at the similarities between whales and cows and never got around to discussing the adaptations that make them so different.

To say that the mind is an evolutionary adaptation is not to say that all behavior is adaptive in Darwin's sense. Natural selection is not a guardian angel that hovers over us making sure that our behavior always maximizes biological fitness. Until recently, scientists with an evolutionary bent felt a responsibility to account for acts that seem like Darwinian suicide, such as celibacy, adoption, and contraception. Perhaps, they ventured, celibate people have more time to raise large broods of nieces and nephews and thereby propagate more copies of their genes than they would if they had their own children. This kind of stretch is unnecessary, however. The reasons, first articulated by the anthropologist Donald Symons, distinguish evolutionary psychology from the school of thought in the 1970s and 1980s called sociobiology (though there is much overlap between the approaches as well).

First, selection operates over thousands of generations. For ninety-nine percent of human existence, people lived as foragers in small nomadic bands. Our brains are adapted to that long-vanished way of life, not to brand-new agricultural and industrial civilizations. They are not wired to cope with anonymous crowds, schooling, written language, government, police, courts, armies, modern medicine, formal social institutions, high technology, and other newcomers to the human experience. Since the modern mind is adapted to the Stone Age, not the computer age, there is no need to strain for adaptive explanations for everything we do. Our ancestral environment lacked the institutions that now entice us to nonadaptive choices, such as religious orders, adoption agencies, and pharmaceutical companies, so until very recently there was never a selection pressure to resist the enticements. Had the Pleistocene savanna contained trees bearing birth-control pills, we might have evolved to find them as terrifying as a venomous spider.

Second, natural selection is not a puppetmaster that pulls the strings of behavior directly. It acts by designing the generator of behavior: the package of information-processing and goal-pursuing mechanisms called the mind. Our minds are designed to generate behavior that would have been adaptive, on average, in our ancestral environment, but any particular deed done today is the effect of dozens of causes. Behavior is the outcome of an internal struggle among many mental modules, and it is played out on the chessboard of opportunities and constraints defined by *other* people's behavior. A recent cover story in *Time* asked, "Adultery: Is It in Our Genes?" The question makes no sense because neither adultery nor any other behavior can be in our genes. Conceivably a *desire* for adultery can be an indirect product of our genes, but the desire may be overridden by *other* desires that are also indirect products of our genes, such as the desire to have a trusting spouse. And the desire, even if it prevails in the rough-and-tumble of the mind, cannot be consummated as overt behavior unless there is a partner around in whom that desire has also prevailed. Behavior itself did not evolve; what evolved was the mind.

Reverse-engineering is possible only when one has a hint of what the device was designed to accomplish. We do not understand the olive-pitter until we catch on that it was designed as a machine for pitting olives rather than as a paperweight or wrist-exerciser. The goals of the designer must be sought for every part of a complex device and for the device as a whole. Automobiles have a component, the carburetor, that is designed to mix air and gasoline, and mixing air and gasoline is a subgoal of the ultimate goal, carting people around. Though the process of natural selection itself has no goal, it evolved entities that (like the automobile) are highly organized to bring about certain goals and subgoals. To reverse-engineer the mind, we must sort them out and identify the ultimate goal in its design. Was the human mind ultimately designed to create beauty? To discover truth? To love and to work? To harmonize with other human beings and with nature?

The logic of natural selection gives the answer. The ultimate goal that the mind was designed to attain is maximizing the number of copies of the genes that created it. Natural selection cares only about the long-term fate of entities that replicate, that is, entities that retain a stable identity across many generations of copying. It predicts only that replicators whose effects tend to enhance the probability of their own replication come to predominate. When we ask questions like "Who or what is supposed to benefit from an adaptation?" and "What is a design in living things a design *for*?" the theory of natural selection provides the answer: the long-term stable replicators, genes. Even our bodies, our selves, are not the ultimate beneficiary of our design. As Gould has said, "What is the 'individual reproductive success' of which Darwin speaks? It cannot be the passage of one's body into the next generation—for, truly, you can't take it with you in this sense above all!" The criterion by which genes get selected is the quality of the bodies they build, but it is the genes making it into the next generation, not the perishable bodies, that are selected to live and fight another day.

Though there are some holdouts (such as Gould himself), the gene's-eye view predominates in evolutionary biology and has been a stunning success. It has asked, and is finding answers to, the deepest questions about life, such as how life arose, why there are cells, why there are bodies, why there is sex, how the genome is structured, why animals interact socially, and why there is communication. It is as indispensable to researchers in animal behavior as Newton's laws are to mechanical engineers.

But almost everyone misunderstands the theory. Contrary to popular belief, the gene-centered theory of evolution does *not* imply that the point of all human striving is to spread our genes. With the exception of the fertility doctor who artificially inseminated patients with his own semen, the donors to the sperm bank for Nobel Prize winners, and other kooks, *no* human being (or animal) strives to spread his or her genes. Dawkins explained the theory in a book called *The Selfish Gene*, and the metaphor was chosen carefully. People don't selfishly spread their genes; genes selfishly spread themselves. They do it by the way they build our brains. By making us enjoy life, health, sex, friends, and children, the genes buy a lottery ticket for representation in the next generation, with odds that were favorable in the environment in which we evolved. Our goals are subgoals of the ultimate goal of the genes, replicating themselves. But the two are different. As far as *we* are concerned, our goals, conscious or unconscious, are not about genes at all, but about health and lovers and children and friends.

The confusion between our goals and our genes' goals has spawned one muddle after another. A reviewer of a book about the evolution of sexuality protests that human adultery, unlike the animal equivalent, cannot be a strategy to spread the genes because adulterers take steps to prevent pregnancy. But whose strategy are we talking about? Sexual desire is *not* people's strategy to propagate their genes. It's people's strategy to attain the pleasures of sex, and the pleasures of sex are the genes' strategy to propagate themselves. If the genes don't get propagated, it's because we are smarter than they are. A book on the emotional life of animals complains that if altruism according to biologists is just helping kin or exchanging favors, both of which serve the interests of one's genes, it would not *really* be altruism after all, but some kind of hypocrisy. This too is a mixup. Just as blueprints don't necessarily specify blue buildings, selfish genes don't necessarily specify selfish organisms. As we shall see, sometimes the most selfish thing a gene can do is to build a selfless brain. Genes are a play within a play, not the interior monologue of the players.

PSYCHOLOGICAL CORRECTNESS

The evolutionary psychology of this book is a departure from the dominant view of the human mind in our intellectual tradition, which Tooby and Cosmides have dubbed the Standard Social Science Model (SSSM). The SSSM proposes a fundamental division between biology and culture. Biology endows humans with the five senses, a few drives like hunger and fear, and a general capacity to learn. But biological evolution, according to the SSSM, has been superseded by cultural evolution. Culture is an autonomous entity that carries out a desire to perpetuate itself by setting up expectations and assigning roles, which can vary arbitrarily from society to society. Even the reformers of the SSSM have accepted its framing of the issues. Biology is "just as important as" culture, say the reformers; biology imposes "constraints" on behavior, and all behavior is a mixture of the two.

The SSSM not only has become an intellectual orthodoxy but has acquired a moral authority. When sociobiologists first began to challenge it, they met with a ferocity that is unusual even by the standards of academic invective. The biologist E. O. Wilson was doused with a pitcher of ice water at a scientific convention, and students yelled for his dismissal over bullhorns and put up posters urging people to bring noisemakers to his lectures. Angry manifestos and book-length denunciations were published by organizations with names like Science for the People and The Campaign Against Racism, IQ, and the Class Society. In *Not in Our Genes*, Richard Lewontin, Steven Rose, and Leon Kamin dropped innuendos about Donald Symons' sex life and doctored a defensible passage of Richard Dawkins' into an insane one. (Dawkins said of the genes, "They created us, body and mind"; the authors have quoted it repeatedly as "They *control* us, body and mind.") When *Scientific American* ran an article on behavior genetics (studies of twins, families, and adoptees), they entitled it "Eugenics Revisited," an allusion to the discredited movement to improve the human genetic stock. When the magazine covered evolutionary psychology, they called the article "The New Social Darwinists," an allusion to the nineteenth-century movement that justified social inequality as part of the wisdom of nature. Even one of sociobiology's distinguished practitioners, the primatologist Sarah Blaffer Hrdy, said, "I question whether sociobiology should be taught at the high school level, or even the undergraduate level. . . . The whole message of sociobiology is oriented toward the success of the individual. It's Machiavellian, and unless a student has a moral framework already in place, we could be producing social monsters by teaching this. It really fits in very nicely with the yuppie 'me first' ethos."

Entire scholarly societies joined in the fun, passing votes on empirical issues that one might have thought would be hashed out in the lab and the field. Margaret Mead's portrayal of an idyllic, egalitarian Samoa was one of the founding documents of the SSSM, and when the anthropologist Derek Freeman showed that she got the facts spectacularly wrong, the American Anthropological Association voted at its business meeting to denounce his finding as unscientific. In 1986, twenty social scientists at a "Brain and Aggression" meeting drafted the Seville Statement on Violence, subsequently adopted by UNESCO and endorsed by several scientific organizations. The statement claimed to "challenge a number of alleged biological findings that have been used, even by some in our disciplines, to justify violence and war":

> It is scientifically incorrect to say that we have inherited a tendency to make war from our animal ancestors.
> It is scientifically incorrect to say that war or any other violent behavior is genetically programmed into our human nature.
> It is scientifically incorrect to say that in the course of human evolution there has been a selection for aggressive behavior more than for other kinds of behavior.
> It is scientifically incorrect to say that humans have a "violent brain."
> It is scientifically incorrect to say that war is caused by "instinct" or any single motivation. . . . We conclude that biology does not condemn humanity to war, and that humanity can be freed from the bondage of biological pessimism and empowered with confidence to undertake the transformative tasks needed in the International Year of Peace and in the years to come.

What moral certainty could have incited these scholars to doctor quotations, censor ideas, attack the ideas' proponents ad hominem, smear them with unwarranted associations to repugnant political movements, and mobilize powerful institutions to legislate what is correct and incorrect? The certainty comes from an opposition to three putative implications of an innate human nature.

First, if the mind has an innate structure, different people (or different classes, sexes, and races) could have different innate structures. That would justify discrimination and oppression.

Second, if obnoxious behavior like aggression, war, rape, clannishness, and the pursuit of status and wealth are innate, that would make them "natural" and hence good. And even if they are deemed objectionable, they are in the genes and cannot be changed, so attempts at social reform are futile.

Third, if behavior is caused by the genes, then individuals cannot be held responsible for their actions. If the rapist is following a biological imperative to spread his genes, it's not his fault.

Aside perhaps from a few cynical defense lawyers and a lunatic fringe who are unlikely to read manifestos in the *New York Review of Books*, no one has actually drawn these mad conclusions. Rather, they are thought to be extrapolations that the untutored masses *might* draw, so the dangerous ideas must themselves be suppressed. In fact, the problem with the three arguments is not that the conclusions are so abhorrent that no one should be allowed near the top of the slippery slope that leads to them. The problem is that there is no such slope; the arguments are non sequiturs. To expose them, one need only examine the logic of the theories and separate the scientific from the moral issues.

My point is not that scientists should pursue the truth in their ivory tower, undistracted by moral and political thoughts. Every human act involving another living being is both the subject matter of psychology and the subject matter of moral philosophy, and both are important. But they are not the same thing. The debate over human nature has been muddied by an intellectual laziness, an unwillingness to make moral arguments when moral issues come up. Rather than reasoning from principles of rights and values, the tendency has been to buy an off-the-shelf moral package (generally New Left or Marxist) or to lobby for a feel-good picture of human nature that would spare us from having to argue moral issues at all.

～

The moral equation in most discussions of human nature is simple: innate equals right-wing equals bad. Now, many hereditarian movements *have* been right-wing and bad, such as eugenics, forced sterilization, genocide, discrimination along racial, ethnic, and sexual lines, and the justification of economic and social castes. The Standard Social Science Model, to its credit, has provided some of the grounds that thoughtful social critics have used to undermine these practices.

But the moral equation is wrong as often as it is right. Sometimes left-wing practices are just as bad, and the perpetrators have tried to justify them using the SSSM's denial of human nature. Stalin's purges, the Gulag, Pol Pot's killing fields, and almost fifty years of repression in China—all have been justified by the doctrine that dissenting ideas reflect not the operation of rational minds that have come to different conclusions, but arbitrary cultural products that can be eradicated by re-engineering the society, "re-educating" those who were tainted by the old upbringing, and, if necessary, starting afresh with a new generation of slates that are still blank.

And sometimes left-wing positions are right *because* the denial of human nature is wrong. In *Hearts and Minds*, the 1974 documentary about the war in Vietnam, an American officer explains that we cannot apply our moral standards to the Vietnamese because their culture does not place a value on individual lives, so they do not suffer as we do when family members are killed. The director plays the quote over footage of wailing mourners at the funeral of a Vietnamese casualty, reminding us that the universality of love and grief refutes the officer's horrifying ratio-nalization. For most of this century, guilty mothers have endured inane theories blaming them for every dysfunction or difference in their chil-dren (mixed messages cause schizophrenia, coldness causes autism, domineering causes homosexuality, lack of boundaries causes anorexia, insufficient "motherese" causes language disorders). Menstrual cramps, pregnancy sickness, and childbirth pain have been dismissed as women's "psychological" reactions to cultural expectations, rather than being treated as legitimate health issues.

The foundation of individual rights is the assumption that people have wants and needs and are authorities on what those wants and needs are. If people's stated desires were just some kind of erasable inscription or repro-grammable brainwashing, any atrocity could be justified. (Thus it is ironic that fashionable "liberation" ideologies like those of Michel Foucault and some academic feminists invoke a socially conditioned "interiorized authority," "false consciousness," or "inauthentic preference" to explain away the inconvenient fact that people enjoy the things that are alleged to oppress them.) A denial of human nature, no less than an emphasis on it, can be warped to serve harmful ends. We should expose whatever ends are harmful and whatever ideas are false, and not confuse the two.

So what about the three supposed implications of an innate human nature? The first "implication"—that an innate human nature implies innate human differences—is no implication at all. The mental machinery I argue for is installed in every neurologically normal human being. The differences among people may have nothing to do with the design of that machinery. They could very well come from random variations in the assembly process or from different life histories. Even if the differences were innate, they could be quantitative variations and minor quirks in equipment present in all of us (how fast a module works, which module prevails in a competition inside the head) and are not necessarily any more pernicious than the kinds of innate differences allowed in the Standard Social Science Model (a faster general-purpose learning process, a stronger sex drive).

A universal structure to the mind is not only logically possible but likely to be true. Tooby and Cosmides point out a fundamental consequence of sexual reproduction: every generation, each person's blueprint is scrambled with someone else's. That means we must be qualitatively alike. If two people's genomes had designs for different kinds of machines, like an electric motor and a gasoline engine, the new pastiche would not specify a working machine at all. Natural selection is a homogenizing force within a species; it eliminates the vast majority of macroscopic design variants because they are not improvements. Natural selection does depend on there having been variation in the past, but it feeds off the variation and uses it up. That is why all normal people have the same physical organs, and why we all surely have the same mental organs as well. There are, to be sure, microscopic variations among people, mostly small differences in the molecule-by-molecule sequence of many of our proteins. But at the level of functioning organs, physical and mental, people work in the same ways. Differences among people, for all their endless fascination to us as we live our lives, are of minor interest when we ask how the mind works. The same is true for differences—whatever their source—between the averages of entire groups of people, such as races.

The sexes, of course, are a different matter. The male and female reproductive organs are a vivid reminder that qualitatively different designs *are* possible for the sexes, and we know that the differences come from the special gadget of a genetic "switch," which triggers a line of biochemical dominoes that activate and deactivate families of genes throughout the brain and body. I will present evidence that some of these effects cause differences in how the mind works. In another of the ironies that run through the academic politics of human nature, this evolution-inspired research has proposed sex differences that are tightly focused on reproduction and related domains, and are far less invidious than the differences proudly claimed by some schools of feminism. Among the claims of "difference feminists" are that women do not engage in abstract linear reasoning, that they do not treat ideas with skepticism or evaluate them through rigorous debate, that they do not argue from general moral principles, and other insults.

But ultimately we cannot just look at who is portrayed more flatteringly; the question is what to make of any group differences we do stumble upon. And here we must be prepared to make a moral argument. Discrimination against individuals on the basis of their race, sex, or ethnicity is wrong. The argument can be defended in various ways that have nothing to do with the average traits of the groups. One might argue that it is unfair to deny a social benefit to individuals because of factors they cannot control, or that a victim of discrimination experiences it as a uniquely painful sting, or that a group of victims is liable to react with rage, or that discrimination tends to escalate into horrors like slavery and genocide. (Those who favor affirmative action could acknowledge that reverse discrimination is wrong but argue that it undoes an even greater wrong.) None of these arguments is affected by anything any scientist will ever claim to discover. The final word on the political non-implications of group differences must go to Gloria Steinem: "There are really not many jobs that actually require a penis or a vagina, and all the other occupations should be open to everyone."

The fallacy of the second supposed implication of a human nature— that if our ignoble motives are innate, they can't be so bad after all—is so obvious it has been given a name: the naturalistic fallacy, that what happens in nature is right. Forget the romantic nonsense in wildlife documentaries, where all creatures great and small act for the greater good and the harmony of the ecosystem. As Darwin said, "What a book a devil's chaplain might write on the clumsy, wasteful, blundering, low, and horribly cruel works of nature!" A classic example is the ichneumon wasp, who paralyzes a caterpillar and lays eggs in its body so her hatchlings can slowly devour its living flesh from the inside.

Like many species, *Homo sapiens* is a nasty business. Recorded history from the Bible to the present is a story of murder, rape, and war, and honest ethnography shows that foraging peoples, like the rest of us, are more savage than noble. The !Kung San of the Kalahari Desert are often held out as a relatively peaceful people, and so they are, compared with other foragers: their murder rate is only as high as Detroit's. A linguist friend of mine who studies the Wari in the Amazon rainforest learned that their language has a term for edible things, which includes anyone who isn't a Wari. Of course humans don't have an "instinct for war" or a "violent brain," as the Seville Statement assures us, but humans don't exactly have an instinct for peace or a nonviolent brain, either. We cannot attribute all of human history and ethnography to toy guns and superhero cartoons.

Does that mean that "biology condemns man to war" (or rape or murder or selfish yuppies) and that any optimism about reducing it should be snuffed out? No one needs a scientist to make the moral point that war is not healthy for children and other living things, or the empirical point that some places and periods are vastly more peaceable than others and that we should try to understand and duplicate what makes them so. And no one needs the bromides of the Seville Statement or its disinformation that war is unknown among animals and that their dominance hierarchies are a form of bonding and affiliation that benefits the group. What could not hurt is a realistic understanding of the psychology of human malevolence. For what it's worth, the theory of a module-packed mind allows both for innate motives that lead to evil acts and for innate motives that can avert them. Not that this is a unique discovery of evolutionary psychology; all the major religions observe that mental life is often a struggle between desire and conscience.

When it comes to the hopes of changing bad behavior, the conventional wisdom again needs to be inverted: a complex human nature may allow *more* scope for change than the blank slate of the Standard Social Science Model. A richly structured mind allows for complicated negotiations inside the head, and one module could subvert the ugly designs of another one. In the SSSM, in contrast, upbringing is often said to have an insidious and irreversible power. "Is it a boy or a girl?" is the first question we ask about a new human being, and from then on parents treat their sons and daughters differently: they touch, comfort, breast-feed, indulge, and talk to boys and girls in unequal amounts. Imagine that this behavior has long-term consequences on the children, which include all the documented sex differences *and* a tendency to treat *their* children differently from birth. Unless we stationed parenting police in the maternity ward, the circle would be complete and irrevocable. Culture would condemn women to inferiority, and we would be enslaved to the bondage of cultural pessimism, disempowered by self-doubt from undertaking transformative tasks.

Nature does not dictate what we should accept or how we should live our lives. Some feminists and gay activists react with fury to the banal observations that natural selection designed women in part for growing and nursing children and that it designed both men and women for heterosexual sex. They see in those observations the sexist and homophobic message that only traditional sexual roles are "natural" and that alternative lifestyles are to be condemned. For example, the novelist Mary Gordon, mocking a historian's remark that what all women have in common is the ability to bear children, wrote, "If the defining quality of being a woman is the ability to bear children, then not bearing children (as, for instance, Florence Nightingale and Greta Garbo did not) is somehow a failure to fulfill your destiny." I'm not sure what "the defining quality of being a woman" and "fulfilling your destiny" even *mean*, but I do know that happiness and virtue have nothing to do with what natural selection designed us to accomplish in the ancestral environment. They are for us to determine. In saying this I am no hypocrite, even though I am a conventional straight white male. Well into my procreating years I am, so far, voluntarily childless, having squandered my biological resources reading and writing, doing research, helping out friends and students, and jogging in circles, ignoring the solemn imperative to spread my genes. By Darwinian standards I am a horrible mistake, a pathetic loser, not one iota less than if I were a card-carrying member of Queer Nation. But I am happy to be that way, and if my genes don't like it, they can go jump in the lake.

～

Finally, what about blaming bad behavior on our genes? The neuroscientist Steven Rose, in a review of a book by E. O. Wilson in which Wilson wrote that men have a greater desire for polygamy than women, accused him of really saying, "Don't blame your mates for sleeping around, ladies, it's not their fault they are genetically programmed." The title of Rose's own book with Lewontin and Kamin, *Not in Our Genes*, is an allusion to *Julius Caesar*:

Men at some time are masters of their fates:
The fault, dear Brutus, lies not in our stars,
But in ourselves . . .

For Cassius, the programming that was thought to excuse human faults
was not genetic but astrological, and that raises a key point. *Any* cause of
behavior, not just the genes, raises the question of free will and responsibility. The difference between explaining behavior and excusing it is an
ancient theme of moral reasoning, captured in the saw "To understand is
not to forgive."

In this scientific age, "to understand" means to try to explain behavior
as a complex interaction among (1) the genes, (2) the anatomy of the
brain, (3) its biochemical state, (4) the person's family upbringing, (5)
the way society has treated him or her, and (6) the stimuli that impinge
upon the person. Sure enough, *every one* of these factors, not just the
stars or the genes, has been inappropriately invoked as the source of our
faults and a claim that we are not masters of our fates.

(1) In 1993 researchers identified a gene that was associated with
uncontrollable violent outbursts. ("Think of the implications," one
columnist wrote. "We may someday have a cure for hockey.") Soon afterward came the inevitable headline: "Man's Genes Have Made Him Kill,
His Lawyers Claim."

(2) In 1982 an expert witness in the insanity defense of John Hinckley, who had shot President Reagan and three other men to impress the
actress Jodie Foster, argued that a CAT scan of Hinckley's brain showed
widened sulci and enlarged ventricles, a sign of schizophrenia and thus
an excusing mental disease or defect. (The judge excluded the evidence,
though the insanity defense prevailed.)

(3) In 1978 Dan White, having resigned from the San Francisco
Board of Supervisors, walked into Mayor George Moscone's office and
begged to be reinstated. When Moscone refused, White shot him dead,
walked down the hall into the office of Supervisor Harvey Milk, and shot
him dead too. White's lawyers successfully argued that at the time of his
crime White had diminished capacity and had not committed a premed-
itated act because his binges on sugary junk food played havoc with his
brain chemistry. White was convicted of voluntary manslaughter and
served five years, thanks to the tactic that lives on in infamy as the
Twinkie Defense. Similarly, in what is now known as the PMS (premen-

strual syndrome) Defense, raging hormones exonerated a surgeon who had assaulted a trooper who stopped her for drunk driving.

(4) In 1989 Lyle and Erik Menendez burst into their millionaire parents' bedroom and killed them with a shotgun. After several months of showing off their new Porsches and Rolexes, they confessed to the shootings. Their lawyers argued the case to a hung jury by claiming self-defense, despite the fact that the victims had been lying in bed, unarmed, eating strawberries and ice cream. The Menendez boys, the lawyers said, had been traumatized into believing that their parents were going to kill them because they had been physically, sexually, and emotionally abused by the father for years. (In a new trial in 1996 they were convicted of murder and sent to prison for life.)

(5) In 1994 Colin Ferguson boarded a train and began to shoot white people at random, killing six. The radical lawyer William Kunstler was prepared to defend him by invoking the Black Rage Syndrome, in which an African American can suddenly burst under the accumulated pressure of living in a racist society. (Ferguson rejected the offer and argued his own case, unsuccessfully.)

(6) In 1992 a death-row inmate asked an appeals court to reduce his sentence for rape and murder because he had committed his crimes under the influence of pornography. The Pornography-Made-Me-Do-It Defense is an irony for the schools of feminism that argue that biological explanations of rape reduce the rapist's responsibility and that a good tactic to fight violence against women is to blame it on pornography.

As science advances and explanations of behavior become less fanciful, the Specter of Creeping Exculpation, as Dennett calls it, will loom larger. Without a clearer moral philosophy, any cause of behavior could be taken to undermine free will and hence moral responsibility. Science is guaranteed to appear to eat away at the will, *regardless* of what it finds, because the scientific mode of explanation cannot accommodate the mysterious notion of uncaused causation that underlies the will. If scientists wanted to show that people had free will, what would they look for? Some random neural event that the rest of the brain amplifies into a signal triggering behavior? But a random event does not fit the concept of free will any more than a lawful one does, and could not serve as the long-sought locus of moral responsibility. We would not find someone guilty if his finger pulled the trigger when it was mechanically connected to a roulette wheel;

why should it be any different if the roulette wheel is inside his skull? The same problem arises for another unpredictable cause that has been suggested as the source of free will, chaos theory, in which, according to the cliché, a butterfly's flutter can set off a cascade of events culminating in a hurricane. A fluttering in the brain that causes a hurricane of behavior, if it were ever found, would still be a cause of behavior and would not fit the concept of uncaused free will that underlies moral responsibility.

Either we dispense with all morality as an unscientific superstition, or we find a way to reconcile causation (genetic or otherwise) with responsibility and free will. I doubt that our puzzlement will ever be completely assuaged, but we can surely reconcile them in part. Like many philosophers, I believe that science and ethics are two self-contained systems played out among the same entities in the world, just as poker and bridge are different games played with the same fifty-two-card deck. The science game treats people as material objects, and its rules are the physical processes that cause behavior through natural selection and neurophysiology. The ethics game treats people as equivalent, sentient, rational, free-willed agents, and its rules are the calculus that assigns moral value to behavior through the behavior's inherent nature or its consequences.

Free will is an idealization of human beings that makes the ethics game playable. Euclidean geometry requires idealizations like infinite straight lines and perfect circles, and its deductions are sound and useful even though the world does not really have infinite straight lines or perfect circles. The world is close enough to the idealization that the theorems can usefully be applied. Similarly, ethical theory requires idealizations like free, sentient, rational, equivalent agents whose behavior is uncaused, and *its* conclusions can be sound and useful even though the world, as seen by science, does not really have uncaused events. As long as there is no outright coercion or gross malfunction of reasoning, the world is close enough to the idealization of free will that moral theory can meaningfully be applied to it.

Science and morality are separate spheres of reasoning. Only by recognizing them as separate can we have them both. If discrimination is wrong only if group averages are the same, if war and rape and greed are wrong only if people are never inclined toward them, if people are responsible for their actions only if the actions are mysterious, then either scientists must be prepared to fudge their data or all of us must be prepared to give up our values. Scientific arguments would turn into the

National Lampoon cover showing a puppy with a gun at its head and the caption "Buy This Magazine or We'll Shoot the Dog."

The knife that separates causal explanations of behavior from moral responsibility for behavior cuts both ways. In the latest twist in the human-nature morality play, a chromosomal marker for homosexuality in some men, the so-called gay gene, was identified by the geneticist Dean Hamer. To the bemusement of Science for the People, this time it is the genetic explanation that is politically correct. Supposedly it refutes right-wingers like Dan Quayle, who had said that homosexuality "is more of a choice than a biological situation. It is a wrong choice." The gay gene has been used to argue that homosexuality is not a choice for which gay people can be held responsible but an involuntary orientation they just can't help. But the reasoning is dangerous. The gay gene could just as easily be said to influence some people to *choose* homosexuality. And like all good science, Hamer's result might be falsified someday, and then where would we be? Conceding that bigotry against gay people is OK after all? The argument against persecuting gay people must be made not in terms of the gay gene or the gay brain but in terms of people's right to engage in private consensual acts without discrimination or harassment.

The cloistering of scientific and moral reasoning in separate arenas also lies behind my recurring metaphor of the mind as a machine, of people as robots. Does this not dehumanize and objectify people and lead us to treat them as inanimate objects? As one humanistic scholar lucidly put it in an Internet posting, does it not render human experience invalid, reifying a model of relating based on an I-It relationship, and delegitimating all other forms of discourse with fundamentally destructive consequences to society? Only if one is so literal-minded that one cannot shift among different stances in conceptualizing people for different purposes. A human being is simultaneously a machine and a sentient free agent, depending on the purpose of the discussion, just as he is also a taxpayer, an insurance salesman, a dental patient, and two hundred pounds of ballast on a commuter airplane, depending on the purpose of the discussion. The mechanistic stance allows us to understand what makes us tick and how we fit into the physical universe. When those discussions wind down for the day, we go back to talking about each other as free and dignified human beings.

The confusion of scientific psychology with moral and political goals, and the resulting pressure to believe in a structureless mind, have rippled perniciously through the academy and modern intellectual discourse. Many of us have been puzzled by the takeover of humanities departments by the doctrines of postmodernism, poststructuralism, and deconstructionism, according to which objectivity is impossible, meaning is self-contradictory, and reality is socially constructed. The motives become clearer when we consider typical statements like "Human beings have constructed and used gender—human beings can deconstruct and stop using gender," and "The heterosexual/homosexual binary is not in nature, but is socially constructed, and therefore deconstructable." Reality is denied to categories, knowledge, and the world itself so that reality can be denied to stereotypes of gender, race, and sexual orientation. The doctrine is basically a convoluted way of getting to the conclusion that oppression of women, gays, and minorities is bad. And the dichotomy between "in nature" and "socially constructed" shows a poverty of the imagination, because it omits a third alternative: that some categories are products of a complex mind designed to mesh with what is in nature.

Mainstream social critics, too, can state any absurdity if it fits the Standard Social Science Model. Little boys are encouraged to argue and fight. Children learn to associate sweets with pleasure because parents use sweets as a reward for eating spinach. Teenagers compete in looks and dress because they follow the example set by spelling bees and award ceremonies. Men are socialized into believing that the goal of sex is an orgasm. Eighty-year-old women are considered less physically attractive than twenty-year-olds because our phallic culture has turned the young girl into the cult object of desire. It's not just that there is no evidence for these astonishing claims, but it is hard to credit that the authors, deep down, believe them themselves. These kinds of claims are uttered without concern for whether they are true; they are part of the secular catechism of our age.

Contemporary social commentary rests on archaic conceptions of the mind. Victims burst under the pressure, boys are conditioned to do this, women are brainwashed to value that, girls are taught to be such-and-such. Where do these explanations come from? From the nineteenth-century hydraulic model of Freud, the drooling dogs and key-pressing vermin of behaviorism, the mind-control plots of bad cold-war movies, the wide-eyed, obedient children of *Father Knows Best*.

But when we look around us, we sense that these simplistic theories just don't ring true. Our mental life is a noisy parliament of competing factions. In dealing with others, we assume they are as complicated as we are, and we guess what they are guessing we are guessing they are guessing. Children defy their parents from the moment they are born, and confound all expectations thereafter: one overcomes horrific circumstances to lead a satisfying life, another is granted every comfort but grows up a rebel without a cause. A modern state loosens its grip, and its peoples enthusiastically take up the vendettas of their grandparents. And there are no robots.

I believe that a psychology of many computational faculties engineered by natural selection is our best hope for a grasp on how the mind works that does justice to its complexity. But I won't convince you with the opening brief in this chapter. The proof must come from insight into problems ranging from how Magic Eye stereograms work to what makes a landscape beautiful to why we find the thought of eating worms disgusting to why men kill their estranged wives. Whether or not you are persuaded by the arguments so far, I hope they have provoked your thoughts and made you curious about the explanations to come.

2

THINKING MACHINES

Like many baby boomers, I was first exposed to problems in philosophy by traveling through another dimension, a dimension not only of sight and sound but of mind, taking a journey into a wondrous land whose boundaries are that of imagination. I am referring to *The Twilight Zone*, the campy television series by Rod Serling that was popular during my childhood. Philosophers often try to clarify difficult concepts using thought experiments, outlandish hypothetical situations that help us explore the implications of our ideas. *The Twilight Zone* actually staged them for the camera.

One of the first episodes was called "The Lonely." James Corry is serving a fifty-year sentence in solitary confinement on a barren asteroid nine million miles from Earth. Allenby, the captain of a supply ship that services the asteroid, takes pity on him and leaves a crate containing "Alicia," a robot that looks and acts like a woman. At first Corry is repulsed, but of course he soon falls deeply in love. A year later Allenby returns with the news that Corry has been pardoned and he has come to get him. Unfortunately Corry can take only fifteen pounds of gear, and Alicia weighs more than that. When Corry refuses to leave, Allenby reluctantly pulls out a gun and shoots Alicia in the face, exposing a tangle of smoking wires. He tells Corry, "All you're leaving behind is loneliness." Corry, devastated, mutters, "I must remember that. I must remember to keep that in mind."

I still remember my horror at the climax, and the episode was much discussed in my pre-teen critics' circle. (Why didn't he just take her head? asked one commentator.) Our pathos came both from sympathy with Corry for his loss and from the sense that a sentient being had been snuffed out. Of course the directors had manipulated the audience by casting a beautiful actress rather than a heap of tin cans to play Alicia. But in evoking our sympathies they raised two vexing questions. Could a mechanical device ever duplicate human intelligence, the ultimate test being whether it could cause a real human to fall in love with it? And if a humanlike machine could be built, would it actually be *conscious*—would dismantling it be the act of murder we felt we had witnessed on the small screen?

The two deepest questions about the mind are "What makes intelligence possible?" and "What makes consciousness possible?" With the advent of cognitive science, intelligence has become intelligible. It may not be too outrageous to say that at a very abstract level of analysis the problem has been solved. But consciousness or sentience, the raw sensation of toothaches and redness and saltiness and middle C, is still a riddle wrapped in a mystery inside an enigma. When asked what consciousness *is*, we have no better answer than Louis Armstrong's when a reporter asked him what jazz is: "Lady, if you have to ask, you'll never know." But even consciousness is not as thoroughgoing a mystery as it used to be. *Parts* of the mystery have been pried off and turned into ordinary scientific problems. In this chapter I will first explore what intelligence is, how a physical being like a robot or a brain could achieve it, and how our brains do achieve it. Then I will turn to what we do and do not understand about consciousness.

THE SEARCH FOR INTELLIGENT LIFE IN THE UNIVERSE

The Search for Intelligent Life in the Universe is the title of a stage act by the comedian Lily Tomlin, an exploration of human follies and foibles. Tomlin's title plays on the two meanings of "intelligence": aptitude (as in the famous tongue-in-cheek definition of intelligence as "whatever IQ tests measure"), and rational, humanlike thought. The second meaning is the one I am writing about here.

We may have trouble defining intelligence, but we recognize it when we see it. Perhaps a thought experiment can clarify the concept. Suppose there was an alien being who in every way looked different from us. What would it have to do to make us think it was intelligent? Science-fiction writers, of course, face this problem as part of their job; what better authority could there be on the answer? The author David Alexander Smith gave as good a characterization of intelligence as I have seen when asked by an interviewer, "What makes a good alien?"

> One, they have to have intelligent but impenetrable responses to situations. You have to be able to observe the alien's behavior and say, "I don't understand the rules by which the alien is making its decisions, but the alien is acting rationally by some set of rules." . . . The second requirement is that they have to care about something. They have to want something and pursue it in the face of obstacles.

To make decisions "rationally," by some set of rules, means to base the decisions on some grounds of truth: correspondence to reality or soundness of inference. An alien who bumped into trees or walked off cliffs, or who went through all the motions of chopping a tree but in fact was hacking at a rock or at empty space, would not seem intelligent. Nor would an alien who saw three predators enter a cave and two leave and then entered the cave as if it were empty.

These rules must be used in service of the second criterion, wanting and pursuing something in the face of obstacles. If we had no fix on what a creature wanted, we could not be impressed when it did something to attain it. For all we know, the creature may have *wanted* to bump into a tree or bang an ax against a rock, and was brilliantly accomplishing what it wanted. In fact, without a specification of a creature's goals, the very idea of intelligence is meaningless. A toadstool could be given a genius award for accomplishing, with pinpoint precision and unerring reliability, the feat of sitting exactly where it is sitting. Nothing would prevent us from agreeing with the cognitive scientist Zenon Pylyshyn that rocks are smarter than cats because rocks have the sense to go away when you kick them.

Finally, the creature has to use the rational rules to attain the goal in different ways, depending on the obstacles to be overcome. As William James explained:

> Romeo wants Juliet as the filings want the magnet; and if no obstacles intervene he moves toward her by as straight a line as they. But Romeo and Juliet, if a wall be built between them, do not remain idiotically pressing their faces against the opposite sides like the magnet and filings with the card. Romeo soon finds a circuitous way, by scaling the wall or otherwise, of touching Juliet's lips directly. With the filings the path is

fixed; whether it reaches the end depends on accidents. With the lover it is the end which is fixed; the path may be modified indefinitely.

Intelligence, then, is the ability to attain goals in the face of obstacles by means of decisions based on rational (truth-obeying) rules. The computer scientists Allen Newell and Herbert Simon fleshed this idea out further by noting that intelligence consists of specifying a goal, assessing the current situation to see how it differs from the goal, and applying a set of operations that reduce the difference. Perhaps reassuringly, by this definition human beings, not just aliens, are intelligent. We have *desires*, and we pursue them using *beliefs*, which, when all goes well, are at least approximately or probabilistically true.

An explanation of intelligence in terms of beliefs and desires is by no means a foregone conclusion. The old theory of stimulus and response from the school of behaviorism held that beliefs and desires have nothing to do with behavior—indeed, that they are as unscientific as banshees and black magic. Humans and animals emit a response to a stimulus either because it was earlier paired with a reflexive trigger for that response (for example, salivating to a bell that was paired with food) or because the response was rewarded in the presence of that stimulus (for example, pressing a bar that delivers a food pellet). As the famous behaviorist B. F. Skinner said, "The question is not whether machines think, but whether men do."

Of course, men and women do think; the stimulus-response theory turned out to be wrong. Why did Sally run out of the building? Because she believed it was on fire and did not want to die. Her fleeing was not a predictable response to some stimulus that can be objectively described in the language of physics and chemistry. Perhaps she left when she saw smoke, but perhaps she left in response to a phone call telling her that the building was on fire, or to the sight of arriving fire trucks, or to the sound of a fire alarm. But none of these stimuli would *necessarily* have sent her out, either. She would *not* have left if she knew that the smoke was from an English muffin in a toaster, or that the phone call was from a friend practicing lines for a play, or that someone had pulled the alarm switch by accident or as a prank, or that the alarms were being tested by an electrician. The light and sound and particles that physicists can measure do not lawfully predict a person's behavior. What does predict Sally's behavior, and predict it well, is whether she *believes* herself to be in danger. Sally's beliefs are, of course, related to the stimuli impinging on her, but only in a tortuous, circuitous way, mediated by all the rest of her beliefs about where she is and how the world works. And Sally's behavior depends just as much on whether she *wants* to escape the danger—if she were a volunteer firefighter, or suicidal, or a zealot who wanted to immolate herself to draw attention to a cause, or had children in the day-care center upstairs, you can bet she would not have fled.

Skinner himself did not pigheadedly insist that measurable stimuli like wavelengths and shapes predicted behavior. Instead, he defined stimuli by his own intuitions. He was perfectly happy calling "danger"—like "praise," "English," and "beauty"—a kind of stimulus. That had the advantage of keeping his theory in line with reality, but it was the advantage of theft over honest toil. We understand what it means for a device to respond to a red light or a loud noise—we can even build one that does—but humans are the only devices in the universe that respond to danger, praise, English, and beauty. The ability of a human to respond to something as physically nebulous as praise is part of the puzzle we are trying to solve, not part of the solution to the puzzle. Praise, danger, English, and all the other things we respond to, no less than beauty, are in the eye of the beholder, and the eye of the beholder is what we want to explain. The chasm between what can be measured by a physicist and what can cause behavior is the reason we must credit people with beliefs and desires.

In our daily lives we all predict and explain other people's behavior from what we think they know and what we think they want. Beliefs and desires are the explanatory tools of our own intuitive psychology, and intuitive psychology is still the most useful and complete science of behavior there is. To predict the vast majority of human acts—going to the refrigerator, getting on the bus, reaching into one's wallet—you don't need to crank through a mathematical model, run a computer simulation of a neural network, or hire a professional psychologist; you can just ask your grandmother.

It's not that common sense should have any more authority in psychology than it does in physics or astronomy. But this part of common sense has so much power and precision in predicting, controlling, and explaining everyday behavior, compared to any alternative ever entertained, that the odds are high that it will be incorporated in some form into our best scientific theories. I call an old friend on the other coast and we agree to meet in Chicago at the entrance of a bar in a certain hotel on a particular day two months hence at 7:45 P.M. I predict, he predicts, and everyone who knows us predicts that on that day at that time we will meet up. And we do meet up. That is amazing! In what other domain could laypeople—

or scientists, for that matter—predict, months in advance, the trajectories of two objects thousands of miles apart to an accuracy of inches and minutes? And do it from information that can be conveyed in a few seconds of conversation? The calculus behind this forecasting is intuitive psychology: the knowledge that I *want* to meet my friend and vice versa, and that each of us *believes* the other will be at a certain place at a certain time and *knows* a sequence of rides, hikes, and flights that will take us there. No science of mind or brain is ever likely to do better. That does not mean that the intuitive psychology of beliefs and desires is itself a science, but it suggests that scientific psychology will have to explain how a hunk of matter, such as a human being, can have beliefs and desires and how the beliefs and desires work so well.

The traditional explanation of intelligence is that human flesh is suffused with a non-material entity, the soul, usually envisioned as some kind of ghost or spirit. But the theory faces an insurmountable problem: How does the spook interact with solid matter? How does an ethereal nothing respond to flashes, pokes, and beeps and get arms and legs to move? Another problem is the overwhelming evidence that the mind is the activity of the brain. The supposedly immaterial soul, we now know, can be bisected with a knife, altered by chemicals, started or stopped by electricity, and extinguished by a sharp blow or by insufficient oxygen. Under a microscope, the brain has a breathtaking complexity of physical structure fully commensurate with the richness of the mind.

Another explanation is that mind comes from some extraordinary form of matter. Pinocchio was animated by a magical kind of wood found by Geppetto that talked, laughed, and moved on its own. Alas, no one has ever discovered such a wonder substance. At first one might think that the wonder substance is brain tissue. Darwin wrote that the brain "secretes" the mind, and recently the philosopher John Searle has argued that the physico-chemical properties of brain tissue somehow produce the mind just as breast tissue produces milk and plant tissue produces sugar. But recall that the same kinds of membranes, pores, and chemicals are found in brain tissue throughout the animal kingdom, not to mention in brain tumors and cultures in dishes. All of these globs of neural tissue have the same physico-chemical properties, but not all of them accomplish humanlike intelligence. Of course, *something* about the tissue in the human brain is necessary for our intelligence, but the physical properties are not sufficient, just as the physical properties of bricks are not sufficient to explain architecture and the physical properties of oxide particles are not sufficient to explain music. Something in the *patterning* of neural tissue is crucial.

Intelligence has often been attributed to some kind of energy flow or force field. Orbs, luminous vapors, auras, vibrations, magnetic fields, and lines of force figure prominently in spiritualism, pseudoscience, and science-fiction kitsch. The school of Gestalt psychology tried to explain visual illusions in terms of electromagnetic force fields on the surface of the brain, but the fields were never found. Occasionally the brain surface has been described as a continuous vibrating medium that supports holograms or other wave interference patterns, but that idea, too, has not panned out. The hydraulic model, with its psychic pressure building up, bursting out, or being diverted through alternative channels, lay at the center of Freud's theory and can be found in dozens of everyday metaphors: anger welling up, letting off steam, exploding under the pressure, blowing one's stack, venting one's feelings, bottling up rage. But even the hottest emotions do not literally correspond to a buildup and discharge of energy (in the physicist's sense) somewhere in the brain. In Chapter 6 I will try to persuade you that the brain does not actually *operate by* internal pressures but *contrives them* as a negotiating tactic, like a terrorist with explosives strapped to his body.

A problem with all these ideas is that even if we *did* discover some gel or vortex or vibration or orb that spoke and plotted mischief like Geppetto's log, or that, more generally, made decisions based on rational rules and pursued a goal in the face of obstacles, we would still be faced with the mystery of *how* it accomplished those feats.

No, intelligence does not come from a special kind of spirit or matter or energy but from a different commodity, *information*. Information is a correlation between two things that is produced by a lawful process (as opposed to coming about by sheer chance). We say that the rings in a stump carry information about the age of the tree because their number correlates with the tree's age (the older the tree, the more rings it has), and the correlation is not a coincidence but is caused by the way trees grow. Correlation is a mathematical and logical concept; it is not defined in terms of the stuff that the correlated entities are made of.

Information itself is nothing special; it is found wherever causes leave effects. What is special is information *processing*. We can regard a piece of matter that carries information about some state of affairs as a symbol; it can "stand for" that state of affairs. But as a piece of matter, it can do other things as well—physical things, whatever that kind of matter in that kind of state can do according to the laws of physics and chemistry. Tree rings carry information about age, but they also reflect light and absorb staining material. Footprints carry information about animal motions, but they also trap water and cause eddies in the wind.

Now here is an idea. Suppose one were to build a machine with parts that are affected by the physical properties of some symbol. Some lever or electric eye or tripwire or magnet is set in motion by the pigment absorbed by a tree ring, or the water trapped by a footprint, or the light reflected by a chalk mark, or the magnetic charge in a bit of oxide. And suppose that the machine then causes something to happen in some other pile of matter. It burns new marks onto a piece of wood, or stamps impressions into nearby dirt, or charges some other bit of oxide. Nothing special has happened so far; all I have described is a chain of physical events accomplished by a pointless contraption.

Here is the special step. Imagine that we now try to interpret the newly arranged piece of matter using the scheme according to which the original piece carried information. Say we *count* the newly burned wood rings and interpret them as the age of some tree at some time, even though they were not caused by the growth of any tree. And let's say that the machine was carefully designed so that the interpretation of its new markings made sense—that is, so that they carried information about something in the world. For example, imagine a machine that scans the rings in a stump, burns one mark on a nearby plank for each ring, moves over to a smaller stump from a tree that was cut down at the same time, scans its rings, and sands off one mark in the plank for each ring. When we count the marks on the plank, we have the age of the first tree at the time that the second one was planted. We would have a kind of *rational* machine, a machine that produces true conclusions from true premises— not because of any special kind of matter or energy, or because of any part that was itself intelligent or rational. All we have is a carefully contrived chain of ordinary physical events, whose first link was a configuration of matter that carries information. Our rational machine owes its rationality to two properties glued together in the entity we call a symbol: a symbol carries information, and it causes things to happen. (Tree rings correlate with the age of the tree, and they can absorb the light beam of a scanner.)

When the caused things themselves carry information, we call the whole system an information processor, or a computer.

Now, this whole scheme might seem like an unrealizable hope. What guarantee is there that *any* collection of thingamabobs can be arranged to fall or swing or shine in just the right pattern so that when their effects are interpreted, the interpretation will make sense? (More precisely, so that it will make sense according to some prior law or relationship we find interesting; any heap of stuff can be given a contrived interpretation after the fact.) How confident can we be that some machine will make marks that actually correspond to some meaningful state of the world, like the age of a tree when another tree was planted, or the average age of the tree's offspring, or anything else, as opposed to being a meaningless pattern corresponding to nothing at all?

The guarantee comes from the work of the mathematician Alan Turing. He designed a hypothetical machine whose input symbols and output symbols could correspond, depending on the details of the machine, to any one of a vast number of sensible interpretations. The machine consists of a tape divided into squares, a read-write head that can print or read a symbol on a square and move the tape in either direction, a pointer that can point to a fixed number of tickmarks on the machine, and a set of mechanical reflexes. Each reflex is triggered by the symbol being read and the current position of the pointer, and it prints a symbol on the tape, moves the tape, and/or shifts the pointer. The machine is allowed as much tape as it needs. This design is called a Turing machine.

What can this simple machine do? It can take in symbols standing for a number or a set of numbers, and print out symbols standing for new numbers that are the corresponding value for any mathematical function that can be solved by a step-by-step sequence of operations (addition, multiplication, exponentiation, factoring, and so on—I am being imprecise to convey the importance of Turing's discovery without the technicalities). It can apply the rules of any useful logical system to derive true statements from other true statements. It can apply the rules of any grammar to derive well-formed sentences. The equivalence among Turing machines, calculable mathematical functions, logics, and grammars, led the logician Alonzo Church to conjecture that *any* well-defined recipe or set of steps that is guaranteed to produce the solution to some problem in a finite amount of time (that is, any algorithm) can be implemented on a Turing machine.

What does this mean? It means that to the extent that the world obeys mathematical equations that can be solved step by step, a machine can be built that simulates the world and makes predictions about it. To the extent that rational thought corresponds to the rules of logic, a machine can be built that carries out rational thought. To the extent that a language can be captured by a set of grammatical rules, a machine can be built that produces grammatical sentences. To the extent that thought consists of applying *any* set of well-specified rules, a machine can be built that, in some sense, thinks.

Turing showed that rational machines—machines that use the physical properties of symbols to crank out new symbols that make some kind of sense—are buildable, indeed, easily buildable. The computer scientist Joseph Weizenbaum once showed how to build one out of a die, some rocks, and a roll of toilet paper. In fact, one doesn't even need a huge warehouse of these machines, one to do sums, another to do square roots, a third to print English sentences, and so on. One kind of Turing machine is called a universal Turing machine. It can take in a *description* of any other Turing machine printed on its tape and thereafter mimic that machine exactly. A single machine can be programmed to do anything that any set of rules can do.

Does this mean that the human brain is a Turing machine? Certainly not. There are no Turing machines in use anywhere, let alone in our heads. They are useless in practice: too clumsy, too hard to program, too big, and too slow. But it does not matter. Turing merely wanted to prove that *some* arrangement of gadgets could function as an intelligent symbol-processor. Not long after his discovery, more practical symbol-processors were designed, some of which became IBM and Univac mainframes and, later, Macintoshes and PCs. But all of them were equivalent to Turing's universal machine. If we ignore size and speed, and give them as much memory storage as they need, we can program them to produce the same outputs in response to the same inputs.

Still other kinds of symbol-processors have been proposed as models of the human mind. These models are often simulated on commercial computers, but that is just a convenience. The commercial computer is first programmed to emulate the hypothetical mental computer (creating what computer scientists call a virtual machine), in much the same way that a Macintosh can be programmed to emulate a PC. Only the virtual mental computer is taken seriously, not the silicon chips that emulate it. Then a program that is meant to model some sort of thinking (solving a problem, understanding a sentence) is run on the virtual mental computer. A new way of understanding human intelligence has been born.

~

Let me show you how one of these models works. In an age when real computers are so sophisticated that they are almost as incomprehensible to laypeople as minds are, it is enlightening to see an example of computation in slow motion. Only then can one appreciate how simple devices can be wired together to make a symbol-processor that shows real intelligence. A lurching Turing machine is a poor advertisement for the theory that the mind is a computer, so I will use a model with at least a vague claim to resembling our mental computer. I'll show you how it solves a problem from everyday life—kinship relations—that is complex enough that we can be impressed when a machine solves it.

The model we'll use is called a production system. It eliminates the feature of commercial computers that is most starkly unbiological: the ordered list of programming steps that the computer follows single-mindedly, one after another. A production system contains a memory and a set of reflexes, sometimes called "demons" because they are simple, self-contained entities that sit around waiting to spring into action. The memory is like a bulletin board on which notices are posted. Each demon is a knee-jerk reflex that waits for a particular notice on the board and responds by posting a notice of its own. The demons collectively constitute a program. As they are triggered by notices on the memory board and post notices of their own, in turn triggering other demons, and so on, the information in memory changes and eventually contains the correct output for a given input. Some demons are connected to sense organs and are triggered by information in the world rather than information in memory. Others are connected to appendages and respond by moving the appendages rather than by posting more messages in memory.

Suppose your long-term memory contains knowledge of the immediate families of you and everyone around you. The content of that knowledge is a set of propositions like "Alex is the father of Andrew." According to the computational theory of mind, that information is embodied in symbols: a collection of physical marks that correlate with the state of the world as it is captured in the propositions.

These symbols cannot be English words and sentences, notwithstanding the popular misconception that we think in our mother tongue. As I showed in *The Language Instinct*, sentences in a spoken language like English or Japanese are designed for vocal communication between impatient, intelligent social beings. They achieve brevity by leaving out any information that the listener can mentally fill in from the context. In contrast, the "language of thought" in which knowledge is couched can leave nothing to the imagination, because it *is* the imagination. Another problem with using English as the medium of knowledge is that English sentences can be ambiguous. When the serial killer Ted Bundy wins a stay of execution and the headline reads "Bundy Beats Date with Chair," we do a double-take because our mind assigns two meanings to the string of words. If one string of words in English can correspond to two meanings in the mind, meanings in the mind cannot be strings of words in English. Finally, sentences in a spoken language are cluttered with articles, prepositions, gender suffixes, and other grammatical boilerplate. They are needed to help get information from one head to another by way of the mouth and the ear, a slow channel, but they are not needed inside a single head where information can be transmitted directly by thick bundles of neurons. So the statements in a knowledge system are not sentences in English but rather inscriptions in a richer language of thought, "mentalese."

In our example, the portion of mentalese that captures family relations comes in two kinds of statements. An example of the first is Alex father-of Andrew: a name, followed by an immediate family relationship, followed by a name. An example of the second is Alex is-male: a name followed by its sex. Do not be misled by my use of English words and syntax in the mentalese inscriptions. This is a courtesy to you, the reader, to help you keep track of what the symbols stand for. As far as the machine is concerned, they are simply different arrangements of marks. As long as we use each one consistently to stand for someone (so the symbol used for Alex is always used for Alex and never for anyone else), and arrange them according to a consistent plan (so they preserve information about who is the father of whom), they could be any marks in any arrangement at all. You can think of the marks as bar codes recognized by a scanner, or keyholes that admit only one key, or shapes that fit only one template. Of course, in a commercial computer they would be patterns of charges in silicon, and in a brain they would be firings in sets of neurons. The key point is that nothing in the machine understands them the way you or I do; parts of the machine respond to their shapes and are triggered to do something, exactly as a gumball machine responds to the shape and weight of a coin by releasing a gumball.

The example to come is an attempt to demystify computation, to get you to see how the trick is done. To hammer home my explanation of the trick—that symbols both stand for some concept and mechanically cause things to happen—I will step through the activity of our production system and describe everything twice: conceptually, in terms of the content of the problem and the logic that solves it, and mechanically, in terms of the brute sensing and marking motions of the system. The system is intelligent because the two correspond exactly, idea-for-mark, logical-step-for-motion.

Let's call the portion of the system's memory that holds inscriptions about family relationships the Long-Term Memory. Let's identify another part as the Short-Term Memory, a scratchpad for the calculations. A part of the Short-Term Memory is an area for goals; it contains a list of questions that the system will "try" to answer. The system wants to know whether Gordie is its biological uncle. To begin with, the memory looks like this:

```
Long-Term Memory          Short-Term Memory        Goal


Abel parent-of Me                                  Gordie uncle-of Me?
Abel is-male
Bella parent-of Me
Bella is-female
Claudia sibling-of Me
Claudia is-female
Duddie sibling-of Me
Duddie is-male
Edgar sibling-of Abel
Edgar is-male
Fanny sibling-of Abel
Fanny is-female
Gordie sibling-of Bella
Gordie is-male
```

Conceptually speaking, our goal is to find the answer to a question; the answer is affirmative if the fact it asks about is true. Mechanically speaking, the system must determine whether a string of marks in the Goal column followed by a question mark (?) has a counterpart with an identical string of marks somewhere in memory. One of the demons is designed to answer these look-up questions by scanning for identical marks in the Goal and Long-Term Memory columns. When it detects a match, it prints a mark next to the question which indicates that it has been answered affirmatively. For convenience, let's say the mark looks like this: Yes.

```
IF: Goal = blah-blah-blah?
      Long-Term Memory = blah-blah-blah
THEN: MARK GOAL
      Yes
```

The conceptual challenge faced by the system is that it does not *explicitly* know who is whose uncle; that knowledge is *implicit* in the other things it knows. To say the same thing mechanically: there is no uncle-of mark in the Long-Term Memory; there are only marks like sibling-of and parent-of. Conceptually speaking, we need to deduce knowledge of unclehood from knowledge of parenthood and knowledge of siblinghood. Mechanically speaking, we need a demon to print an uncle-of inscription flanked by appropriate marks found in sibling-of and parent-of inscriptions. Conceptually speaking, we need to find out who our parents are, identify their siblings, and then pick the males. Mechanically speaking, we need the following demon, which prints new inscriptions in the Goal area that trigger the appropriate memory searches:

```
IF: Goal = Q uncle-of P
THEN: ADD GOAL
      Find P's Parents
      Find Parents' Siblings
      Distinguish Uncles/Aunts
```

This demon is triggered by an uncle-of inscription in the Goal column. The Goal column indeed has one, so the demon goes to work and adds some new marks to the column:

Long-Term Memory	Short-Term Memory	Goal
Abel parent-of Me		Gordie uncle-of Me?
Abel is-male		Find Me's Parents
Bella parent-of Me		Find Parents' Siblings
Bella is-female		Distinguish Uncles/Aunts
Claudia sibling-of Me		
Claudia is-female		
Duddie sibling-of Me		

```
Long-Term Memory       Short-Term Memory       Goal


Duddie is-male
Edgar sibling-of Abel
Edgar is-male
Fanny sibling-of Abel
Fanny is-female
Gordie sibling-of Bella
Gordie is-male
   ...
```

There must also be a device—some other demon, or extra machinery inside this demon—that minds its Ps and Qs. That is, it replaces the P label with a list of the actual labels for names: Me, Abel, Gordie, and so on. I'm hiding these details to keep things simple.

The new Goal inscriptions prod other dormant demons into action. One of them (conceptually speaking) looks up the system's parents, by (mechanically speaking) copying all the inscriptions containing the names of the parents into Short-Term Memory (unless the inscriptions are already there, of course; this proviso prevents the demon from mindlessly making copy after copy like the Sorcerer's Apprentice):

```
IF: Goal = Find P's Parents
    Long-Term Memory = X parent-of P
    Short-Term Memory ≠ X parent-of P
THEN: COPY TO Short-Term Memory
    X parent-of P
    ERASE GOAL
```

Our bulletin board now looks like this:

```
Long-Term Memory       Short-Term Memory       Goal


Abel parent-of Me      Abel parent-of Me       Gordie uncle-of Me?
Abel is-male           Bella parent-of Me      Find Parents' Siblings
Bella parent-of Me                             Distinguish Uncles/Aunts
Bella is-female
Claudia sibling-of Me
Claudia is-female
Duddie sibling-of Me
```

```
Long-Term Memory        Short-Term Memory        Goal


Duddie is-male
Edgar sibling-of Abel
Edgar is-male
Fanny sibling-of Abel
Fanny is-female
Gordie sibling-of Bella
Gordie is-male
  ...
```

Now that we know the parents, we can find the parents' siblings. Mechanically speaking: now that the names of the parents are written in Short-Term Memory, a demon can spring into action that copies inscriptions about the parents' siblings:

```
IF: Goal = Find Parent's Siblings
    Short-Term Memory = X parent-of Y
    Long-Term Memory = Z sibling-of X
    Short-Term Memory ≠ Z sibling-of X
THEN: COPY TO SHORT-TERM MEMORY
    Z sibling-of X
    ERASE GOAL
```

Here is its handiwork:

```
Long-Term Memory        Short-Term Memory        Goal


Abel parent-of Me       Abel parent-of Me        Gordie uncle-of Me?
Abel is-male            Bella parent-of Me       Distinguish Uncles/Aunts
Bella parent-of Me      Edgar sibling-of Abel
Bella is-female         Fanny sibling-of Abel
Claudia sibling-of Me   Gordie sibling-of Bella
Claudia is-female
Duddie sibling-of Me
Duddie is-male
Edgar sibling-of Abel
Edgar is-male
Fanny sibling-of Abel
Fanny is-female
Gordie sibling-of Bella
Gordie is-male
  ...
```

As it stands, we are considering the aunts and uncles collectively. To separate the uncles from the aunts, we need to find the males. Mechanically speaking, the system needs to see which inscriptions have counterparts in Long-Term Memory with `is-male` marks next to them. Here is the demon that does the checking:

```
IF: Goal = Distinguish Uncles/Aunts
    Short-Term Memory = X parent-of Y
    Long-Term Memory = Z sibling-of X
    Long-Term Memory = Z is-male
THEN: STORE IN LONG-TERM MEMORY
    Z uncle-of Y
    ERASE GOAL
```

This is the demon that most directly embodies the system's knowledge of the meaning of "uncle": a male sibling of a parent. It adds the unclehood inscription to Long-Term Memory, not Short-Term Memory, because the inscription represents a piece of knowledge that is permanently true:

Long-Term Memory	Short-Term Memory	Goal
Edgar uncle-of-Me		
Gordie uncle-of-Me		
Abel parent-of Me	Abel parent-of Me	Gordie uncle-of Me?
Abel is-male	Bella parent-of Me	Distinguish Uncles/Aunts
Bella parent-of Me	Edgar sibling-of Abel	
Bella is-female	Fanny sibling-of Abel	
Claudia sibling-of Me	Gordie sibling-of Bella	
Claudia is-female		
Duddie sibling-of Me		
Duddie is-male		
Edgar sibling-of Abel		
Edgar is-male		
Fanny sibling-of Abel		
Fanny is-female		
Gordie sibling-of Bella		
Gordie is-male		
...		

Conceptually speaking, we have just deduced the fact that we inquired about. Mechanically speaking, we have just created mark-for-mark identical inscriptions in the Goal column and the Long-Term Memory column. The very first demon I mentioned, which scans for such duplicates, is triggered to make the mark that indicates the problem has been solved:

```
Long-Term Memory        Short-Term Memory        Goal


Edgar uncle-of-Me
Gordie uncle-of-Me
Abel parent-of Me       Abel parent-of Me        Gordie uncle-of Me?
Abel is-male            Bella parent-of Me       Distinguish Uncles/Aunts
Bella parent-of Me      Edgar sibling-of Abel
Bella is-female         Fanny sibling-of Abel
Claudia sibling-of Me   Gordie sibling-of Bella
Claudia is-female
Duddie sibling-of Me
Duddie is-male
Edgar sibling-of Abel
Edgar is-male
Fanny sibling-of Abel
Fanny is-female
Gordie sibling-of Bella
Gordie is-male

   ...
```

What have we accomplished? We have built a system out of lifeless gumball-machine parts that did something vaguely mindlike: it deduced the truth of a statement that it had never entertained before. From ideas about particular parents and siblings and a knowledge of the meaning of unclehood, it manufactured true ideas about particular uncles. The trick, to repeat, came from the processing of symbols: arrangements of matter that have both *representational* and *causal* properties, that is, that simultaneously carry information about something and take part in a chain of physical events. Those events make up a computation, because the machinery was crafted so that if the interpretation of the symbols that trigger the machine is a true statement, then the interpretation of the symbols created by the machine is also a true statement. The computational theory of mind is the hypothesis that intelligence is computation in this sense.

"This sense" is broad, and it shuns some of the baggage found in other definitions of computation. For example, we need not assume that the computation is made up of a sequence of discrete steps, that the symbols must be either completely present or competely absent (as opposed to being stronger or weaker, more active or less active), that a correct answer is guaranteed in a finite amount of time, or that the truth value be "absolutely true" or "absolutely false" rather than a probability or a degree of certainty. The computational theory thus embraces an alternative kind of computer with many elements that are active to a degree corresponding to the *probability* that some statement is true or false, and in which the activity levels change smoothly to register new and roughly accurate probabilities. (As we shall see, that may be the way the brain works.) The key idea is that the answer to the question "What makes a system smart?" is not the kind of stuff it is made of or the kind of energy flowing through it, but what the parts of the machine stand for and how the patterns of changes inside it are designed to mirror truth-preserving relationships (including probabilistic and fuzzy truths).

NATURAL COMPUTATION

Why should you buy the computational theory of mind? Because it has solved millennia-old problems in philosophy, kicked off the computer revolution, posed the significant questions of neuroscience, and provided psychology with a magnificently fruitful research agenda.

Generations of thinkers have banged their heads against the problem of how mind can interact with matter. As Jerry Fodor has put it, "Self-pity can make one weep, as can onions." How can our intangible beliefs, desires, images, plans, and goals reflect the world around us and pull the levers by which we, in turn, shape the world? Descartes became the laughingstock of scientists centuries after him (unfairly) because he proposed that mind and matter were different kinds of stuff that somehow interacted in a part of the brain called the pineal gland. The philosopher Gilbert Ryle ridiculed the general idea by calling it the Doctrine of the Ghost in the Machine (a phrase that was later co-opted for book titles by the writer Arthur Koestler and the psychologist Stephen Kosslyn and for an album title by the rock group The Police). Ryle and other philosophers argued that mentalistic terms such as "beliefs," "desires," and "images" are meaningless and come from sloppy misunderstandings of language, as if someone heard the expression "for Pete's sake" and went around looking for Pete. Simpatico behaviorist psychologists claimed that these invisible entities were as unscientific as the Tooth Fairy and tried to ban them from psychology.

And then along came computers: fairy-free, fully exorcised hunks of metal that could not be explained without the full lexicon of mentalistic taboo words. "Why isn't my computer printing?" "Because the program doesn't *know* you replaced your dot-matrix printer with a laser printer. It still *thinks* it is *talking to* the dot-matrix and is *trying* to print the document by *asking* the printer to *acknowledge* its *message*. But the printer doesn't *understand* the message; it's *ignoring* it because it *expects* its input to begin with '%!' The program *refuses* to *give up control* while it *polls* the printer, so you have to *get the attention* of the *monitor* so that it can *wrest control* back from the program. Once the program *learns* what printer is connected to it, they can *communicate*." The more complex the system and the more expert the users, the more their technical conversation sounds like the plot of a soap opera.

Behaviorist philosophers would insist that this is all just loose talk. The machines aren't really understanding or trying anything, they would say; the observers are just being careless in their choice of words and are in danger of being seduced into grave conceptual errors. Now, what is wrong with this picture? The *philosophers* are accusing the *computer scientists* of fuzzy thinking? A computer is the most legalistic, persnickety, hard-nosed, unforgiving demander of precision and explicitness in the universe. From the accusation you'd think it was the befuddled computer scientists who call a philosopher when their computer stops working rather than the other way around. A better explanation is that computation has finally demystified mentalistic terms. Beliefs are inscriptions in memory, desires are goal inscriptions, thinking is computation, perceptions are inscriptions triggered by sensors, trying is executing operations triggered by a goal.

(You are objecting that we humans *feel* something when we have a belief or a desire or a perception, and a mere inscription lacks the power to create such feelings. Fair enough. But try to separate the problem of explaining intelligence from the problem of explaining conscious feelings. So far I'm trying to explain intelligence; we'll get to consciousness later in the chapter.)

The computational theory of mind also rehabilitates once and for all the infamous homunculus. A standard objection to the idea that thoughts are internal representations (an objection popular among scientists trying to show how tough-minded they are) is that a representation would require a little man in the head to look at it, and the little man would require an even littler man to look at the representations inside him, and so on, ad infinitum. But once more we have the spectacle of the theoretician insisting to the electrical engineer that if the engineer is correct his workstation must contain hordes of little elves. Talk of homunculi is indispensable in computer science. Data structures are read and interpreted and examined and recognized and revised all the time, and the subroutines that do so are unashamedly called "agents," "demons," "supervisors," "monitors," "interpreters," and "executives." Why doesn't all this homunculus talk lead to an infinite regress? Because an internal representation is not a lifelike photograph of the world, and the homunculus that "looks at it" is not a miniaturized copy of the entire system, requiring its entire intelligence. That indeed would have explained nothing. Instead, a representation is a set of symbols corresponding to *aspects* of the world, and each homunculus is required only to react in a few circumscribed ways to some of the symbols, a feat far simpler than what the system as a whole does. The intelligence of the system emerges from the activities of the not-so-intelligent mechanical demons inside it. The point, first made by Jerry Fodor in 1968, has been succinctly put by Daniel Dennett:

> Homunculi are *bogeymen* only if they duplicate *entire* the talents they are rung in to explain. . . . If one can get a team or committee of *relatively* ignorant, narrow-minded, blind homunculi to produce the intelligent behavior of the whole, this is progress. A flow chart is typically the organizational chart of a committee of homunculi (investigators, librarians, accountants, executives); each box specifies a homunculus by prescribing a function *without saying how it is accomplished* (one says, in effect: put a little man in there to do the job). If we then look closer at the individual boxes we see that the function of each is accomplished by subdividing it via another flow chart into still smaller, more stupid homunculi. Eventually this nesting of boxes within boxes lands you with homunculi so stupid (all they have to do is remember whether to say yes or no when asked) that they can be, as one says, "replaced by a machine." One *discharges* fancy homunculi from one's scheme by organizing armies of idiots to do the work.

You still might wonder how the marks being scribbled and erased by demons inside the computer are supposed to *represent* or *stand for* things in the world. Who decides that this mark in the system corresponds to that bit of the world? In the case of a computer, the answer is obvious: *we* get to decide what the symbols mean, because we built the machine. But who means the meaning of the symbols allegedly inside *us*? Philosophers call this the problem of "intentionality" (confusingly, because it has nothing to do with intentions). There are two common answers. One is that a symbol is connected to its referent in the world by our sense organs. Your mother's face reflects light, which stimulates your eye, which triggers a cascade of templates or similar circuits, which inscribe the symbol mother in your mind. The other answer is that the unique pattern of symbol manipulations triggered by the first symbol mirrors the unique pattern of relationships between the referent of the first symbol and the referents of the triggered symbols. Once we agree, for whatever reason, to say that mother means mother, uncle means uncle, and so on, the new interlocking kinship statements generated by the demons turn out to be uncannily true, time and again. The device prints Bella mother-of Me, and sure enough, Bella *is* my mother. Mother means "mother" because it plays a role in inferences about mothers.

These are called the "causal" and the "inferential-role" theories, and philosophers hostile to each have had fun thinking up preposterous thought experiments to refute them. Oedipus didn't want to marry his mother, but he did so anyway. Why? Because his mother triggered the symbol Jocasta in him rather than the symbol Mom, and his desire was couched as "If it's Mom, don't marry her." The causal effects of Jocasta, the woman who really was Oedipus' mother, were irrelevant; all that mattered was the inferential role that the *symbols* Jocasta and Mom played inside Oedipus' head. A lightning bolt hits a dead tree in the middle of a swamp, and by an amazing coincidence the slime coalesces into a molecule-for-molecule replica of me at this moment, memories included. Swampman has never been in contact with my mother, but most people would say that his mother thoughts are about my mother, just as mine are. Again we conclude that causation by something in the world is not necessary for a symbol to be about something; its inferential role is enough.

But, but, but! Suppose the sequence of information-processing steps in a chess-playing computer turns out, by a remarkable coincidence, to be identical to the battlefield events in the Six-Day War (King's knight = Moshe Dayan, Rook to c7 = Israeli army captures the Golan Heights, and so on). Would the program be "about" the Six-Day War every bit as much as it is "about" the chess game? Suppose that someday we discovered that cats are not animals after all, but lifelike robots controlled from Mars. Any inference rule that computed "If it's a cat, then it must be an animal" would be inoperative. The inferential role of our mental symbol cat would have changed almost beyond recognition. But surely the *meaning* of cat would be unchanged: you'd still be thinking "cat" when Felix the Robot slunk by. Score two points for the causal theory.

A third view is summarized by the television ad parody on *Saturday Night Live*: You're both right—it's a floor wax *and* a dessert topping. *Together* the causal and inferential roles of a symbol determine what it represents. (On this view, Swampman's thoughts would be about my mother because he has a *future*-oriented causal connection with her: he can recognize her when he meets her.) Causal and inferential roles tend to be in sync because natural selection designed both our perceptual systems and our inference modules to work accurately, most of the time, in this world. Not all philosophers agree that causation plus inference plus natural selection are enough to nail down a concept of "meaning" that would work perfectly in all worlds. ("Suppose Swampman has an identical twin on another planet . . .") But if so, one might respond, so much the worse for that concept of meaning. Meaning might make sense only relative to a device that was designed (by engineers or by natural selection) to function in a particular kind of world. In other worlds—Mars, Swampland, the Twilight Zone—all bets are off. Whether or not the causal-plus-inferential theory is completely philosopher-proof, it takes the mystery out of how a symbol in a mind or a machine can mean something.

Another sign that the computational theory of mind is on the right track is the existence of artificial intelligence: computers that perform human-like intellectual tasks. Any discount store can sell you a computer that surpasses a human's ability to calculate, store and retrieve facts, draft drawings, check spelling, route mail, and set type. A well-stocked software house can sell you programs that play excellent chess and that recognize alphabetic characters and carefully pronounced speech. Clients with deeper pockets can buy programs that respond to questions in English about restricted topics, control robot arms that weld and spray-paint, and duplicate human expertise in hundreds of areas such as picking stocks, diagnosing diseases, prescribing drugs, and troubleshooting equipment breakdowns. In 1996 the computer Deep Blue defeated the world chess champion Gary Kasparov in one game and played him to a draw in two others before losing the match, and it is only a matter of time before a computer defeats a world champion outright. Though there are no Terminator-class robots, there are thousands of smaller-scale artificial intelligence programs in the world, including some hidden in your personal computer, car, and television set. And progress continues.

These low-key successes are worth pointing out because of the emotional debate over What Computers Will-Soon/Won't-Ever Do. One side says robots are just around the corner (showing that the mind is a computer); the other side says it will never happen (showing that it isn't). The debate seems to come right out of the pages of Christopher Cerf and Victor Navasky's *The Experts Speak*:

> Well-informed people know it is impossible to transmit the voice over wires and that were it possible to do so, the thing would be of no practical value.
>
> —Editorial, *The Boston Post*, 1865

> Fifty years hence . . . [w]e shall escape the absurdity of growing a whole chicken in order to eat the breast or wing, by growing these parts separately under a suitable medium.
>
> —Winston Churchill, 1932

> Heavier-than-air flying machines are impossible.
> —Lord Kelvin, pioneer in thermodynamics and electricity, 1895

> [By 1965] the deluxe open-road car will probably be 20 feet long, powered by a gas turbine engine, little brother of the jet engine.
> —Leo Cherne, editor-publisher of The Research Institute of America, 1955

> Man will never reach the moon, regardless of all future scientific advances.
> —Lee Deforest, inventor of the vacuum tube, 1957

> Nuclear powered vacuum cleaners will probably be a reality within 10 years.
> —Alex Lewyt, manufacturer of vacuum cleaners, 1955

The one prediction coming out of futurology that is undoubtedly correct is that in the future today's futurologists will look silly. The ultimate attainments of artificial intelligence are unknown, and will depend on countless practical vicissitudes that will be discovered only as one goes along. What is indisputable is that computing machines can be intelligent.

Scientific understanding and technological achievement are only loosely connected. For some time we have understood much about the hip and the heart, but artificial hips are commonplace while artificial hearts are elusive. The pitfalls between theory and application must be kept in mind when we look to artificial intelligence for clues about computers and minds. The proper label for the study of the mind informed by computers is not Artificial Intelligence but Natural Computation.

The computational theory of mind has quietly entrenched itself in neuroscience, the study of the physiology of the brain and nervous system. No corner of the field is untouched by the idea that information processing is the fundamental activity of the brain. Information processing is what makes neuroscientists more interested in neurons than in glial cells, even though the glia take up more room in the brain. The axon (the long output fiber) of a neuron is designed, down to the molecule, to propagate information with high fidelity across long separations, and when its electrical signal is transduced to a chemical one at the synapse (the junction between neurons), the physical format of the information changes while the information itself remains the same. And as we shall see, the tree of dendrites (input fibers) on each neuron appears to perform the basic logical and statistical operations underlying computation. Information-theoretic terms such as "signals," "codes," "representations," "transformations," and "processing" suffuse the language of neuroscience.

Information processing even defines the legitimate questions of the field. The retinal image is upside down, so how do we manage to see the world right-side up? If the visual cortex is in the back of the brain, why doesn't it feel like we are seeing in the back of our heads? How is it possible that an amputee can feel a phantom limb in the space where his real limb used to be? How can our experience of a green cube arise from neurons that are neither colored green nor in the shape of a cube? Every neuroscientist knows that these are pseudo-questions, but why? Because they are about properties of the brain that make no difference to the transmission and processing of information.

~

If a scientific theory is only as good as the facts it explains and the discoveries it inspires, the biggest selling point for the computational theory of mind is its impact on psychology. Skinner and other behaviorists insisted that all talk about mental events was sterile speculation; only stimulus-response connections could be studied in the lab and the field. Exactly the opposite turned out to be true. Before computational ideas were imported in the 1950s and 1960s by Newell and Simon and the psychologists George Miller and Donald Broadbent, psychology was dull, dull, dull. The psychology curriculum comprised physiological psychology, which meant reflexes, and perception, which meant beeps, and learning, which meant rats, and memory, which meant nonsense syllables, and intelligence, which meant IQ, and personality, which meant personality tests. Since then psychology has brought the questions of history's deepest thinkers into the laboratory and has made thousands of discoveries, on every aspect of the mind, that could not have been dreamed of a few decades ago.

The blossoming came from a central agenda for psychology set by the computational theory: discovering the form of mental representations (the symbol inscriptions used by the mind) and the processes (the demons) that access them. Plato said that we are trapped inside a cave and know the world only through the shadows it casts on the wall. The skull is our cave, and mental representations are the shadows. The information in an internal representation is all that we can know about the world. Consider, as an analogy, how *external* representations work. My bank statement lists each deposit as a single sum. If I deposited several checks and some cash, I cannot verify whether a particular check was among them; that information was obliterated in the representation. What's more, the *form* of a representation determines what can easily be inferred from it, because the symbols and their arrangement are the only things a homunculus stupid enough to be replaced by a machine can respond to. Our representation of numbers is valuable because addition can be performed on the numbers with a few dronelike operations: looking up entries in the addition table and carrying digits. Roman numerals have not survived, except as labels or decorations, because addition operations are far more complicated with them, and multiplication and division operations are practically impossible.

Pinning down mental representations is the route to rigor in psychology. Many explanations of behavior have an airy-fairy feel to them because they explain psychological phenomena in terms of other, equally mysterious psychological phenomena. Why do people have more trouble with this task than with that one? Because the first one is "more difficult." Why do people generalize a fact about one object to another object? Because the objects are "similar." Why do people notice this event but not that one? Because the first event is "more salient." These explanations are scams. Difficulty, similarity, and salience are in the mind of the beholder, which is what we should be trying to explain. A computer finds it more difficult to remember the gist of *Little Red Riding Hood* than to remember a twenty-digit number; you find it more difficult to remember the number than the gist. You find two crumpled balls of newspaper to be similar, even though their shapes are completely different, and find two people's faces to be different, though their shapes are almost the same. Migrating birds that navigate at night by the stars in the sky find the positions of the constellations at different times of night quite salient; to a typical person, they are barely noticeable.

But if we hop down to the level of representations, we find a firmer sort of entity, which can be rigorously counted and matched. If a theory of psychology is any good, it should predict that the representations required by the "difficult" task contain more symbols (count 'em) or trigger a longer chain of demons than those of the "easy" task. It should predict that the representations of two "similar" things have more shared symbols and fewer nonshared symbols than the representations of "dissimilar" things. The "salient" entities should have different representations from their neighbors; the "nonsalient" entities should have the same ones.

Research in cognitive psychology has tried to triangulate on the mind's internal representations by measuring people's reports, reaction times, and errors as they remember, solve problems, recognize objects, and generalize from experience. The way people generalize is perhaps the most telltale sign that the mind uses mental representations, and lots of them.

Suppose it takes a while for you to learn to read a fancy new typeface, festooned with curlicues. You have practiced with some words and are now as quick as you are for any other typeface. Now you see a familiar word that was not in your practice set—say, *elk*. Do you have to relearn that the word is a noun? Do you have to relearn how to pronounce it? Relearn that the referent is an animal? What the referent looks like? That it has mass and breathes and suckles its young? Surely not. But this banal talent of yours tells a story. Your knowledge about the word *elk* could not have been connected directly to the physical shapes of printed letters. If it had, then when new letters were introduced, your knowledge would have no connection to them and would be unavailable until you learned the connections anew. In reality, your knowledge must have been connected to a node, a number, an address in memory, or an entry in a mental dictionary representing the abstract word *elk*, and that entry must be neutral with respect to how it is printed or pronounced. When you learned the new typeface, you created a new visual trigger for the letters of the alphabet, which in turn triggered the old *elk* entry, and everything hooked up to the entry was instantly available, without your having to reconnect, piece by piece, everything you know about elks to the new way of printing *elk*. This is how we know that your mind contains mental representations specific to abstract entries for words, not just the shapes of the words when they are printed.

These leaps, and the inventory of internal representations they hint at, are the hallmark of human cognition. If you learned that *wapiti* was another name for an elk, you could take all the facts connected to the word *elk* and instantly transfer them to *wapiti*, without having to solder new connections to the word one at a time. Of course, only your zoological knowledge would transfer; you would not expect *wapiti* to be *pronounced* like *elk*. That suggests you have a level of representation specific to the concepts behind the words, not just the words themselves. Your knowledge of facts about elks hangs off the concept; the words *elk* and *wapiti* also hang off the concept; and the spelling *e-l-k* and pronunciation [ɛlk] hang off the word *elk*.

We have moved upward from the typeface; now let's move downward. If you had learned the typeface as black ink on white paper, you wouldn't have to relearn it for white ink on red paper. This unmasks a representation for visual edges. Any color abutting any other color is seen as an edge; edges define strokes; an arrangement of strokes makes up an alphanumeric character.

The various mental representations connected with a concept like an elk can be shown in a single diagram, sometimes called a semantic network, knowledge representation, or propositional database.

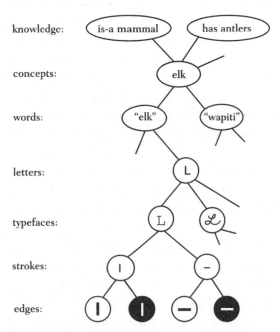

This is a fragment of the immense multimedia dictionary, encyclopedia, and how-to manual we keep in our heads. We find these layers upon layers of representations everywhere we look in the mind. Say I asked you to print the word *elk* in any typeface you wanted, but with your left hand (if you are a righty), or by writing it in the sand with your toe, or by tracing it with a penlight held in your teeth. The printing would be messy but recognizable. You might have to practice to get the motions to be smoother, but you would not have to relearn the strokes composing each letter, let alone the alphabet or the spelling of every English word. This transfer of skill must tap into a level of representation for motor control that specifies a geometric trajectory, not the muscle contractions or limb movements that accomplish it. The trajectory would be translated into actual motions by lower-level control programs for each appendage.

Or recall Sally escaping from the burning building earlier in this chapter. Her desire must have been couched as the abstract representation flee-from-danger. It could not have been couched as run-from-smoke, because the desire could have been triggered by signs other than smoke (and sometimes smoke would not trigger it), and her flight could have been accomplished by many kinds of action, not just running. Yet her behavioral response was put together for the first time there and then. Sally must be modular: one part of her assesses danger, another decides whether to flee, yet another figures out how to flee.

The combinatorics of mentalese, and of other representations composed of parts, explain the inexhaustible repertoire of human thought and action. A few elements and a few rules that combine them can generate an unfathomably vast number of different representations, because the number of possible representations grows exponentially with their size. Language is an obvious example. Say you have ten choices for the word to begin a sentence, ten choices for the second word (yielding a hundred two-word beginnings), ten choices for the third word (yielding a thousand three-word beginnings), and so on. (Ten is in fact the approximate geometric mean of the number of word choices available at each point in assembling a grammatical and sensible sentence.) A little arithmetic shows that the number of sentences of twenty words or less (not an unusual length) is about 10^{20}: a one followed by twenty zeros, or a hundred million trillion, or a hundred times the number of seconds since the birth of the universe. I bring up the example to impress you not with the vastness of language but with the vastness of thought. Language, after all, is not scat-singing: every sentence expresses a distinct idea. (There are no truly synonymous sentences.) So in addition to whatever ineffable thoughts people might have, they can entertain something like a hundred million trillion different effable thoughts.

The combinatorial immensity of thinkable structures is found in many spheres of human activity. The young John Stuart Mill was alarmed to discover that the finite number of musical notes, together with the maximum practical length of a musical piece, meant that the world would soon run out of melodies. At the time he sank into this melancholy, Brahms, Tchaikovsky, Rachmaninoff, and Stravinsky had not yet been born, to say nothing of the entire genres of ragtime, jazz, Broadway musicals, electric blues, country and western, rock and roll, samba, reggae, and punk. We are unlikely to have a melody shortage anytime soon because music is combinatorial: if each note of a melody can be selected from, say, eight notes on average, there are 64 pairs of notes, 512 motifs of three notes, 4,096 phrases of four notes, and so on, multiplying out to trillions and trillions of musical pieces.

Our everyday ease in generalizing our knowledge is one class of evidence that we have several kinds of data representations inside our heads. Mental representations also reveal themselves in the psychology laboratory. With clever techniques, psychologists can catch a mind in the act of flipping from representation to representation. A nice demonstration comes from the psychologist Michael Posner and colleagues. Volunteers sit in front of a video screen and see pairs of letters flashed briefly: A A, for example. They are asked to press one button if the letters are the same, another button if they are different (say, A B). Sometimes the matching letters are both uppercase or both lowercase (A A or a a); that is, they are physically identical. Sometimes one is uppercase and one is lowercase (A a or a A); they are the same letter of the alphabet, but physically different. When the letters are physically identical, people press the buttons more quickly and accurately than when they are physically different, presumably because the people are processing the letters as visual forms and can simply match them by their geometry, template-style. When one letter is A and the other letter is a, people have to convert them into a format in which they are equivalent, namely "the letter *a*"; this conversion adds about a tenth of a second to the reaction time. But if one letter is flashed and the other follows seconds later, it doesn't matter whether they were physically identical or not; A-then-A is as slow as A-then-a. Quick template-matching is no longer possible. Apparently after a few seconds the mind automatically converts a visual representation into an alphabetic one, discarding the information about its geometry.

Such laboratory legerdemain has revealed that the human brain uses at least four major formats of representation. One format is the visual image, which is like a template in a two-dimensional, picturelike mosaic. (Visual images are discussed in Chapter 4.) Another is a phonological representation, a stretch of syllables that we play in our minds like a tape loop, planning out the mouth movements and imagining what the syllables sound like. This stringlike representation is an important component of our short-term memory, as when we look up a phone number and silently repeat it to ourselves just long enough to dial the number. Phonological short-term memory lasts between one and five seconds and can hold from four to seven "chunks." (Short-term memory is measured in chunks rather than sounds because each item can be a label that points to a much bigger information structure in long-term memory, such as the content of a phrase or sentence.) A third format is the grammatical representation: nouns and verbs, phrases and clauses, stems and roots, phonemes and syllables, all arranged into hierarchical trees. In *The Language Instinct* I explained how these representations determine what goes into a sentence and how people communicate and play with language.

The fourth format is mentalese, the language of thought in which our conceptual knowledge is couched. When you put down a book, you forget almost everything about the wording and typeface of the sentences and where they sat on the page. What you take away is their content or gist. (In memory tests, people confidently "recognize" sentences they never saw if they are paraphrases of the sentences they did see.) Mentalese is the medium in which content or gist is captured; I used bits of it in the bulletin board of the production system that identified uncles, and in the "knowledge" and "concept" levels of the semantic network shown in the last diagram. Mentalese is also the mind's lingua franca, the traffic of information among mental modules that allows us to describe what we see, imagine what is described to us, carry out instructions, and so on. This traffic can actually be seen in the anatomy of the brain. The hippocampus and connected structures, which put our memories into long-term storage, and the frontal lobes, which house the circuitry for decision making, are not directly connected to the brain areas that process raw sensory input (the mosaic of edges and colors and the ribbon of changing pitches). Instead, most of their input fibers carry what neuroscientists call "highly processed" input coming from regions one or more stops downstream from the first sensory areas. The input consists of codes for objects, words, and other complex concepts.

⌒

Why so many kinds of representations? Wouldn't it be simpler to have an Esperanto of the mind? In fact, it would be hellishly complicated. The modular organization of mental software, with its packaging of knowledge into separate formats, is a nice example of how evolution and engineering converge on similar solutions. Brian Kernighan, a wizard in the software world, wrote a book with P. J. Plauger called *The Elements of Programming Style* (a play on Strunk and White's famous writing manual, *The Elements of Style*). They give advice on what makes a program work powerfully, run efficiently, and evolve gracefully. One of their maxims is "Replace repetitive expressions by calls to a common function." For example, if a program has to compute the areas of three triangles, it should not have three different commands, each with the coordinates of one of the triangles embedded in its own copy of the formula for the area of a triangle. Instead, the program should have the formula spelled out *once*. There should be a "calculate-triangle-area" function, and it should have slots labeled X, Y, and Z that can stand for *any* triangle's coordinates. That function can be invoked three times, with the coordinates from the input plugged into the X, Y, and Z slots. This design principle becomes even more important as the function grows from a one-line formula to a multistep subroutine, and it inspired these related maxims, all of which seem to have been followed by natural selection as it designed our modular, multiformat minds:

> Modularize.
> Use subroutines.
> Each module should do one thing well.
> Make sure every module hides something.
> Localize input and output in subroutines.

A second principle is captured in the maxim

> Choose the data representation that makes the program simple.

Kernighan and Plauger give the example of a program that reads in a line of text and then has to print it out centered within a border. The line of text could be stored in many formats (as a string of characters, a list of coordinates, and so on), but one format makes the centering child's play: allocate eighty consecutive memory slots that mirror the eighty positions in the input-output display. The centering can be accomplished in a few steps, without error, for an input of any size; with any other format, the program would have to be more complicated. Presumably the distinct formats of representation used by the human mind—images, phonological loops, hierarchical trees, mentalese—evolved because they allow simple programs (that is, stupid demons or homunculi) to compute useful things from them.

And if you like the intellectual stratosphere in which "complex systems" of all kinds are lumped together, you might be receptive to Herbert

Simon's argument that modular design in computers and minds is a special case of modular, hierarchical design in *all* complex systems. Bodies contain tissues made of cells containing organelles; armed forces comprise armies which contain divisions broken into battalions and eventually platoons; books contain chapters divided into sections, subsections, paragraphs, and sentences; empires are assembled out of countries, provinces, and territories. These "nearly decomposable" systems are defined by rich interactions among the elements belonging to the same component and few interactions among elements belonging to different components. Complex systems are hierarchies of modules because only elements that hang together in modules can remain stable long enough to be assembled into larger and larger modules. Simon gives the analogy of two watchmakers, Hora and Tempus:

> The watches the men made consisted of about 1,000 parts each. Tempus had so constructed his that if he had one partly assembled and had to put it down—to answer the phone, say—it immediately fell to pieces and had to be reassembled from the elements. . . .
>
> The watches that Hora made were no less complex than those of Tempus. But he had designed them so that he could put together subassemblies of about ten elements each. Ten of these subassemblies, again, could be put together into a larger subassembly; and a system of ten of the latter subassemblies constituted the whole watch. Hence, when Hora had to put down a partly assembled watch in order to answer the phone, he lost only a small part of his work, and he assembled his watches in only a fraction of the man-hours it took Tempus.

Our complex mental activity follows the wisdom of Hora. As we live our lives, we don't have to attend to every squiggle or plan out every muscle twitch. Thanks to word symbols, any typeface can awaken any bit of knowledge. Thanks to goal symbols, any sign of danger can trigger any means of escape.

The payoff for the long discussion of mental computation and mental representation I have led you through is, I hope, an understanding of the complexity, subtlety, and flexibility that the human mind is capable of *even if* it is nothing but a machine, nothing but the on-board computer of a robot made of tissue. We don't need spirits or occult forces to explain intelligence. Nor, in an effort to look scientific, do we have to ignore the evidence of our own eyes and claim that human beings are bundles of conditioned associations, puppets of the genes, or followers of brutish instincts. We can have both the agility and discernment of human thought *and* a mechanistic framework in which to explain it. The later chapters, which try to explain common sense, the emotions, social relations, humor, and the arts, build on the foundation of a complex computational psyche.

THE DEFENDING CHAMPION

Of course, if it was *unimaginable* that the computational theory of mind was false, that would mean it had no content. In fact, it has been attacked head-on. As one would expect of a theory that has become so indispensable, pea-shooting is not enough; nothing less than undermining the foundations could bring it down. Two flamboyant writers have taken on the challenge. Both have chosen weapons suitable to the occasion, though the weapons are as opposite as can be: one is an appeal to down-home common sense, the other to esoteric physics and mathematics.

The first attack comes from the philosopher John Searle. Searle believes that he refuted the computational theory of mind in 1980 with a thought experiment he adapted from another philosopher, Ned Block (who, ironically, is a major proponent of the computational theory). Searle's version has become famous as the Chinese Room. A man who knows no Chinese is put in a room. Pieces of paper with squiggles on them are slipped under the door. The man has a long list of complicated instructions such as "Whenever you see [squiggle squiggle squiggle], write down [squoggle squoggle squoggle]." Some of the rules tell him to slip his scribbles back out under the door. He gets good at following the instructions. Unknown to him, the squiggles and squoggles are Chinese characters, and the instructions are an artificial intelligence program for answering questions about stories in Chinese. As far as a person on the other side of the door knows, there is a native Chinese speaker in the room. Now, if understanding consists of running a suitable computer program, the guy must understand Chinese, because he is running such a program. But the guy doesn't understand Chinese, not a word of it; he's just manipulating symbols. Therefore, understanding—and, by extension, any aspect of intelligence—is not the same as symbol manipulation or computation.

Searle says that what the program is missing is intentionality, the connection between a symbol and what it means. Many people have interpreted him as saying that the program is missing *consciousness*, and indeed Searle believes that consciousness and intentionality are closely related because we are conscious of what we mean when we have a thought or use a word. Intentionality, consciousness, and other mental phenomena are caused not by information processing, Searle concludes, but by the "actual physical-chemical properties of actual human brains" (though he never says what those properties are).

The Chinese Room has kicked off a truly unbelievable amount of commentary. More than a hundred published articles have replied to it, and I have found it an excellent reason to take my name off all Internet discussion-group lists. To people who say that *the whole room* (man plus rule sheet) understands Chinese, Searle replies: Fine, let the guy memorize the rules, do the calculations in his head, and work outdoors. The room is gone, and our symbol-manipulator still does not understand Chinese. To those who say the man lacks any sensorimotor connection to the world, and that is the crucial missing factor, Searle replies: Suppose that the incoming squiggles are the outputs of a television camera and the outgoing squoggles are the commands to a robot arm. He has the connections, but he still doesn't speak the language. To those who say his program does not mirror what the brain does, Searle can invoke Block's parallel distributed counterpart to the Chinese Room, the Chinese Gym: millions of people in a huge gym act as if they are neurons and shout signals to each other over walkie-talkies, duplicating a neural network that answers questions about stories in Chinese. But the *gym* does not understand Chinese any more than the guy did.

Searle's tactic is to appeal over and over to our common sense. You can almost hear him saying, "Aw, c'mon! You mean to claim that *the guy understands Chinese*??!!! Geddadahere! He doesn't understand a word!! He's lived in Brooklyn all his life!!" and so on. But the history of science has not been kind to the simple intuitions of common sense, to put it mildly. The philosophers Patricia and Paul Churchland ask us to imagine how Searle's argument might have been used against Maxwell's theory that light consists of electromagnetic waves. A guy holds a magnet in his hand and waves it up and down. The guy is creating electromagnetic radiation, *but no light comes out*; therefore, light is not an electromagnetic wave. The thought experiment slows down the waves to a range in which we humans no longer see them as light. By trusting our intuitions in the thought experiment, we falsely conclude that *rapid* waves cannot be light, either. Similarly, Searle has slowed down the mental computation to a range in which we humans no longer think of it as understanding (since understanding is ordinarily much faster). By trusting our intuitions in the thought experiment, we falsely conclude that rapid computation cannot be understanding, either. But if a speeded-up version of Searle's preposterous story could come true, and we met a person who seemed to converse intelligently in Chinese but was really deploying millions of memorized rules in fractions of a second, it is not so clear that we would deny that he understood Chinese.

My own view is that Searle is merely exploring facts about the English word *understand*. People are reluctant to use the word unless certain stereotypical conditions apply: the rules of the language are used rapidly and unconsciously, and the content of the language is connected to the beliefs of the whole person. If people balk at using the vernacular word *understand* to embrace exotic conditions that violate the stereotype but preserve the essence of the phenomenon, then nothing, scientifically speaking, is really at stake. We can look for another word, or agree to use the old one in a technical sense; who cares? The explanation of *what makes understanding work* is the same. Science, after all, is about the principles that make things work, not which things are "really" examples of a familiar word. If a scientist explains the functioning of the human elbow by saying it is a second-class lever, it is no refutation to describe a guy holding a second-class lever made of steel and proclaim, "But look, *the guy doesn't have three elbows*!!!"

As for the "physical-chemical properties" of the brain, I have already mentioned the problem: brain tumors, the brains of mice, and neural tissue kept alive in a dish don't understand, but their physical-chemical properties are the same as the ones of our brains. The computational theory explains the difference: those hunks of neural tissue are not *arranged* into patterns of connectivity that carry out the right kind of information processing. For example, they do not have parts that distinguish nouns from verbs, and their activity patterns do not carry out the rules of syntax, semantics, and common sense. Of course, we can always *call* that a difference in physical-chemical properties (in the same sense that two books differ in their physical-chemical properties), but then the term is meaningless because it can no longer be defined in the language of physics and chemistry.

With thought experiments, turnabout is fair play. Perhaps the ultimate reply to Searle's Chinese Room may be found in a story by the science-fiction writer Terry Bisson, widely circulated on the Internet, which has the incredulity going the other way. It reports a conversation between the leader of an interplanetary explorer fleet and his commander in chief, and begins as follows:

"They're made out of meat."

"Meat?" . . . "There's no doubt about it. We picked several from different parts of the planet, took them aboard our recon vessels, probed them all the way through. They're completely meat."

"That's impossible. What about the radio signals? The messages to the stars?"

"They use the radio waves to talk, but the signals don't come from them. The signals come from machines."

"So who made the machines? That's who we want to contact."

"They made the machines. That's what I'm trying to tell you. Meat made the machines."

"That's ridiculous. How can meat make a machine? You're asking me to believe in sentient meat."

"I'm not asking you, I'm telling you. These creatures are the only sentient race in the sector and they're made out of meat."

"Maybe they're like the Orfolei. You know, a carbon-based intelligence that goes through a meat stage."

"Nope. They're born meat and they die meat. We studied them for several of their life spans, which didn't take too long. Do you have any idea [of] the life span of meat?"

"Spare me. Okay, maybe they're only part meat. You know, like the Weddilei. A meat head with an electron plasma brain inside."

"Nope, we thought of that, since they do have meat heads like the Weddilei. But I told you, we probed them. They're meat all the way through."

"No brain?"

"Oh, there is a brain all right. It's just that the brain is made out of meat!"

"So . . . what does the thinking?"

"You're not understanding, are you? The brain does the thinking. The meat."

"Thinking meat! You're asking me to believe in thinking meat!"

"Yes, thinking meat! Conscious meat! Loving meat. Dreaming meat. The meat is the whole deal! Are you getting the picture?"

The other attack on the computational theory of mind comes from the mathematical physicist Roger Penrose, in a best-seller called *The Emperor's New Mind* (how's *that* for an in-your-face impugnment!). Penrose draws not on common sense but on abstruse issues in logic and physics. He argues that Gödel's famous theorem implies that mathematicians—and, by extension, all humans—are not computer programs. Roughly, Gödel proved that any formal system (such as a computer program or a set of axioms and rules of inference in mathematics) that is even moderately powerful (powerful enough to state the truths of arithmetic) and consistent (it does not generate contradictory statements) can generate statements that are true but that the system cannot prove to be true. Since we human mathematicians can just *see* that those statements are true, we are not formal systems like computers. Penrose believes that the mathematician's ability comes from an aspect of consciousness that cannot be explained as computation. In fact, it cannot be explained by the operation of neurons; they're too big. It cannot be explained by Darwin's theory of evolution. It cannot even be explained by physics as we currently understand it. Quantum-mechanical effects, to be explained in an as yet nonexistent theory of quantum gravity, operate in the microtubules that make up the miniature skeleton of neurons. Those effects are so strange that they might be commensurate with the strangeness of consciousness.

Penrose's mathematical argument has been dismissed as fallacious by logicians, and his other claims have been reviewed unkindly by experts in the relevant disciplines. One big problem is that the gifts Penrose attributes to his idealized mathematician are not possessed by real-life mathematicians, such as the certainty that the system of rules being relied on is consistent. Another is that quantum effects almost surely cancel out in nervous tissue. A third is that microtubules are ubiquitous among cells and appear to play no role in how the brain achieves intelligence. A fourth is that there is not even a hint as to how consciousness might arise from quantum mechanics.

The arguments from Penrose and Searle have something in common other than their target. Unlike the theory they attack, they are so unconnected to discovery and explanation in scientific practice that they have been empirically sterile, contributing no insight and inspiring no discoveries on how the mind works. In fact, the most interesting implication of

The Emperor's New Mind was pointed out by Dennett. Penrose's denunciation of the computational theory of mind turns out to be a backhanded compliment. The computational theory fits so well into our understanding of the world that, in trying to overthrow it, Penrose had to reject most of contemporary neuroscience, evolutionary biology, and physics!

REPLACED BY A MACHINE

In Lewis Carroll's story "What the Tortoise Said to Achilles," the swift-footed warrior has caught up with the plodding tortoise, defying Zeno's paradox in which any head start given to the tortoise should make him uncatchable. (In the time it would take for Achilles to close the gap, the tortoise would have progressed a small amount; in the time it took to close *that* gap, the tortoise would have moved a bit farther, ad infinitum.) The tortoise offers Achilles a similar paradox from logic. Achilles pulls an enormous notebook and a pencil from his helmet, and the tortoise dictates Euclid's First Proposition:

(A) Things that are equal to the same are equal to each other.
(B) The two sides of this Triangle are things that are equal to the same.
(Z) The two sides of this Triangle are equal to each other.

The tortoise gets Achilles to agree that anyone who accepts A and B and "If A and B then Z" must also accept Z. But now the tortoise disagrees with Achilles' logic. He says he is entitled to reject conclusion Z, because no one ever wrote down the if-then rule on the list of premises he must accept. He challenges Achilles to *force* him to conclude Z. Achilles replies by adding C to the list in his notebook:

(C) If A and B are true, Z must be true.

The tortoise replies that he fails to see why he should assume that just because A and B and C are true, Z is true. Achilles adds one more statement—

(D) If A and B and C are true, Z must be true.

—and declares that "Logic [must] take you by the throat, and *force* you" to accept Z. The tortoise replies,

"Whatever *Logic* is good enough to tell me is worth *writing down*. So enter it in your book, please. We will call it

(E) If A and B and C and D are true, Z must be true."

"I see," said Achilles; and there was a touch of sadness in his tone.

Here the narrator, having pressing business at the Bank, was obliged to leave the happy pair, and did not again pass the spot until some months afterwards. When he did so, Achilles was still seated on the back of the much-enduring tortoise, and was writing in his notebook, which appeared to be nearly full. The tortoise was saying, "Have you got that last step written down? Unless I've lost count, that makes a thousand and one. There are several millions more to come."

The solution to the paradox, of course, is that no inference system follows explicit rules all the way down. At some point the system must, as Jerry Rubin (and later the Nike Corporation) said, just do it. That is, the rule must simply be *executed* by the reflexive, brute-force operation of the system, no more questions asked. At that point the system, if implemented as a machine, would not be following rules but obeying the laws of physics. Similarly, if representations are read and written by demons (rules for replacing symbols with symbols), and the demons have smaller (and stupider) demons inside them, eventually you have to call Ghostbusters and replace the smallest and stupidest demons with machines—in the case of people and animals, machines built from neurons: neural networks. Let's see how our picture of how the mind works can be grounded in simple ideas of how the brain works.

The first hints came from the mathematicians Warren McCulloch and Walter Pitts, who wrote about the "neuro-logical" properties of connected neurons. Neurons are complicated and still not understood, but McCulloch and Pitts and most neural-network modelers since have identified one thing neurons do as the most significant thing. Neurons, in effect, add up a set of quantities, compare the sum to a threshold, and indicate whether the threshold is exceeded. That is a conceptual description of what they do; the corresponding physical description is that a firing neuron is active to varying degrees, and its activity level is influenced by the activity levels of the incoming axons from other neurons attached at synapses to the neuron's dendrites (input structures). A synapse has a strength ranging from positive (excitatory) through zero (no effect) to negative (inhibitory). The activation level of each incoming axon is multiplied by the strength of the synapse. The neuron sums these incoming levels; if the total exceeds a threshold, the neuron will become more active, sending a signal in turn to any neuron connected to *it*. Though neurons are always firing and incoming signals merely cause it to fire at a detectably faster or slower rate, it is sometimes convenient to describe them as being either off (resting rate) or on (elevated rate).

McCulloch and Pitts showed how these toy neurons could be wired up to make logic gates. Logic gates implement the basic logical relations "and," "or," and "not" that underlie simple inferences. "A and B" (conceptually) is true if A is true and if B is true. An AND-gate (mechanically) produces an output if both of its inputs are on. To make an AND-gate out of toy neurons, set the threshold of the output unit to be greater than each of the incoming weights but less than their sum, as in the mini-network on the left below. "A or B" (conceptually) is true if A is true or if B is true. An OR-gate (mechanically) produces an output if either of its inputs is on. To make one, set the threshold to be less than each incoming weight, as in the middle mini-network below. Finally, "not A" (conceptually) is true if A is false, and vice versa. A NOT-gate (mechanically) produces an output when it receives no input, and vice versa. To make one, set the threshold at zero, so the neuron will fire when it gets no input, and make the incoming weight negative, so that an incoming signal will turn the neuron off, as in the mini-network on the right.

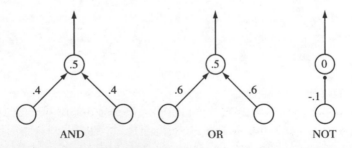

Suppose that each toy neuron represents a simple proposition. The mini-networks can be wired together, with the output of one feeding the input to another, to evaluate the truth of a complex proposition. For example, a neural network could evaluate the proposition {[(X chews its cud) and (X has cloven hooves)] or [(X has fins) and (X has scales)]}, a summary of what it takes for an animal to be kosher. In fact, if a network of toy neurons is connected to some kind of extendable memory (such as a roll of paper moving under a rubber stamp and an eraser), it would be a Turing machine, a full-powered computer.

It is utterly impractical, though, to represent propositions, or even the concepts composing them, in logic gates, whether those logic gates are made out of neurons or semiconductors. The problem is that every concept and proposition has to be hard-wired in advance as a separate unit. Instead, both computers and brains represent concepts as *patterns* of activity over *sets* of units. A simple example is the lowly byte, which represents an alphanumeric character in your computer. The representation of the letter *B* is 01000010, where the digits (bits) correspond to tiny pieces of silicon laid out in a row. The second and seventh pieces are charged, corresponding to the ones, and the other pieces are uncharged, corresponding to the zeros. A byte can also be built out of toy neurons, and a circuit for recognizing the *B* pattern can be built as a simple neural network:

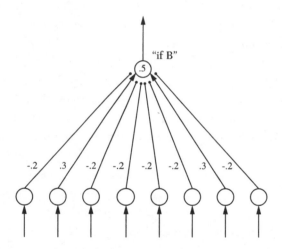

You can imagine that this network is one of the parts making up a demon. If the bottom row of toy neurons is connected to short-term memory, the top one detects whether short-term memory contains an instance of the symbol *B*. And on page 102 is a network for a demon-part that *writes* the symbol *B* into memory.

We are on our way to building a conventional digital computer out of toy neurons, but let's change direction a bit and make a more biomorphic computer. First, we can use the toy neurons to implement not classical logic but fuzzy logic. In many domains people do not have all-or-none convictions about whether something is true. A thing can be a better or a worse example of a category rather than being either in or out. Take the category "vegetable." Most people agree that celery is a full-fledged

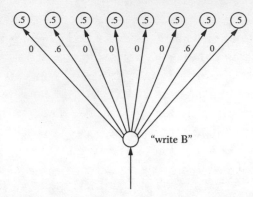

vegetable but that garlic is only a so-so example. And if we are to believe the Reagan administration when it justified its parsimonious school lunch program, even ketchup is a kind of vegetable—though after a firestorm of criticism the administration conceded that it is not a very good example of one. Conceptually speaking, we eschew the idea that something either is or is not a vegetable and say that things can be better or worse examples of a vegetable. Mechanically speaking, we no longer insist that a unit representing vegetablehood be either on or off, but allow it to have a value ranging from 0 (for a rock) through 0.1 (for ketchup) through .4 (for garlic) to 1.0 (for celery).

We can also scrap the arbitrary code that relates each concept to a meaningless string of bits. Each bit can earn its keep by representing something. One bit might represent greenness, another leafiness, another crunchiness, and so on. Each of these vegetable-property units could be connected with a small weight to the vegetable unit itself. Other units, representing features that vegetables lack, such as "magnetic" or "mobile," could be connected with negative weights. Conceptually speaking, the more vegetable properties something has, the better an example it is of a vegetable. Mechanically speaking, the more vegetable-property units are turned on, the higher the activation level of the vegetable unit.

Once a network is allowed to be squishy, it can represent degrees of evidence and probabilities of events and can make statistical decisions. Suppose each unit in a network represents a piece of evidence implicating the butler (fingerprints on the knife, love letters to the victim's wife, and so on). Suppose the top node represents the conclusion that the butler did it. Conceptually speaking, the more clues there are that the butler might have done it, the higher our estimate would be that the butler did

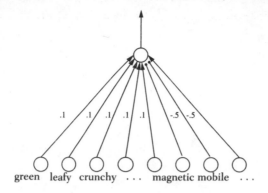

green leafy crunchy . . . magnetic mobile . . .

do it. Mechanically speaking, the more clue units there are that are turned on, the greater the activation of the conclusion unit. We could implement different statistical procedures in the network by designing the conclusion unit to integrate its inputs in different ways. For example, the conclusion unit could be a threshold unit like the ones in crisp logic gates; that would implement a policy to put out a decision only if the weight of evidence exceeded a critical value (say, "beyond a reasonable doubt"). Or the conclusion unit could increase its activity gradually; its degree of confidence could increase slowly with the first clues trickling in, build quickly as more and more are amassed, and level off at a point of diminishing returns. These are two of the kinds of unit that neural-network modelers like to use.

We can get even more adventurous, and take inspiration from the fact that with neurons, unlike silicon chips, connections are cheap. Why not connect every unit to every other unit? Such a network would embody not only the knowledge that greenness predicts vegetablehood and crunchiness predicts vegetablehood, but that greenness predicts crunchiness, crunchiness predicts leafiness, greenness predicts lack of mobility, and so on:

With this move, interesting things begin to happen. The network begins to resemble human thought processes in ways that sparsely connected networks do not. For this reason psychologists and artificial intelligence researchers have been using everything-connected-to-everything networks to model many examples of simple pattern recognition. They have built networks for the lines that co-occur in letters, the letters that co-occur in words, the animal parts that co-occur in animals, and the pieces of furniture that co-occur in rooms. Often the decision node at the top is thrown away and only the correlations among the properties are calculated. These networks, sometimes called auto-associators, have five nifty features.

First, an auto-associator is a reconstructive, content-addressable memory. In a commercial computer, the bits themselves are meaningless, and the bytes made out of them have arbitrary addresses, like houses on a street, which have nothing to do with their contents. Memory locations are accessed by their addresses, and to determine whether a pattern has been stored somewhere in memory you have to search them all (or use clever shortcuts). In a content-addressable memory, on the other hand, specifying an item automatically lights up any location in memory containing a copy of the item. Since an item is represented in an auto-associator by turning on the units that represent its properties (in this case celery, greenness, leafiness, and so on), and since those units are connected to one another with strong weights, the activated units will reinforce one another, and after a few rounds in which activation reverberates through the network, all the units pertaining to the item will lock into the "on" position. That indicates that the item has been recognized. In fact, a single auto-associator can accommodate many sets of weights in its battery of connections, not just one, so it can store many items at a time.

Better yet, the connections are redundant enough that even if only a *part* of the pattern for an item is presented to the auto-associator, say, greenness and crunchiness alone, the rest of the pattern, leafiness, gets completed automatically. In some ways this is reminiscent of the mind. We do not need predefined retrieval tags for items in memory; almost any *aspect* of an object can bring the entire object to mind. For example, we can recall "vegetable" upon thinking about things that are green and leafy *or* green and crunchy *or* leafy and crunchy. A visual example is our ability to complete a word from a few of its fragments. We do not see this figure as random line segments or even as an arbitrary sequence of letters like MIHB, but as something more probable:

A second selling point, called "graceful degradation," helps deal with noisy input or hardware failure. Who isn't tempted to throw a shoe through the computer screen when it responds to the command `pritn file` with the error message `pritn: command not found`? In Woody Allen's *Take the Money and Run*, the bank robber Virgil Starkwell is foiled by his penmanship when the teller asks him why he wrote that he is pointing a gub at her. In a Gary Larson cartoon that adorns the office door of many a cognitive psychologist, a pilot flying over a castaway on a desert island reads the message scratched in the sand and shouts into his radio, "Wait! Wait! . . . Cancel that, I guess it says 'HELF'." Real-life humans do better, perhaps because we are fitted with auto-associators that use a preponderance of mutually consistent pieces of information to override one unusual piece. "Pritn" would activate the more familiar pattern "print"; "gub" would be warped to "gun," "HELF" to "HELP." Similarly, a computer with a single bad bit on its disk, a smidgen of corrosion in one of its sockets, or a brief dip in its supply of power can lock up and crash. But a human being who is tired, hung over, or brain-damaged does not lock up and crash; usually he or she is slower and less accurate but can muster an intelligible response.

A third advantage is that auto-associators can do a simple version of the kind of computation called constraint satisfaction. Many problems that humans solve have a chicken-and-egg character. An example from Chapter 1 is that we compute the lightness of a surface from a guess about its angle *and* compute the angle of the surface from a guess about its lightness, without knowing either for sure beforehand. These problems abound in perception, language, and common-sense reasoning. Am I looking at a fold or at an edge? Am I hearing the vowel [I] (as in *pin*) or the vowel [ɛ] (as in *pen*) with a southern accent? Was I the victim of an act of malice or an act of stupidity? These ambiguities can sometimes be resolved by choosing the interpretation that is consistent with the greatest number of interpretations of *other* ambiguous events, if they could all be resolved at once. For example, if one speech sound can be interpreted as either *send* or *sinned*, and another as either *pen* or *pin*, I can resolve the uncertainties if I hear one speaker utter both words with the same vowel sound. He must have intended *send* and *pen*, I would reason, because *send a pen* is the only guess that does not violate some constraint. *Sinned* and *pin* would give me *sinned a pin*, which violates the rules of grammar and plausible meaning; *send* and *pin* can be ruled out by the constraint that the two vowels were pronounced identically; *sinned* and *pen* can be ruled out because they violate both these constraints.

This kind of reasoning takes a long time if all the compatibilities must be tested one at a time. But in an auto-associator, they are coded beforehand in the connections, and the network can evaluate them all at once. Suppose each interpretation is a toy neuron, one for *sinned*, one for *send*, and so on. Suppose that pairs of units whose interpretations are consistent are connected with positive weights and pairs of units whose interpretations are inconsistent are connected with negative weights. Activation will ricochet around the network, and if all goes well, it will settle into a state in which the greatest number of mutually consistent interpretations are active. A good metaphor is a soap bubble that wobbles in eggy and amoeboid shapes as the tugs among its neighboring molecules pull it into a sphere.

Sometimes a constraint network can have mutually inconsistent but equally stable states. That captures the phenomenon of global ambiguity, in which an entire object, not just its parts, can be interpreted in two ways. If you stare at the drawing of a cube on page 107 (called a Necker cube), your perception will flip from a downward view of its top face to an upward view of its bottom face. When the global flip occurs, the interpretations of all of the local parts are dragged with it. Every near edge becomes a far edge, every convex corner becomes a concave corner, and so on. Or vice versa: if you *try* to see a convex corner as concave, you can sometimes nudge the whole cube into flipping. The dynamics can be captured in a network, shown below the cube, in which the units represent the interpretations of the parts, and the interpretations that are consistent in a 3-D object excite each other while the ones that are inconsistent inhibit each other.

A fourth advantage comes from a network's ability to generalize automatically. If we had connected our letter-detector (which funneled a bank of input units into a decision unit) to our letter-printer (which had an intention unit fanning out into a bank of output units), we would have made a simple read-write or lookup demon—for example, one that responds to a *B* by printing a *C*. But interesting things happen if you skip the middleman and connect the input units directly to the output units.

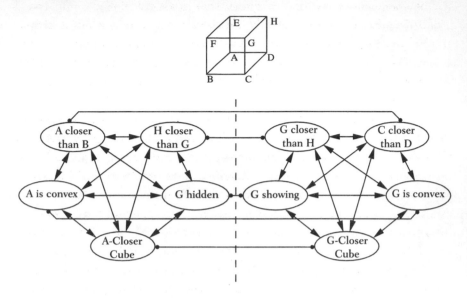

Instead of a faithful-to-the-letter lookup demon, you have one that can generalize a bit. The network is called a pattern associator.

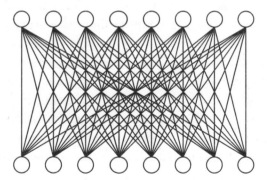

Suppose the input units at the bottom represent the appearance of animals: "hairy," "quadrupedal," "feathered," "green," "long-necked," and so on. With enough units, every animal can be represented by turning on the units for its unique set of properties. A parrot is represented by turning the "feathered" unit on, the "hairy" unit off, and so on. Now suppose the output units at the top stand for zoological facts. One represents the fact that the animal is herbivorous, another that it is warm-blooded, and so on. With no units standing for a particular animal (that is, with no unit for "parrot"), the weights automatically represent statistical knowledge about *classes* of animals. They embody the knowledge that feathered things tend to be warm-blooded, animals with hair tend to bear live young, and so on. Any fact stored in the connections for one animal (parrots are warm-blooded) automatically transfers to similar animals (budgies are warm-blooded), because the network does not care that the connections belong to an animal at all. The connections merely say which visible properties predict which invisible properties, skipping ideas about species of animals altogether.

Conceptually speaking, a pattern associator captures the idea that if two objects are similar in some ways, they are probably similar in other ways. Mechanically speaking, similar objects are represented by some of the very same units, so any piece of information connected to the units for one object will ipso facto be connected to many of the units for the other. Moreover, classes of different degrees of inclusiveness are superimposed in the same network, because any subset of the units implicitly defines a class. The fewer the units, the larger the class. Say there are input units for "moves," "breathes," "hairy," "barks," "bites," and "lifts-leg-at-hydrants." The connections emanating out of all six trigger facts about dogs. The connections emanating out of the first three trigger facts about mammals. The connections emanating out of the first two trigger facts about animals. With suitable weights, the knowledge programmed in for one animal can be shared with both its immediate and its distant family members.

A fifth trick of neural networks is that they learn from examples, where learning consists of changes in the connection weights. The model-builder (or evolution) does not have to hand-set the thousands of weights needed to get the outputs right. Suppose a "teacher" feeds a pattern associator with an input and *also* with the correct output. A learning mechanism compares the network's actual output—which at first will be pretty random—with the correct one, and adjusts the weights to minimize the difference between the two. If the network leaves an output node off that the teacher says ought to be on, we want to make it more likely that the current funnel of active inputs will turn it on in the future. So the weights on the active inputs to the recalcitrant output unit are increased slightly. In addition, the output node's own threshold is lowered slightly, to make it more trigger-happy across the board. If the network turns an output node on and the teacher says it should be off, the opposite happens: the weights of the currently active input lines are taken down a notch (possibly driving the weight past zero to a negative value), and the target node's threshold is raised. This all makes the hyperactive output node more likely to turn off in response to those inputs in the future. A whole series of inputs and their outputs is presented to the network, over and over, causing waves of little adjustments of the connection weights, until it gets every output right for every input, at least as well as it can manage to.

A pattern associator equipped with this learning technique is called a perceptron. Perceptrons are interesting but have a big flaw. They are like the chef from hell: they think that if a little of each ingredient is good, a lot of everything must be better. In deciding whether a set of inputs justifies turning on an output, the perceptron weights them and adds them up. Often that gives the wrong answer, even on very simple problems. A textbook example of this flaw is the perceptron's handling of the simple logical operation called exclusive-or ("xor"), which means "A or B, but not both."

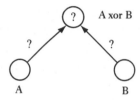

When A is on, the network should turn A-xor-B on. When B is on, the network should turn A-xor-B on. These facts will coax the network into increasing the weight for the connection from A (say, to .6) and increasing the weight for the connection from B (say, to .6), making each one high enough to overcome the output unit's threshold (say, .5). But when A and B are *both* on, we have too much of a good thing—A-xor-B is screaming its head off just when we want it to shut up. If we try smaller weights or a higher threshold, we can keep it quiet when A and B are both on, but then, unfortunately, it will be quiet when *just* A or *just* B is on. You can experiment with your own weights and you will see that nothing works. Exclusive-or is just one of many demons that cannot be built out of perceptrons; others include demons to determine whether an even or an odd number of units are on, to determine whether a string of active units is symmetrical, and to get the answer to a simple addition problem.

The solution is to make the network less of a stimulus-response creature and give it an *internal representation* between the input and output layers. It needs a representation that makes the crucial kinds of information about the inputs explicit, so that each output unit really *can* just add up its inputs and get the right answer. Here is how it can be done for exclusive-or:

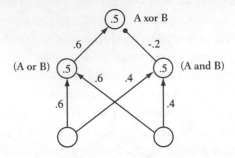

The two hidden units between the input and the output calculate useful intermediate products. The one on the left computes the simple case of "A or B," which in turn simply excites the output node. The one on the right computes the vexing case of "A and B," and it *inhibits* the output node. The output node can simply compute "(A or B) and not (A and B)," which is well within its feeble powers. Note that even at the microscopic level of building the simplest demons out of toy neurons, internal representations are indispensable; stimulus-response connections are not enough.

Even better, a hidden-layer network can be trained to set its own weights, using a fancier version of the perceptron learning procedure. As before, a teacher gives the network the correct output for every input, and the network adjusts the connection weights up or down to try to reduce the difference. But that poses a problem the perceptron did not have to worry about: how to adjust the connections from the input units to the hidden units. It is problematic because the teacher, unless it is a mind reader, has no way of knowing the "correct" states for the hidden units, which are sealed inside the network. The psychologists David Rumelhart, Geoffrey Hinton, and Ronald Williams hit on a clever solution. The output units propagate back to each hidden unit a signal that represents the *sum* of the hidden unit's errors across *all* the output units it connects to ("you're sending too much activation," or "you're sending too little activation," and by what amount). That signal can serve as a surrogate teaching signal which may be used to adjust the hidden layer's inputs. The connections from the input layer to each hidden unit can be nudged up or down to reduce the hidden unit's tendency to overshoot or undershoot, given the current input pattern. This procedure, called "error back-propagation" or simply "backprop," can be iterated backwards to any number of layers.

We have reached what many psychologists treat as the height of the neural-network modeler's art. In a way, we have come full circle, because a hidden-layer network is like the arbitrary road map of logic gates that McCulloch and Pitts proposed as their neuro-logical computer. Conceptually speaking, a hidden-layer network is a way to compose a set of propositions, which can be true or false, into a complicated logical function held together by ands, ors, and nots—though with two twists. One is that the values can be continuous rather than on or off, and hence they can represent the degree of truth or the probability of truth of some statement rather than dealing only with statements that are absolutely true or absolutely false. The second twist is that the network can, in many cases, be trained to take on the right weights by being fed with inputs and their correct outputs. On top of these twists there is an attitude: to take inspiration from the many connections among neurons in the brain and feel no guilt about going crazy with the number of gates and connections put into a network. That ethic allows one to design networks that compute many probabilities and hence that exploit the statistical redundancies among the features of the world. And that, in turn, allows neural networks to generalize from one input to similar inputs without further training, as long as the problem is one in which similar inputs yield similar outputs.

Those are a few ideas on how to implement our smallest demons and their bulletin boards as vaguely neural machines. The ideas serve as a bridge, rickety for now, along the path of explanation that begins in the conceptual realm (Grandma's intuitive psychology and the varieties of knowledge, logic, and probability theory that underlie it), continues on to rules and representations (demons and symbols), and eventually arrives at real neurons. Neural networks also offer some pleasant surprises. In figuring out the mind's software, ultimately we may use only demons stupid enough to be replaced by machines. If we seem to need a smarter demon, someone has to figure out how to build *him* out of stupider ones. It all goes faster, and sometimes goes differently, when neural-network modelers working from the neurons upward can build an inventory of stock demons that do handy things, like a content-addressable memory or an automatically generalizing pattern associator. The mental software engineers (actually, reverse-engineers) have a good parts catalogue from which they can order smart demons.

CONNECTOPLASM

Where do the rules and representations in mentalese leave off and the neural networks begin? Most cognitive scientists agree on the extremes. At the highest levels of cognition, where we consciously plod through steps and invoke rules we learned in school or discovered ourselves, the mind is something like a production system, with symbolic inscriptions in memory and demons that carry out procedures. At a lower level, the inscriptions and rules are implemented in something like neural networks, which respond to familiar patterns and associate them with other patterns. But the boundary is in dispute. Do simple neural networks handle the bulk of everyday thought, leaving only the products of book-learning to be handled by explicit rules and propositions? Or are the networks more like building blocks that aren't humanly smart until they are assembled into structured representations and programs?

A school called connectionism, led by the psychologists David Rumelhart and James McClelland, argues that simple networks by themselves can account for most of human intelligence. In its extreme form, connectionism says that the mind is one big hidden-layer back-propagation network, or perhaps a battery of similar or identical ones, and intelligence emerges when a trainer, the environment, tunes the connection weights. The only reason that humans are smarter than rats is that our networks have more hidden layers between stimulus and response and we live in an environment of other humans who serve as network trainers. Rules and symbols might be useful as a rough-and-ready approximation to what is happening in a network for a psychologist who can't keep track of the millions of streams of activation flowing through the connections, but they are no more than that.

The other view—which I favor—is that those neural networks alone cannot do the job. It is the *structuring* of networks into programs for manipulating symbols that explains much of human intelligence. In particular, symbol manipulation underlies human language and the parts of reasoning that interact with it. That's not all of cognition, but it's a lot of it; it's everything we can talk about to ourselves and others. In my day job as a psycholinguist I have gathered evidence that even the simplest of talents that go into speaking English, such as forming the past tense of verbs (*walk* into *walked*, *come* into *came*), is too computationally sophisticated to be handled in a single neural network. In this section, I will present a more general class of evidence. Does the content of our common-sense thoughts (the kind of information we exchange in conversation) require a computational device designed to implement a highly structured mentalese, or can it be handled by generic neural-network stuff—what one wag has called connectoplasm? I will show you that our thoughts have a delicate logical structuring that no simple network of homogeneous layers of units can handle.

Why should you care? Because these demonstrations cast doubt on the most influential theory of how the mind works that has ever been proposed. By itself, a perceptron or a hidden-layer network is a high-tech implementation of an ancient doctrine: the association of ideas. The British philosophers John Locke, David Hume, George Berkeley, David Hartley, and John Stuart Mill proposed that thought is governed by two laws. One is contiguity: ideas that are frequently experienced together get associated in the mind. Thereafter, when one is activated, the other is activated too. The other law is resemblance: when two ideas are similar, whatever has been associated with the first idea is automatically associated with the second. As Hume summed up the theory in 1748:

> Experience shows us a number of uniform effects, resulting from certain objects. When a new object, endowed with similar sensible qualities, is produced, we expect similar powers and forces, and look for a like effect. From a body of like color and consistence with bread we expect like nourishment and support.

Association by contiguity and resemblance was also thought to be the scrivener that fills the famous blank slate, Locke's metaphor for the neonate mind. The doctrine, called associationism, dominated British and American views of the mind for centuries, and to a large extent still does. When the "ideas" were replaced by stimuli and responses, associationism became behaviorism. The blank slate and the two general-purpose laws of learning are also the psychological underpinnings of the Standard Social Science Model. We hear it in clichés about how our upbringing leads us to "associate" food with love, wealth with happiness, height with power, and so on.

Until recently, associationism was too vague to test. But neural-network models, which are routinely simulated on computers, make the ideas precise. The learning scheme, in which a teacher presents the network with an input and the correct output and the network strives to duplicate the pairing in the future, is a good model of the law of contiguity. The distributed input representation, in which a concept does not get its own unit ("parrot") but is represented by a pattern of activity over units for its properties ("feathered," "winged," and so on), allows for automatic generalization to similar concepts and thus nicely fits the law of association by resemblance. And if all parts of the mind start off as the same kind of network, we have an implementation of the blank slate. Connectionism thus offers an opportunity. In seeing what simple neural-network models can and cannot do, we can put the centuries-old doctrine of the association of ideas to a rigorous test.

Before we begin, we need to set aside some red herrings. Connectionism is not an alternative to the computational theory of mind, but a variety of it, which claims that the main kind of information processing done by the mind is multivariate statistics. Connectionism is not a necessary corrective to the theory that the mind is like a commercial computer, with a high-speed, error-free, serial central processing unit; no one holds that theory. And there is no real-life Achilles who claims that every form of thinking consists of cranking through thousands of rules from a logic textbook. Finally, connectionist networks are not particularly realistic models of the brain, despite the hopeful label "neural networks." For example, the "synapse" (connection weight) can switch from excitatory to inhibitory, and information can flow in both directions along an "axon" (connection), both anatomically impossible. When there is a choice between getting a job done and mirroring the brain, connectionists often opt for getting the job done; that shows that the networks are used as a form of artificial intelligence based loosely on the metaphor of neurons, and are not a form of neural modeling. The question is, do they perform the right kinds of computations to model the workings of human thought?

~

Raw connectoplasm has trouble with five feats of everyday thinking. The feats appear to be subtle at first, and were not even suspected of existing until logicians, linguists, and computer scientists began to put the meanings of sentences under a microscope. But the feats give human thought its distinctive precision and power and are, I think, an important part of the answer to the question, How does the mind work?

One feat is entertaining the concept of an individual. Let's go back to the first departure of neural networks from computerlike representations. Rather than symbolizing an entity as an arbitrary pattern in a string of bits, we represented it as a pattern in a layer of units, each standing for one of the entity's properties. An immediate problem is that there is no longer a way to tell apart two individuals with identical properties. They are represented in one and the same way, and the system is blind to the fact that they are not the same hunk of matter. We have lost the individual: we can represent vegetableness or horsehood, but not a particular vegetable or a particular horse. Whatever the system learns about one horse melds into what it knows about another, identical one. And there is no natural way to represent *two* horses. Making the horsey nodes twice as active won't do it, because that is indistinguishable from being twice as confident that the properties of a horse are present or from thinking that the properties of a horse are present to twice the degree.

It is easy to confuse the relationship between a class and a subclass, such as "animal" and "horse" (which a network handles easily), with the relationship between a subclass and an individual, such as "horse" and "Mr. Ed." The two relationships are, to be sure, similar in one way. In both, any property of the higher entity is inherited by the lower entity. If animals breathe, and horses are animals, then horses breathe; if horses have hooves, and Mr. Ed is a horse, then Mr. Ed has hooves. This can lure a modeler into treating an individual as a very, very specific subclass, using some slight difference between the two entities—a freckle unit that is on for one individual but off for the other—to distinguish near-doppelgangers.

Like many connectionist proposals, the idea dates back to British associationism. Berkeley wrote, "Take away the sensations of softness, moisture, redness, tartness, and you take away the cherry, since it is not a being distinct from sensations. A cherry, I say, is nothing but a congeries of sensible impressions." But Berkeley's suggestion never did work. Your knowledge of the properties of two objects can be identical and still you can know they are distinct. Imagine a room with two identical chairs. Someone comes in and switches them around. Is the room the same as or different from before? Obviously, everyone understands that it is different. But you know of no feature that distinguishes one chair from the other—except that you can think of one as Chair Number One and the other as Chair Number Two. We are back to arbitrary labels for memory slots, as in the despised digital computer! The same point underlies a joke from the comedian Stephen Wright: "While I was gone, someone stole everything in my apartment and replaced it with an exact replica. When I told my roommate, he said, 'Do I know you?'"

There is, admittedly, one feature that always distinguishes individuals: they cannot be in the same place at the same time. Perhaps the mind could stamp every object with the time and place and constantly update those coordinates, allowing it to distinguish individuals with identical properties. But even that fails to capture our ability to keep individuals apart in our minds. Suppose an infinite white plane contains nothing but two identical circles. One of them slides over and superimposes itself on the second one for a few moments, then proceeds on its way. I don't think anyone has trouble conceiving of the circles as distinct entities even in the moments in which they are in the same place at the same time. That shows that being in a certain place at a certain time is not our mental definition of "individual."

The moral is not that individuals cannot be represented in neural networks. It's easy; just dedicate some units to individuals' *identities* as individuals, independent of the individuals' *properties*. One could give each individual its own unit, or give each individual the equivalent of a serial number, coded in a pattern of active units. The moral is that the networks of the mind have to be crafted to implement the abstract logical notion of the individual, analogous to the role played by an arbitrarily labeled memory location in a computer. What does not work is a pattern associator restricted to an object's observable properties, a modern instantiation of the Aristotelian dictum that "there is nothing in the intellect that was not previously in the senses."

Is this discussion just an exercise in logic? Not at all: the concept of the individual is the fundamental particle of our faculties of social reasoning. Let me give you two real-life examples, involving those grand arenas of human interaction, love and justice.

Monozygotic twins share most of their properties. Apart from the physical resemblance, they think alike, feel alike, and act alike. Not identically, of course, and that is a loophole through which one might try to represent them as very narrow subclasses. But any creature representing them as subclasses should at least *tend* to treat identical twins alike. The creature should transfer its opinions from one to the other, at least probabilistically or to some extent—remember, that is a selling point of associationism and its implementation in connectoplasm. For example, whatever attracts you to one twin—the way he walks, the way he talks, the way he looks, and so on—should attract you to the other. And this should cast identical twins in tales of jealousy and betrayal of truly gothic proportions. In fact, nothing happens. The spouse of one identical twin feels no romantic attraction toward the other twin. Love locks our feelings in to another person *as that person*, not as a *kind* of person, no matter how narrow the kind.

On March 10, 1988, someone bit off half the ear of Officer David J. Storton. No one doubts who did it: either Shawn Blick, a twenty-one-year-old man living in Palo Alto, California, or Jonathan Blick, his identical twin brother. Both were scuffling with the officer, and one of them bit off part of his ear. Both were charged with mayhem, attempted burglary, assaulting a police officer, and aggravated mayhem. The aggravated mayhem charge, for the ear biting, carries a life sentence. Officer Storton testified that one of the twins had short hair and the other long, and it was the long-haired man who bit him. Unfortunately, by the time the men surrendered three days later they sported identical crew cuts and weren't talking. Their lawyers argued that neither one could be given the severe sentence for aggravated mayhem. For each brother there is a reasonable doubt as to whether he did it, because it could have been the other. The argument is compelling because our sense of justice picks out the *individual* who did a deed, not the characteristics of that individual.

Our obsession with individual personhood is not an inexplicable quirk, but probably evolved because every human being we meet, quite apart from any property we can observe, is guaranteed to house an unreplicable collection of memories and desires owing to a unique embryological and biographical history. In Chapter 6, when we reverse-engineer the sense of justice and the emotion of romantic love, we will see that the mental act of registering individual persons is at the heart of their design.

Human beings are not the only class of confusable individuals we have to keep distinct; a shell game is another real-life example. Many animals have to play shell games and thus keep track of individuals. One example is the mother who has to track her offspring, which may look like everyone else's but invisibly carries her genes. Another is the predator of herding animals, who has to track one member of the herd, following the tag-in-the-swimming-pool strategy: if you're "It," don't switch quarries, giving everyone but yourself time to catch their breath. When zoologists in Kenya tried to make their data collection easier by color-coding the horns of wildebeests they had tranquilized, they found that no matter how carefully they restored the marked animal to vigor before reintroducing it to the herd, it was killed within a day or so by hyenas. One explanation is that the colored marker made it easy for the hyenas to individuate the wildebeest and chase it to the point of exhaustion. Recent thinking about zebra stripes is that they are not for blending in with stripey tall grass—always a dubious explanation—but for turning the zebras into a living shell game, baffling lions and other predators as they try to keep their attention on just one zebra. Of course, we do not know that hyenas or lions have the concept of an individual; perhaps an odd man out just looks more appetizing. But the examples illustrate the computational problem of distinguishing individuals from classes, and they underscore the human mind's facility in solving it.

A second problem for associationism is called compositionality: the ability of a representation to be built out of parts and to have a meaning that comes from the meanings of the parts *and* from the way they are combined. Compositionality is the quintessential property of all human languages. The meaning of *The baby ate the slug* can be calculated from the meanings of *baby*, *ate*, *the*, and *slug* and from their positions in the sentence. The whole is not the sum of the parts; when the words are rearranged into *The slug ate the baby*, a different idea is conveyed. Since you have never heard either sentence before, you must have interpreted them by applying a set of algorithms (incorporating the rules of syntax) to the strings of words. The end product in each case is a novel thought you assembled on the fly. Equipped with the concepts of babies, slugs, and eating, and with an ability to arrange symbols for them on a mental bulletin board according to a scheme that can be registered by the demons that read it, you can think the thought for the first time in your life.

Journalists say that when a dog bites a man, that is not news, but when a man bites a dog, that is news. The compositionality of mental representations is what allows us to understand news. We can entertain wild and wonderful new ideas, no matter how outlandish. The cow jumped over the moon; the Grinch stole Christmas; the universe began with a big bang; aliens land at Harvard; Michael Jackson married Elvis' daughter. Thanks to the mathematics of combinatorics, we will never run out of news. There are hundreds of millions of trillions of thinkable thoughts.

You might think it is easy to put compositionality in a neural network: just turn on the units for "baby," "eats," and "slug." But if that was all that happened in your mind, you would be in a fog as to whether the baby ate the slug, the slug ate the baby, or the baby and the slug ate. The concepts must be assigned to *roles* (what logicians call "arguments"): who is the eater, who is the eaten.

Perhaps, then, one could dedicate a node to each *combination* of concepts and roles. There would be a baby-eats-slug node and a slug-eats-baby node. The brain contains a massive number of neurons, one might think, so why not do it that way? One reason not to is that there is massive and then there is *really* massive. The number of combinations grows exponentially with their allowable size, setting off a combinatorial explosion whose numbers surpass even our most generous guess of the brain's capacity. According to legend, the vizier Sissa Ben Dahir claimed a humble reward from King Shirham of India for inventing the game of chess. All he asked for was a grain of wheat to be placed on the first square of a chessboard, two grains of wheat on the second, four on the third, and so on. Well before they reached the sixty-fourth square the king discovered he had unwittingly committed all the wheat in his kingdom. The reward amounted to four trillion bushels, the world's wheat production for two thousand years. Similarly, the combinatorics of thought can overwhelm the number of neurons in the brain. A hundred million trillion sentence meanings cannot be squeezed into a brain with a hundred billion neurons if each meaning must have its own neuron.

But even if they did fit, a complex thought is surely not stored whole, one thought per neuron. The clues come from the way our thoughts are related to one another. Imagine that each thought had its own unit. There would have to be separate units for the baby eating the slug, the slug eating the baby, the chicken eating the slug, the chicken eating the baby, the slug eating the chicken, the baby seeing the slug, the slug seeing the baby, the chicken seeing the slug, and so on. Units have to be assigned to all of these thoughts and many more; any human being capable of thinking the thought that the baby saw the chicken is also capable of thinking the thought that the chicken saw the baby. But there is something suspicious about this inventory of thought-units; it is shot through with coincidences. Over and over again we have babies eating, slugs eating, babies seeing, slugs seeing, and so on. The thoughts perfectly slot themselves into the rows, columns, layers, hyper-rows, hyper-columns, and hyper-layers of a vast matrix. But this striking pattern is baffling if thoughts are just a very big collection of separate units; the units could just as easily have represented an inventory of isolated factoids that had nothing do with one another. When nature presents us with objects that perfectly fill a rectangular bank of pigeonholes, it's telling us that the objects must be built out of smaller components which correspond to the rows and the columns. That's how the periodic table of the elements led to an understanding of the structure of the atom. For similar reasons we can conclude that the warp and weft of our thinkable thoughts are the concepts composing them. Thoughts are assembled out of concepts; they are not stored whole.

Compositionality is surprisingly tricky for connectoplasm. All the obvious tricks turn out to be inadequate halfway measures. Suppose we dedicate each unit to a combination of one concept and one role. Perhaps one unit would stand for baby-eats and another for slug-is-eaten, or perhaps one unit would stand for baby-does-something and another for slug-has-something-done-to-it. This cuts down the number of combinations considerably—but at the cost of reintroducing befuddlement about who did what to whom. The thought "The baby ate the chicken when the poodle ate the slug" would be indistinguishable from the thought "The baby ate the slug when the poodle ate the chicken." The problem is that a unit for baby-eats does not say *what* it ate, and a unit for slug-is-eaten does not say who ate it.

A step in the right direction is to build into the hardware a distinction between the concepts (baby, slug, and so on) and the roles they play (actor, acted upon, and so on). Suppose we set up separate *pools* of units, one for the role of actor, one for the action, one for the role of acted upon. To represent a proposition, each pool of units is filled with the pattern for the concept currently playing the role, shunted in from a separate memory store for concepts. If we connected every node to every other node, we would have an auto-associator for propositions, and it could achieve a modicum of facility with combinatorial thoughts. We could store "baby ate slug," and then when any two of the components were presented as a question (say, "baby" and "slug," representing the question "What is the relationship between the baby and the slug?"), the network would complete the pattern by turning on the units for the third component (in this case, "ate").

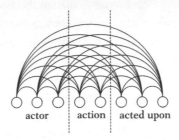

actor   action   acted upon

Or would it? Alas, it would not. Consider these thoughts:

Baby same-as baby.
Baby different-from slug.
Slug different-from baby.
Slug same-as slug.

No set of connection weights that allow "baby" in the first slot and "same-as" in the middle to turn on "baby" in the third slot, *and* that allow "baby" and "different-from" to turn on "slug," *and* that allow "slug" and "different-from" to turn on "baby," will *also* allow "slug" and "same-as" to turn on "slug." It's the exclusive-or problem in a different guise. If the baby-to-baby and same-to-baby links are strong, they will turn on "baby" in response to "baby same-as ____" (which is good), but they will also turn on "baby" in response to "baby different-from ____" (which is bad) and in response to "slug same-as ____" (also bad). Jigger the weights all you want; you will never find ones that work for all four sentences. Since any human can understand the four sentences without getting confused, the human mind must represent propositions with something more sophisticated than a set of concept-to-concept or concept-to-role associations. The mind needs a representation for the proposition itself. In this example, the model needs an *extra* layer of units—most straightforwardly, a layer dedicated to representing the entire proposition, separately from the concepts and their

proposition

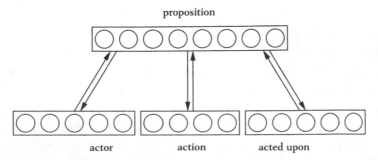

actor   action   acted upon

roles. The bottom of page 121 shows, in simplified form, a model devised by Geoffrey Hinton that does handle the sentences.

The bank of "proposition" units light up in arbitrary patterns, a bit like serial numbers, that label complete thoughts. It acts as a superstructure keeping the concepts in each proposition in their proper slots. Note how closely the architecture of the network implements standard, language-like mentalese! There have been other suggestions for compositional networks that aren't such obvious mimics, but they all have to have *some* specially engineered parts that separate concepts from their roles and that bind each concept to its role properly. The ingredients of logic such as predicate, argument, and proposition, and the computational machinery to handle them, have to be snuck back in to get a model to do mind-like things; association-stuff by itself is not enough.

Another mental talent that you may never have realized you have is called quantification, or variable-binding. It arises from a combination of the first problem, individuals, with the second, compositionality. Our compositional thoughts are, after all, often about individuals, and it makes a difference how those individuals are linked to the various parts of the thought. The thought that a particular baby ate a particular slug is different from the thought that a particular baby eats slugs in general, or that babies in general eat slugs in general. There is a family of jokes whose humor depends on the listener appreciating that difference. "Every forty-five seconds someone in the United States sustains a head injury." "Omigod! That poor guy!" When we hear that "Hildegard wants to marry a man with big muscles," we wonder whether she has a particular he-man lined up or if she is just hanging hopefully around the gym. Abraham Lincoln said, "You may fool all the people some of the time; you can even fool some of the people all the time; but you can't fool all of the people all the time." Without an ability to compute quantification, we could not understand what he said.

In these examples, we have several sentences, or several readings of an ambiguous sentence, in which the same concepts play the same roles but the ideas as a whole are very different. Hooking up concepts to their roles is not enough. Logicians capture these distinctions with variables and quantifiers. A variable is a place-holding symbol like x or y which stands for the same entity across different propositions or different parts of one proposition. A quantifier is a symbol that can express "There exists a particular *x* who . . ." and "For all *x* it is true that . . ." A thought can then be captured in a proposition built out of symbols for concepts, roles, quantifiers, and variables, all precisely ordered and bracketed. Compare, for example, "Every forty-five seconds {there exists an X [who gets injured]}" with "There exists an X {who every forty-five seconds [gets injured]}." Our mentalese must have machinery that does something similar. But so far, we have no hint as to how this can be done in an associative network.

Not only can a proposition be about an individual, it must be treated as a kind of individual itself, and that gives rise to a new problem. Connectoplasm gets its power from superimposing patterns in a single set of units. Unfortunately, that can breed bizarre chimeras or make a network fall between two stools. It is part of a pervasive bugaboo for connectoplasm called interference or cross-talk.

Here are two examples. The psychologists Neal Cohen and Michael McCloskey trained a network to add two digits. They first trained it to add 1 to the other numbers: when the inputs were "1" and "3," the network learned to put out "4," and so on. Then they trained it to add 2 to any other number. Unfortunately, the add-2 problem sucked the connection weights over to values that were optimal for adding 2, and because the network had no hardware set aside to anchoring the knowledge of how to add 1, it became amnesic for how to add 1! The effect is called "catastrophic forgetting" because it is unlike the mild forgetting of everyday life. Another example comes from a network designed by McClelland and his collaborator Alan Kawamoto to assign meanings to ambiguous sentences. For example, *A bat broke the window* can mean either that a baseball bat was hurled at it or that a winged mammal flew through it. The network came up with the one interpretation that humans do not make: a winged mammal broke the window using a baseball bat!

As with any other tool, the features that make connectoplasm good for some things make it bad for other things. A network's ability to generalize comes from its dense interconnectivity and its superposition of inputs. But if you're a unit, it's not always so great to have thousands of other units yammering in your ear and to be buffeted by wave after wave of inputs. Often different hunks of information should be packaged and stored separately, not blended. One way to do this is to give each proposition its own storage slot and address—once again showing that not all aspects of computer design can be dismissed as silicon curiosities. Computers, after all, were not designed as room heaters; they were designed to process information in a way that is meaningful to human users.

The psychologists David Sherry and Dan Schacter have pushed this line of reasoning farther. They note that the different engineering demands on a memory system are often at cross-purposes. Natural selection, they argue, responded by giving organisms *specialized* memory systems. Each has a computational structure optimized for the demands of one of the tasks the mind of the animal must fulfill. For example, birds that cache seeds to retrieve in leaner times have evolved a capacious memory for the hiding places (ten thousand places, in the case of the Clark's Nutcracker). Birds whose males sing to impress the females or to intimidate other males have evolved a capacious memory for songs (two hundred, in the case of the nightingale). The memory for caches and the memory for songs are in different brain structures and have different patterns of wiring. We humans place two very different demands on our memory system at the same time. We have to remember individual episodes of who did what to whom, when, where, and why, and that requires stamping each episode with a time, a date, and a serial number. But we also must extract generic knowledge about how people work and how the world works. Sherry and Schacter suggest that nature gave us one memory system for each requirement: an "episodic" or autobiographical memory, and a "semantic" or generic-knowledge memory, following a distinction first made by the psychologist Endel Tulving.

The trick that multiplies human thoughts into truly astronomical numbers is not the slotting of concepts into three or four roles but a kind of mental fecundity called recursion. A fixed set of units for each role is not enough. We humans can take an entire proposition and give it a role in some larger proposition. Then we can take the larger proposition and embed it in a still-larger one, creating a hierarchical tree structure of propositions inside propositions. Not only did the baby eat the slug, but the father saw the baby eat the slug, and I wonder whether the father saw the baby eat the slug, and the father knows that I wonder whether he saw the baby eat the slug, and I can guess that the father knows that I wonder whether he saw the baby eat the slug, and so on. Just as an ability to add 1 to a number bestows the ability to generate an infinite set of numbers, the ability to embed a proposition inside another proposition bestows the ability to think an infinite number of thoughts.

To get propositions-inside-propositions out of the network displayed in the preceding diagram, one could add a new layer of connections to the top of the diagram, connecting the bank of units for the whole proposition to the role slot in some bigger proposition; the role might be something like "event-observed." If we continue to add enough layers, we could accommodate an entire multiply nested proposition by etching a full tree diagram for it in connectoplasm. But this solution is clumsy and raises suspicions. For every kind of recursive structure, there would have to be a different network hard-wired in: one network for a person thinking about a proposition, another for a person thinking about a proposition about a person thinking about a proposition, a third for a person communicating a proposition about some person to another person, and so on.

In computer science and psycholinguistics, a more powerful and flexible mechanism is used. Each simple structure (for a person, an action, a proposition, and so on) is represented in long-term memory *once*, and a processor shuttles its attention from one structure to another, storing the itinerary of visits in short-term memory to thread the proposition together. This dynamic processor, called a recursive transition network, is especially plausible for sentence understanding, because we hear and read words one at a time rather than inhaling an entire sentence at once. We also seem to chew our complex thoughts piece by piece rather swallowing or regurgitating them whole, and that suggests that the mind is equipped with a recursive proposition-cruncher for thoughts, not just for sentences. The psychologists Michael Jordan and Jeff Elman have built networks whose output units send out connections that loop back into a set of short-term memory units, triggering a new cycle of activation flow. That looping design provides a glimpse of how iterative information processing might be implemented in neural networks, but it is not enough to interpret or assemble structured propositions. More recently, there have been attempts to combine a looping network with a propositional network to implement a kind of recursive transition network out of pieces of connectoplasm. These attempts show that unless neural networks are specially assembled into a recursive processor, they cannot handle our recursive thoughts.

The human mind must be given credit for one more cognitive feat that is difficult to wring out of connectoplasm, and therefore difficult to explain by associationism. Neural networks easily implement a fuzzy logic in which everything is a kind-of something to some degree. To be sure, many common-sense concepts really are fuzzy at their edges and have no clear definitions. The philosopher Ludwig Wittgenstein offered the example of "a game," whose exemplars (jigsaw puzzles, roller derby, curling, Dungeons and Dragons, cockfighting, and so on) have nothing in common, and earlier I gave you two others, "bachelor" and "vegetable." The members of a fuzzy category lack a single defining feature; they overlap in many features, much like the members of a family or the strands of a rope, none of which runs the entire length. In the comic strip *Bloom County*, Opus the Penguin, temporarily amnesic, objects when he is told he is a bird. Birds are svelte and aerodynamic, he points out; he is not. Birds can fly; he cannot. Birds can sing; his performance of "Yesterday" left his listeners gagging. Opus suspects he is really Bullwinkle the Moose. So even concepts like "bird" seem to be organized not around necessary and sufficient conditions but around prototypical members. If you look up *bird* in the dictionary, it will be illustrated not with a penguin but with Joe Bird, typically a sparrow.

Experiments in cognitive psychology have shown that people are bigots about birds, other animals, vegetables, and tools. People share a stereotype, project it to all the members of a category, recognize the stereotype more quickly than the nonconformists, and even claim to have seen the stereotype when all they really saw were examples similar to it. These responses can be predicted by tallying up the properties that a member shares with other members of the category: the more birdy properties, the better the bird. An auto-associator presented with examples from a category pretty much does the same thing, because it computes correlations among properties. That's a reason to believe that parts of human memory are wired something like an auto-associator.

But there must be more to the mind than that. People are not *always* fuzzy. We laugh at Opus because a part of us knows that he really *is* a bird. We may agree on the prototype of a grandmother—the kindly, gray-haired septuagenarian dispensing blueberry muffins or chicken soup (depending on whose stereotype we're talking about)—but at the same time we have no trouble understanding that Tina Turner and Elizabeth

Taylor are grandmothers (indeed, a *Jewish* grandmother, in Taylor's case). When it comes to bachelors, many people—such as immigration authorities, justices of the peace, and health care bureaucrats—are notoriously *un*fuzzy about who belongs in the category; as we all know, a lot can hinge on a piece of paper. Examples of unfuzzy thinking are everywhere. A judge may free an obviously guilty suspect on a technicality. Bartenders deny beer to a responsible adult the day before his twenty-first birthday. We joke that you can't be a little bit pregnant or a little bit married, and after a Canadian survey reported that married women have sex 1.57 times a week, the cartoonist Terry Mosher drew a woman sitting up in bed beside her dozing husband and muttering, "Well, that was .57."

In fact, fuzzy and crisp versions of *the same category* can live side by side in a single head. The psychologists Sharon Armstrong, Henry Gleitman, and Lila Gleitman mischievously gave the standard tests for fuzzy categories to university students but asked them about knife-edged categories like "odd number" and "female." The subjects happily agreed to daft statements such as that 13 is a better example of an odd number than 23 is, and that a mother is a better example of a female than a comedienne is. Moments later the subjects also claimed that a number either is odd or is even, and that a person either is female or is male, with no gray areas.

People think in two modes. They can form fuzzy stereotypes by uninsightfully soaking up correlations among properties, taking advantage of the fact that things in the world tend to fall into clusters (things that bark also bite and lift their legs at hydrants). But people can also create systems of rules—intuitive theories—that define categories in terms of the rules that apply to them, and that treat all the members of the category equally. All cultures have systems of formal kinship rules, often so precise that one can prove theorems in them. Our own kinship system gives us a crisp version of "grandmother": the mother of a parent, muffins be damned. Law, arithmetic, folk science, and social conventions (with their rites of passage sharply delineating adults from children and husbands from bachelors) are other rule systems in which people all over the planet reckon. The grammar of a language is yet another.

Rule systems allow us to rise above mere similarity and reach conclusions based on explanations. Hinton, Rumelhart, and McClelland wrote: "People are good at generalizing newly acquired knowledge. If, for example, you learn that chimpanzees like onions you will probably raise your estimate of the probability that gorillas like onions. In a network that uses distributed representations, this kind of generalization is automatic." Their boast is a twentieth-century echo of Hume's remark that from a body similar to bread in color and consistency we expect a similar degree of nourishment. But the assumption breaks down in any domain in which a person actually knows something. The onion-loving gorilla was intended only as an example, of course, but it is interesting to see how even this simple example underestimates us. Knowing a bit of zoology and not much about gorillas, I would definitely *not* raise my estimate of the probability that gorillas like onions. Animals can be cross-classified. They may be grouped by genealogy and resemblance into a taxon, such as the great apes, but they also may be grouped into "guilds" that specialize in certain ways of getting food, such as omnivores, herbivores, and carnivores. Knowing this principle leads me to reason as follows. Chimpanzees are omnivores, and it is not surprising that they eat onions; after all, we are omnivores, and we eat them. But gorillas are herbivores, who spend their days munching wild celery, thistles, and other plants. Herbivores are often finicky about which species they feed on, because their digestive systems are optimized to detoxify the poisons in some kinds of plants and not others (the extreme example being koalas, who specialize in eating eucalyptus leaves). So it would not surprise me if gorillas avoided the pungent onion, regardless of what chimpanzees do. Depending on which system of explanation I call to mind, chimpanzees and gorillas are either highly similar category-mates or as different as people and cows.

In associationism and its implementation in connectoplasm, the way an object is represented (namely, as a set of properties) automatically commits the system to generalizing in a certain way (unless it is trained out of the generalization with specially provided contrary examples). The alternative I am pushing is that humans can mentally *symbolize* kinds of objects, and those symbols can be referred to in a number of rule systems we carry around in our heads. (In artificial intelligence, this technique is called explanation-based generalization, and connectionist designs are an example of the technique called similarity-based generalization.) Our rule systems couch knowledge in compositional, quantified, recursive propositions, and collections of these propositions interlock to form modules or intuitive theories about particular domains of experience, such as kinship, intuitive science, intuitive psychology, number, language, and law. Chapter 5 explores some of those domains.

What good are crisp categories and systems of rules? In the social world they can adjudicate between haggling parties each pointing at the fuzzy boundary of a category, one saying something is inside and the other saying it is outside. Rites of passage, the age of majority, diplomas, licenses, and other pieces of legal paper draw sharp lines that all parties can mentally represent, lines that let everyone know where everyone else stands. Similarly, all-or-none rules are a defense against salami tactics, in which a person tries to take advantage of a fuzzy category by claiming one borderline case after another to his advantage.

Rules and abstract categories also help in dealing with the natural world. By sidestepping similarity, they allow us to get beneath the surface and ferret out hidden laws that make things tick. And because they are, in a sense, digital, they give representations stability and precision. If you make a chain of analog copies from an analog tape, the quality declines with each generation of copying. But if you make a chain of digital copies, the last can be as good as the first. Similarly, crisp symbolic representations allow for chains of reasoning in which the symbols are copied verbatim in successive thoughts, forming what logicians call a sorites:

All ravens are crows.
All crows are birds.
All birds are animals.
All animals need oxygen.

A sorites allows a thinker to draw conclusions with confidence despite meager experience. For example, a thinker can conclude that ravens need oxygen even if no one has ever actually deprived a raven of oxygen to see what happens. The thinker can reach that conclusion even if he or she has never witnessed an experiment depriving *any* animal of oxygen but only heard the statement from a credible expert. But if each step in the deduction were fuzzy or probabilistic or cluttered with the particulars of the category members one step before, the slop would accumulate. The last statement would be as noisy as an *n*th-generation bootleg tape or as unrecognizable as the last whisper in a game of broken telephone. People in all cultures carry out long chains of reasoning built from links whose truth they could not have observed directly. Philosophers have often pointed out that science is made possible by that ability.

~

Like many issues surrounding the mind, the debate over connectionism is often cast as a debate between innateness and learning. And as always, that makes it impossible to think clearly. Certainly learning plays an enormous role in connectionist modeling. Often a modeler, sent back to the drawing board by the problems I have mentioned, will take advantage of a hidden-layer network's ability to learn a set of inputs and outputs and generalize them to new, similar ones. By training the living daylights out of a generic hidden-layer network, one can sometimes get it to do approximately the right thing. But heroic training regimes cannot, by themselves, be the salvation of connectoplasm. That is not because the networks have too little innate structure and too much environmental input. It is because raw connectoplasm is so underpowered that networks must often be built with the worst combination: too much innate structure combined with too much environmental input.

For example, Hinton devised a three-layer network to compute family relationships. (He intended it as a demonstration of how networks work, but other connectionists have treated it as a real theory of psychology.) The input layer had units for a name and units for a relationship, such as "Colin" and "mother." The output layer had units for the name of the person so related, such as "Victoria." Since the units and connections are the innate structure of a network, and only the connection weights are learned, taken literally the network corresponds to an innate module in the brain just for spitting out answers to questions about who is related to a named person in a given way. It is not a system for reasoning about kinship in general, because the knowledge is smeared across the connection weights linking the question layer to the answer layer, rather than being stored in a database that can be accessed by different retrieval processes. So the knowledge is useless if the question is changed slightly, such as asking how two people are related or asking for the names and relationships in a person's family. In this sense, the model has too much innate structure; it is tailored to a specific quiz.

After training the model to reproduce the relationships in a small, made-up family, Hinton called attention to its ability to generalize to new pairs of kin. But in the fine print we learn that the network had to be trained on 100 of the 104 possible pairs in order to generalize to the remaining 4. And each of the 100 pairs in the training regime had to be fed into the network 1,500 times (150,000 lessons in all)! Obviously children do not learn family relationships in a manner even remotely like this. The numbers are typical of connectionist networks, because they do not cut to the solution by means of rules but need to have most of the examples pounded into them and merely interpolate between the examples. Every substantially different kind of example must be in the training set, or the network will interpolate spuriously, as in the story of the statisticians on a duck hunt: One shoots a yard too high, the second shoots a yard too low, and the third shouts, "We got him!"

Why put connectoplasm under such strong lights? Certainly not because I think neural-network modeling is unimportant—quite the contrary! Without it, my whole edifice on how the mind works would be left levitating in midair. Nor do I think that network modeling is merely subcontracting out the work of building demons and data structures from neural hardware. Many connectionist models offer real surprises about what the simplest steps of mental computation can accomplish. I do think that connectionism has been oversold. Because networks are advertised as soft, parallel, analogical, biological, and continuous, they have acquired a cuddly connotation and a diverse fan club. But neural networks don't perform miracles, only some logical and statistical operations. The choices of an input representation, of the number of networks, of the wiring diagram chosen for each one, and of the data pathways and control structures that interconnect them explain more about what makes a system smart than do the generic powers of the component connectoplasm.

But my main intent is not to show what certain kinds of models cannot do but what the mind *can* do. The point of this chapter is to give you a feel for the stuff our minds are made of. Thoughts and thinking are no longer ghostly enigmas but mechanical processes that can be studied, and the strengths and weaknesses of different theories can be examined and debated. I find it particularly illuminating to see the shortcomings of the venerable doctrine of the association of ideas, because they highlight the precision, subtlety, complexity, and open-endedness of our everyday thinking. The computational power of human thought has real consequences. It is put to good use in our capacity for love, justice, creativity, literature, music, kinship, law, science, and other activities we will explore in later chapters. But before we get to them, we must return to the other question that opened this chapter.

ALADDIN'S LAMP

What about consciousness? What makes us actually *suffer* the pain of a toothache or see the blue of the sky as *blue*? The computational theory of mind, even with complete neural underpinnings, offers no clear answer. The symbol blue is inscribed, goal states change, some neurons fire; so what? Consciousness has struck many thinkers as not just a problem but almost a miracle:

> Matter can differ from matter only in form, bulk, density, motion and direction of motion: to which of these, however varied or combined, can consciousness be annexed? To be round or square, to be solid or fluid, to be great or little, to be moved slowly or swiftly one way or another, are modes of material existence, all equally alien from the nature of cogitation.
>
> —Samuel Johnson

> How it is that anything so remarkable as a state of consciousness comes about as a result of irritating nervous tissue, is just as unaccountable as the appearance of the Djin, when Aladdin rubbed his lamp.
>
> —Thomas Huxley

> Somehow, we feel, the water of the physical brain is turned into the wine of consciousness, but we draw a total blank on the nature of this conversion. Neural transmissions just seem like the wrong kind of materials with which to bring consciousness into the world.
>
> —Colin McGinn

Consciousness presents us with puzzle after puzzle. How can a neural event cause consciousness to happen? What good is consciousness? That is, what does the raw sensation of redness add to the train of billiard-ball events taking place in our neural computers? Any *effect* of perceiving something as red—noticing it against a sea of green, saying out loud, "That's red," reminiscing about Santa Claus and fire engines, becoming agitated—could be accomplished by pure information processing triggered by a sensor for long-wavelength light. Is consciousness an impotent side effect hovering over the symbols, like the lights flashing on a computer or the thunder that accompanies lightning? And if consciousness is useless—if a creature without it could negotiate the world as well as a creature with it—why would natural selection have favored the conscious one?

Consciousness has recently become the circle that everyone wants to square. Almost every month an article announces that consciousness has been explained at last, often with a raspberry blown at the theologians and humanists who would put boundaries on science and another one for the scientists and philosophers who dismiss the topic as too subjective or muddled to be studyable.

Unfortunately, many of the things that people write about consciousness are almost as puzzling as consciousness itself. Stephen Jay Gould wrote, "*Homo sapiens* is one small twig [on the tree of life]. . . . Yet our twig, for better or worse, has developed the most extraordinary new quality in all the history of multicellular life since the Cambrian explosion. We have invented consciousness with all its sequelae from Hamlet to Hiroshima." Gould has denied consciousness to all nonhuman animals; other scientists grant it to some animals but not all. Many test for consciousness by seeing whether an animal recognizes that the image in a mirror is itself and not another animal. By this standard, monkeys, young chimpanzees, old chimpanzees, elephants, and human toddlers are unconscious. The only conscious animals are gorillas, orangutans, chimpanzees in their prime, and, according to Skinner and his student Robert Epstein, properly trained pigeons. Others are even more restrictive than Gould: not even all people are conscious. Julian Jaynes claimed that consciousness is a recent invention. The people of early civilizations, including the Greeks of Homer and the Hebrews of the Old Testament, were unconscious. Dennett is sympathetic to the claim; he believes that consciousness "is largely a product of cultural evolution that gets imparted to brains in early training" and that it is "a huge complex of memes," *meme* being Dawkins' term for a contagious feature of culture, such as a catchy jingle or the latest fashion craze.

Something about the topic of consciousness makes people, like the White Queen in *Through the Looking Glass*, believe six impossible things before breakfast. Could most animals really be *unconscious*—sleepwalkers, zombies, automata, out cold? Hath not a dog senses, affections, passions? If you prick them, do they not feel pain? And was Moses really unable to taste salt or see red or enjoy sex? Do children learn to become conscious in the same way that they learn to wear baseball caps turned around?

People who write about consciousness are not crazy, so they must have something different in mind when they use the word. One of the best observations about the concept of consciousness came from Woody Allen in his hypothetical college course catalogue:

Introduction to Psychology: The theory of human behavior. . . . Is there a split between mind and body, and, if so, which is better to have?

. . . Special consideration is given to a study of consciousness as opposed to unconsciousness, with many helpful hints on how to remain conscious.

Verbal humor sets readers up with one meaning of an ambiguous word and surprises them with another. Theoreticians also trade on the ambiguity of the word *consciousness*, not as a joke but as a bait-and-switch: the reader is led to expect a theory for one sense of the word, the hardest to explain, and is given a theory for another sense, the easiest to explain. I don't like to dwell on definitions, but when it comes to consciousness we have no choice but to begin by disentangling the meanings.

Sometimes "consciousness" is just used as a lofty synonym for "intelligence." Gould, for example, must have been using it in this way. But there are three more-specialized meanings, nicely distinguished by the linguist Ray Jackendoff and the philosopher Ned Block.

One is *self-knowledge*. Among the various people and objects that an intelligent being can have information about is the being itself. Not only can I feel pain and see red, I can think to myself, "Hey, here I am, Steve Pinker, feeling pain and seeing red!" Oddly enough, this recondite sense of the word is the one that most academic discussions have in mind. Consciousness is typically defined as "building an internal model of the world that contains the self," "reflecting back on one's own mode of understanding," and other kinds of navel-gazing that have nothing to do with consciousness as it is commonly understood: being alive and awake and aware.

Self-knowledge, including the ability to use a mirror, is no more mysterious than any other topic in perception and memory. If I have a mental database for people, what's to prevent it from containing an entry for myself? If I can learn to raise my arm and crane my neck to sight a hidden spot on my back, why couldn't I learn to raise a mirror and look up at it to sight a hidden spot on my forehead? And access to information about the self is perfectly easy to model. Any beginning programmer can write a short piece of software that examines, reports on, and even modifies itself. A robot that could recognize itself in a mirror would not be much more difficult to build than a robot that could recognize anything at all. There are, to be sure, good questions to ask about the evolution of self-knowledge, its development in children, and its advantages (and, more interesting, disadvantages, as we shall see in Chapter 6). But self-knowledge is an everyday topic in cognitive science, not the paradox of water becoming wine. Because it is so easy to say something about self-knowledge, writers can crow about their "theory of consciousness."

A second sense is *access to information*. I ask, "A penny for your thoughts?" You reply by telling me the content of your daydreams, your plans for the day, your aches and itches, and the colors, shapes, and sounds in front of you. But you cannot tell me about the enzymes secreted by your stomach, the current settings of your heart and breathing rate, the computations in your brain that recover 3-D shapes from the 2-D retinas, the rules of syntax that order the words as you speak, or the sequence of muscle contractions that allow you to pick up a glass. That shows that the mass of information processing in the nervous system falls into two pools. One pool, which includes the products of vision and the contents of short-term memory, can be accessed by the systems underlying verbal reports, rational thought, and deliberate decision making. The other pool, which includes autonomic (gut-level) responses, the internal calculations behind vision, language, and movement, and repressed desires or memories (if there are any), cannot be accessed by those systems. Sometimes information can pass from the first pool to the second or vice versa. When we first learn how to use a stick shift, every motion has to be thought out, but with practice the skill becomes automatic. With intense concentration and biofeedback, we can focus on a hidden sensation like our heartbeat.

This sense of consciousness, of course, also embraces Freud's distinction between the conscious and the unconscious mind. As with self-knowledge, there is nothing miraculous or even mysterious about it. Indeed, there are obvious analogues in machines. My computer has access to information about whether the printer is working or not working (it is "conscious" of it, in this particular sense) and can print out an error message, `Printer not responding`. But it has no access to information about *why* the printer is not working; the signal carried back along the cable from printer to computer does not include the information. The chip inside the printer, in contrast, *does* have access to that information (it is conscious of it, in this sense); the sensors in different parts of the printer feed into the chip, and the chip can turn on a yellow light if the toner supply is low and a red light if the paper is jammed.

Finally, we come to the most interesting sense of all, *sentience*: subjective experience, phenomenal awareness, raw feels, first-person present tense, "what it is like" to be or do something, if you have to ask you'll never know. Woody Allen's joke turned on the difference between this sense of consciousness and Freud's sense of it as access to information by the deliberative, language-using parts of the mind. And this sense, sentience, is the one in which consciousness seems like a miracle.

The remainder of the chapter is about consciousness in these last two senses. First I will look at access, at what kinds of information the different parts of the mind make available to one another. In this sense of the word, we really are coming to understand consciousness. Interesting things can be said about how it is implemented in the brain, the role it plays in mental computation, the engineering specs it is designed to meet (and hence the evolutionary pressures that gave rise to it), and how those specs explain the main features of consciousness—sensory awareness, focal attention, emotional coloring, and the will. Finally, I will turn to the problem of sentience.

Someday, probably sooner rather than later, we will have a fine understanding of what in the brain is responsible for consciousness in the sense of access to information. Francis Crick and Christof Koch, for example, have set out straightforward criteria for what we should look for. Most obviously, information from sensation and memory guides behavior only in an awake animal, not an anesthetized one. Therefore some of the neural bases of access-consciousness can be found in whatever brain structures act differently when an animal is awake and when it is in a dreamless sleep or out cold. The lower layers of the cerebral cortex are one candidate for that role. Also, we know that information about an object being perceived is scattered across many parts of the cerebral cortex. Therefore information access requires a mechanism that binds together geographically separated data. Crick and Koch suggest that synchronization of neural firing might be one such mechanism, perhaps entrained by loops from the cortex to the thalamus, the cerebrum's central way-station. They also note that voluntary, planned behavior requires activity in the frontal lobes. Therefore access-consciousness may be determined by the anatomy of the fiber tracts running from various parts of the brain to the frontal lobes. Whether or not they are right, they have shown that the problem can be addressed in the lab.

Access-consciousness is also a mere problem, not a mystery, in our grasp of the computations carried out by the brain. Recall our uncle-detecting production system. It has a communal short-term memory: a workspace or bulletin board visible to all of the demons in the system. In a separate part of the system lies a larger repository of information, a long-term memory, that cannot be read by the demons until pieces of it are copied to the short-term memory. Many cognitive psychologists have pointed out that in these models the short-term memory (communal bulletin board, global workspace) acts just like consciousness. When we are aware of a piece of information, many parts of the mind can act on it. We not only see a ruler in front of us but can describe it, reach for it, deduce that it can prop up a window, or count its markings. As the philosopher Stephen Stich has put it, conscious information is inferentially *promiscuous*; it makes itself available to a large number of information-processing agents rather than committing itself to one alone. Newell and Simon have made headway in understanding human problem-solving simply by asking a person to think aloud when working on a puzzle. They have nicely simulated the mental activity using a production system where the contents of the bulletin board correspond step for step with the person's report of what he is consciously thinking.

The engineering specs of information access, and thus the selection pressures that probably gave rise to it, are also becoming clearer. The general principle is that any information processor must be given limited access to information because information has costs as well as benefits.

One cost is space: the hardware to hold the information. The limitation is all too clear to microcomputer owners deciding whether to invest in more RAM. Of course the brain, unlike a computer, comes with vast amounts of parallel hardware for storage. Sometimes theorists infer that the brain can store *all* contingencies in advance and that thought can be reduced to one-step pattern recognition. But the mathematics of a combinatorial explosion bring to mind the old slogan of MTV: Too much is never enough. Simple calculations show that the number of humanly graspable sentences, sentence meanings, chess games, melodies, seeable objects, and so on can exceed the number of particles in the universe. For example, there are thirty to thirty-five possible moves at each point in a chess game, each of which can be followed by thirty to thirty-five responses, defining about a thousand complete turns. A typical chess game lasts forty turns, yielding 10^{120} different chess games. There are about 10^{70} particles in the visible universe. So no one can play chess by memorizing all the games and recognizing every sequence of moves. The same is true for sentences, stories, melodies, and so on. Of course, *some* combinations can be stored, but pretty soon either you run out of brain or you start to superimpose the patterns and get useless chimeras and blends. Rather than storing googols of inputs and their outputs or questions and their answers, an information processor needs rules or algorithms that operate on a subset of information at a time and calculate an answer just when it is needed.

A second cost of information is time. Just as one couldn't store all the chess games in a brain less than the size of the universe, one can't mentally play out all the chess games in a lifetime less than the age of the universe (10^{18} seconds). Solving a problem in a hundred years is, practically speaking, the same as not solving it at all. In fact, the requirements on an intelligent agent are even more stringent. Life is a series of deadlines. Perception and behavior take place in real time, such as in hunting an animal or keeping up one's end of a conversation. And since computation itself takes time, information processing can be part of the problem rather than part of the solution. Think about a hiker planning the quickest route back to camp before it gets dark and taking twenty minutes to plot out a path that saves her ten minutes.

A third cost is resources. Information processing requires energy. That is obvious to anyone who has stretched out the battery life of a laptop computer by slowing down the processor and restricting its access to information on the disk. Thinking, too, is expensive. The technique of functional imaging of brain activity (PET and MRI) depends on the fact that working brain tissue calls more blood its way and consumes more glucose.

Any intelligent agent incarnated in matter, working in real time, and subject to the laws of thermodynamics must be restricted in its access to information. Only information *relevant* to the problem at hand should be allowed in. That does not mean that the agent should wear blinkers or become an amnesiac. Information that is irrelevant at one time for one purpose might be relevant at another time for another purpose. So information must be *routed*. Information that is always irrelevant to a kind of computation should be permanently sealed off from it. Information that is sometimes relevant and sometimes irrelevant should be accessible to a computation when it is relevant, insofar as that can be predicted in advance. This design specification explains why access-consciousness exists in the human mind and also allows us to understand some of its details.

Access-consciousness has four obvious features. First, we are aware, to varying degrees, of a rich field of sensation: the colors and shapes of the world in front of us, the sounds and smells we are bathed in, the pressures and aches of our skin, bone, and muscles. Second, portions of this information can fall under the spotlight of attention, get rotated into and out of short-term memory, and feed our deliberative cogitation. Third, sensations and thoughts come with an emotional flavoring: pleasant or unpleasant, interesting or repellent, exciting or soothing. Finally, an executive, the "I," appears to make choices and pull the levers of behavior. Each of these features discards some information in the nervous system, defining the highways of access-consciousness. And each has a clear role in the adaptive organization of thought and perception to serve rational decision making and action.

Let's begin with the perceptual field. Jackendoff, after reviewing the levels of mental representation used by various modules, asked which level corresponds to the rich field of present-tense awareness. For example, visual processing runs from the rods and cones in the retina, through intermediate levels representing edges, depths, and surfaces, to a recognition of the objects in front of us. Language understanding proceeds from raw sound up through representations of syllables, words, and phrases, to an understanding of the content of the message.

Jackendoff observed that access-consciousness seems to tap the intermediate levels. People are unaware of the lowest levels of sensation. We do not spend our lives in Proustian contemplation of every crumb of the madeleine and every nuance of the decoction of lime flowers. We literally cannot see the lightness of the coal in the sun, the darkness of the snowball inside, the pale green-gray of the "black" areas on the television screen, or the rubbery parallelograms that a moving square projects on our retinas. What we "see" is a highly processed product: the surfaces of objects, their intrinsic colors and textures, and their depths, slants, and tilts. In the sound wave arriving at our ears, syllables and words are warped and smeared together, but we don't hear that seamless acoustic ribbon; we "hear" a chain of well-demarcated words. Our immediate awareness does not exclusively tap the *highest* level of representation, either. The highest levels—the contents of the world, or the gist of a message—tend to stick in long-term memory days and years after an experience, but as the experience is unfolding, we are aware of the sights and sounds. We do not just abstractly think "Face!" when we see a face; the shadings and contours are available for scrutiny.

The advantages of intermediate-level awareness are not hard to find. Our perception of a constant shape and lightness across changes in viewing conditions tracks the object's inherent properties: the lump of coal itself stays rigid and black as we move around it or raise the lights, and we experience it as looking the same. The lower levels are not needed, and the higher levels are not enough. The raw data and computational steps behind these constancies are sealed off from our awareness, no doubt because they use the eternal laws of optics and neither need advice from, nor have any insights to offer to, the rest of cognition. The products of the computation are released for general consumption well before the identities of objects are established, because we need more than a terse *mise en scène* to make our way around the world. Behavior is a game of inches, and the geometry and composition of surfaces must be available to the decision processes that plan the next step or grasp. Similarly, while we are understanding a sentence there is nothing to be gained in peering all the way down to the hisses and hums of the sound wave; they have to be decoded into syllables before they match up with anything meaningful in the mental dictionary. The speech decoder uses a special key with lifelong validity and should be left to do its job without interference from kibbitzers in the rest of the mind. But as with vision, the rest of the mind cannot be satisfied with only the final product, either—in this case the speaker's gist. The choice of words and the tone of voice carry information that allows us to hear between the lines.

The next noteworthy feature of conscious access is the spotlight of attention. It serves as the quintessential demonstration that unconscious parallel processing (in which many inputs are processed at the same time, each by its own mini-processor) can go only so far. An early stage of parallel processing does what it can, and passes along a representation from which a more cramped and plodding processor must select the information it needs. The psychologist Anne Treisman thought up a few simple, now classic demonstrations of where unconscious processing leaves off and conscious processing begins. People are shown a display of colored shapes, like X's and O's, and are asked to press a button if they see a specified target. If the search target is an O and the display shows one O in a sea of X's, the person responds quickly. It doesn't matter how many X's there are; people say the O just pops out. (Pop-out, as the effect is now called, is a nice sign of unconscious parallel processing.) Similarly, a green O pops out from a sea of red O's. But if the experimenter asks the person to find a letter that is both green *and* an O, and the letter sits somewhere in a mixed sea of green X's and red O's, the person must consciously search the display, letter by letter, checking each one to see if it meets the two-part criterion. The task becomes like the children's comic strip *Where's Waldo?*, in which the hero in the red-and-white-striped jersey hides in a throng of people wearing red, white, or stripes.

What exactly is happening? Imagine that the visual field is sprinkled with thousands of little processors, each of which detects a color or a simple shape like a curve, an angle, or a line whenever it appears at the processor's location. The output of one set of processors looks like this: red red red red green red red red, and so on. The output of another set looks like this: straight straight straight curved straight straight straight, and so on. Superimposed on these processors is a layer of odd-man-out detectors. Each stands astride a group of line or color detectors and "marks" any spot on the visual field that differs from its neighbors in color or in contour. The green surrounded by reds acquires a little extra flag. All it takes to see a green among reds is to spot the flag, a task within the powers of even the simplest demon. An O among X's can be detected in the same way. But the thousands of processors tiled across the field are too stupid to calculate *conjunctions* of features: a patch that is green *and* curved, or red *and* straight. The conjunctions are detected only by a programmable logic machine that looks at one part of the visual field at a time through a narrow, movable window, and passes on its answer to the rest of cognition.

Why is visual computation divided into an unconscious parallel stage and a conscious serial stage? Conjunctions are combinatorial. It would be impossible to sprinkle conjunction detectors at every location in the visual field because there are too many kinds of conjunctions. There are a million visual locations, so the number of processors needed would be a million multiplied by the number of logically possible conjunctions: the number of colors we can discriminate times the number of contours times the number of depths times the number of directions of motion times the number of velocities, and so on, an astronomical number. Parallel, unconscious computation stops after it labels each location with a color, contour, depth, and motion; the combinations then have to be computed, consciously, at one location at a time.

The theory makes a surprising prediction. If the conscious processor is focused at one location, the features at other locations should float around unglued. For example, a person not deliberately attending to a region should not know whether it contains a red X and a green O or a green X and a red O—the color and shape should float in separate planes until the conscious processor binds them together at a particular spot. Treisman found that that is what happens. When people are distracted from some colored letters, they can report the letters and they can report the colors, but they misreport which color went with which letter. These illusory combinations are a striking demonstration of the limits of unconscious visual computation, and they are not uncommon in everyday life. When words are glimpsed absent-mindedly or out of the corner of the eye, the letters sometimes rearrange themselves. One psychologist began to study the phenomenon after he walked past a coffee machine and wondered why it claimed to be dispensing the World's Worst Coffee. The sign, of course, really said "World's Best Coffee." One time I did a double-take when driving past a billboard advertising a brothel (actually the Brothers' Hotel). When flipping through a magazine I once caught sight of a headline about anti-semitic cameras (they were semi-antique).

There are bottlenecks constricting the flow of information from inside the person as well as from outside. When we try to retrieve a memory, the items drip into awareness one at a time, often with agonizing delays if the information is old or uncommon. Ever since Plato invoked the metaphor of soft wax, psychologists have assumed that the neural medium must be inherently resistant to retaining information, fading with time unless the information is pounded in. But the brain *can* record indelible memories, such as the content of shocking news and a few of the details of the time and place at which one hears it. So the neural medium itself is not necessarily to blame.

The psychologist John Anderson has reverse-engineered human memory retrieval, and has shown that the limits of memory are not a by-product of a mushy storage medium. As programmers like to say, "It's not a bug, it's a feature." In an optimally designed information-retrieval system, an item should be recovered only when the relevance of the item outweighs the cost of retrieving it. Anyone who has used a computerized library retrieval system quickly comes to rue the avalanche of titles spilling across the screen. A human expert, despite our allegedly feeble powers of retrieval, vastly outperforms any computer in locating a piece of information from its content. When I need to find articles on a topic in an unfamiliar field, I don't use the library computer; I send email to a pal in the field.

What would it mean for an information-retrieval system to be optimally designed? It should cough up the information most likely to be useful at the time of the request. But how could that be known in advance? The probabilities could be estimated, using general laws about what kinds of information are most likely to be needed. If such laws exist, we should be able to find them in information systems in general, not just human memory; for example, the laws should be visible in the statistics of books requested at a library or the files retrieved in a computer. Information scientists have discovered several of these laws. A piece of information that has been requested many times in the past is more likely to be needed now than a piece that has been requested only rarely. A piece that has been requested recently is more likely to be needed now than a piece that has not been requested for a while. An optimal information-retrieval system should therefore be biased to fetch frequently and recently encountered items. Anderson notes that that is exactly what human memory retrieval does: we remember common and recent events better than rare and long-past events. He found four other classic phenomena in memory research that meet the optimal design criteria independently established for computer information-retrieval systems.

A third notable feature of access-consciousness is the emotional coloring of experience. We not only register events but register them as pleasurable or painful. That makes us take steps to have more of the former and less of the latter, now and in the future. None of this is a mystery. Computationally speaking, representations trigger goal states, which in turn trigger information-gathering, problem-solving, and behavior-selecting demons that calculate how to attain, shun, or modify the charged situation. And evolutionarily speaking, there is seldom any mystery in why we seek the goals we seek—why, for example, people would rather make love with an attractive partner than get a slap on the belly with a wet fish. The things that become objects of desire are the kinds of things that led, on average, to enhanced odds of survival and reproduction in the environment in which we evolved: water, food, safety, sex, status, mastery over the environment, and the well-being of children, friends, and kin.

The fourth feature of consciousness is the funneling of control to an executive process: something we experience as the self, the will, the "I." The self has been under assault lately. The mind is a society of agents, according to the artificial intelligence pioneer Marvin Minsky. It's a large collection of partly finished drafts, says Daniel Dennett, who adds, "It's a mistake to look for the President in the Oval Office of the brain."

The society of mind is a wonderful metaphor, and I will use it with gusto when explaining the emotions. But the theory can be taken too far if it outlaws any system in the brain charged with giving the reins or the floor to one of the agents at a time. The agents of the brain might very well be organized hierarchically into nested subroutines with a set of master decision rules, a computational demon or agent or good-kind-of-homunculus, sitting at the top of the chain of command. It would not be a ghost in the machine, just another set of if-then rules or a neural network that shunts control to the loudest, fastest, or strongest agent one level down.

We even have hints about the brain structures that house the decision-making circuitry. The neurologist Antonio Damasio has noted that damage to the anterior cingulate sulcus, which receives input from many higher perceptual areas and is connected to the higher levels of the motor system, leaves a patient in a seemingly alert but strangely unresponsive state. The report led Francis Crick to proclaim, only partly in jest, that the seat of the will had been discovered. And for many decades neurologists have known that exercising the will—forming and carrying out plans—is a job of the frontal lobes. A sad but typical example came to me from a man who called about his fifteen-year-old son, who had suffered an injury to his frontal lobes in a car accident. The boy would stay in the shower for hours at a time, unable to decide when to get out, and could not leave the house because he kept looping back to his room to check whether he had turned off the lights.

Why would a society of mental agents need an executive at the top? The reason is as clear as the old Yiddish expression "You can't dance at two weddings with only one *tuches*." No matter how many agents we have in our minds, we each have exactly one body. Custody of each major part must be granted to a controller that selects a plan from the hubbub of competing agents. The eyes have to point at one object at a time; they can't fixate on the empty space halfway between two interesting objects or wobble between them in a tug-of-war. The limbs must be choreographed to pull the body or objects along a path that attains the goal of just one of the mind's agents. The alternative, a truly egalitarian society of mind, is shown in the wonderfully silly movie *All of Me*. Lily Tomlin is a hypochondriac heiress who hires a swami to transfer her soul into the body of a woman who doesn't want hers. During the transfer, a chamberpot containing her soul falls out the window and conks a passerby, played by Steve Martin, on the head. Tomlin's dybbuk comes to rest in the right half of his body while he retains control of the left half. He lurches in a zigzag as first his left half strides in one direction and then his right half, pinkie extended, minces in the other.

⚊

So, consciousness in the sense of access is coming to be understood. What about consciousness in the sense of sentience? Sentience and access may be two sides of a single coin. Our subjective experience is also the grist for our reasoning, speech, and action. We do not just experience a toothache; we complain about it and head to the dentist.

Ned Block has tried to clarify the distinction between access and sentience by thinking up scenarios in which access could occur without sentience and vice versa. An example of access without sentience might be found in the strange syndrome called blindsight. When a person has a large blind spot because of damage to his visual cortex, he will adamantly deny that he can see a thing there, but when forced to guess where an object is, he performs well above chance. One interpretation is that the blindsighter has access to the objects but is not sentient of them. Whether or not this is correct, it shows that it is possible to *conceive of* a difference between access and sentience. Sentience without access might occur when you are engrossed in a conversation and suddenly realize that there is a jackhammer outside the window and that you have been hearing it, but not noticing it, for some time. Prior to the epiphany you were sentient of the noise but had no access to it. But Block admits that the examples are a bit strained, and suspects that in reality access and sentience go together.

So we may not need a separate theory of where sentience occurs in the brain, how it fits into mental computation, or why it evolved. It seems to be an extra quality of some kinds of information access. What we do need is a theory of how the subjective qualities of sentience emerge out of mere information access. To complete the story, then, I must present a theory that addresses questions like these:

• If we could ever duplicate the information processing in the human mind as an enormous computer program, would a computer running the program be conscious?

• What if we took that program and trained a large number of people, say, the population of China, to hold in mind the data and act out the steps? Would there be one gigantic consciousness hovering over China, separate from the consciousnesses of the billion individuals? If they were implementing the brain state for agonizing pain, would there be some entity that really was in pain, even if every citizen was cheerful and light-hearted?

• Suppose the visual receiving area at the back of your brain was surgically severed from the rest and remained alive in your skull, receiving input from the eyes. By every behavioral measure you are blind. Is there a mute but fully aware visual consciousness sealed off in the back of your head? What if it was removed and kept alive in a dish?

• Might your experience of red be the same as my experience of green? Sure, you might *label* grass as "green" and tomatoes as "red," just as I do, but perhaps you actually *see* the grass as having the color that I would describe, if I were in your shoes, as red.

• Could there be zombies? That is, could there be an android rigged up to act as intelligently and as emotionally as you and me, but in which there is "no one home" who is actually *feeling* or *seeing* anything? How do I know that *you're* not a zombie?

• If someone could download the state of my brain and duplicate it in another collection of molecules, would it have my consciousness? If someone destroyed the original, but the duplicate continued to live my life and think my thoughts and feel my feelings, would I have been murdered? Was Captain Kirk snuffed out and replaced by a twin every time he stepped into the transporter room?

• What is it like to be a bat? Do beetles enjoy sex? Does a worm scream silently when a fisherman impales it on a hook?

• Surgeons replace one of your neurons with a microchip that duplicates its input-output functions. You feel and behave exactly as before. Then they replace a second one, and a third one, and so on, until more and more of your brain becomes silicon. Since each microchip does exactly what the neuron did, your behavior and memory never change. Do you even notice the difference? Does it feel like dying? Is some *other* conscious entity moving in with you?

Beats the heck out of me! I have some prejudices, but no idea of how to begin to look for a defensible answer. And neither does anyone else. The computational theory of mind offers no insight; neither does any finding in neuroscience, once you clear up the usual confusion of sentience with access and self-knowledge.

How can a book called *How the Mind Works* evade the responsibility of explaining where sentience comes from? I could, I suppose, invoke the doctrine of logical positivism, which holds that if a statement cannot be verified it is literally meaningless. The imponderables in my list ask about the quintessentially unverifiable. Many thinkers, such as Dennett, conclude that worrying about them is simply flaunting one's confusion: sentient experiences (or, as philosophers call them, *qualia*) are a cognitive illusion. Once we have isolated the computational and neurological correlates of access-consciousness, there is nothing left to explain. It's just irrational to insist that sentience remains unexplained after all the manifestations of sentience have been accounted for, just because the computations don't have anything sentient in them. It's like insisting that wetness remains unexplained even after all the manifestations of wetness have been accounted for, because moving molecules aren't wet.

Most people are uncomfortable with the argument, but it is not easy to find anything wrong with it. The philosopher Georges Rey once told me that he has no sentient experiences. He lost them after a bicycle accident when he was fifteen. Since then, he insists, he has been a zombie. I assume he is speaking tongue-in-cheek, but of course I have no way of knowing, and that is his point.

The qualia-debunkers do have a point. At least for now, we have no scientific purchase on the special extra ingredient that gives rise to sentience. As far as scientific explanation goes, it might as well not exist. It's not just that claims about sentience are perversely untestable; it's that testing them would make no difference to anything anyway. Our incomprehension of sentience does not impede our understanding of how the mind works in the least. Generally the parts of a scientific problem fit together like a crossword puzzle. To reconstruct human evolution, we need physical anthropology to find the bones, archeology to understand the tools, molecular biology to date the split from chimpanzees, and paleobotany to reconstruct the environment from fossil pollen. When any part of the puzzle is blank, such as a lack of chimpanzee fossils or an uncertainty about whether the climate was wet or dry, the gap is sorely felt and everyone waits impatiently for it to be filled. But in the study of the mind, sentience floats in its own plane, high above the causal chains of psychology and neuroscience. If we ever could trace all the neurocomputational steps from perception through reasoning and emotion to behavior, the only thing left missing by the lack of a theory of sentience would be an understanding of sentience itself.

But saying that we have no scientific explanation of sentience is not the same as saying that sentience does not exist at all. I am as certain that I am sentient as I am certain of *anything*, and I bet you feel the same. Though I concede that my curiosity about sentience may never be satisfied, I refuse to believe that I am just confused when I think I am sentient at all! (Dennett's analogy of unexplained wetness is not decisive: wetness is itself a subjective feeling, so the observer's dissatisfaction is just the problem of sentience all over again.) And we cannot banish sentience from our discourse or reduce it to information access, because moral reasoning depends on it. The concept of sentience underlies our certainty that torture is wrong and that disabling a robot is the destruction of property but disabling a person is murder. It is the reason that the death of a loved one does not impart to us just self-pity at our loss but the uncomprehending pain of knowing that the person's thoughts and pleasures have vanished forever.

If you bear with me to the end of the book, you will learn my own hunch about the mystery of sentience. But the mystery remains a mystery, a topic not for science but for ethics, for late-night dorm-room bull sessions, and, of course, for one other realm:

On a microscopic piece of sand that floats through space is a fragment of a man's life. Left to rust is the place he lived in and the machines he used. Without use, they will disintegrate from the wind and the sand and the years that act upon them; all of Mr. Corry's machines—including the one made in his image, kept alive by love, but now obsolete . . . in the Twilight Zone.

3

REVENGE OF THE NERDS

Somewhere beyond the edge of our solar system, hurtling into interstellar space, is a phonograph and a golden record with hieroglyphic instructions on the sleeve. They are attached to the *Voyager 2* space probe, launched in 1977 to transmit photographs and data back to us from the outer planets in our solar system. Now that it has flown by Neptune and its thrilling scientific mission is over, it serves as an interplanetary calling card from us to any spacefaring extraterrestrial that might snag it.

The astronomer Carl Sagan was the record producer, and he chose sights and sounds that captured our species and its accomplishments. He included greetings in fifty-five human languages and one "whale language," a twelve-minute sound essay made up of a baby's cry, a kiss, and an EEG record of the meditations of a woman in love, and ninety minutes of music sampled from the world's idioms: Mexican mariachi, Peruvian panpipes, Indian raga, a Navajo night chant, a Pygmy girl's initiation song, a Japanese *shakuhachi* piece, Bach, Beethoven, Mozart, Stravinsky, Louis Armstrong, and Chuck Berry singing "Johnny B. Goode."

The disk also bore a message of peace from our species to the cosmos. In an unintended bit of black comedy, the message was recited by the secretary-general of the United Nations at the time, Kurt Waldheim. Years later historians discovered that Waldheim had spent World War II as an intelligence officer in a German army unit that carried out brutal reprisals against Balkan partisans and deported the Jewish population of Salonika to Nazi death camps. It is too late to call *Voyager* back, and this mordant joke on us will circle the center of the Milky Way galaxy forever.

GET SMART

The *Voyager* phonograph record, in any case, was a fine idea, if only because of the questions it raised. Are we alone? If not, do alien life forms have the intelligence and the desire to develop space travel? If so, would they interpret the sounds and images as we intended, or would they hear the voice as the whine of a modem and see the line drawings of people on the cover as showing a race of wire frames? If they understood it, how would they respond? By ignoring us? By coming over to enslave us or eat us? Or by starting an interplanetary dialogue? In a *Saturday Night Live* skit, the long-awaited reply from outer space was "Send more Chuck Berry."

These are not just questions for late-night dorm-room bull sessions. In the early 1990s NASA allocated a hundred million dollars to a ten-year Search for Extraterrestrial Intelligence (SETI). Scientists were to listen with radio antennas for signals that could have come only from intelligent extraterrestrials. Predictably, some congressmen objected. One said it was a waste of federal money "to look for little green men with mis-shapen heads." To minimize the "giggle factor," NASA renamed the project the High-Resolution Microwave Survey, but it was too late to save the project from the congressional ax. Currently it is funded by donations from private sources, including Steven Spielberg.

The opposition to SETI came not just from the know-nothings but from some of the world's most distinguished biologists. Why did they join the discussion? SETI depends on assumptions from evolutionary theory, not just astronomy—in particular, about the evolution of intelligence. Is intelligence inevitable, or was it a fluke? At a famous conference in 1961, the astronomer and SETI enthusiast Frank Drake noted that the number of extraterrestrial civilizations that might contact us can be estimated with the following formula:

(1) (The number of stars in the galaxy) ×

(2) (The fraction of stars with planets) ×

(3) (The number of planets per solar system with a life-supporting environment) ×

(4) (The fraction of these planets on which life actually appears) ×

(5) (The fraction of life-bearing planets on which intelligence emerges) ×

(6) (The fraction of intelligent societies willing and able to communicate with other worlds) ×

(7) (The longevity of each technology in the communicative   state).

The astronomers, physicists, and engineers at the conference felt unable to estimate factor (6) without a sociologist or a historian. But they felt confident in estimating factor (5), the proportion of life-bearing planets on which intelligence emerges. They decided it was one hundred percent.

Finding intelligent life elsewhere in the cosmos would be the most exciting discovery in human history. So why are the biologists being such grinches? It is because they sense that the SETI enthusiasts are reasoning from a pre-scientific folk belief. Centuries-old religious dogma, the Victorian ideal of progress, and modern secular humanism all lead people to misunderstand evolution as an internal yearning or unfolding toward greater complexity, climaxing in the appearance of man. The pressure builds up, and intelligence emerges like popcorn in a pan.

The religious doctrine was called the Great Chain of Being—amoeba to monkey to man—and even today many scientists thoughtlessly use words like "higher" and "lower" life forms and the evolutionary "scale" and "ladder." The parade of primates, from gangly-armed gibbon through stoop-shouldered caveman to upright modern man, has become an icon of pop culture, and we all understand what someone means when she says she turned down a date because the guy is not very evolved. In science fiction like H. G. Wells' *The Time Machine*, episodes of *Star Trek*, and stories from *Boy's Life*, the momentum is extrapolated to our descendants, shown as bald, varicose-veined, bulbous-brained, spindly-bodied homunculi. In *The Planet of the Apes* and other stories, after we have blown ourselves to smithereens or choked in our pollutants, apes or dolphins rise to the occasion and take on our mantle.

Drake expressed these assumptions in a letter to *Science* defending SETI against the eminent biologist Ernst Mayr. Mayr had noted that only one of the fifty million species on earth had developed civilizations, so the probability that life on a given planet would include an intelligent species might very well be small. Drake replied:

> The first species to develop intelligent civilizations will discover that it is the only such species. Should it be surprised? Someone must be first, and being first says nothing about how many other species had or have the potential to evolve into intelligent civilizations, or may do so in the future. . . . Similarly, among many civilizations, one will be the first, and temporarily the only one, to develop electronic technology. How else

could it be? The evidence does suggest that planetary systems need to exist in sufficiently benign circumstances for a few billion years for a technology-using species to evolve.

To see why this thinking runs so afoul of the modern theory of evolution, consider an analogy. The human brain is an exquisitely complex organ that evolved only once. The elephant's trunk, which can stack logs, uproot trees, pick up a dime, remove thorns, powder the elephant with dust, siphon water, serve as a snorkel, and scribble with a pencil, is another complex organ that evolved only once. The brain and the trunk are products of the same evolutionary force, natural selection. Imagine an astronomer on the Planet of the Elephants defending SETT, the Search for Extraterrestrial Trunks:

> The first species to develop a trunk will discover that it is the only such species. Should it be surprised? Someone must be first, and being first says nothing about how many other species had or have the potential to evolve trunks, or may do so in the future. . . . Similarly, among many trunk-bearing species, one will be the first, and temporarily the only one, to powder itself with dust. The evidence does suggest that planetary systems need to exist in sufficiently benign circumstances for a few billion years for a trunk-using species to evolve. . . .

This reasoning strikes us as cockeyed because the elephant is assuming that evolution did not just *produce* the trunk in a species on this planet but was *striving* to produce it in some lucky species, each waiting and hoping. The elephant is merely "the first," and "temporarily" the only one; other species have "the potential," though a few billion years will have to pass for the potential to be realized. Of course, we are not chauvinistic about trunks, so we can see that trunks evolved, but not because a rising tide made it inevitable. Thanks to fortuitous preconditions in the elephants' ancestors (large size and certain kinds of nostrils and lips), certain selective forces (the problems posed by lifting and lowering a huge head), and luck, the trunk evolved as a workable solution for those organisms at that time. Other animals did not and will not evolve trunks because in their bodies and circumstances it is of no great help. Could it happen again, here or elsewhere? It could, but the proportion of planets on which the necessary hand has been dealt in a given period of time is presumably small. Certainly it is less than one hundred percent.

We *are* chauvinistic about our brains, thinking them to be the goal of evolution. And that makes no sense, for reasons articulated over the years by Stephen Jay Gould. First, natural selection does nothing even close to striving for intelligence. The process is driven by differences in the survival and reproduction rates of replicating organisms in a particular environment. Over time the organisms acquire designs that adapt them for survival and reproduction in that environment, period; nothing pulls them in any direction other than success there and then. When an organism moves to a new environment, its lineage adapts accordingly, but the organisms who stayed behind in the original environment can prosper unchanged. Life is a densely branching bush, not a scale or a ladder, and living organisms are at the tips of the branches, not on lower rungs. Every organism alive today has had the same amount of time to evolve since the origin of life—the amoeba, the platypus, the rhesus macaque, and, yes, Larry on the answering machine asking for another date.

But, a SETI fan might ask, isn't it true that animals become more complex over time? And wouldn't intelligence be the culmination? In many lineages, of course, animals have become more complex. Life began simple, so the complexity of the *most* complex creature alive on earth at any time has to increase over the eons. But in many lineages they have not. The organisms reach an optimum and stay put, often for hundreds of millions of years. And those that do become more complex don't always become smarter. They become bigger, or faster, or more poisonous, or more fecund, or more sensitive to smells and sounds, or able to fly higher and farther, or better at building nests or dams—whatever works for them. Evolution is about ends, not means; becoming smart is just one option.

Still, isn't it inevitable that *many* organisms would take the route to intelligence? Often different lineages converge on a solution, like the forty different groups of animals that evolved complex designs for eyes. Presumably you can't be too rich, too thin, or too smart. Why wouldn't humanlike intelligence be a solution that many organisms, on this planet and elsewhere, might converge on?

Evolution could indeed have converged on humanlike intelligence several times, and perhaps that point could be developed to justify SETI. But in calculating the odds, it is not enough to think about how great it is to be smart. In evolutionary theory, that kind of reasoning merits the accusation that conservatives are always hurling at liberals: they specify a benefit but neglect to factor in the costs. Organisms don't evolve toward every imaginable advantage. If they did, every creature would be faster than a speeding bullet, more powerful than a locomotive, and able to leap tall buildings in a single bound. An organism that devotes some of its matter and energy to one organ must take it away from another. It must have thinner bones or less muscle or fewer eggs. Organs evolve only when their benefits outweigh their costs.

Do you have a Personal Digital Assistant, like the Apple Newton? These are the hand-held devices that recognize handwriting, store phone numbers, edit text, send faxes, keep schedules, and many other feats. They are marvels of engineering and can organize a busy life. But I don't have one, though I am a gadget-lover. Whenever I am tempted to buy a PDA, four things dissuade me. First, they are bulky. Second, they need batteries. Third, they take time to learn to use. Fourth, their sophistication makes simple tasks, like looking up a phone number, slow and cumbersome. I get by with a notebook and a fountain pen.

The same disadvantages would face any creature pondering whether to evolve a humanlike brain. First, the brain is bulky. The female pelvis barely accommodates a baby's outsize head. That design compromise kills many women during childbirth and requires a pivoting gait that makes women biomechanically less efficient walkers than men. Also, a heavy head bobbing around on a neck makes us more vulnerable to fatal injuries in accidents such as falls. Second, the brain needs energy. Neural tissue is metabolically greedy; our brains take up only two percent of our body weight but consume twenty percent of our energy and nutrients. Third, brains take time to learn to use. We spend much of our lives either being children or caring for children. Fourth, simple tasks can be slow. My first graduate advisor was a mathematical psychologist who wanted to model the transmission of information in the brain by measuring reaction times to loud tones. Theoretically, the neuron-to-neuron transmission times should have added up to a few milliseconds. But there were seventy-five milliseconds unaccounted for between stimulus and response—"There's all this cogitation going on, and we just want him to push his finger down," my advisor grumbled. Lower-tech animals can be much quicker; some insects can bite in less than a millisecond. Perhaps this answers the rhetorical question in the sporting equipment ad: The average man's IQ is 107. The average brown trout's IQ is 4. So why can't a man catch a brown trout?

Intelligence isn't for everyone, any more than a trunk is, and this should give SETI enthusiasts pause. But I am not arguing against the search for extraterrestrial intelligence; my topic is terrestrial intelligence. The fallacy that intelligence is some exalted ambition of evolution is part of the same fallacy that treats it as a divine essence or wonder tissue or all-encompassing mathematical principle. The mind is an organ, a biological gadget. We have our minds because their design attains outcomes whose benefits outweighed the costs in the lives of Plio-Pleistocene African primates. To understand ourselves, we need to know the how, why, where, and when of this episode in history. They are the subject of this chapter.

LIFE'S DESIGNER

One evolutionary biologist *has* made a prediction about extraterrestrial life—not to help us look for life on other planets, but to help us understand life on this planet. Richard Dawkins has ventured that life, anywhere it is found in the universe, will be a product of Darwinian natural selection. That may seem like the most overreaching prognosis ever made from an armchair, but in fact it is a straightforward consequence of the argument for the theory of natural selection. Natural selection is the only explanation we have of how complex life *can* evolve, putting aside the question of how it *did* evolve. If Dawkins is right, as I think he is, natural selection is indispensable to understanding the human mind. If it is the only explanation of the evolution of little green men, it certainly is the only explanation of the evolution of big brown and beige ones.

The theory of natural selection—like the other foundation of this book, the computational theory of mind—has an odd status in modern intellectual life. Within its home discipline, it is indispensable, explaining thousands of discoveries in a coherent framework and constantly inspiring new ones. But outside its home, it is misunderstood and reviled. As in Chapter 2, I want to spell out the case for this foundational idea: how it explains a key mystery that its alternatives cannot explain, how it has been verified in the lab and the field, and why some famous arguments against it are wrong.

Natural selection has a special place in science because it alone explains what makes life special. Life fascinates us because of its *adaptive complexity* or *complex design*. Living things are not just pretty bits of bric-a-brac, but do amazing things. They fly, or swim, or see, or digest food, or catch prey, or manufacture honey or silk or wood or poison. These are rare accomplishments, beyond the means of puddles, rocks, clouds, and other nonliving things. We would call a heap of extraterrestrial matter "life" only if it achieved comparable feats.

Rare accomplishments come from special structures. Animals can see and rocks can't because animals have eyes, and eyes have precise arrangements of unusual materials capable of forming an image: a cornea that focuses light, a lens that adjusts the focus to the object's depth, an iris that opens and closes to let in the right amount of light, a sphere of transparent jelly that maintains the eye's shape, a retina at the focal plane of the lens, muscles that aim the eyes up-and-down, side-to-side, and in-and-out, rods and cones that transduce light into neural signals, and more, all exquisitely shaped and arranged. The odds are mind-bogglingly stacked against these structures' being assembled out of raw materials by tornados, landslides, waterfalls, or the lightning bolt vaporizing swamp goo in the philosopher's thought experiment.

The eye has so many parts, arranged so precisely, that it appears to have been designed in advance with the *goal* of putting together something that sees. The same is true for our other organs. Our joints are lubricated to pivot smoothly, our teeth meet to sheer and grind, our hearts pump blood—every organ seems to have been designed with a function in mind. One of the reasons God was invented was to *be* the mind that formed and executed life's plans. The laws of the world work forwards, not backwards: rain causes the ground to be wet; the ground's benefiting from being wet cannot cause the rain. What else but the plans of God could effect the teleology (goal-directedness) of life on earth?

Darwin showed what else. He identified a forward-causation physical process that mimics the paradoxical appearance of backward causation or teleology. The trick is *replication*. A replicator is something that can make a copy of itself, with most of its traits duplicated in the copy, including the ability to replicate in turn. Consider two states of affairs, A and B. B can't cause A if A comes first. (Seeing well can't cause an eye to have a clear lens.)

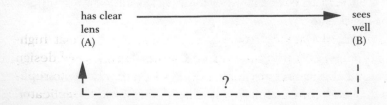

But let's say that A causes B, and B in turn causes the protagonist of A to make a copy of itself—let's call it AA. AA looks just like A, so it appears as if B has caused A. But it hasn't; it has only caused AA, the *copy* of A. Suppose there are three animals, two with a cloudy lens, one with a clear lens. Having a clear lens (A) causes an eye to see well (B); seeing well causes the animal to reproduce by helping it avoid predators and find mates. The offspring (AA) have clear lenses and can see well, too. It looks as if the offspring have eyes *so that* they can see well (bad, teleological, backward causation), but that's an illusion. The offspring have eyes because *their parents'* eyes *did* see well (good, ordinary, forward causation). Their eyes *look like* their parents' eyes, so it's easy to mistake what happened for backward causation.

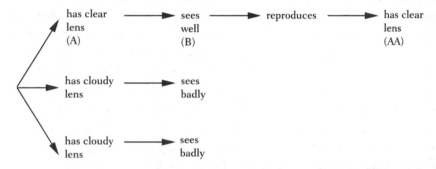

There's more to an eye than a clear lens, but the special power of a replicator is that its copies can replicate, too. Consider what happens when the clear-lensed daughter of our hypothetical animal reproduces. Some of her offspring will have rounder eyeballs than others, and the round-eyed versions see better because the images are focused from center to edge. Better vision leads to better reproduction, and the next generation has both clear lenses and round eyeballs. They, too, are replicators, and the sharper-visioned of their offspring are more likely to leave a new generation with sharp vision, and so on. In every generation, the traits that lead to good vision are disproportionately passed down to the next generation. That is why a late generation of replicators will have traits that seem to have been designed by an intelligent engineer (see figure on page 158).

I have introduced Darwin's theory in an unorthodox way that highlights its extraordinary contribution: explaining the appearance of design without a designer, using ordinary forward causation as it applies to replicators. The full story runs as follows. In the beginning was a replicator.

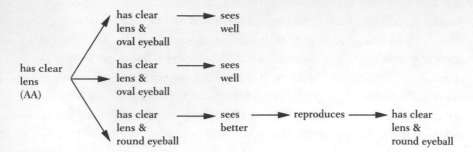

This molecule or crystal was a product not of natural selection but of the laws of physics and chemistry. (If it were a product of selection, we would have an infinite regress.) Replicators are wont to multiply, and a single one multiplying unchecked would fill the universe with its great-great-great-. . .-great-grandcopies. But replicators use up materials to make their copies and energy to power the replication. The world is finite, so the replicators will compete for its resources. Because no copying process is one hundred percent perfect, errors will crop up, and not all of the daughters will be exact duplicates. Most of the copying errors will be changes for the worse, causing a less efficient uptake of energy and materials or a slower rate or lower probability of replication. But by dumb luck a few errors will be changes for the better, and the replicators bearing them will proliferate over the generations. Their descendants will accumulate any subsequent errors that are changes for the better, including ones that assemble protective covers and supports, manipulators, catalysts for useful chemical reactions, and other features of what we call bodies. The resulting replicator with its apparently well-engineered body is what we call an organism.

Natural selection is not the only process that changes organisms over time. But it is the only process that seemingly *designs* organisms over time. Dawkins stuck out his neck about extraterrestrial evolution because he reviewed every alternative to selection that has been proposed in the history of biology and showed that they are impotent to explain the signature of life, complex design.

The folk theory that organisms respond to an urge to unfold into more complex and adaptive forms obviously won't do. The urge—and, more important, the power to achieve its ambitions—is a bit of magic that is left unexplained.

The two principles that have come to be associated with Darwin's predecessor Jean Baptiste Lamarck—use and disuse, and the inheritance of acquired characteristics—are also not up to the job. The problem goes beyond the many demonstrations that Lamarck was wrong in fact. (For example, if acquired traits really could be inherited, several hundred generations of circumcision should have caused Jewish boys today to be born without foreskins.) The deeper problem is that the theory would not be able to explain adaptive complexity even if it *had* turned out to be correct. First, *using* an organ does not, by itself, make the organ function better. The photons passing through a lens do not somehow wash it clear, and using a machine does not improve it but wears it out. Now, many parts of organisms do adjust adaptively to use: exercised muscle bulks up, rubbed skin thickens, sunlit skin darkens, rewarded acts increase and punished ones decrease. But these responses are themselves part of the evolved design of the organism, and we need to explain how *they* arose: no law of physics or chemistry makes rubbed things thicken or illuminated surfaces darken. The inheritance of acquired characteristics is even worse, for most acquired characteristics are cuts, scrapes, scars, decay, weathering, and other assaults by the pitiless world, not improvements. And even if a blow did lead to an improvement, it is mysterious how the size and shape of the helpful wound could be read off the affected flesh and encoded back into DNA instructions in the sperm or egg.

Yet another failed theory is the one that invokes the macromutation: a mammoth copying error that begets a new kind of adapted organism in one fell swoop. The problem here is that the laws of probability astronomically militate against a large random copying error creating a complex functioning organ like the eye out of homogeneous flesh. Small random errors, in contrast, can make an organ *a bit more* like an eye, as in our example where an imaginable mutation might make a lens a tiny bit clearer or an eyeball a tiny bit rounder. Indeed, way before our scenario begins, a long sequence of small mutations must have accumulated to give the organism an eye at all. By looking at organisms with simpler eyes, Darwin reconstructed how that could have happened. A few mutations made a patch of skin cells light-sensitive, a few more made the underlying tissue opaque, others deepened it into a cup and then a spherical hollow. Subsequent mutations added a thin translucent cover, which subsequently was thickened into a lens, and so on. Each step offered a small improvement in vision. Each mutation was improbable, but not astronomically so. The entire sequence was not astronomically impossible because the mutations were not dealt all at once like a big gin rummy hand; each beneficial mutation was added to a set of prior ones that had been selected over the eons.

A fourth alternative is random genetic drift. Beneficial traits are beneficial only on average. Actual creatures suffer the slings and arrows of outrageous fortune. When the number of individuals in a generation is small enough, an advantageous trait can vanish if its bearers are unlucky, and a disadvantageous or neutral one can take over if its bearers are lucky. Genetic drift can, in principle, explain why a population has a simple trait, like being dark or light, or an inconsequential trait, like the sequence of DNA bases in a part of the chromosome that doesn't do anything. But because of its very randomness, random drift cannot explain the appearance of an improbable, useful trait like an ability to see or fly. The required organs need hundreds or thousands of parts to work, and the odds are astronomically stacked against the required genes accumulating by sheer chance.

Dawkins' argument about extraterrestrial life is a timeless claim about the logic of evolutionary theories, about the power of an *explanans* to cause the *explanandum*. And indeed his argument works against two subsequent challenges. One is a variant of Lamarckism called *directed* or *adaptive* mutation. Wouldn't it be nice if an organism could react to an environmental challenge with a slew of new mutations, and not wasteful, random ones, but mutations for traits that would allow it to cope? Of course it would be nice, and that's the problem—chemistry has no sense of niceness. The DNA inside the testes and ovaries cannot peer outside and considerately mutate to make fur when it's cold and fins when it's wet and claws when there are trees around, or to put a lens in front of the retina as opposed to between the toes or inside the pancreas. That is why a cornerstone of evolutionary theory—indeed, a cornerstone of the scientific worldview—is that mutations are indifferent overall to the benefits they confer on the organism. They cannot be adaptive in general, though of course a tiny few can be adaptive by chance. The periodic announcements of discoveries of "adaptive mutations" inevitably turn out to be laboratory curiosities or artifacts. No mechanism short of a guardian angel can guide mutations to respond to organisms' needs in general, there being billions of kinds of organisms, each with thousands of needs.

The other challenge comes from the fans of a new field called the theory of complexity. The theory looks for mathematical principles of order underlying many complex systems: galaxies, crystals, weather systems, cells, organisms, brains, ecosystems, societies, and so on. Dozens of new books have applied these ideas to topics such as AIDS, urban decay, the Bosnian war, and, of course, the stock market. Stuart Kauffman, one of the movement's leaders, suggested that feats like self-organization, order, stability, and coherence may be an "innate property of some complex systems." Evolution, he suggests, may be a "marriage of selection and self-organization."

Complexity theory raises interesting issues. Natural selection presupposes that a replicator arose somehow, and complexity theory might help explain the "somehow." Complexity theory might also pitch in to explain other assumptions. Each body has to hang together long enough to function rather than fly apart or melt into a puddle. And for evolution to happen at all, mutations have to change a body enough to make a difference in its functioning but not so much as to bring it to a chaotic crash. If there are abstract principles that govern whether a web of interacting parts (molecules, genes, cells) has such properties, natural selection would have to work within those principles, just as it works within other constraints of physics and mathematics like the Pythagorean theorem and the law of gravitation.

But many readers have gone much further and conclude that natural selection is now trivial or obsolete, or at best of unknown importance. (Incidentally, the pioneers of complexity theory themselves, such as Kauffman and Murray Gell-Mann, are appalled by that extrapolation.) This letter to the *New York Times Book Review* is a typical example:

> Thanks to recent advances in nonlinear dynamics, nonequilibrium thermodynamics and other disciplines at the boundary between biology and physics, there is every reason to believe that the origin and evolution of life will eventually be placed on a firm scientific footing. As we approach the 21st century, those other two great 19th century prophets—Marx and Freud—have finally been deposed from their pedestals. It is high time we freed the evolutionary debate from the anachronistic and unscientific thrall of Darwin worship as well.

The letter-writer must have reasoned as follows: complexity has always been treated as a fingerprint of natural selection, but now it can be explained by complexity theory; therefore natural selection is obsolete. But the reasoning is based on a pun. The "complexity" that so impresses biologists is not just any old order or stability. Organisms are not just cohesive blobs or pretty spirals or orderly grids. They are machines, and their "complexity" is *functional, adaptive design*: complexity in the service of accomplishing some interesting outcome. The digestive tract is not just patterned; it is patterned as a factory line for extracting nutrients from ingested tissues. No set of equations applicable to everything from galaxies to Bosnia can explain why teeth are found in the mouth rather than in the ear. And since organisms are collections of digestive tracts, eyes, and other systems organized to attain goals, general laws of complex systems will not suffice. Matter simply does not have an innate tendency to organize itself into broccoli, wombats, and ladybugs. Natural selection remains the only theory that explains how *adaptive* complexity, not just any old complexity, can arise, because it is the only nonmiraculous, forward-direction theory in which *how well something works* plays a causal role in *how it came to be*.

❧

Because there are no alternatives, we would almost *have* to accept natural selection as the explanation of life on this planet even if there were no evidence for it. Thankfully, the evidence is overwhelming. I don't just mean evidence that life evolved (which is way beyond reasonable doubt, creationists notwithstanding), but that it evolved by natural selection. Darwin himself pointed to the power of selective breeding, a direct analogue of natural selection, in shaping organisms. For example, the differences among dogs—Chihuahuas, greyhounds, Scotties, Saint Bernards, shar-peis—come from selective breeding of wolves for only a few thousand years. In breeding stations, laboratories, and seed company greenhouses, artificial selection has produced catalogues of wonderful new organisms befitting Dr. Seuss.

Natural selection is also readily observable in the wild. In a classic example, the white peppered moth gave way in nineteenth-century Manchester to a dark mutant form after industrial soot covered the lichen on which the moth rested, making the white form conspicuous to birds. When air pollution laws lightened the lichen in the 1950s, the then-rare white form reasserted itself. There are many other examples, perhaps the most pleasing coming from the work of Peter and Rosemary Grant. Darwin was inspired to the theory of natural selection in part by the thirteen species of finches on the Galápagos islands. They clearly were related to a species on the South American mainland, but differed from them and from one another. In particular, their beaks resembled different kinds of pliers: heavy-duty lineman's pliers, high-leverage diagonal pliers, straight needle-nose pliers, curved needle-nose pliers, and so on. Darwin eventually reasoned that one kind of bird was blown to the islands and then differentiated into the thirteen species because of the demands of different ways of life on different parts of the islands, such as stripping bark from trees to get at insects, probing cactus flowers, or cracking tough seeds. But he despaired of ever seeing natural selection happen in real time: "We see nothing of these slow changes in progress, until the hand of time has marked the lapse of ages." The Grants painstakingly measured the size and toughness of the seeds in different parts of the Galápagos at different times of the year, the length of the finches' beaks, the time they took to crack the seeds, the numbers and ages of the finches in different parts of the islands, and so on—every variable relevant to natural selection. Their measurements showed the beaks evolving to track changes in the availability of different kinds of seeds, a frame-by-frame analysis of the movie that Darwin could only imagine. Selection in action is even more dramatic among faster-breeding organisms, as the world is discovering to its peril in the case of pesticide-resistant insects, drug-resistant bacteria, and the AIDS virus in a single patient.

And two of the prerequisites of natural selection—enough variation and enough time—are there for the having. Populations of naturally living organisms maintain an enormous reservoir of genetic variation that can serve as the raw material for natural selection. And life has had more than three billion years to evolve on earth, complex life a billion years, according to a recent estimate. In *The Ascent of Man*, Jacob Bronowski wrote:

> I remember as a young father tiptoeing to the cradle of my first daughter when she was four or five days old, and thinking, "These marvelous fingers, every joint so perfect, down to the fingernails. I could not have designed that detail in a million years." But of course it is exactly a million years that it took me, a million years that it took mankind . . . to reach its present stage of evolution.

Finally, two kinds of formal modeling have shown that natural selection can work. Mathematical proofs from population genetics show how genes combining according to Gregor Mendel's laws can change in frequency under the pressure of selection. These changes can occur impressively fast. If a mutant produces just 1 percent more offspring than its rivals, it can increase its representation in a population from 0.1 percent to 99.9 percent in just over four thousand generations. A hypothetical mouse subjected to a selection pressure for increased size that is so weak it cannot be measured could nonetheless evolve to the size of an elephant in only twelve thousand generations.

More recently, computer simulations from the new field of Artificial Life have shown the power of natural selection to evolve organisms with complex adaptations. And what better demonstration than everyone's favorite example of a complex adaptation, the eye? The computer scientists Dan Nilsson and Susanne Pelger simulated a three-layer slab of virtual skin resembling a light-sensitive spot on a primitive organism. It was a simple sandwich made up of a layer of pigmented cells on the bottom, a layer of light-sensitive cells above it, and a layer of translucent cells forming a protective cover. The translucent cells could undergo random mutations of their refractive index: their ability to bend light, which in real life often corresponds to density. All the cells could undergo small mutations affecting their size and thickness. In the simulation, the cells in the slab were allowed to mutate randomly, and after each round of mutation the program calculated the spatial resolution of an image projected onto the slab by a nearby object. If a bout of mutations improved the resolution, the mutations were retained as the starting point for the next bout, as if the slab belonged to a lineage of organisms whose survival depended on reacting to looming predators. As in real evolution, there was no master plan or project scheduling. The organism could not put up with a less effective detector in the short run even if its patience would have been rewarded by the best conceivable detector in the long run. Every change it retained had to be an improvement.

Satisfyingly, the model evolved into a complex eye right on the computer screen. The slab indented and then deepened into a cup; the transparent layer thickened to fill the cup and bulged out to form a cornea. Inside the clear filling, a spherical lens with a higher refractive index emerged in just the right place, resembling in many subtle details the excellent optical design of a fish's eye. To estimate how long it would take in real time, rather than in computer time, for an eye to unfold, Nilsson and Pelger built in pessimistic assumptions about heritability, variation in the population, and the size of the selective advantage, and even forced the mutations to take place in only one part of the "eye" each generation. Nonetheless, the entire sequence in which flat skin became a complex eye took only four hundred thousand generations, a geological instant.

❧

I have reviewed the modern case for the theory of natural selection because so many people are hostile to it. I don't mean fundamentalists from the Bible Belt, but professors at America's most distinguished universities from coast to coast. Time and again I have heard the objections: the theory is circular, what good is half an eye, how can structure arise from random mutation, there hasn't been enough time, Gould has disproved it, complexity just emerges, physics will make it obsolete someday.

People desperately want Darwinism to be wrong. Dennett's diagnosis in *Darwin's Dangerous Idea* is that natural selection implies there is no plan to the universe, including human nature. No doubt that is a reason, though another is that people who study the mind would rather not have to think about how it evolved because it would make a hash of cherished theories. Various scholars have claimed that the mind is innately equipped with fifty thousand concepts (including "carburetor" and "trombone"), that capacity limitations prevent the human brain from solving problems that are routinely solved by bees, that language is designed for beauty rather than for use, that tribal people kill their babies to protect the ecosystem from human overpopulation, that children harbor an unconscious wish to copulate with their parents, and that people could just as easily be conditioned to enjoy the thought of their spouse being unfaithful as to be upset by the thought. When advised that these claims are evolutionarily improbable, they attack the theory of evolution rather than rethinking the claim. The efforts that academics have made to impugn Darwinism are truly remarkable.

One claim is that reverse-engineering, the attempt to discover the functions of organs (which I am arguing should be done to the human mind), is a symptom of a disease called "adaptationism." Apparently if you believe that any aspect of an organism has a function, you absolutely must believe that *every* aspect has a function, that monkeys are brown to hide amongst the coconuts. The geneticist Richard Lewontin, for example, has defined adaptationism as "that approach to evolutionary studies which assumes without further proof that all aspects of the morphology, physiology and behavior of organisms are adaptive optimal solutions to problems." Needless to say, there is no such madman. A sane person can believe that a complex organ is an adaptation, that is, a product of natural selection, while also believing that features of an organism that are *not* complex organs are a product of drift or a by-product of some other adaptation. Everyone acknowledges that the redness of blood was not selected for itself but is a by-product of selection for a molecule that carries oxygen, which just happens to be red. That does not imply that the ability of the eye to see could easily be a by-product of selection for something else.

There also are no benighted fools who fail to realize that animals carry baggage from their evolutionary ancestors. Readers young enough to have had sex education or old enough to be reading articles about the prostate may have noticed that the seminal ducts in men do not lead directly from the testicles to the penis but snake up into the body and pass over the ureter before coming back down. That is because the testes of our reptilian ancestors were inside their bodies. The bodies of mammals are too hot for the production of sperm, so the testes gradually descended into a scrotum. Like a gardener who snags a hose around a tree, natural selection did not have the foresight to plan the shortest route. Again, that does not mean that the entire eye could very well be useless phylogenetic baggage.

Similarly, because adaptationists believe that the laws of physics are not *enough* to explain the design of animals, they are also imagined to be prohibited from *ever* appealing to the laws of physics to explain *anything*. A Darwin critic once defiantly asked me, "Why has no animal evolved the ability to disappear and instantly reappear elsewhere, or to turn into King Kong at will (great for frightening predators)?" I think it is fair to say that "not being able to turn into King Kong at will" and "being able to see" call for different kinds of explanations.

Another accusation is that natural selection is a sterile exercise in after-the-fact storytelling. But if that were true, the history of biology would be a quagmire of effete speculation, with progress having to wait for today's enlightened anti-adaptationists. Quite the opposite has happened. Mayr, the author of a definitive history of biology, wrote,

> The adaptationist question, "What is the function of a given structure or organ?" has been for centuries the basis of every advance in physiology. If it had not been for the adaptationist program, we probably would still not yet know the functions of thymus, spleen, pituitary, and pineal. Harvey's question "Why are there valves in the veins?" was a major stepping stone in his discovery of the circulation of blood.

From the shape of an organism's body to the shape of its protein molecules, everything we have learned in biology has come from an understanding, implicit or explicit, that the organized complexity of an organism is in the service of its survival and reproduction. This includes what we have learned about the nonadaptive by-products, because they can be found only in the course of a search for the adaptations. It is the bald claim that a feature is a lucky product of drift or of some poorly understood dynamic that is untestable and post hoc.

Often I have heard it said that animals are not well engineered after all. Natural selection is hobbled by shortsightedness, the dead hand of the past, and crippling constraints on what kinds of structures are biologically and physically possible. Unlike a human engineer, selection is incapable of good design. Animals are clunking jalopies saddled with ancestral junk and occasionally blunder into barely serviceable solutions.

People are so eager to believe this claim that they seldom think it through or check the facts. Where do we find this miraculous human engineer who is *not* constrained by availability of parts, manufacturing practicality, and the laws of physics? Of course, natural selection does not have the foresight of engineers, but that cuts both ways: it does not have their mental blocks, impoverished imagination, or conformity to bourgeois sensibilities and ruling-class interests, either. Guided only by what works, selection can home in on brilliant, creative solutions. For millennia, biologists have discovered to their astonishment and delight the ingenious contrivances of the living world: the biomechanical perfection of cheetahs, the infrared pinhole cameras of snakes, the sonar of bats, the superglue of barnacles, the steel-strong silk of spiders, the dozens of grips of the human hand, the DNA repair machinery in all complex organisms. After all, entropy and more malevolent forces like predators and parasites are constantly gnawing at an organism's right to life and do not forgive slapdash engineering.

And many of the examples of bad design in the animal kingdom turn out to be old spouses' tales. Take the remark in a book by a famous cognitive psychologist that natural selection has been powerless to eliminate the wings of any bird, which is why penguins are stuck with wings even though they cannot fly. Wrong twice. The moa had no trace of a wing, and penguins do use their wings to fly—under water. Michael French makes the point in his engineering textbook using a more famous example:

> It is an old joke that a camel is a horse designed by a committee, a joke which does grave injustice to a splendid creature and altogether too

much honour to the creative power of committees. For a camel is no chimera, no odd collection of bits, but an elegant design of the tightest unity. So far as we can judge, every part is contrived to suit the difficult role of the whole, a large herbivorous animal to live in harsh climates with much soft going, sparse vegetation and very sparse water. The specification for a camel, if it were ever written down, would be a tough one in terms of range, fuel economy and adaptation to difficult terrains and extreme temperatures, and we must not be surprised that the design that meets it appears extreme. Nevertheless, every feature of the camel is of a piece: the large feet to diffuse load, the knobbly knees that derive from some of the design principles of Chapter 7 [bearings and pivots], the hump for storing food and the characteristic profile of the lips have a congruity that derives from function and invests the whole creation with a feeling of style and a certain bizarre elegance, borne out by the beautiful rhythms of its action at a gallop.

Obviously, evolution is constrained by the legacies of ancestors and the kinds of machinery that can be grown out of protein. Birds could not have evolved propellers, even if that had been advantageous. But many claims of biological constraints are howlers. One cognitive scientist has opined that "many properties of organisms, like symmetry, for example, do not really have anything to do with specific selection but just with the ways in which things can exist in the physical world." In fact, most things that exist in the physical world are *not* symmetrical, for obvious reasons of probability: among all the possible arrangements of a volume of matter, only a tiny fraction are symmetrical. Even in the living world, the molecules of life are asymmetrical, as are livers, hearts, stomachs, flounders, snails, lobsters, oak trees, and so on. Symmetry has *everything* to do with selection. Organisms that move in straight lines have bilaterally symmetrical external forms because otherwise they would go in circles. Symmetry is so improbable and difficult to achieve that any disease or defect can disrupt it, and many animals size up the health of prospective mates by checking for minute asymmetries.

Gould has emphasized that natural selection has only limited freedom to alter basic body plans. Much of the plumbing, wiring, and architecture of the vertebrates, for example, has been unchanged for hundreds of millions of years. Presumably they come from embryological recipes that cannot easily be tinkered with. But the vertebrate body plan accommodates eels, cows, hummingbirds, aardvarks, ostriches, toads, gerbils, seahorses, giraffes, and blue whales. The similarities are important, but the differences are important, too! Developmental constraints only rule out broad classes of options. They cannot, by themselves, force a functioning organ to come into being. An embryological constraint like "Thou shalt grow wings" is an absurdity. The vast majority of hunks of animal flesh do not meet the stringent engineering demands of powered flight, so it is infinitesimally unlikely that the creeping and bumping cells in the microscopic layers of the developing embryo are obliged to align themselves into bones, skin, muscles, and feathers with just the right architecture to get the bird aloft—unless, of course, the developmental program had been shaped to bring about that outcome by the history of successes and failures of the whole body.

Natural selection should not be pitted against developmental, genetic, or phylogenetic constraints, as if the more important one of them is, the less important the others are. Selection versus constraints is a phony dichotomy, as crippling to clear thinking as the dichotomy between innateness and learning. Selection can only select from alternatives that are growable as carbon-based living stuff, but in the absence of selection that stuff could just as easily grow into scar tissue, scum, tumors, warts, tissue cultures, and quivering amorphous protoplasm as into functioning organs. Thus selection and constraints are both important but are answers to different questions. The question "Why does this creature have such-and-such an organ?" by itself is meaningless. It can only be asked when followed by a compared-to-what phrase. Why do birds have wings (as opposed to propellers)? Because you can't grow a vertebrate with propellers. Why do birds have wings (as opposed to forelegs or hands or stumps)? Because selection favored ancestors of birds that could fly.

Another widespread misconception is that if an organ changed its function in the course of evolution, it did not evolve by natural selection. One discovery has been cited over and over in support of the misconception: the wings of insects were not originally used for locomotion. Like a friend-of-a-friend legend, that discovery has mutated in the retelling: wings evolved for something else but happened to be perfectly adapted for flight, and one day the insects just decided to fly with them; the evolution of insect wings refutes Darwin because they would have had to evolve gradually and half a wing is useless; the wings of *birds* were not originally used for locomotion (probably a misremembering of another fact, that the first feathers evolved not for flight but for insulation). All one has to do is say "the evolution of wings" and audiences will nod knowingly, completing the anti-adaptationist argument for themselves.

How can anyone say that any organ was selected for its current function? Maybe it evolved for something else and the animal is only *using* it for that function now, like the nose holding up spectacles and all that stuff about insect wings that everyone knows about (or was it bird wings?).

Here is what you find when you check the facts. Many organs that we see today have maintained their original function. The eye was always an eye, from light-sensitive spot to image-focusing eyeball. Others changed their function. That is not a new discovery. Darwin gave many examples, such as the pectoral fins of fishes becoming the forelimbs of horses, the flippers of whales, the wings of birds, the digging claws of moles, and the arms of humans. In Darwin's day the similarities were powerful evidence for the fact of evolution, and they still are. Darwin also cited changes in function to explain the problem of "the incipient stages of useful structures," perennially popular among creationists. How could a complex organ gradually evolve when only the final form is usable? Most often the premise of unusability is just wrong. For example, partial eyes have partial sight, which is better than no sight at all. But sometimes the answer is that before an organ was selected to assume its current form, it was adapted for something else and then went through an intermediate stage in which it accomplished both. The delicate chain of middle-ear bones in mammals (hammer, anvil, stirrup) began as parts of the jaw hinge of reptiles. Reptiles often sense vibrations by lowering their jaws to the ground. Certain bones served both as jaw hinges and as vibration transmitters. That set the stage for the bones to specialize more and more as sound transmitters, causing them to shrink and move into their current shape and role. Darwin called the earlier forms "pre-adaptations," though he stressed that evolution does not somehow anticipate next year's model.

There is nothing mysterious about the evolution of birds' wings. Half a wing will not let you soar like an eagle, but it will let you glide or parachute from trees (as many living animals do), and it will let you leap or take off in bursts while running, like a chicken trying to escape a farmer. Paleontologists disagree about which intermediate stage is best supported by the fossil and aerodynamic evidence, but there is nothing here to give comfort to a creationist or a social scientist.

The theory of the evolution of insect wings proposed by Joel Kingsolver and Mimi Koehl, far from being a refutation of adaptationism, is one of its finest moments. Small cold-blooded animals like insects struggle to regulate their temperature. Their high ratio of surface area to volume makes them heat up and cool down quickly. (That is why there are no bugs outside in cold months; winter is the best insecticide.) Perhaps the incipient wings of insects first evolved as adjustable solar panels, which soak up the sun's energy when it is colder out and dissipate heat when it's warmer. Using thermodynamic and aerodynamic analyses, Kingsolver and Koehl showed that proto-wings too small for flight are effective heat exchangers. The larger they grow, the more effective they become at heat regulation, though they reach a point of diminishing returns. That point is in the range of sizes in which the panels could serve as effective wings. Beyond that point, they become more and more useful for flying as they grow larger and larger, up to their present size. Natural selection could have pushed for bigger wings throughout the range from no wings to current wings, with a gradual change of function in the middle sizes.

So how did the work get garbled into the preposterous story that one day an ancient insect took off by flapping unmodified solar panels and the rest of them have been doing it ever since? Partly it is a misunderstanding of a term introduced by Gould, *exaptation*, which refers to the adaptation of an old organ to a new function (Darwin's "pre-adaptation") or the adaptation of a non-organ (bits of bone or tissue) to an organ with a function. Many readers have interpreted it as a new theory of evolution that has replaced adaptation and natural selection. It's not. Once again, complex design is the reason. Occasionally a machine designed for a complicated, improbable task can be pressed into service to do something simpler. A book of cartoons called *101 Uses for a Dead Computer* showed PCs being used as a paperweight, an aquarium, a boat anchor, and so on. The humor comes from the relegation of sophisticated technology to a humble function that cruder devices can fulfill. But there will never be a book of cartoons called *101 Uses for a Dead Paperweight* showing one being used as a computer. And so it is with exaptation in the living world. On engineering grounds, the odds are against an organ designed for one purpose being usable out of the box for some other purpose, unless the new purpose is quite simple. (And even then the nervous system of the animal must often be adapted for it to find and keep the new use.) If the new function is at all difficult to accomplish, natural selection must have revamped and retrofitted the part considerably, as it did to give modern insects their wings. A housefly dodging a crazed human can decelerate from rapid flight, hover, turn in its own length, fly upside down, loop, roll, and land on the ceiling, all in less than a second. As an article entitled "The Mechanical Design of Insect Wings" notes,

"Subtle details of engineering and design, which no man-made airfoil can match, reveal how insect wings are remarkably adapted to the acrobatics of flight." The evolution of insect wings is an argument *for* natural selection, not against it. A *change* in selection pressure is not the same as *no* selection pressure.

Complex design lies at the heart of all these arguments, and that offers a final excuse to dismiss Darwin. Isn't the whole idea a bit squishy? Since no one knows the number of kinds of possible organisms, how can anyone say that an infinitesimal fraction of them have eyes? Perhaps the idea is circular: the things one calls "adaptively complex" are just the things that one believes couldn't have evolved any other way than by natural selection. As Noam Chomsky wrote,

> So the thesis is that natural selection is the only physical explanation of design that fulfills a function. Taken literally, that cannot be true. Take my physical design, including the property that I have positive mass. That fulfills some function—namely, it keeps me from drifting into outer space. Plainly, it has a physical explanation which has nothing to do with natural selection. The same is true of less trivial properties, which you can construct at will. So you can't mean what you say literally. I find it hard to impose an interpretation that doesn't turn it into the tautology that where systems have been *selected* to satisfy some function, then the process is selection.

Claims about functional design, because they cannot be stated in exact numbers, do leave an opening for a skeptic, but a little thought about the magnitudes involved closes it. Selection is not invoked to explain mere usefulness; it's invoked to explain *improbable* usefulness. The mass that keeps Chomsky from floating into outer space is not an improbable condition, no matter how you measure the probabilities. "Less trivial properties"—to pick an example at random, the vertebrate eye—*are* improbable conditions, no matter how you measure the probabilities. Take a dip net and scoop up objects from the solar system; go back to life on the planet a billion years ago and sample the organisms; take a collection of molecules and calculate all their physically possible configurations; divide the human body into a grid of one-inch cubes. Calculate the proportion of samples that have positive mass. Now calculate the proportion of samples that can form an optical image. There will be a statistically significant difference in the proportions, and it needs to be explained.

At this point the critic can say that the criterion—seeing versus not seeing—is set *a posteriori*, after we know what animals can do, so the probability estimates are meaningless. They are like the infinitesimal probability that I would have been dealt whatever poker hand I happened to have been dealt. Most hunks of matter cannot see, but then most hunks of matter cannot flern either, where I hereby define *flern* as the ability to have the exact size and shape and composition of the rock I just picked up.

Recently I visited an exhibition on spiders at the Smithsonian. As I marveled at the Swiss-watch precision of the joints, the sewing-machine motions by which it drew silk from its spinnerets, the beauty and cunning of the web, I thought to myself, "How could anyone see this and not believe in natural selection!" At that moment a woman standing next to me exclaimed, "How could anyone see this and not believe in God!" We agreed *a priori* on the facts that need to be explained, though we disagreed about how to explain them. Well before Darwin, theologians such as William Paley pointed to the engineering marvels of nature as proof of the existence of God. Darwin did not invent the facts to be explained, only the explanation.

But what, exactly, are we all so impressed by? Everyone might agree that the Orion constellation looks like a big guy with a belt, but that does not mean we need a special explanation of why stars align themselves into guys with belts. But the intuition that eyes and spiders show "design" and that rocks and Orion don't can be unpacked into explicit criteria. There has to be a heterogeneous structure: the parts or aspects of an object are unpredictably different from one another. And there has to be a unity of function: the different parts are organized to cause the system to achieve some special effect—special because it is improbable for objects lacking that structure, and special because it benefits someone or something. If you can't state the function more economically than you can describe the structure, you don't have design. A lens is different from a diaphragm, which in turn is different from a photopigment, and no unguided physical process would deposit the three in the same object, let alone align them perfectly. But they do have something in common—all are needed for high-fidelity image formation—and that makes sense of why they are found together in an eye. For the flerning rock, in contrast, describing the structure and stating the function are one and the same. The notion of *function* adds nothing.

And most important, attributing adaptive complexity to natural selection is not just a recognition of design excellence, like the expensive appliances in the Museum of Modern Art. Natural selection is a falsifiable hypothesis about the origin of design and imposes onerous empirical requirements. Remember how it works: from competition among replicators. Anything that showed signs of design but did not come from a long line of replicators could not be explained by—in fact, would refute—the theory of natural selection: natural species that lacked reproductive organs, insects growing like crystals out of rocks, television sets on the moon, eyes spewing out of vents on the ocean floor, caves shaped like hotel rooms down to the details of hangers and ice buckets. Moreover, the beneficial functions all have to be in the ultimate service of reproduction. An organ can be designed for seeing or eating or mating or nursing, but it had better not be designed for the beauty of nature, the harmony of the ecosystem, or instant self-destruction. Finally, the beneficiary of the function has to be the replicator. Darwin pointed out that if horses had evolved saddles, his theory would immediately be falsified.

Rumors and folklore notwithstanding, natural selection remains the heart of explanation in biology. Organisms can be understood only as interactions among adaptations, by-products of adaptations, and noise. The by-products and noise don't rule out the adaptations, nor do they leave us staring blankly, unable to tell them apart. It is exactly what makes organisms so fascinating—their improbable adaptive design—that calls for reverse-engineering them in the light of natural selection. The by-products and noise, because they are defined negatively as un-adaptations, also can be discovered only via reverse-engineering.

This is no less true for human intelligence. The major faculties of the mind, with their feats no robot can duplicate, show the handiwork of selection. That does not mean that every aspect of the mind is adaptive. From low-level features like the sluggishness and noisiness of neurons, to momentous activities like art, music, religion, and dreams, we should expect to find activities of the mind that are not adaptations in the biologists' sense. But it *does* mean that our understanding of how the mind works will be woefully incomplete or downright wrong unless it meshes with our understanding of how the mind evolved. That is the topic of the rest of the chapter.

THE BLIND PROGRAMMER

Why did brains evolve to start with? The answer lies in the value of information, which brains have been designed to process.

Every time you buy a newspaper, you are paying for information. Economic theorists have explained why you should: information confers a benefit that is worth paying for. Life is a choice among gambles. One turns left or right at the fork in the road, stays with Rick or leaves with Victor, knowing that neither choice guarantees fortune or happiness; the best one can do is play the odds. Stripped to its essentials, every decision in life amounts to choosing which lottery ticket to buy. Say a ticket costs $1.00 and offers a one-in-four chance of winning $10.00. On average, you will net $1.50 per play ($10.00 divided by 4 equals $2.50, minus $1.00 for the ticket). The other ticket costs $1.00 and offers a one-in-five chance of winning $12.00. On average, you will net $1.40 per play. The two kinds of tickets come in equal numbers, and neither has the odds or winnings marked on it. How much should you pay for someone to tell you which is which? You should pay up to four cents. With no information, you would have to choose at random, and you could expect to make $1.45 on average ($1.50 half the time, $1.40 half the time). If you knew which had the better average payoff, you would make an average of $1.50 each play, so even if you paid four cents you would be ahead by one cent each play.

Most organisms don't buy lottery tickets, but they all choose between gambles every time their bodies can move in more than one way. They should be willing to "pay" for information—in tissue, energy, and time—if the cost is lower than the expected payoff in food, safety, mating opportunities, and other resources, all ultimately valued in the expected number of surviving offspring. In multicellular animals the information is gathered and translated into profitable decisions by the nervous system.

Often, more information brings a greater reward and earns back its extra cost. If a treasure chest has been buried somewhere in your neighborhood, the single bit of information that locates it in the north or the south half is helpful, because it cuts your digging time in half. A second bit that told you which quadrant it was in would be even more useful, and so on. The more digits there are in the coordinates, the less time you will waste digging fruitlessly, so you should be willing to pay for more bits, up to the point where you are so close that further subdivision would not be worth the cost. Similarly, if you were trying to crack a combination lock, every number you bought would cut down the number of possibilities to try, and could be worth its cost in the time saved. So very often more information is better, up to a point of diminishing returns, and that is why some lineages of animals have evolved more and more complex nervous systems.

Natural selection cannot directly endow an organism with information about its environment, or with the computational networks, demons, modules, faculties, representations, or mental organs that process the information. It can only select among genes. But genes build brains, and different genes build brains that process information in different ways. The evolution of information processing has to be accomplished at the nuts-and-bolts level by selection of genes that affect the brain-assembly process.

Many kinds of genes could be the targets of selection for better information processing. Altered genes could lead to different numbers of proliferative units along the walls of the ventricles (the cavities in the center of the brain), which beget the cortical neurons making up the gray matter. Other genes could allow the proliferative units to divide for different numbers of cycles, creating different numbers and kinds of cortical areas. Axons connecting the neurons can be re-routed by shifting the chemical trails and molecular guideposts that coax the axons in particular directions. Genes can change the molecular locks and keys that encourage neurons to connect with other ones. As in the old joke about how to carve a statue of an elephant (remove all the bits that don't look like an elephant), neural circuits can be sculpted by programming certain cells and synapses to commit suicide on cue. Neurons can become active at different points in embryogenesis, and their firing patterns, both spontaneous and programmed, can be interpreted downstream as information about how to wire together. Many of these processes interact in cascades. For example, increasing the size of one area allows it to compete better for real estate downstream. Natural selection does not care how baroque the brain-assembly process is, or how ugly the resulting brain. Modifications are evaluated strictly on how well the brain's algorithms work in guiding the perception, thought, and action of the whole animal. By these processes, natural selection can build a better and better functioning brain.

But could the selection of random variants really improve the design of a nervous system? Or would the variants crash it, like a corrupted byte in a computer program, and the selection merely preserve the systems that do not crash? A new field of computer science called genetic algorithms has shown that Darwinian selection can create increasingly intelligent software. Genetic algorithms are programs that are duplicated to make multiple copies, though with random mutations that make each one a tiny bit different. All the copies have a go at solving a problem, and the ones that do best are allowed to reproduce to furnish the copies for the next round. But first, parts of each program are randomly mutated again, and pairs of programs have sex: each is split in two, and the halves are exchanged. After many cycles of computation, selection, mutation, and reproduction, the surviving programs are often better than anything a human programmer could have designed.

More apropos of how a mind can evolve, genetic algorithms have been applied to neural networks. A network might be given inputs from simulated sense organs and outputs to simulated legs and placed in a virtual environment with scattered "food" and many other networks competing for it. The ones that get the most food leave the most copies before the next round of mutation and selection. The mutations are random changes in the connection weights, sometimes followed by sexual recombination between networks (swapping some of their connection weights). During the early iterations, the "animals"—or, as they are sometimes called, "animats"—wander randomly over the terrain, occasionally bumping into a food source. But as they evolve they come to zip directly from food source to food source. Indeed, a population of networks that is allowed to evolve innate connection weights often does better than a single neural network that is allowed to learn them. That is especially true for networks with multiple hidden layers, which complex animals, especially humans, surely have. If a network can only learn, not evolve, the environmental teaching signal gets diluted as it is propagated backward to the hidden layers and can only nudge the connection weights up and down by minuscule amounts. But if a population of networks can evolve, even if they cannot learn, mutations and recombinations can reprogram the hidden layers directly, and can catapult the network into a combination of innate connections that is much closer to the optimum. Innate structure is selected for.

Evolution and learning can also go on simultaneously, with innate structure evolving in an animal that also learns. A population of networks can be equipped with a generic learning algorithm *and* can be allowed to evolve the innate parts, which the network designer would ordinarily have built in by guesswork, tradition, or trial and error. The innate specs include how many units there are, how they are connected, what the initial connection weights are, and how much the weights should be nudged up and down on each learning episode. Simulated evolution gives the networks a big head start in their learning careers.

So evolution can guide learning in neural networks. Surprisingly, learning can guide evolution as well. Remember Darwin's discussion of "the incipient stages of useful structures"—the what-good-is-half-an-eye problem. The neural-network theorists Geoffrey Hinton and Steven Nowlan invented a fiendish example. Imagine an animal controlled by a neural network with twenty connections, each either excitatory (on) or neutral (off). But the network is utterly useless unless all twenty connections are correctly set. Not only is it no good to have half a network; it is no good to have ninety-five percent of one. In a population of animals whose connections are determined by random mutation, a fitter mutant, with all the right connections, arises only about once every million (2^{20}) genetically distinct organisms. Worse, the advantage is immediately lost if the animal reproduces sexually, because after having finally found the magic combination of weights, it swaps half of them away. In simulations of this scenario, no adapted network ever evolved.

But now consider a population of animals whose connections can come in three forms: innately on, innately off, or settable to on or off by learning. Mutations determine which of the three possibilities (on, off, learnable) a given connection has at the animal's birth. In an average animal in these simulations, about half the connections are learnable, the other half on or off. Learning works like this. Each animal, as it lives its life, tries out settings for the learnable connections at random until it hits upon the magic combination. In real life this might be figuring out how to catch prey or crack a nut; whatever it is, the animal senses its good fortune and retains those settings, ceasing the trial and error. From then on it enjoys a higher rate of reproduction. The earlier in life the animal acquires the right settings, the longer it will have to reproduce at the higher rate.

Now with these evolving learners, or learning evolvers, there *is* an advantage to having less than one hundred percent of the correct network. Take all the animals with ten innate connections. About one in a thousand (2^{10}) will have all ten correct. (Remember that only one in a million *non*learning animals had all twenty of its innate connections correct.) That well-endowed animal will have some probability of attaining the completely correct network by learning the other ten connections; if it has a thousand occasions to learn, success is fairly likely. The successful animal will reproduce earlier, hence more often. And among its descendants, there are advantages to mutations that make more and more of the connections innately correct, because with more good connections to begin with, it takes less time to learn the rest, and the chances of going through life without having learned them get smaller. In Hinton and Nowlan's simulations, the networks thus evolved more and more innate connections. The connections never became completely innate, however. As more and more of the connections were fixed, the selection pressure to fix the remaining ones tapered off, because with only a few connections to learn, every organism was guaranteed to learn them quickly. Learning leads to the evolution of innateness, but not complete innateness.

Hinton and Nowlan submitted the results of their computer simulations to a journal and were told that they had been scooped by a hundred years. The psychologist James Mark Baldwin had proposed that learning could guide evolution in precisely this way, creating an illusion of Lamarckian evolution without there really being Lamarckian evolution. But no one had shown that the idea, known as the Baldwin effect, would really work. Hinton and Nowlan showed why it can. The ability to learn alters the evolutionary problem from looking for a needle in a haystack to looking for the needle with someone telling you when you are getting close.

The Baldwin effect probably played a large role in the evolution of brains. Contrary to standard social science assumptions, learning is not some pinnacle of evolution attained only recently by humans. All but the simplest animals learn. That is why mentally uncomplicated creatures like fruit flies and sea slugs have been convenient subjects for neuroscientists searching for the neural incarnation of learning. If the ability to learn was in place in an early ancestor of the multicellular animals, it could have guided the evolution of nervous systems toward their specialized circuits even when the circuits are so intricate that natural selection could not have found them on its own.

INSTINCT AND INTELLIGENCE

Complex neural circuitry has evolved in many animals, but the common image of animals climbing up some intelligence ladder is wrong. The common view is that lower animals have a few fixed reflexes, and that in higher ones the reflexes can be associated with new stimuli (as in Pavlov's experiments) and the responses can be associated with rewards (as in Skinner's). On this view, the ability to associate gets better in still higher organisms, and eventually it is freed from bodily drives and physical stimuli and responses and can associate ideas directly to each other, reaching an apex in man. But the distribution of intelligence in real animals is nothing like this.

The Tunisian desert ant leaves its nest, travels some distance, and then wanders over the burning sands looking for the carcass of an insect that has keeled over from the heat. When it finds one, it bites off a chunk, turns, and makes a beeline for the nest, a hole one millimeter in diameter as much as fifty meters away. How does it find its way back? The navigation depends on information gathered during the outward journey, not on sensing the nest like a beacon. If someone lifts the ant as it emerges from the nest and plunks it down some distance away, the ant wanders in random circles. If someone moves the ant *after* it finds food, it runs in a line within a degree or two of the direction of its nest with respect to the abduction site, slightly overshoots the point where the nest should be, does a quick U-turn, and searches for the nonexistent nest. This shows that the ant has somehow measured and stored the direction and distance back to the nest, a form of navigation called path integration or dead reckoning.

This example of information processing in animals, discovered by the biologist Rudiger Wehner, is one of many that the psychologist Randy Gallistel has used to try to get people to stop thinking about learning as the formation of associations. He explains the principle:

> Path integration is the integration of the velocity vector with respect to time to obtain the position vector, or some discrete equivalent of this computation. The discrete equivalent in traditional marine navigation is to record the direction and speed of travel (the velocity) at intervals, multiply each recorded velocity by the interval since the previous recording to get interval-by-interval displacements (e.g., making 5 knots on a northeast course for half an hour puts the ship 2.5 nautical miles northeast of where it was), and sum the successive displacements (changes in position) to get the net change in position. These running sums of the longitudinal and latitudinal displacements are the deduced reckoning of the ship's position.

Audiences are incredulous. All that computation inside the little bitty pinhead of an ant? Actually, as computation goes, this is pretty simple stuff; you could build a device to do it for a few dollars out of little parts hanging on the pegboard at Radio Shack. But intuitions about the nervous system have been so impoverished by associationism that a psychologist would be accused of wild, profligate speculation if she were to attribute this machinery to a *human* brain, let alone an ant brain. Could an ant really do calculus, or even arithmetic? Not overtly, of course, but then neither do we when we exercise our own faculty of dead reckoning, our "sense of direction." The path integration calculations are done unconsciously, and their output pokes into our awareness—and the ant's, if it has any—as an abstract feeling that home is thataway, yea far.

Other animals execute even more complicated sequences of arithmetic, logic, and data storage and retrieval. Many migratory birds fly thousands of miles at night, maintaining their compass direction by looking at the constellations. As a Cub Scout I was taught how to find the North Star: locate the tip of the handle of the Little Dipper, or extrapolate from the front lip of the Big Dipper a distance seven times its depth. Birds are not born with this knowledge, not because it is unthinkable that it could be innate, but because if it were innate it would soon be obsolete. The earth's axis of rotation, and hence the celestial pole (the point in the sky corresponding to north), wobbles in a 27,000-year cycle called the precession of the equinoxes. The cycle is rapid in an evolutionary timetable, and the birds have responded by evolving a special algorithm for *learning* where the celestial pole is in the night sky. It all happens while they are still in the nest and cannot fly. The nestlings gaze up at the night sky for hours, watching the slow rotation of the constellations. They find the point around which the stars appear to move, and record its position with respect to several nearby constellations, acquiring the information imparted to me by the Cub Scout manual. Months later they can use any of these constellations to maintain a constant heading—say, keeping north behind them while flying south, or flying into the celestial pole the next spring to return north.

Honeybees perform a dance that tells their hivemates the direction and distance of a food source with respect to the sun. As if that weren't impressive enough, the bees have evolved a variety of calibrations and backup systems to deal with the engineering complexities of solar navigation. The dancer uses an internal clock to compensate for the movement of the sun between the time she discovered the source and the time she passes on the information. If it's cloudy, the other bees estimate the direction using the polarization of light in the sky. These feats are the tip of an iceberg of honeybee ingenuity, documented by Karl Von Frisch, James Gould, and others. A psychologist colleague of mine once thought that bees offered a good pedagogical opportunity to convey the sophistication of neural computation to our undergraduates. He devoted the first week of his entry-level course in cognitive science to some of the ingenious experiments. The next year the lectures spilled over to the second week, then the third, and so on, until the students complained that the course had become an Introduction to Bee Cognition.

There are dozens of comparable examples. Many species compute how much time to forage at each patch so as to optimize their rate of return of calories per energy expended in foraging. Some birds learn the emphemeris function, the path of the sun above the horizon over the course of the day and the year, necessary for navigating by the sun. The barn owl uses sub-millisecond discrepancies between the arrival times of a sound at its two ears to swoop down on a rustling mouse in pitch blackness. Cacheing species place nuts and seeds in unpredictable hiding places to foil thieves, but months later must recall them all. I mentioned in the preceding chapter that the Clark's Nutcracker can remember ten thousand hiding places. Even Pavlovian and operant conditioning, the textbook cases of learning by association, turn out to be not a general stickiness of coinciding stimuli and responses in the brain, but complex algorithms for multivariate, nonstationary time series analysis (predicting when events will occur, based on their history of occurrences).

The moral of this animal show is that animals' brains are just as specialized and well engineered as their bodies. A brain is a precision instrument that allows a creature to use information to solve the problems presented by its lifestyle. Since organisms' lifestyles differ, and since they are related to one another in a great bush, not a great chain, species cannot be ranked in IQ or by the percentage of human intelligence they have achieved. Whatever is special about the human mind cannot be just more, or better, or more flexible animal intelligence, because there is no such thing as generic animal intelligence. Each animal has evolved information-processing machinery to solve its problems, and we evolved machinery to solve ours. The sophisticated algorithms found in even the tiniest dabs of nervous tissue serve as yet another eye-opener—joining the difficulty of building a robot, the circumscribed effects of brain damage, and the similarities between twins reared apart—for the hidden complexity we should expect to find in the human mind.

~

The brains of mammals, like the bodies of mammals, follow a common general plan. Many of the same cell types, chemicals, tissues, sub-organs, way-stations, and pathways are found throughout the class, and the major visible differences come from inflating or shrinking the parts. But under the microscope, differences appear. The number of cortical areas differs widely, from twenty or fewer in rats to fifty or more in humans. Primates differ from other mammals in the number of visual areas, their interconnections, and their hookup to the motor and decision regions of the frontal lobes. When a species has a noteworthy talent, it is reflected in the gross anatomy of its brain, sometimes in ways visible to the naked eye. The takeover of monkeys' brains by visual areas (about one-half the territory) reflects—more accurately, allows—their aptitude for depth, color, motion, and visually guided grasping. Bats that rely on sonar have additional brain areas dedicated to their ultrasonic hearing, and desert mice that cache seeds are born with a bigger hippocampus— a seat of the cognitive map—than closely related species that don't cache.

The human brain, too, tells an evolutionary story. Even a quick side-by-side comparison shows that the primate brain must have been considerably re-engineered to end up as a human brain. Our brains are about three times too big for a generic monkey or ape of our body size. The inflation is accomplished by prolonging fetal brain growth for a year after birth. If our bodies grew proportionally during that period, we would be ten feet tall and weigh half a ton.

The major lobes and patches of the brain have been revamped as well. The olfactory bulbs, which underlie the sense of smell, have shriveled to a third of the expected primate size (already puny by mammalian standards), and the main cortical areas for vision and movement have shrunk proportionally as well. Within the visual system, the first stop for information, the primary visual cortex, takes up a smaller proportion of the whole brain, while the later areas for complex-form processing expand, as do the temporo-parietal areas that shunt visual information to the language and conceptual regions. The areas for hearing, especially for understanding speech, have grown, and the prefrontal lobes, the seat of deliberate thought and planning, have ballooned to twice what a primate our size should have. While the brains of monkeys and apes are subtly asymmetrical, the human brain, especially in the areas devoted to language, is so lopsided that the two hemispheres can be distinguished by shape in the jar. And there have been takeovers of primate brain areas for new functions. Broca's area, involved in speech, has a homologue (evolutionary counterpart) in monkeys, but they obviously don't use it for speech, and they don't even seem to use it to produce shrieks, barks, and other calls.

It's interesting to find these differences, but the human brain could be radically different from an ape's brain even if one looked like a perfect scale model of the other. The real action is in the patterns of connections among neurons, just as the differences in content among different computer programs, microchips, books, or videocassettes lie not in their gross shapes but in the combinatorial arrangements of their tiny constituents. Virtually nothing is known about the functioning microcircuitry of the human brain, because there is a shortage of volunteers willing to give up their brains to science before they are dead. If we could somehow read the code in the neural circuitry of growing humans and apes, we would surely find substantial differences.

Are the marvelous algorithms of animals mere "instincts" that we have lost or risen above? Humans are often said to have no instincts beyond the vegetative functions; we are said to reason and behave flexibly, freed from specialized machinery. The featherless biped surely understands astronomy in a sense that the feathered biped does not! True enough, but it is not because we have fewer instincts than other animals; it is because we have *more*. Our vaunted flexibility comes from scores of instincts assembled into programs and pitted in competitions. Darwin called human language, the epitome of flexible behavior, "an instinct to acquire an art" (giving me the title for *The Language Instinct*), and his follower William James pressed the point:

> *Now, why do the various animals do what seem to us such strange things*, in the presence of such outlandish stimuli? Why does the hen, for example,

submit herself to the tedium of incubating such a fearfully uninteresting set of objects as a nestful of eggs, unless she have some sort of a prophetic inkling of the result? The only answer is *ad hominem*. We can only interpret the instincts of brutes by what we know of instincts in ourselves. Why do men always lie down, when they can, on soft beds rather than on hard floors? Why do they sit round the stove on a cold day? Why, in a room, do they place themselves, ninety-nine times out of a hundred, with their faces towards its middle rather than to the wall? Why do they prefer saddle of mutton and champagne to hard-tack and pond-water? Why does the maiden interest the youth so that everything about her seems more important and significant than anything else in the world? Nothing more can be said than that these are human ways, and that every creature *likes* its own ways, and takes to the following them as a matter of course. Science may come and consider these ways, and find that most of them are useful. But it is not for the sake of their utility that they are followed, but because at the moment of following them we feel that that is the only appropriate and natural thing to do. Not one man in a billion, when taking his dinner, ever thinks of utility. He eats because the food tastes good and makes him want more. If you ask him *why* he should want to eat more of what tastes like that, instead of revering you as a philosopher we will probably laugh at you for a fool. . . .

And so, probably, does each animal feel about the particular things it tends to do in presence of particular objects. To the broody hen the notion would probably seem monstrous that there should be a creature in the world to whom a nestful of eggs was not the utterly fascinating and precious and never-to-be-too-much sat-upon object which it is to her.

The human reactions described in the passage still may strike you as versions of animal instincts. What about our rational, flexible thought? Can it be explained as a set of instincts? In the preceding chapter I showed how our precision intelligence can be broken down into smaller and smaller agents or networks of information processing. At the lowest levels, the steps have to be as automatic and unanalyzed as the reactions of the most brutish animal. Remember what the tortoise said to Achilles. No rational creature can consult rules all the way down; that way infinite regress lies. At some point a thinker must *execute* a rule, because he just can't help it: it's the human way, a matter of course, the only appropriate and natural thing to do—in short, an instinct. When all goes well, our reasoning instincts link up into complex programs for rational analysis, but that is not because we somehow commune with a realm of truth and reason. The same instincts can be seduced by sophistry, bump up against paradoxes like Zeno's beguiling demonstrations that motion is impossible, or make us dizzy as they ponder mysteries like sentience and free will. Just as an ethologist unmasks an animal's instincts with clever manipulations of its world, such as slipping a mechanical bee into a hive or rearing a chick in a planetarium, psychologists can unmask human reasoning instincts by couching problems in devilish ways, as we shall see in Chapter 5.

THE COGNITIVE NICHE

Ambrose Bierce's *Devil's Dictionary* defines our species as follows:

> **Man,** *n.* An animal so lost in rapturous contemplation of what he thinks he is as to overlook what he indubitably ought to be. His chief occupation is extermination of other animals and his own species, which, however, multiplies with such insistent rapidity as to infest the whole habitable earth and Canada.

Homo sapiens sapiens is indeed an unprecedented animal, with many zoologically unique or extreme traits. Humans achieve their goals by complex chains of behavior, assembled on the spot and tailored to the situation. They plan the behavior using cognitive models of the causal structure of the world. They learn these models in their lifetimes and communicate them through language, which allows the knowledge to accumulate within a group and over generations. They manufacture and depend upon many kinds of tools. They exchange goods and favors over long periods of time. Food is transported long distances, processed extensively, stored, and shared. Labor is divided between the sexes. Humans form large, structured coalitions, especially among males, and coalitions wage war against each other. Humans use fire. Kinship systems are complex and vary with other aspects of their lifestyles. Mating relations are negotiated by kin, often by groups exchanging daughters. Ovulation is concealed, and females may choose to have sex at any time rather than at certain points in a reproductive cycle.

A few of these traits are found among some of the great apes, but to a much lesser degree, and most are not found at all. And humans have rediscovered traits that are rare among primates but found in other animals. They are bipedal. They live longer than other apes, and bear help-less offspring who stay children (that is, sexually immature) for a sub-stantial part of their lives. Hunting is important, and meat a large part of the diet. Males invest in their offspring: they tote children around, pro-tect them against animals and other humans, and give them food. And as *The Devil's Dictionary* points out, humans occupy every ecozone on earth.

Aside from the retooling of the skeleton that gives us upright posture and precision manipulation, what makes us unusual is not our body but our behavior and the mental programs that organize it. In the comic strip *Calvin and Hobbes*, Calvin asks his tiger companion why people are never content with what they have. Hobbes replies, "Are you kidding? Your fingernails are a joke, you've got no fangs, you can't see at night, your pink hides are ridiculous, your reflexes are nil, and you don't even have tails! Of course people aren't content!" But despite these handi-caps, humans control the fate of tigers, rather than vice versa. Human evolution is the original revenge of the nerds.

Perhaps recoiling from the image of the pasty-faced, pocket-protected, polyester-clad misfits, theorists on human evolution have looked far and wide for alternative theories. Human ingenuity has been explained away as a by-product of blood vessels in the skull that radiate heat, as a run-away courtship device like the peacock's tail, as a stretching of chim-panzee childhood, and as an escape hatch that saved the species from the evolutionary dead end of bearing fewer and fewer offspring. Even in theo-ries that acknowledge that intelligence itself was selected for, the causes are badly underpowered in comparison with the effects. In various stories the full human mind sprang into existence to solve narrow problems like chipping tools out of stone, cracking open nuts and bones, throwing rocks at animals, keeping track of toddlers, following herds to scavenge their dead, and maintaining social bonds in a large group.

There are grains of truth in these accounts, but they lack the leverage of good reverse-engineering. Natural selection for success in solving a particular problem tends to fashion an idiot savant like the dead-reckon-ing ants and stargazing birds. We need to know what the more general kinds of intelligence found in our species are good for. That requires a good description of the improbable feats the human mind accomplishes, not just one-word compliments like "flexibility" or "intelligence." That description must come from the study of the modern mind, cognitive sci-ence. And because selection is driven by the fate of the whole individual, it is not enough to explain the evolution of a brain in a vat. A good theory has to connect all the parts of the human lifestyle—all ages, both sexes, anatomy, diet, habitat, and social life. That is, it has to characterize the ecological niche that humans entered.

The only theory that has risen to this challenge comes from John Tooby and the anthropologist Irven DeVore. Tooby and DeVore begin by noting that species evolve at one another's expense. We fantasize about the land of milk and honey, the big rock candy mountain, and tangerine trees with marmalade skies, but real ecosystems are different. Except for fruits (which trick hungry animals into dispersing seeds), virtually every food is the body part of some other organism, which would just as soon keep that part for itself. Organisms evolve defenses against being eaten, and would-be diners evolve weapons to overcome these defenses, prodding the would-be meals to evolve better defenses, and so on, in an evolutionary arms race. These weapons and defenses are genetically based and relatively fixed within the lifetime of the individual; therefore they change slowly. The balance between eater and eaten develops only over evolutionary time.

Humans, Tooby and DeVore suggest, entered the "cognitive niche." Remember the definition of intelligence from Chapter 2: using knowledge of how things work to attain goals in the face of obstacles. By learning which manipulations achieve which goals, humans have mastered the art of the surprise attack. They use novel, goal-oriented courses of action to overcome the Maginot Line defenses of other organisms, which can respond only over evolutionary time. The manipulations can be novel because human knowledge is not just couched in concrete instructions like "how to catch a rabbit." Humans analyze the world using intuitive theories of objects, forces, paths, places, manners, states, substances, hidden biochemical essences, and, for other animals and people, beliefs and desires. (These intuitive theories are the topic of Chapter 5.) People compose new knowledge and plans by mentally playing out combinatorial interactions among these laws in their mind's eye.

Many theorists have wondered what illiterate foragers do with their capacity for abstract intelligence. The foragers would have better grounds for asking the question about modern couch potatoes. Life for foragers (including our ancestors) is a camping trip that never ends, but without the space blankets, Swiss Army knives, and freeze-dried pasta al pesto. Living by their wits, human groups develop sophisticated technologies and bodies of folk science. All human cultures ever documented have words for the elements of space, time, motion, speed, mental states, tools, flora, fauna, and weather, and logical connectives (not, and, same, opposite, part-whole, and general-particular). They combine the words into grammatical sentences and use the underlying propositions to reason about invisible entities like diseases, meteorological forces, and absent animals. Mental maps represent the locations of thousands of note-worthy sites, and mental calendars represent nested cycles of weather, ani-mal migrations, and the life histories of plants. The anthropologist Louis Liebenberg recounts a typical experience with the !Xõ of the central Kalahari Desert:

> While tracking down a solitary wildebeest spoor [tracks] of the previous evening !Xõ trackers pointed out evidence of trampling which indicated that the animal had slept at that spot. They explained consequently that the spoor leaving the sleeping place had been made early that morning and was therefore relatively fresh. The spoor then followed a straight course, indicating that the animal was on its way to a specific destina-tion. After a while, one tracker started to investigate several sets of foot-prints in a particular area. He pointed out that these footprints all belonged to the same animal, but were made during the previous days. He explained that the particular area was the feeding ground of that par-ticular wildebeest. Since it was, by that time, about mid-day, it could be expected that the wildebeest may be resting in the shade in the near vicinity.

All foraging peoples manufacture cutters, pounders, containers, cordage, nets, baskets, levers, and spears and other weapons. They use fire, shelters, and medicinal drugs. Their engineering is often ingenious, exploiting poisons, smokeouts, glue traps, gill nets, baited lines, snares, corrals, weirs, camouflaged pits and clifftops, blowguns, bows and arrows, and kites trailing sticky fishing lines made out of spider silk.

The reward is an ability to crack the safes of many other living things: burrowing animals, plants' underground storage organs, nuts, seeds, bone marrow, tough-skinned animals and plants, birds, fish, shellfish, turtles, poisonous plants (detoxified by peeling, cooking, soaking, par-boiling, fermenting, leaching, and other tricks of the kitchen magician), quick animals (which can be ambushed), and large animals (which coop-erating groups can drive, exhaust, surround, and dispatch with weapons). Ogden Nash wrote:

> The hunter crouches in his blind
> 'Neath camouflage of every kind,
> And conjures up a quacking noise
> To lend allure to his decoys.
> This grown-up man, with pluck and luck
> Is hoping to outwit a duck.

And outwit it he does. Humans have the unfair advantage of attacking in this lifetime organisms that can beef up their defenses only in subsequent ones. Many species cannot evolve defenses rapidly enough, even over evolutionary time, to defend themselves against humans. That is why species drop like flies whenever humans first enter an ecosystem. And it's not just the snail darters and snowy owls recently threatened by dams and loggers. The reason you have never seen a living mastodon, saber-tooth, giant woolly rhinoceros, or other fantastic Ice Age animal is that humans apparently extinguished them thousands of years ago.

The cognitive niche embraces many of the zoologically unusual features of our species. Tool manufacture and use is the application of knowledge about causes and effects among objects in the effort to bring about goals. Language is a means of exchanging knowledge. It multiplies the benefit of knowledge, which can not only be used but exchanged for other resources, and lowers its cost, because knowledge can be acquired from the hard-won wisdom, strokes of genius, and trial and error of others rather than only from risky exploration and experimentation. Information can be shared at a negligible cost: if I give you a fish, I no longer possess the fish, but if I give you information on how to fish, I still possess the information myself. So an information-exploiting lifestyle goes well with living in groups and pooling expertise—that is, with culture. Cultures differ from one another because they pool bodies of expertise fashioned in different times and places. A prolonged childhood is an apprenticeship for knowledge and skills. That shifts the balance of payoffs for males toward investing time and resources in their offspring and away from competing over sexual access to females (see Chapter 7). And that in turn makes kinship a concern of both sexes and all ages. Human lives are long to repay the investment of a long apprenticeship. New habitats can be colonized because even if their local conditions differ, they obey the laws of physics and biology that are already within humans' ken, and can be exploited and outsmarted in their turn.

WHY US?

Why did some miocene ape first enter the cognitive niche? Why not a groundhog, or a catfish, or a tapeworm? It only happened once, so no one knows. But I would guess that our ancestors had four traits that made it especially easy and worth their while to evolve better powers of causal reasoning.

First, primates are visual animals. In monkeys such as the rhesus macaque, half the brain is dedicated to sight. Stereoscopic vision, the use of differences in the vantage points of the two eyes to give a sense of depth, developed early in the primate lineage, allowing early nocturnal primates to move among treacherous fine branches and to grab insects with their hands. Color vision accompanied the switch of the ancestors of monkeys and apes to the day shift and their new taste for fruits, which advertise their ripeness with gaudy hues.

Why would the vision thing make such a difference? Depth perception defines a three-dimensional space filled with movable solid objects. Color makes objects pop out from their backgrounds, and gives us a sensation that corresponds to the stuff an object is made of, distinct from our perception of the shape of the stuff. Together they have pushed the primate brain into splitting the flow of visual information into two streams: a "what" system, for objects and their shapes and compositions, and a "where" system, for their locations and motions. It can't be a coincidence that the human mind grasps the world—even the most abstract, ethereal concepts—as a space filled with movable things and stuff (see Chapters 4 and 5). We say that John *went from* being sick *to* being well, even if he didn't move an inch; he could have been in bed the whole time. Mary can *give* him *many pieces of* advice, even if they merely talked on the phone and nothing changed hands. Even scientists, when they try to grasp abstract mathematical relationships, plot them in graphs that show them as two- and three-dimensional shapes. Our capacity for abstract thought has co-opted the coordinate system and inventory of objects made available by a well-developed visual system.

It is harder to see how a standard mammal could have moved in that direction. Most mammals hug the ground sniffing the rich chemical tracks and trails left behind by other living things. Anyone who has walked a frisky cocker spaniel as it explores the invisible phantasmagoria on a sidewalk knows that it lives in an olfactory world beyond our understanding. Here is an exaggerated way of stating the difference. Rather than living in a three-dimensional coordinate space hung with movable objects, standard mammals live in a two-dimensional flatland which they explore through a zero-dimensional peephole. Edwin Abbott's *Flatland*, a mathematical novel about the denizens of a plane, showed that a two-dimensional world differs from our own in ways other than just lacking one third of the usual dimensions. Many geometric arrangements are simply impossible. A full-faced human figure has no way of getting food into his mouth, and a profiled one would be divided into two pieces by his digestive tract. Simple devices like tubes, knots, and wheels with axles are unbuildable. If most mammals think in a cognitive flatland, they would lack the mental models of movable solid objects in 3-D spatial and mechanical relationships that became so essential to our mental life.

A second possible prerequisite, this one found in the common ancestor of humans, chimpanzees, and gorillas, is group living. Most apes and monkeys are gregarious, though most mammals are not. Living together has advantages. A cluster of animals is not much more detectable to a predator than a single animal, and if it is detected, the likelihood that any individual will be picked off is diluted. (Drivers feel less vulnerable speeding when they are in a group of speeders, because chances are the traffic cop will stop someone else.) There are more eyes, ears, and noses to detect a predator, and the attacker can sometimes be mobbed. A second advantage is in foraging efficiency. The advantage is most obvious in cooperative hunting of large animals, such as in wolves and lions, but it also helps in sharing and defending other ephemeral food resources too big to be consumed by the individual who found it, such as a tree laden with ripe fruit. Primates that depend on fruit, and primates that spend time on the ground (where they are more vulnerable to predators), tend to hang out in groups.

Group living could have set the stage for the evolution of humanlike intelligence in two ways. With a group already in place, the value of having better information is multiplied, because information is the one commodity that can be given away and kept at the same time. Therefore a smarter animal living in a group enjoys a double advantage: the benefit of the knowledge and the benefit of whatever it can get in trade for the knowledge.

The other way in which a group can be a crucible of intelligence is that group living itself poses new cognitive challenges. There are also disadvantages to the madding crowd. Neighbors compete over food, water, mates, and nest sites. And there is the risk of exploitation. Hell is other people, said Jean-Paul Sartre, and if baboons were philosophers no doubt they would say that hell is other baboons. Social animals risk theft, cannibalism, cuckoldry, infanticide, extortion, and other treachery.

Every social creature is poised between milking the benefits and suffering the costs of group living. That creates a pressure to stay on the right side of the ledger by becoming smarter. In many kinds of animals, the largest-brained and smartest-behaving species are social: bees, parrots, dolphins, elephants, wolves, sea lions, and, of course, monkeys, gorillas, and chimpanzees. (The orangutan, smart but almost solitary, is a puzzling exception.) Social animals send and receive signals to coordinate predation, defense, foraging, and collective sexual access. They exchange favors, repay and enforce debts, punish cheaters, and join coalitions.

The collective expression for hominoids, "a shrewdness of apes," tells a story. Primates are sneaky baldfaced liars. They hide from rivals' eyes to flirt, cry wolf to attract or divert attention, even manipulate their lips into a poker face. Chimpanzees monitor one another's goals, at least crudely, and sometimes appear to use them in pedagogy and deception. One chimp, shown a set of boxes with food and one with a snake, led his companions to the snake, and after they fled screaming, feasted in peace. Vervet monkeys are yentas who keep close track of everyone's comings and goings and friends and enemies. But they are so dense about the nonsocial world that they ignore the tracks of a python and the ominous sight of a carcass in a tree, the unique handiwork of a leopard.

Several theorists have proposed that the human brain is the outcome of a cognitive arms race set in motion by the Machiavellian intelligence of our primate forebears. There's only so much brain power you need to subdue a plant or a rock, the argument goes, but the other guy is about as smart as you are and may use that intelligence against your interests. You had better think about what he is thinking about what you are thinking he is thinking. As far as brain power goes, there's no end to keeping up with the Joneses.

My own guess is that a cognitive arms race by itself was not enough to launch human intelligence. Any social species can begin a never-ending escalation of brain power, but none except ours has, probably because without some other change in lifestyle, the costs of intelligence (brain size, extended childhood, and so on) would damp the positive feedback loop. Humans are exceptional in mechanical and biological, not just social, intelligence. In a species that runs on information, each faculty multiplies the value of the others. (Incidentally, the expansion of the human brain is no evolutionary freak crying out for a runaway positive feedback loop. The brain tripled in size in five million years, but that is leisurely by evolutionary timekeeping. There was enough time in hominid evolution for the brain to shoot up to human size, shrink back down, and shoot up again several times over.)

A third pilot of intelligence, alongside good vision and big groups, is the hand. Primates evolved in trees and have hands to grasp the branches. Monkeys use all four limbs to run along the tops of branches, but apes hang from them, mainly by their arms. They have put their well-developed hands to use in manipulating objects. Gorillas meticulously dissect tough or thorny plants to pick out the edible matter, and chimpanzees use simple tools such as stems to fish out termites, rocks to bash open nuts, and mashed leaves to sponge up water. As Samuel Johnson said about dogs walking on their hind legs, while it is not done well, you are surprised to find it done at all. Hands are levers of influence on the world that make intelligence worth having. Precision hands and precision intelligence co-evolved in the human lineage, and the fossil record shows that hands led the way.

Finely tooled hands are useless if you have to walk on them all the time, and they could not have evolved by themselves. Every bone in our bodies has been reshaped to give us our upright posture, which frees the hands for carrying and manipulating. Once again we have our ape ancestors to thank. Hanging from trees calls for a body plan that is different from the horizontal four-wheel-drive design of most mammals. Apes' bodies are already tilted upward with arms that differ from their legs, and chimpanzees (and even monkeys) walk upright for short distances to carry food and objects.

Fully upright posture may have evolved under several selection pressures. Bipedal walking is a biomechanically efficient way to retool a tree-hanging body to cover distance on the flat ground of the newly entered savanna. Upright posture also allows one to peer over grass like a marmot. Hominids go out in the midday sun; this zoologically unusual work shift brought in several human adaptations for keeping cool, such as hairlessness and profuse sweating. Upright posture might be another; it is the opposite of lying down to get a tan. But carrying and manipulation must have been crucial inducements. With the hands free, tools could be assembled out of materials from different locations and brought to where they were most useful, and food and children could be carried to safe or productive areas.

A final usher of intelligence was hunting. Hunting, tool use, and bipedalism were for Darwin the special trinity that powered human evolution. "Man the Hunter" was the major archetype in both serious and pop accounts through the 1960s. But the macho image that resonated with the decade of John Glenn and James Bond lost its appeal in the feminist-influenced small planet of the 1970s. A major problem for Man the Hunter was that it credited the growth of intelligence to the teamwork and foresight needed by men in groups to fell large game. But natural selection sums over the lives of both sexes. Women did not wait in the kitchen to cook the mastodon that Dad brought home, nor did they forgo the expansion of intelligence enjoyed by evolving men. The ecology of modern foraging peoples suggests that Woman the Gatherer provided a substantial portion of the calories in the form of highly processed plant foods, and that requires mechanical and biological acumen. And, of course, in a group-living species, social intelligence is as important a weapon as spears and clubs.

But Tooby and DeVore have argued that hunting was nonetheless a major force in human evolution. The key is to ask not what the mind can do for hunting, but what hunting can do for the mind. Hunting provides sporadic packages of concentrated nutrients. We did not always have tofu, and the best natural material for building animal flesh is animal flesh. Though plant foods supply calories and other nutrients, meat is a complete protein containing all twenty amino acids, and provides energy-rich fat and indispensable fatty acids. Across the mammals, carnivores have larger brains for their body size than herbivores, partly because of the greater skill it takes to subdue a rabbit than to subdue grass, and partly because meat can better feed ravenous brain tissue. Even in the most conservative estimates, meat makes up a far greater proportion of foraging humans' diet than of any other primate's. That may have been one of the reasons we could afford our expensive brains.

Chimpanzees collectively hunt small animals like monkeys and bush pigs, so our common ancestor probably hunted as well. The move to the savanna must have made hunting more appealing. Notwithstanding the teeming wildlife in the Save-the-Rainforest posters, real forests have few large animals. Only so much solar energy falls on a patch of ground, and if the biomass it supports is locked up in wood it is not available to make animals. But grass is like the legendary self-replenishing goblet, growing back as soon as it is grazed. Grasslands can feed vast herds of herbivores, who in turn feed carnivores. Evidence of butchery appears in the fossil record almost two million years ago, the time of *Homo habilis*. Hunting must be even older, since we know that chimpanzees do it, and their activities would not leave evidence in the fossil record. Once our ancestors increased their hunting, the world opened up. Plant foods are scarce during the winter at higher altitudes and latitudes, but hunters can survive there. There are no vegetarian Eskimos.

Our ancestors have sometimes been characterized as meek scavengers rather than brave hunters, in keeping with today's machismo-puncturing ethos. But while hominids may occasionally have scavenged, they probably could not have made a living from it, and if they did, they were no wimps. Vultures get away with scavenging because they can scan large territories for carcasses and flee on short notice when more formidable competitors show up. Otherwise, scavenging is not for the faint of heart. A carcass is jealously guarded by its hunter or an animal fierce enough to have stolen it. It is attractive to microorganisms, who quickly poison the meat to repel other would-be scavengers. So when modern primates or hunter-gatherers come across a carcass, they usually leave it alone. In a poster widely available in head shops in the early 1970s, one vulture says to the other, "Patience, my ass! I'm going to *kill* something." The poster got it right, except for the vulture part: mammals that do scavenge, such as hyenas, also hunt.

Meat is also a major currency of our social life. Imagine a cow who tries to win the favors of a neighbor by dropping a clump of grass at its feet. One could forgive the second cow for thinking, "Thanks, but I can get my own grass." The nutritional jackpot of a felled animal is another matter. Miss Piggy once advised, "Never eat anything bigger than you can lift." A hunter with a dead animal larger than he can eat and about to become a putrefying mass is faced with a unique opportunity. Hunting is largely a matter of luck. In the absence of refrigeration, a good place to store meat for leaner times is in the bodies of other hunters who will return the favor when fortunes reverse. This eases the way for the male coalitions and the extensive reciprocity that are ubiquitous in foraging societies.

And there are other markets for a hunter's surplus. Having concentrated food to offer one's offspring changes the relative payoffs for males between investing in their young and competing with other males for access to females. The robin bringing a worm to the nestlings reminds us that most animals that provision their young do so with prey, the only food that repays the effort to obtain it and transport it.

Meat also figures into sexual politics. In all foraging societies, presumably including our ancestors', hunting is overwhelmingly a male activity. Women are encumbered with children, which makes hunting inconvenient, and men are bigger and more adept at killing because of their evolutionary history of killing each other. As a result, males can invest surplus meat in their children by provisioning the children's pregnant or nursing mothers. They also can trade meat with females for plant foods or for sex. Brazen bartering of the carnal for the carnal has been observed in baboons and chimpanzees and is common in foraging peoples. Though people in modern societies are ever-so-more discreet, an exchange of resources for sexual access is still an important part of the interactions between men and women all over the world. (Chapter 7 explores these dynamics and how they originated in differences in reproductive anatomy, though of course anatomy is not destiny in modern ways of life.) In any case, we have not lost the association completely. *Miss Manners' Guide to Excruciatingly Correct Behavior* advises:

> There are three possible parts to a date, of which at least two must be offered: entertainment, food, and affection. It is customary to begin a series of dates with a great deal of entertainment, a moderate amount of food, and the merest suggestion of affection. As the amount of affection increases, the entertainment can be reduced proportionately. When the affection *is* the entertainment, we no longer call it dating. Under no circumstances can the food be omitted.

Of course no one really knows whether these four habits formed the base camp for the ascent of human intelligence. And no one knows whether there are other, untried gradients to intelligence in biological design space. But if these traits do explain why our ancestors were the only species out of fifty million to follow that route, it would have sobering implications for the search for extraterrestrial intelligence. A planet with life may not be enough of a launching pad. Its history might have to include a nocturnal predator (to get stereo vision), with descendants that switched to a diurnal lifestyle (for color) in which they depended on fruit and were vulnerable to predators (for group living), which then changed their means of locomotion to swinging beneath branches (for hands and for precursors to upright posture), before a climate shift sent them from the forest into grasslands (for upright posture and hunting). What is the probability that a given planet, even a planet with life, has such a history?

Species	Date	Height	Physique	Brain
Chimp-hominid ancestor (if similar to modern chimps)	8–6 million years ago	1–1.7 meters	long arms, short thumbs, curved fingers and toes; adapted for knuckle-walking and tree-climbing	450 cc
Ardipithecus ramidus	4.4 million years ago	?	probably bipedal	?
Australopithecus anamensis	4.2–3.9 million years ago	?	bipedal	?
Australopithecus afarensis (Lucy)	4–2.5 million years ago	1–1.2 meters	fully bipedal with modified hands but ape-like features: thorax, long arms, curved fingers and toes	400–500 cc
Homo habilis (Handyman)	2.3–1.6 million years ago	1–1.5 meters	some specimens: small with long arms; others: robust but human	500–800 cc
Homo erectus	1.9 million–300,000 (maybe 27,000) years ago	1.3–1.5 meters	robust but human	750–1250 cc
Archaic Homo sapiens	400,000–100,000 years ago	?	robust but modern	1100–1400 cc
Early Homo sapiens	130,000–60,000 years ago	1.6–1.85 meters	robust but modern	1200–1700 cc
Homo sapiens (Cro-Magnon)	45,000–12,000 years ago	1.6–1.8 meters	modern	1300–1600 (cf. today: 1000–2000, average 1350)

THE MODERN STONE AGE FAMILY

The dry bones of the fossil record tell of a gradual entry into the cognitive niche. A summary of the current evidence on the species thought to be our direct ancestors is shown in the table below.

Skull	Teeth	Tools	Distribution
very low forehead; projecting face; huge brow ridges	large canines	stone hammers, leaf sponges, stem probes, branch levers	West Africa
?	chimplike molars but not canines	?	East Africa
apelike fragments	chimplike size and placement; humanlike enamel	?	East Africa
low flat forehead; projecting face; big brow ridges	large canines and molars	none? flakes?	East Africa (maybe also west)
smaller face; rounder skull	smaller molars	flakes, choppers, scrapers	East and South Africa
thick; large brow ridges (Asia); smaller, protruding face	smaller teeth	symmetrical hand axes	Africa (may be separate species), Asia, Europe
higher skull; smaller, protruding face; large brow ridges	smaller teeth	better hand axes, retouched flakes	Africa, Asia, Europe
high skull; medium brow ridges; slightly protruding face; chin	smaller teeth	retouched flakes; flake-blades; points	Africa, Western Asia
modern	modern	blades; drills; spear throwers; needles; engravers; bone	worldwide

Millions of years before our brains billowed out, some descendants of the common ancestor of chimpanzees and humans walked upright. In the 1920s that discovery came as a shock to human chauvinists who imagined that our glorious brains led us up the ladder, perhaps as our ancestors decided at each rung what use to make of their newfound smarts. But natural selection could not have worked that way. Why bulk up your brain if you can't put it to use? The history of paleoanthropology is the discovery of earlier and earlier birthdays for upright posture. The most recent discoveries put it at four or even four and a half million years ago. With hands freed, subsequent species ratchet upward, click by click, in the features that distinguish us: the dexterity of hands, the sophistication of tools, the dependence on hunting, the size of brains, the range of habitats. The teeth and jaw become smaller. The face that opposes it becomes less muzzle-like. The brow ridges that anchor the muscles that close the jaw shrink and disappear. Our delicate faces differ from the brutes' because tools and technology have taken over from teeth. We slaughter and skin animals with blades, and soften plants and meats with fire. That eases the mechanical demands on the jaw and skull, allowing us to shave bone from our already heavy heads. The sexes come to differ less in size, suggesting that males spent less of their resources beating each other up and perhaps more on their children and the children's mothers.

The stepwise growth of the brain, propelled by hands and feet and manifested in tools, butchered bones, and increased range, is good evidence, if evidence were needed, that intelligence is a product of natural selection for exploitation of the cognitive niche. The package was not an inexorable unfolding of hominid potential. Other species, omitted from the table, spun off in every epoch to occupy slightly different niches: nutcracking and root-gnawing australopithecines, perhaps one of the two habiline subtypes, quite possibly the Asian branches of *erectus* and archaic *sapiens*, and probably the Ice Age-adapted Neanderthals. Each species might have been outcompeted when a neighboring, more *sapiens*-like population had entered far enough into the cognitive niche to duplicate the species' more specialized feats and do much else besides. The package was also not the gift of a macromutation or random drift—for how could such luck have held up in one lineage for millions of years, over hundreds of thousands of generations, in species after bigger-brained species? Moreover, the bigger brains were no mere ornaments but allowed their owners to make finer tools and infest more of the planet.

∿

According to the standard timetable in paleoanthropology, the human brain evolved to its modern form in a window that began with the appearance of *Homo habilis* two million years ago and ended with the appearance of "anatomically modern humans," *Homo sapiens sapiens*, between 200,000 and 100,000 years ago. I suspect that our ancestors were penetrating the cognitive niche for far longer than that. Both ends of the R&D process might have to be stretched beyond the textbook dates, providing even more time for our fantastic mental adaptations to have evolved.

At one end of the timetable is the four-million-year-old australopithecines-like *afarensis* (the species of the charismatic fossil called Lucy). They are often described as chimpanzees with upright posture because their brain size was in the chimpanzee ballpark and they left no clear evidence of tool use. That implies that cognitive evolution did not begin till two million years later, when larger-brained habilines earned their "handyman" name by chipping choppers.

But that can't be right. First, it is ecologically improbable that a treedweller could have moved onto open ground and retooled its anatomy for upright walking without repercussions on every other aspect of its lifestyle and behavior. Modern chimps use tools and transport objects, and would have had much more incentive and success if they could carry them around freely. Second, though australopithecines' hands retain some apelike curvature of the fingers (and may have been used at times to run up trees for safety), the hands visibly evolved for manipulation. Compared to chimps' hands, their thumbs are longer and more opposable to the other fingers, and their index and middle fingers are angled to allow cupping the palm to grasp a hammerstone or a ball. Third, it's not so clear that they had a chimp-sized brain, or that they lacked tools. The paleoanthropologist Yves Coppens argues that their brains are thirty to forty percent bigger than expected for a chimpanzee of their body size, and that they left behind modified quartz flakes and other tools. Fourth, skeletons of the tool-using habilines (handymen) have now been found, and they do not look so different from the australopithecines'.

Most important, hominids did not arrange their lives around the convenience of anthropologists. We are lucky that a rock can be carved into a cutter and that it lasts for millions of years, so some of our ancestors inadvertently left us time capsules. But it's much harder to carve a rock into a basket, a baby sling, a boomerang, or a bow and arrow. Contemporary hunter-gatherers use many self-composting implements for every lasting one, and that must have been true of hominids at every stage. The archeological record is bound to underestimate tool use.

So the standard timetable for human brain evolution begins the story too late; I think it also ends the story too early. Modern humans (us) are said to have first arisen between 200,000 and 100,000 years ago in Africa. One kind of evidence is that the mitochondrial DNA (mDNA) of everyone on the planet (which is inherited only from one's mother) can be traced back to an African woman living sometime in that period. (The claim is controversial, but the evidence is growing.) Another is that anatomically modern fossils first appear in Africa more than 100,000 years ago and in the Middle East shortly afterward, around 90,000 years ago. The assumption is that human biological evolution had pretty much stopped then. This leaves an anomaly in the timeline. The anatomically modern early humans had the same toolkit and lifestyle as their doomed Neanderthal neighbors. The most dramatic change in the archeological record, the Upper Paleolithic transition—also called the Great Leap Forward and the Human Revolution—had to wait another 50,000 years. Therefore, it is said, the human revolution must have been a cultural change.

Calling it a revolution is no exaggeration. All other hominids come out of the comic strip *B.C.*, but the Upper Paleolithic people were the Flintstones. More than 45,000 years ago they somehow crossed sixty miles of open ocean to reach Australia, where they left behind hearths, cave paintings, the world's first polished tools, and today's aborigines. Europe (home of the Cro-Magnons) and the Middle East also saw unprecedented arts and technologies, which used new materials like antler, ivory, and bone as well as stone, sometimes transported hundreds of miles. The toolkit included fine blades, needles, awls, many kinds of axes and scrapers, spear points, spear throwers, bows and arrows, fishhooks, engravers, flutes, maybe even calendars. They built shelters, and they slaughtered large animals by the thousands. They decorated everything in sight—tools, cave walls, their bodies—and carved knick-knacks in the shapes of animals and naked women, which archeologists euphemistically call "fertility symbols." They were us.

Ways of life certainly can shoot off without any biological change, as in the more recent agricultural, industrial, and information revolutions. That is especially true when populations grow to a point where the insights of thousands of inventors can be pooled. But the first human revolution was not a cascade of changes set off by a few key inventions. Ingenuity itself was the invention, manifested in hundreds of innovations tens of thousands of miles and years apart. I find it hard to believe that the people of 100,000 years ago had the same minds as those of the Upper Paleolithic revolutionaries to come—indeed, the same minds as ours—and sat around for 50,000 years without it dawning on a single one of them that you could carve a tool out of bone, or without a single one feeling the urge to make anything look pretty.

And there is no need to believe it—the 50,000-year gap is an illusion. First, the so-called anatomically modern humans of 100,000 years ago may have been more modern than their Neanderthal contemporaries, but no one would mistake them for contemporary humans. They had brow ridges, protruding faces, and heavily built skeletons outside the contemporary range. Their bodies had to evolve to become us, and their brains surely did as well. The myth that they are completely modern grew out of the habit of treating species labels as if they were real entities. When applied to evolving organisms, they are no more than a convenience. No one wants to invent a new species every time a tooth is found, so intermediate forms tend to get shoehorned into the nearest available category. The reality is that hominids must always have come in dozens or hundreds of variants, scattered across a large network of occasionally interacting subpopulations. The tiny fraction of individuals immortalized as fossils at any point were not necessarily our direct ancestors. The "anatomically modern" fossils are closer to us than to anyone else, but either they had more evolving to do or they were away from the hotbed of change.

Second, the revolution probably began well before the commonly cited watershed of 40,000 years ago. That's when fancy artifacts begin to appear in European caves, but Europe has always attracted more attention than it deserves, because it has lots of caves and lots of archeologists. France alone has three hundred well-excavated paleolithic sites, including one whose cave paintings were scrubbed off by an overenthusiastic boy scout troop that mistook them for graffiti. The entire continent of Africa has only two dozen. But one, in Zaire, contains finely crafted bone implements including daggers, shafts, and barbed points, together with grindstones brought from miles away and the remains of thousands of catfish, presumably the victims of these instruments. The collection looks postrevolutionary but is dated at 75,000 years ago. One commentator said it was like finding a Pontiac in Leonardo da Vinci's attic. But as archeologists are starting to explore this continental attic and date its contents, they are finding more and more Pontiacs: fine stone blades, decorated tools, useless but colorful minerals transported hundreds of miles.

Third, the mitochondrial Eve of 200,000 to 100,000 years ago was not a party to any evolutionary event. Contrary to some fantastic misunderstandings, she did not undergo some mutation that left her descendants smarter or more talkative or less brutish. Nor did she mark the end of human evolution. She is merely a mathematical necessity: the most recent common ancestor of all living people along the female-female-female line of great-great-. . .-great-grandmothers. For all the definition says, Eve could have been a fish.

Eve, of course, turned out to be not a fish but an African hominid. Why would anyone assume that she was a special hominid, or even that she lived in special times? One reason is that she made many other times and places non-special. If twentieth-century Europeans' and Asians' mDNA is a variant of 200,000-year-old African mDNA, they must be descendants of an African population at the time. Eve's contemporary Europeans and Asians left no mDNA in today's Europeans and Asians, and thus presumably were not their ancestors (at least—and this is a big proviso—not their all-maternal-line ancestors).

But that says nothing about evolution's having stopped with Eve. We can assume that most evolution was done with by the time the ancestors of the modern races separated and stopped exchanging genes, since today we are birds of a feather. But that did not happen as soon as Eve breathed her last. The diaspora of the races, and the end of significant human evolution, must have occurred much later. Eve is not our most recent common ancestor, only our most recent common ancestor in the all-maternal line. The most recent common ancestor along a mixed-sex line of descendants lived much later. You and a first cousin share an ancestor of just two generations ago, your common grandmother or grandfather. But in looking for a shared all-female-line ancestor (your mother's mother's mother, and so on), then except for one kind of cousin (the child of your mother's sister), there's almost no limit to how far back you might have to go. So if someone were to guess the degree of relatedness between you and your cousin based on your most recent ancestor, he would say you were closely related. But if he could check only the most recent all-female-line ancestor, he might guess that you are not related at all! Similarly, the birthday of humanity's most recent common all-female-line ancestor, mitochondrial Eve, overestimates how long ago all of humanity was still interbreeding.

Well after Eve's day, some geneticists think, our ancestors passed through a population bottleneck. According to their scenario, which is based on the remarkable sameness of genes across modern human populations, around 65,000 years ago our ancestors dwindled to a mere ten thousand people, perhaps because of a global cooling triggered by a volcano in Sumatra. The human race was as endangered as mountain gorillas are today. The population then exploded in Africa and spun off small bands that moved to other corners of the world, possibly mating now and again with other early humans in their path. Many geneticists believe that evolution is especially rapid when scattered populations exchange occasional migrants. Natural selection can quickly adapt each group to local conditions, so one or more can cope with any new challenge that arises, and their handy genes will then be imported by the neighbors. Perhaps this period saw a final flowering in the evolution of the human mind.[0]

All reconstructions of our evolutionary history are controversial, and the conventional wisdom changes monthly. But I predict that the closing date of our biological evolution will creep later, and the opening date of the archeological revolution will creep earlier, until they coincide. Our minds and our way of life evolved together.

WHAT NOW?

Are we still evolving? Biologically, probably not much. Evolution has no momentum, so we will not turn into the creepy bloat-heads of science fiction. The modern human condition is not conducive to real evolution either. We infest the whole habitable and not-so-habitable earth, migrate at will, and zigzag from lifestyle to lifestyle. This makes us a nebulous, moving target for natural selection. If the species is evolving at all, it is happening too slowly and unpredictably for us to know the direction.

But Victorian hopes spring eternal. If genuine natural selection cannot improve us, maybe a human-made substitute can. The social sciences are filled with claims that new kinds of adaptation and selection have extended the biological kind. But the claims, I think, are misleading.

The first claim is that the world contains a wonderful process called "adaptation" that causes organisms to solve problems. Now, in Darwin's strict sense, adaptation in the present is caused by selection in the past. Remember how natural selection gives an illusion of teleology: selection may *look like* it is adapting each organism to its needs in the present, but really it is just favoring the descendants of organisms that were adapted to their own needs in the past. The genes that built the most adaptive bodies and minds among our ancestors got passed down to build the innate bodies and minds of today (including innate abilities to track certain kinds of environmental variation, as in tanning, callusing, and learning).

But for some, that does not go far enough; adaptation happens daily. "Darwinian social scientists" such as Paul Turke and Laura Betzig believe that "modern Darwinian theory predicts that human behavior will be adaptive, that is, designed to promote maximum reproductive success . . . through available descendent and nondescendent relatives." "Functionalists" such as the psychologists Elizabeth Bates and Brian MacWhinney say that they "view the selectional processes operating during evolution and the selectional processes operating during [learning] as part of one seamless natural fabric." The implication is that there is no need for specialized mental machinery: if adaptation simply *makes* organisms do the right thing, who could ask for anything more? The optimal solution to a problem—eating with one's hands, finding the right mate, inventing tools, using grammatical language—is simply inevitable.

The problem with functionalism is that it is Lamarckian. Not in the sense of Lamarck's second principle, the inheritance of acquired characteristics—the giraffes who stretched their necks and begat baby giraffes with necks pre-stretched. Everyone knows to stay away from that. (Well, almost everyone: Freud and Piaget stuck to it long after it was abandoned by biologists.) It is Lamarckian in the sense of his first principle, "felt need"—the giraffes growing their necks when they hungrily eyed the leaves just out of reach. As Lamarck put it, "New needs which establish a necessity for some part really bring about the existence of that part as a result of efforts." If only it were so! As the saying goes, if wishes were horses, beggars would ride. There are no guardian angels seeing to it that every need is met. They are met only when mutations appear that are capable of building an organ that meets the need, when the organism finds itself in an environment in which meeting the need translates into more surviving babies, and in which that selection pressure persists over thousands of generations. Otherwise, the need goes unmet. Swimmers do not grow webbed fingers; Eskimos do not grow fur. I have studied three-dimensional mirror-images for twenty years, and though I know mathematically that you can convert a left shoe into a right shoe by turning it around in the fourth dimension, I have been unable to grow a 4-D mental space in which to visualize the flip.

Felt need is an alluring idea. Needs really do feel like they bring forth their own solutions. You're hungry, you have hands, the food's in front of you, you eat with your hands; how else could it be? Ah, but you're the last one we should ask. Your brain was fashioned by natural selection so that it *would* find such problems obvious. Change the mind (to a robot's, or to another animal's, or to a neurological patient's), or change the problem, and it's no longer so obvious what's obvious. Rats can't learn to drop a piece of food for a larger reward. When chimpanzees try to imitate someone raking in an inaccessible snack, they don't notice that the rake has to be held business-end down, even if the role model makes a conspicuous show of aligning it properly. Lest you feel smug, the chapters to come will show how the design of our own minds gives rise to paradoxes, brain-teasers, myopias, illusions, irrationalities, and self-defeating strategies that prevent, rather than guarantee, the meeting of our everyday needs.

But what about the Darwinian imperative to survive and reproduce? As far as day-to-day behavior is concerned, there is no such imperative. People watch pornography when they could be seeking a mate, forgo food to buy heroin, sell their blood to buy movie tickets (in India), postpone childbearing to climb the corporate ladder, and eat themselves into an early grave. Human vice is proof that biological adaptation is, speaking literally, a thing of the past. Our minds are adapted to the small foraging bands in which our family spent ninety-nine percent of its existence, not to the topsy-turvy contingencies we have created since the agricultural and industrial revolutions. Before there was photography, it was adaptive to receive visual images of attractive members of the opposite sex, because those images arose only from light reflecting off fertile bodies. Before opiates came in syringes, they were synthesized in the brain as natural analgesics. Before there were movies, it was adaptive to witness people's emotional struggles, because the only struggles you could witness were among people you had to psych out every day. Before there was contraception, children were unpostponable, and status and wealth could be converted into more children and healthier ones. Before there was a sugar bowl, salt shaker, and butter dish on every table, and when lean years were never far away, one could never get too much sweet, salty, and fatty food. People do not divine what is adaptive for them or their genes; their genes give them thoughts and feelings that were adaptive in the environment in which the genes were selected.

⁓

The other extension of adaptation is the seemingly innocuous cliché that "cultural evolution has taken over from biological evolution." For millions of years, genes were transmitted from body to body and were selected to confer adaptations on organisms. But after humans emerged, units of culture were transmitted from mind to mind and were selected to confer adaptations on cultures. The torch of progress has been passed to a swifter runner. In *2001: A Space Odyssey*, a hairy arm hurls a bone into the air, and it fades into a space station.

The premise of cultural evolution is that there is a single phenomenon—the march of progress, the ascent of man, apes to Armageddon—that Darwin explained only up to a point. My own view is that human brains evolved by one set of laws, those of natural selection and genetics, and now interact with one another according to other sets of laws, those of cognitive and social psychology, human ecology, and history. The reshaping of the skull and the rise and fall of empires may have little in common.

Richard Dawkins has drawn the clearest analogy between the selection of genes and the selection of bits of culture, which he dubbed memes. Memes such as tunes, ideas, and stories spread from brain to brain and sometimes mutate in the transmission. New features of a meme that make its recipients more likely to retain and disseminate it, such as being catchy, seductive, funny, or irrefutable, will lead to the meme's becoming more common in the meme pool. In subsequent rounds of retelling, the most spreadworthy memes will spread the most and will eventually take over the population. Ideas will therefore evolve to become better adapted to spreading themselves. Note that we are talking about *ideas* evolving to become more spreadable, not *people* evolving to become more knowledgeable.

Dawkins himself used the analogy to illustrate how natural selection pertains to anything that can replicate, not just DNA. Others treat it as a genuine theory of cultural evolution. Taken literally, it predicts that cultural evolution works like this. A meme impels its bearer to broadcast it, and it mutates in some recipient: a sound, a word, or a phrase is randomly altered. Perhaps, as in *Monty Python's Life of Brian*, the audience of the Sermon on the Mount mishears "Blessed are the peacemakers" as "Blessed are the cheesemakers." The new version is more memorable and comes to predominate in the majority of minds. It too is mangled by typos and speakos and hearos, and the most spreadable ones accumulate, gradually transforming the sequence of sounds. Eventually they spell out, "That's one small step for a man, one giant leap for mankind."

I think you'll agree that this is not how cultural change works. A complex meme does not arise from the retention of copying errors. It arises because some person knuckles down, racks his brain, musters his ingenuity, and composes or writes or paints or invents something. Granted, the fabricator is influenced by ideas in the air, and may polish draft after draft, but neither of these progressions is like natural selection. Just compare the input and the output—draft five and draft six, or an artist's inspiration and her oeuvre. They do not differ by a few random substitutions. The value added with each iteration comes from focusing brainpower on improving the product, not from retelling or recopying it hundreds of thousands of times in the hope that some of the malaprops or typos will be useful.

Stop being so literal-minded! respond the fans of cultural evolution. Of course cultural evolution is not an exact replica of the Darwinian version. In cultural evolution, the mutations are directed and the acquired characteristics are inherited. Lamarck, while being wrong about biological evolution, turned out to be right about cultural evolution.

But this won't do. Lamarck, recall, was not just unlucky in his guess about life on this planet. As far as explaining complex design goes, his theory was, and is, a non-starter. It is mute about the beneficent force in the universe or all-knowing voice in the organism that bestows the useful mutations. And it's that force or voice that's doing all the creative work. To say that cultural evolution is Lamarckian is to confess that one has no idea how it works. The striking features of cultural products, namely their ingenuity, beauty, and truth (analogous to organisms' complex adaptive design), come from the mental computations that "direct"—that is, invent—the "mutations," and that "acquire"—that is, understand—the "characteristics."

Models of cultural transmission do offer insight on other features of cultural change, particularly their demographics—how memes can become popular or unpopular. But the analogy is more from epidemiology than from evolution: ideas as contagious diseases that cause epidemics, rather than as advantageous genes that cause adaptations. They explain how ideas become popular, but not where ideas come from.

Many people unfamiliar with cognitive science see cultural evolution as the only hope for grounding wispy notions like ideas and culture in rigorous evolutionary biology. To bring culture into biology, they reason, one shows how it evolved by its own version of natural selection. But that is a non sequitur; the products of evolution don't have to look like evolution. The stomach is firmly grounded in biology, but it does not randomly secrete variants of acids and enzymes, retain the ones that break down food a bit, let them sexually recombine and reproduce, and so on for hundreds of thousands of meals. Natural selection already went through such trial and error in designing the stomach, and now the stomach is an efficient chemical processor, releasing the right acids and enzymes on cue. Likewise, a group of minds does not have to recapitulate the process of natural selection to come up with a good idea. Natural selection designed the mind to be an information processor, and now it perceives, imagines, simulates, and plans. When ideas are passed around, they aren't merely copied with occasional typographical errors; they are evaluated, discussed, improved on, or rejected. Indeed, a mind that passively accepted ambient memes would be a sitting duck for exploitation by others and would have quickly been selected against.

The geneticist Theodosius Dobzhansky famously wrote that nothing in biology makes sense except in the light of evolution. We can add that nothing in culture makes sense except in the light of psychology. Evolution created psychology, and that is how it explains culture. The most important relic of early humans is the modern mind.

4

THE MIND'S EYE

To gaze is to think.
—Salvador Dali

Past decades had hula hoops, black-light posters, CB radios, and Rubik's cube. The craze of the 1990s is the autostereogram, also called Magic Eye, Deep Vision, and Superstereogram. These are the computer-generated squiggles that when viewed with crossed eyes or a distant gaze spring into a vivid illusion of three-dimensional, razor-edged objects majestically suspended in space. The fad is now five years old and autostereograms are everywhere, from postcards to Web pages. They have been featured in editorial cartoons, in the *Blondie* comic strip, and in situation comedies like *Seinfeld* and *Ellen*. In one episode, the comedian Ellen DeGeneres belongs to a reading club that has chosen a stereogram book as its weekly selection. Ashamed that she cannot see the illusions, she sets aside an evening to train herself, without success. In desperation she joins a support group for people who cannot "get" stereograms.

Visual illusions fascinated people long before the psychologist Christopher Tyler inadvertently created this sensation in his research on binocular (two-eyed) vision. Simpler illusions made up of parallel lines that seem to converge and congruent lines that look unequal have long appeared in cereal-box reading material, Crackerjack prizes, children's museums, and psychology courses. Their fascination is obvious. "Who are you going to believe, me or your own eyes?" says Groucho Marx to Margaret Dumont, playing on our faith that vision is a certain route to knowledge. As the sayings go: I call them as I see them; Seeing is believing; We have an eyewitness; I saw it with my own eyes. But if a devilish display can make us see things that aren't there, how can we trust our eyes at other times?

Illusions are no mere curiosities; they set the intellectual agenda for centuries of Western thought. Skeptical philosophy, as old as philosophy itself, impugns our ability to know *anything* by rubbing our faces in illusions: the oar in the water that appears bent, the round tower that from a distance looks flat, the cold finger that perceives tepid water as hot while the hot finger perceives it as cold. Many of the great ideas of the Enlightenment were escape hatches from the depressing conclusions skeptical philosophers drew from illusions. We can know by faith, we can know by science, we can know by reason, we can know that we think and therefore that we are.

Perception scientists take a lighter view. Vision may not work all the time, but we should marvel that it works at all. Most of the time we don't bump into walls, bite into plastic fruit, or fail to recognize our mothers. The robot challenge shows that this is no mean feat. The medieval philosophers were wrong when they thought that objects conveniently spray tiny copies of themselves in all directions and the eye captures a few and grasps their shape directly. We can imagine a science-fiction creature that embraces an object with calipers, prods it with probes and dipsticks, makes rubber molds, drills core samples, and snips off bits for biopsies. But real organisms don't have these luxuries. When they apprehend the world by sight, they have to use the splash of light reflected off its objects, projected as a two-dimensional kaleidoscope of throbbing, heaving streaks on each retina. The brain somehow analyzes the moving collages and arrives at an impressively accurate sense of the objects out there that gave rise to them.

The accuracy is impressive because the problems the brain is solving are literally unsolvable. Recall from Chapter 1 that inverse optics, the deduction of an object's shape and substance from its projection, is an "ill-posed problem," a problem that, as stated, has no unique solution. An elliptical shape on the retina could have come from an oval viewed head-on or a circle viewed at a slant. A patch of gray could have come from a snowball in the shade or a lump of coal in the sun. Vision has evolved to convert these ill-posed problems into solvable ones by adding premises: assumptions about how the world we evolved in is, on average, put together. For example, I will explain how the human visual system "assumes" that matter is cohesive, surfaces are uniformly colored, and objects don't go out of their way to line up in confusing arrangements.

When the current world resembles the average ancestral environment, we see the world as it is. When we land in an exotic world where the assumptions are violated—because of a chain of unlucky coincidences or because a sneaky psychologist concocted the world to violate the assumptions—we fall prey to an illusion. That is why psychologists are obsessed with illusions. They unmask the assumptions that natural selection installed to allow us to solve unsolvable problems and know, much of the time, what is out there.

Perception is the only branch of psychology that has been consistently adaptation-minded, seeing its task as reverse-engineering. The visual system is not there to entertain us with pretty patterns and colors; it is contrived to deliver a sense of the true forms and materials in the world. The selective advantage is obvious: animals that know where the food, the predators, and the cliffs are can put the food in their stomachs, keep themselves out of the stomachs of others, and stay on the right side of the clifftop.

The grandest vision of vision has come from the late artificial intelligence researcher David Marr. Marr was the first to describe vision as solving ill-posed problems by adding assumptions about the world, and was a forceful defender of the computational theory of mind. He also offered the clearest statement of what vision is *for*. Vision, he said, "is a process that produces from images of the external world a description that is useful to the viewer and not cluttered with irrelevant information."

It may seem strange to read that the goal of vision is a "description." After all, we don't walk around muttering a play-by-play narration of everything we see. But Marr was referring not to a publicly spoken description in English but to an internal, abstract one in mentalese. What does it mean to see the world? We *can* describe it in words, of course, but we can also negotiate it, manipulate it physically and mentally, or file it away in memory for future reference. *All* these feats depend on construing the world as real things and stuff, not as the psychedelia of the retinal image. We call a book "rectangular," not "trapezoidal," though it projects a trapezoid on the retina. We mold our fingers into a rectangular (not trapezoidal) posture as we reach for it. We build rectangular (not trapezoidal) shelves to hold it, and we deduce that it can support a broken couch by fitting into the rectangular space beneath it. Somewhere in the mind there must be a mental symbol for "rectangle," delivered by vision but available at once to the rest of the verbal and non-verbal mind. That mental symbol, and the mental propositions that capture the spatial relations among objects ("book lying face down on shelf near door"), are examples of the "description" that Marr charged vision with computing.

If vision did not deliver a description, every mental faculty—language, walking, grasping, planning, imagining—would need its *own* procedure for deducing that the trapezoid on the retina is a rectangle in the world. That alternative predicts that a person who can call a slanted rectangle a "rectangle" may still have to learn how to hold it as a rectangle, how to predict that it will fit into rectangular spaces, and so on. That seems unlikely. When vision deduces the shape of an object that gave rise to a pattern on the retina, all parts of the mind can exploit the discovery. Though some parts of the visual system siphon off information to motor-control circuits that need to react quickly to moving targets, the system as a whole is not dedicated to any one kind of behavior. It creates a description or representation of the world, couched in objects and 3-D coordinates rather than retinal images, and inscribes it on a blackboard readable by all the mental modules.

This chapter explores how vision turns retinal depictions into mental descriptions. We will work our way up from splashes of light to concepts of objects, and beyond them to a kind of interaction between seeing and thinking known as mental imagery. The repercussions reach to the rest of the psyche. We are primates—highly visual creatures—with minds that evolved around this remarkable sense.

DEEP EYE

Let's begin with the stereograms. How do they work, and why, for some people, don't they work? Despite all the posters, books, and jigsaw puzzles, I have not seen a single attempt at explaining them to the millions of curious consumers. Understanding stereograms is not only a good way to grasp the workings of perception but it is also a treat for the intellect. Stereograms are yet another example of the marvelous contrivances of natural selection, this one inside our own heads.

Autostereograms exploit not one but four discoveries on how to trick the eye. The first, strange to say, is the picture. We are so jaded by photographs, drawings, television, and movies that we forget that they are a benign illusion. Smears of ink or flickering phosphor dots can make us laugh, cry, even become sexually aroused. Humans have made pictures for at least thirty thousand years, and contrary to some social-science folklore, the ability to see them as depictions is universal. The psychologist Paul Ekman created a furor in anthropology by showing that isolated New Guinean highlanders could recognize the facial expressions in photographs of Berkeley students. (Emotions, like everything else, were thought to be culturally relative.) Lost in the brouhaha was a more basic discovery: that the New Guineans were seeing things in the photographs at all rather than treating them as blotchy gray paper.

The picture exploits projection, the optical law that makes perception such a hard problem. Vision begins when a photon (unit of light energy) is reflected off a surface and zips along a line through the pupil to stimulate one of the photoreceptors (rods and cones) lining the curved inner surface of the eyeball. The receptor passes a neural signal up to the brain, and the brain's first task is to figure out where in the world that photon came from. Unfortunately, the ray defining the photon's path extends out to infinity, and all the brain knows is that the originating patch lies somewhere along the ray. For all the brain knows, it could be a foot away, a mile away, or many light-years away; information about the third dimension, distance from the eye, has been lost in the process of projection. The ambiguity is multiplied combinatorially by the million other receptors in the retina, each fundamentally confused about how far away its stimulating patch lies. Any retinal image, then, could have been produced by an infinite number of arrangements of three-dimensional surfaces in the world (see the diagram on p. 9).

Of course, we don't *perceive* infinite possibilities; we home in on one, generally close to the correct one. And here is an opening for a crafter of illusions. Arrange some matter so that it projects the same retinal image as an object the brain is biased to recognize, and the brain should have no way of telling the difference. A simple example is the Victorian novelty in which a peephole in a door revealed a sumptuously furnished room, but when the door was opened the room was empty. The sumptuous room was in a dollhouse nailed to the door over the peephole.

The painter-turned-psychologist Adelbert Ames, Jr., made a career out of carpentering even stranger illusory rooms. In one, rods and slabs were suspended from wires higgledy-piggledy throughout the room. But when the room was seen from outside through a peephole in a wall, the rods and slabs lined up into a projection of a kitchen chair. In another room, the rear wall slanted away from left to right, but it had crazy angles that made its left side just short enough to cancel its expansion in perspective, and its right side just tall enough to cancel its contraction. Through a peephole on the opposite side, the wall projected a rectangle. The visual system hates coincidences: it assumes that a regular image comes from something that really *is* regular and that it doesn't just look that way because of the fortuitous alignment of an irregular shape. Ames *did* align an irregular shape to give a regular image, and he reinforced his cunning trick with crooked windows and floor tiles. When a child stands in the near corner and her mother stands in the far one, the child projects a larger retinal image. The brain takes depth into account when assessing size; that's why a looming toddler never seems to dominate her distant parent in everyday life. But now the viewer's sense of depth is a victim of its distaste for coincidence. Every inch of the wall appears the same distance away, so the retinal images of the bodies are interpreted at face value, and Junior towers over Mom. When they change places by walking along the rear wall, Junior shrinks to lapdog size and Mom becomes Wilt Chamberlain. Ames' room has been built in several museums of science, such as the Exploratorium in San Francisco, and you can see (or be seen in) this astonishing illusion for yourself.

Now, a *picture* is nothing but a more convenient way of arranging matter so that it projects a pattern identical to real objects. The mimicking matter sits on a flat surface, rather than in a dollhouse or suspended by wires, and it is formed by smearing pigments rather than by cutting shapes out of wood. The shapes of the smears can be determined without the twisted ingenuity of an Ames. The trick was stated succinctly by Leonardo da Vinci: "Perspective is nothing else than seeing a place behind a pane of glass, quite transparent, on the surface of which the objects behind the glass are drawn." If the painter sights the scene from a fixed viewing position and copies the contours faithfully, down to the last hair of the dog, a person who then views the painting from the position of the painter would have his eye impaled by the same sheaf of light rays that the original scene projected. In that part of the visual field the painting and the world would be indistinguishable. Whatever assumptions impel the brain to see the world as the world and not as smeared pigment will impel it to see the *painting* as the world and not as smeared pigment.

What are those assumptions? We'll explore them later, but here is a preview. Surfaces are evenly colored and textured (that is, covered with regular grain, weave, or pockmarking), so a gradual change in the markings on a surface is caused by lighting and perspective. The world often contains parallel, symmetrical, regular, right-angled figures lying on the flat ground, which only *appear* to taper in tandem; the tapering is written off as an effect of perspective. Objects have regular, compact silhouettes, so if Object A has a bite taken out that is filled by Object B, A is behind B; accidents don't happen in which a bulge in B fits flush into the bite in A. You can feel the force of the assumptions in these line drawings, which convey an impression of depth.

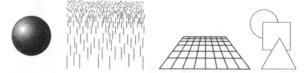

In practice, realist painters do not daub paint on windows but use visual images from memory and a host of tricks to accomplish the same thing on a canvas. They use grids made of wire or etched in glass, taut strings running from the scene through pinholes in the canvas to a viewing reticle, the camera obscura, the camera lucida, and now the camera Nikon. And, of course, no painter reproduces every hair of the dog. Brush strokes, the texture of the canvas, and the shape of the frame make a painting depart from the idealization of Leonardo's window. Also, we almost always see a painting from a vantage point different from the one the painter assumed in front of his window, and this makes the sheaf of light rays impaling the eye different from the one the real scene would send out. That is why paintings are only partly illusory: we see what the painting depicts, but we simultaneously see it *as* a painting, not as reality. The canvas and frame tip us off, and remarkably, we use these very clues about picturehood to ascertain our vantage point relative to the painting and to compensate for its difference from the painter's. We undo the distortion of the picture as if seeing it from the painter's perspective, and interpret the adjusted shapes correctly. The compensation works only up to a point. When we arrive late to a movie and sit in the front row, the difference between our vantage point and the camera's (analogous to the painter at Leonardo's window) is too much of a stretch, and we see warped actors slithering across a trapezoid.

There is another difference between art and life. The painter had to sight the scene from a single vantage point. The viewer peeps at the world from *two* vantage points: his left eye's and his right eye's. Hold out a finger and remain still while you close one eye, then the other. The finger obscures different parts of the world behind it. The two eyes have slightly different views, a fact of geometry called binocular parallax.

Many kinds of animals have two eyes, and whenever they aim forward, so that their fields overlap (rather than aiming outward for a panoramic view), natural selection must have faced the problem of combining their pictures into a unified image that the rest of the brain can use. That hypothetical image is named after a mythical creature with a single eye in the middle of its forehead: the Cyclops, a member of a race of monocular giants encountered by Odysseus in his travels. The problem in making a cyclopean image is that there is no direct way to overlay the views of the two eyes. Most objects fall on different places in the two images, and the difference depends on how far away they are: the closer the object, the farther apart its facsimiles lie in the two eyes' projections. Imagine looking at an apple on a table, with a lemon behind it and cherries in front.

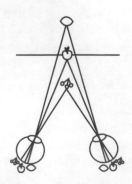

Your eyes are aimed at the apple, so its image lands on each eye's fovea (the dead center of the retina, where vision is sharpest). The apple is at six o'clock in both retinas. Now look at the projections of the cherries, which are nearer. In the left eye they sit at seven o'clock, but in the right eye they sit at *five* o'clock, not seven. The lemon, which is farther, projects an image at five-thirty in the left eye but at six-thirty in the right eye. Objects closer than the point of fixation wander outward toward the temples; objects that are farther squeeze inward toward the nose.

But the impossibility of a simple overlay presented evolution with an opportunity. With a bit of high school trigonometry, one can *use* the difference in an object's projection in the two eyes, together with the angle formed by the two eyes' gaze and their separation in the skull, to calculate how far away the object is. If natural selection could wire up a neural computer to do the trig, a two-eyed creature could shatter Leonardo's window and sense an object's depth. The mechanism is called stereoscopic vision, stereo for short.

Incredibly, for thousands of years no one noticed. Scientists thought that animals have two eyes for the same reason they have two kidneys: as a by-product of a bilaterally symmetrical body plan, and perhaps so that one could serve as a spare if the other got damaged. The possibility of stereo vision escaped Euclid, Archimedes, and Newton, and even Leonardo did not fully appreciate it. He did notice that the two eyes have different views of a sphere, the left eye seeing slightly farther around it on the left and the right eye seeing farther around it on the right. If only he had used a cube in his example instead of a sphere, he would have noticed that the shapes on the retinas are different. Stereo vision was not discovered until 1838, by Charles Wheatstone, a physicist and inventor after whom the "Wheatstone bridge" electrical circuit is named. Wheatstone wrote:

> It will now be obvious why it is impossible for the artist to give a faithful representation of any near solid object, that is, to produce a painting which shall not be distinguished in the mind from the object itself. When the painting and the object are seen with both eyes, in the case of the painting two *similar* pictures are projected on the retinae, in the case of the solid object the two pictures are *dissimilar*; there is therefore an essential difference between the impressions on the organs of sensation in the two cases, and consequently between the perceptions formed in the mind; the painting therefore cannot be confounded with the solid object.

The late discovery of stereo vision is surprising, because it is not hard to notice in everyday experience. Keep one eye closed for a few minutes as you walk around. The world is a flatter place, and you might find your-self grazing doorways and spooning sugar into your lap. Of course, the world does not flatten completely. The brain still has the kinds of infor-mation that are present in pictures and television, like tapering, occlu-sion, placement on the ground, and gradients of texture. Most important, it has motion. As you move around, your vantage point changes continu-ously, making nearby objects whiz by and farther ones budge more slowly. The brain interprets the flow pattern as a three-dimensional world going by. The perception of structure from optical flow is obvious in *Star Trek, Star Wars,* and popular computer screen-savers where white dots fleeing the center of the monitor convey a vivid impression of flying through space (though real stars would be too far away to give that impression to a real-life starfleet crew). All these monocular cues to depth allow people who are blind in one eye to get around pretty well, including the aviator Wiley Post and a wide receiver for the New York Giants football team in the 1970s. The brain is an opportunistic and mathematically adroit consumer of information, and perhaps that is why its use of one cue, binocular disparity, eluded scientists for so long.

Wheatstone proved that the mind turns trigonometry into conscious-ness when he designed the first fully three-dimensional picture, the stereogram. The idea is simple. Capture a scene using *two* of Leonardo's windows, or, more practically, two cameras, each positioned where one eye would be. Place the right picture in front of a person's right eye and the left picture in front of his left eye. If the brain assumes that the two eyes look at one three-dimensional world, with differences in the views coming from binocular parallax, it should be fooled by the pictures and combine them into a cyclopean image in which objects appear at differ-ent depths.

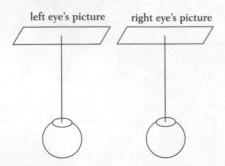

left eye's picture   right eye's picture

But here Wheatstone ran into a problem, one that still challenges all stereoscopic gadgets. The brain physically adjusts the eyes to the depth of a surface in two ways. First, though I have been describing the pupil as if it were a pinhole, in fact it has a lens to accumulate many rays of light emanating from a point in the world and to focus them all at a point on the retina. The closer the object, the more the rays have to be bent for them to converge to a point rather than to a blurry disk, and the fatter the lens of the eye has to be. Muscles inside the eyeball have to thicken the lens to focus nearby objects and flatten it to focus distant objects.

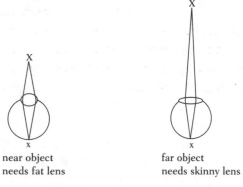

near object
needs fat lens

far object
needs skinny lens

The squeezing is controlled by the focusing reflex, a feedback loop that adjusts the shape of the lens until the fine detail on the retina is at a maximum. (The circuit is similar to the one used in some autofocus cameras.) Poorly focused movies are annoying to watch because the brain keeps trying to eliminate the blur by accommodating the lens, a futile gesture.

The second physical adjustment is to aim the two eyes, which are about two and a half inches apart, at the same spot in the world. The closer the object, the more the eyes must be crossed.

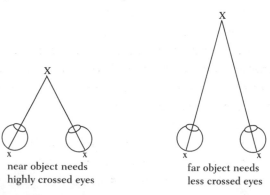

near object needs
highly crossed eyes

far object needs
less crossed eyes

The eyes are crossed and uncrossed by muscles attached to their sides; the muscles are controlled by a brain circuit that tries to eliminate double images. (Seeing double is often a sign that the brain has been poisoned, suffocated, or bruised.) The circuit is similar to the rangefinders in old cameras, in which a prism superimposes the views from two viewfinder windows and the photographer angles the prism (which is geared to the camera lens) until the images line up. The brain uses the rangefinder principle as another source of information about depth, perhaps an indispensable one. Stereo vision gives information only about *relative* depth—depth in front of or behind the point on which the eyes have converged—and feedback from eyeball direction must be used to anchor a sense of absolute depth.

Now here's the problem for the stereoscope designer. The focusing reflex and the eye-crossing reflex are coupled. If you focus on a nearby point to eliminate blur, the eyes converge; if you focus on a distant one, they become parallel. If you converge your eyes on a nearby point to eliminate double vision, the eyes squeeze the lens to close-up focus; if you diverge your eyes on a distant point, they relax for distant focus. The coupling defeats the most straightforward design for a stereoscope, in which a small picture is placed in front of each eye and both eyes point straight ahead, each at its own picture. Pointing the eyes straight ahead is what you do for distant objects, and it drags the focus of each eye to distance vision, blurring the pictures. Focusing the pictures then brings the eyes together, so the eyes are pointing at the same picture rather than each eye aiming at a different one, and that's no good, either. The eyes bob in and out and the lenses thicken and flatten, but not at the right times. To get a stereoscopic illusion, something has to give.

One solution is to uncouple the responses. Many experimental psychologists have trained themselves like fakirs to wrest control of their reflexes and to "free-fuse" stereograms by an act of will. Some cross their eyes at an imaginary point in front of the picture, so that the left eye is staring at the right picture and vice versa, while they focus each eye on the picture behind the imaginary point. Others lock their eyes straight ahead to infinity while maintaining focus. I once took an afternoon out to train myself to do this after I learned that William James said it was a skill every good psychologist should master. But people with lives cannot be expected to show such dedication.

Wheatstone's invention was a bit ungainly because he faced a second problem: the drawings and daguerreotypes of his age were too big to fit in front of the eyes without overlapping, and people could not point their eyes outward to gaze at one on each side like fish. So he put one picture off to each side, the two facing each other like bookends, and between them he placed two mirrors glued together like the cover of an open book, each mirror reflecting a picture. He then put a prism in front of each mirror and adjusted them so that the two mirrors appeared to be superimposed. When people looked through the prisms and saw the superimposed reflections of the two pictures, the scene in the pictures leapt into three-dimensionality. The advent of better cameras and smaller film led to a simpler, hand-held design that is still with us. Small pictures—as always, photographed from two vantage points positioned like the eyes—are placed side by side with a perpendicular blinker between them and a glass lens in front of each eye. The glass lens relieves the eye of having to focus its nearby picture, and the eye can relax to its infinity setting. That spreads the eyes so they are pointing straight ahead, one at each picture, and the pictures easily fuse.

The stereoscope became the television of the nineteenth century. Victorian-era families and friends spent cozy hours taking turns to view stereo photographs of Parisian boulevards, Egyptian pyramids, or Niagara Falls. Beautiful wooden stereoscopes and the software for them (cards with side-by-side photographs) are still sold in antique stores to avid collectors. A modern version is the ViewMaster, available at tourist traps the world over: an inexpensive viewer that displays a ring of stereo slides of the local attractions.

A different technique, the anaglyph, overlays the two images on one surface and uses clever tricks so that each eye sees only the image intended for it. A familiar example is the notorious red-and-green cardboard eyeglasses associated with the 3-D movie craze of the early 1950s. The left eye's image is projected in red and the right eye's image is projected in green onto a single white screen. The left eye peers at the screen through a green filter, which makes the white background look green and the green lines intended for the other eye invisible; the red lines intended for the left eye stand out as black. Similarly, the red filter over the right eye makes the background red, the red lines invisible, and the green lines black. Each eye gets its own image, and the Sludge Monsters from Alpha Centauri rise out of the screen. An unfortunate side effect is that when the two eyes see very different patterns like the red and green backgrounds, the brain cannot fuse them. It carves the visual field into a patchwork and seesaws between seeing each patch as green or red, a disconcerting effect called binocular rivalry. You can experience a milder case by holding a finger a few inches in front of you with both eyes open gazing into the distance so you get a double image. If you pay attention to one of the double images, you will notice that portions slowly become opaque, dissolve into transparency, fill in again, and so on.

A better kind of anaglyph puts polarizing filters, rather than colored filters, over two projector lenses and in the cardboard glasses. The image intended for the left eye is projected from the left projector in light waves that oscillate in a diagonal plane, like this: /. The light can pass through a filter in front of the left eye which has microscopic slits that are also in that orientation, but cannot pass through a filter in front of the right eye with slits in the opposite orientation, like this: \. Conversely, the filter in front of the right eye allows in only the light coming from the right projector. The superimposed images can be in color, and they do not incite rivalry between the eyes. The technique was used to excellent effect by Alfred Hitchcock in *Dial "M" for Murder* in the scene in which Grace Kelly reaches out for the scissors to stab her would-be strangler. The same cannot be said for the film adaptation of Cole Porter's *Kiss Me Kate*, in which a dancer belts out "Too Darn Hot" on a coffee table while flinging scarves at the camera.

Modern anaglyph glasses have panes made of liquid crystal displays (like the numbers on a digital watch) which act as silent, electrically controlled shutters. At any moment one shutter is transparent and the other is opaque, forcing the eyes to take turns at seeing a computer screen in front of them. The glasses are synchronized with the screen, which shows the left eye's image while the left shutter is open and the right eye's image while the right shutter is open. The views alternate too quickly for the eyes to notice the flicker. The technology is used in some virtual reality displays. But the state of the art in virtual reality is a high-tech version of the Victorian stereoscope. A computer displays each image on a little LCD screen with a lens in front of it, mounted in front of each eye on the inside of a helmet or visor.

~

These technologies all force the viewer to don or peer through some kind of apparatus. The illusionist's dream is a stereogram that can be seen with the naked eye—an autostereogram.

The principle was discovered a century and a half ago by David Brewster, the Scottish physicist who also studied polarized light and invented the kaleidoscope and the Victorian-era stereoscope. Brewster noticed that the repeating patterns on wallpaper can leap out in depth. Adjacent copies of the pattern, say a flower, can each lure one eye into fixating on it. That can happen because identical flowers are positioned at the same places on the two retinas, so the double image looks like a single image. In fact, like a misbuttoned shirt, a whole parade of double images can falsely mesh into a single image, except for the unpaired members at each end. The brain, seeing no double image, is prematurely satisfied that it has converged the eyes properly, and locks them into the false alignment. This leaves the eyes aimed at an imaginary point behind the wall, and the flowers seem to float in space at that distance. They also seem inflated, because the brain does its trigonometry and calculates how big the flower would have to be at that depth to project its current retinal image.

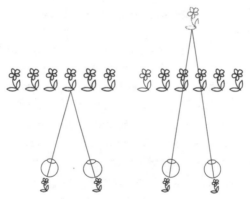

what the eyes should do   what the repeating pattern
                             tricks the eyes into doing

An easy way to experience the wallpaper effect is to stare at a tile wall a few inches away, too close to focus and converge on comfortably. (Many men rediscover the effect as they stand at a urinal.) The tiles in front of each eye easily fuse, creating the surreal impression of a very large tile wall a great distance away. The wall bows outward, and as the head moves from side to side the wall rocks in the opposite direction. Both would have to happen in the world if the wall were really at that distance while projecting the current retinal image. The brain creates those illusions in its headlong attempt to keep the geometry of the whole hallucination consistent.

Brewster also noticed that any irregularity in the spacing of a pair of copies makes them protrude or recess from the rest. Imagine that the flowers pierced by the lines of sight in the diagram are printed a bit closer to each other. The lines of sight are brought together and cross each other closer to the eyes. The images on the retina will splay out to the temples, and the brain sees the imaginary flower as being nearer. Similarly, if the flowers had been printed a bit farther apart, the lines of sight will cross farther away, and their retinal projections will crowd toward the nose. The brain hallucinates the ghost object at a slightly greater distance.

We have now arrived at a simple kind of "magic eye" illusion, the wallpaper autostereogram. Some of the stereograms in the books and greeting cards show rows of repeating figures—trees, clouds, mountains, people. When you view the stereogram, each tier of objects drifts in or out and lands at its own depth (although in these autostereograms, unlike the squiggly ones, no new shape emerges; we'll come to those soon). Here is an example, designed by Ilavenil Subbiah.

It is like Brewster's wallpaper, but with the unequal separations put in deliberately rather than by a paperhanger's sloppiness. The picture accommodates seven sailboats because they are closely packed, but only five arches because they are spaced farther apart. When you look behind the picture, the sailboats seem closer than the arches because their misbuttoned lines of sight meet in a nearer plane.

If you don't already know how to fuse stereograms, try holding the book right up to your eyes. It is too close to focus; just let your eyes point straight ahead, seeing double. Slowly move the book away while keeping your eyes relaxed and "looking through" the book to an imaginary point beyond it. (Some people place a pane of glass or a transparency on top of the stereogram, so they can focus on the reflections of distant objects.) You should still be seeing double. The trick is to let one of the double images drift on top of the other, and then to keep them there as if they were magnets. Try to keep the images aligned. The superimposed shapes should gradually come into focus and pop in or out to different depths. As Tyler has noted, stereo vision is like love: if you're not sure, you're not experiencing it.

Some people have better luck holding a finger a few centimeters in front of the stereogram, focusing on the finger, and then removing it while keeping the eyes converged to that depth. With this technique, the false fusion comes from the eyes crossing so that the left eye sights a boat on the right while the right eye sights a boat on the left. Don't worry about what your mother said; your eyes will not freeze into that position forever. Whether you can fuse stereograms with your eyes crossed too much or not enough probably depends on whether you are slightly cross-eyed or wall-eyed to begin with.

With practice, most people can fuse wallpaper autostereograms. They do not need the yogi-like concentration of the psychologists who free-fuse the two-picture stereograms, because they do not have to uncouple their focusing reflex from their convergence reflex to the same degree. Free-fusing a two-picture stereogram requires jamming your eyes far enough apart that each eye remains aimed at one of the pictures. Fusing a wallpaper stereogram requires merely keeping the eyes far enough apart that each eye remains aimed at neighboring clones *inside a single picture*. The clones are close enough together that the convergence angle is not too far out of line from what the focusing reflex wants it to be. It shouldn't be too hard for you to exploit this small wiggle in the mesh between the two reflexes and focus a wee bit closer than your eyes converge. If it is, Ellen DeGeneres may be able to get you into her support group.

The trick behind the wallpaper stereogram—identical drawings luring the eyes into mismatching their views—uncovers a fundamental problem the brain has to solve to see in stereo. Before it can measure the positions of a spot on the two retinas, the brain has to be sure that the spot on one retina came from the same mark in the world as the spot on the other retina. If the world had only one mark in it, it would be easy. But add a second mark, and their retinal images can be matched in two ways: spot 1 in the left eye with spot 1 in the right eye, and spot 2 in the left eye with spot 2 in the right eye—the correct matchup—or spot 1 in the left eye with spot 2 in the right eye, and spot 2 in the left eye with spot 1 in the right eye—a mismatch that would lead to the hallucination of two ghost marks instead.

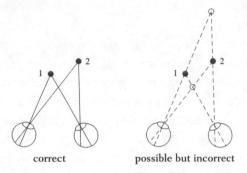

correct   possible but incorrect

Add more marks, and the matching problems multiply. With three marks, there are six ghost matches; with ten marks, ninety; with a hundred marks, almost ten thousand. This "correspondence problem" was noticed in the sixteenth century by the astronomer Johannes Kepler, who thought about how stargazing eyes match up their thousands of white dots and how an object's position in space could be determined from its multiple projections. The wallpaper stereogram works by coaxing the brain to accept a plausible but false solution to the correspondence problem.

Until recently, everyone thought that the brain solved the correspondence problem in everyday scenes by first *recognizing* the objects in each eye and then matching up images of the same object. Lemon in left eye goes with lemon in right eye, cherries in left eye go with cherries in right eye. Stereo vision, guided by the intelligence of the whole person, could head off the mismatches by only joining up points that came from the same kind of object. A typical scene may contain millions of dots, but it will contain far fewer lemons, maybe only one. So if the brain matched whole objects, there would be fewer ways for it to go wrong.

But nature did not opt for that solution. The first hint came from another of Ames' wacky rooms. This time the indefatigable Ames built an ordinary rectangular room but glued leaves on every inch of its floor, walls, and ceiling. When the room was viewed with one eye through a peephole, it looked like an amorphous sea of green. But when it was viewed with both eyes, it sprang into its correct three-dimensional shape. Ames had built a world that could be seen only by the mythical cyclopean eye, not by the left eye or the right eye alone. But how could the brain have matched up the two eyes' views if it had to depend on recognizing and linking the objects in each one? The left eye's view was "leaf leaf leaf leaf leaf leaf leaf leaf." The right eye's view was "leaf leaf leaf leaf leaf leaf leaf leaf." The brain was faced with the hardest correspondence problem imaginable. Nonetheless it effortlessly coupled the views and conjured up a cyclopean vision.

The demonstration is not airtight. What if the edges and corners of the room were not perfectly masked by the leaves? Perhaps each eye had a rough inkling of the room's shape, and when the brain fused the two images it became more confident that the inklings were accurate. The airtight proof that the brain can solve the correspondence problem without recognizing objects came from an ingenious early use of computer graphics by the psychologist Bela Julesz. Before he fled Hungary for the United States in 1956, Julesz was a radar engineer with an interest in aerial reconnaissance. Spying from the air uses a clever trick: stereo views penetrate camouflage. A camouflaged object is covered with markings resembling the background it lies on, making the boundary between the object and its background invisible. But as long as the object is not pancake-flat, when it is viewed from *two* vantage points its markings will appear in slightly different positions in the two views, whereas the background markings will not have moved quite as much because they are farther away. The trick in aerial reconnaissance is to photograph the land, let the plane fly a bit, and photograph it again. The pictures are placed side by side and then fed into a hypersensitive detector of disparity in two images: a human being. A person literally looks at the photographs with a stereo viewer, as if he were a giant peering down from the sky with one eye at each position from which the airplane took a picture, and the camouflaged objects pop out in depth. Since a camouflaged object, by definition, is near-invisible in a single view, we have another example of the cyclopean eye seeing what neither real eye can see.

The proof had to come from *perfect* camouflage, and here Julesz went to the computer. For the left eye's view, he had the computer make a square covered with random dots, like television snow. Julesz then had the computer make a copy for the right eye, but with one twist: he shifted a patch of dots a bit over to the left, and inserted a new stripe of random dots into the gap at the right so the shifted patch would be perfectly camouflaged. Each picture on its own looked like pepper. But when put in the stereoscope, the patch levitated into the air.

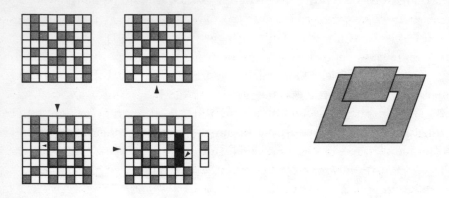

Many authorities on stereo vision at the time refused to believe it because the correspondence problem the brain had to solve was just too hard. They suspected that Julesz had somehow left little cut marks behind in one of the pictures. But of course the computer did no such thing. Anyone who sees a random-dot stereogram is immediately convinced.

All it took for Julesz' occasional collaborator, Christopher Tyler, to invent the magic-eye autostereogram was to combine the wallpaper autostereogram with the random-dot stereogram. The computer generates a vertical stripe of dots and lays copies of it side by side, creating random-dot wallpaper. Say each stripe is ten dots wide, and we number the dots from 1 to 10 (using "0" for 10):

```
12345678901234567890123456789012345678901234567890123456789 0
12345678901234567890123456789012345678901234567890123456789 0
12345678901234567890123456789012345678901234567890123456789 0
```

—and so on. Any clump of dots—say, "5678"—repeats itself every ten spaces. When the eyes fixate on neighboring stripes, they falsely fuse, just as they do with a wallpaper stereogram, except that the brain is superimposing stretches of random dots rather than flowers. Remember that in a wallpaper stereogram, copies of a pattern that have been squashed closer together will float above the rest because their lines of sight cross closer to the viewer. To make a patch float out of a magic-eye autostereogram, the designer identifies the patch and makes each clump of dots inside it closer to the nearest copy of itself. In the picture below, I want to make a floating rectangle. So I snip out two copies of dot 4 in the stretch between the arrows; you can spot the snipped rows because they are now two spaces shorter. Inside the rectangle, every clump of dots, like "5678," repeats itself every *nine* spaces instead of every ten. The brain interprets copies that are closer together as coming from nearer objects, so the rectangle levitates. The diagram, by the way, not only shows how autostereograms are made, but it works as a passable autostereogram itself. If you fuse it like wallpaper, a rectangle should arise. (The asterisks at the top are there to help you fuse it; let your eyes drift until you have a double image with four asterisks and slowly try to bring the images together until the middle two asterisks fuse and you are seeing three asterisks in a row rather than four. Carefully look down at the diagram without re-aiming your eyes, and you may see the floating rectangle.)

```
        *              *
        ↓              ↓
12345678901234567890123456789012345678901234567890123456789012345678901234567890
12345678901234567890123456789012345678901234567890123456789012345678901234567890
12345678901234567890123456789012345678901234567890123456789012345678901234567890
123456789012345678901234567890123567890123567890123456789012345678901234567890
123456789012345678901234567890123567890123567890123456789012345678901234567890
123456789012345678901234567890123567890123567890123456789012345678901234567890
123456789012345678901234567890123567890123567890123456789012345678901234567890
123456789012345678901234567890123567890123567890123456789012345678901234567890
123456789012345678901234567890123567890123567890123456789012345678901234567890
12345678901234567890123456789012345678901234567890123456789012345678901234567890
12345678901234567890123456789012345678901234567890123456789012345678901234567890
12345678901234567890123456789012345678901234567890123456789012345678901234567890
1234567890123456789012345678901234567890123X4567890123X4567890123456789012345678901234567890
1234567890123456789012345678901234567890123X4567890123X4567890123456789012345678901234567890
1234567890123456789012345678901234567890123X4567890123X4567890123456789012345678901234567890
1234567890123456789012345678901234567890123X4567890123X4567890123456789012345678901234567890
1234567890123456789012345678901234567890123X4567890123X4567890123456789012345678901234567890
1234567890123456789012345678901234567890123X4567890123X4567890123456789012345678901234567890
12345678901234567890123456789012345678901234567890123456789012345678901234567890
12345678901234567890123456789012345678901234567890123456789012345678901234567890
12345678901234567890123456789012345678901234567890123456789012345678901234567890
```

You should also see a cutout window lower in the picture. I made it by picking out a rectangular patch and doing the opposite of what I did before: I stuffed an *extra* dot (labeled "X") next to every copy of dot 4 inside the patch. That pushes the clumps of dots farther apart, so they repeat themselves every *eleven* spaces. (The stuffed rows, you will notice, are longer than the rest.) Copies that are more widely spaced equals a surface that is more distant. A real random-dot autostereogram, of course, is made of dots, not numbers, so you don't notice the snipped-out or stuffed-in material, and the uneven lines are filled out with extra dots. Here is an example. The fun in viewing a real random-dot autostereogram is that the moment of pop-out surprises the viewer with previously invisible shapes:

*   *

When the autostereogram craze hit Japan, it soon developed into an art form. Dots are not necessary; any tapestry of small contours rich enough to fool the brain into locking the eyes on neighboring stripes will do. The first commercial autostereograms used colored squiggles, and the Japanese ones use flowers, ocean waves, and, taking a leaf out of Ames' book, leaves. Thanks to the computer, the shapes don't have to be flat cutouts like in a diorama. By reading in the three-dimensional coordinates of the points on a surface, the computer can shift every dot by a slightly different amount to sculpt the solid shape in cyclopean space, rather than shifting the entire patch rigidly. Smooth, bulbous shapes materialize, looking as if they are shrink-wrapped in leaves or flowers.

Why did natural selection equip us with true cyclopean vision—an ability to see shapes in stereo that neither eye can see in mono—rather than with a simpler stereo system that would match up the lemons and cherries that are seeable by each eye? Tyler points out that our ancestors really did live in Ames' leaf room. Primates evolved in trees and had to negotiate a network of branches masked by a veil of foliage. The price of failure was a long drop to the forest floor below. Building a stereo computer into these two-eyed creatures must have been irresistible to natural selection, but it could have worked only if the disparities were calculated over thousands of bits of visual texture. Single objects that allow unambiguous matches were just too few and far between.

Julesz points out another advantage of cyclopean vision. Camouflage was discovered by animals long before it was discovered by armies. The earliest primates were similar to today's prosimians, the lemurs and tarsiers of Madagascar, who snatch insects off trees. Many insects hide from predators by freezing, which defeats the hunter's motion detectors, and by camouflage, which defeats its contour detectors. Cyclopean vision is an effective countermeasure, revealing the prey just as aerial reconnaissance reveals tanks and planes. Advances in weaponry spawn arms races in nature no less than in war. Some insects have outwitted their predators' stereo vision by flattening their bodies and lying flush against the background, or by turning into living sculptures of leaves and twigs, a kind of three-dimensional camouflage.

~

How does the cyclopean eye work? The correspondence problem— matching up the marks in one eye with their counterparts in the other— is a fearsome chicken-and-egg riddle. You can't measure the stereo disparity of a pair of marks until you have picked a pair of marks to measure. But in a leaf room or a random-dot stereogram, there are thousands of candidates for the matchmaker. If you knew how far away the surface was, you would know where to look on the left retina to find the mate of a mark on the right. But if you knew that, there would be no need to do the stereo computation; you would already have the answer. How does the mind do it?

David Marr noted that built-in assumptions about the world we evolved in can come to the rescue. Among the n^2 possible matches of n points, not all are likely to have come from this goodly frame, the earth. A well-engineered matcher should consider only the matchups that are physically likely.

First, every mark in the world is anchored to one position on one surface at one time. So a legitimate match must pair up identical points in the two eyes that came from a single splotch in the world. A black dot in one eye should match a black dot in the other, not a white dot, because the matchup has to represent a single position on some surface, and that position cannot be a black splotch and a white splotch at the same time. Conversely, if a black dot does match a black dot, they must come from a single position on some surface in the world. (That is the assumption violated by autostereograms: each of their splotches appears in *several* positions.)

Second, a dot in one eye should be matched with no more than one dot in the other. That means that a line of sight from one eye is assumed to end at a splotch on one and only one surface in the world. At first glance it looks as if the assumption rules out a line of sight passing through a transparent surface to an opaque one, like the bottom of a shallow lake. But the assumption is more subtle; it only rules out the coincidence in which two identical splotches, one on the lake's surface and one on the bottom, line up one behind the other from the left eye's vantage point while both being visible from the right eye's.

Third, matter is cohesive and smooth. Most of the time a line of sight will end up on a surface in the world that is not drastically closer or farther than the surface hit by the neighboring line of sight. That is, neighboring patches of the world tend to lie on the same smooth surface. Of course, at the boundary of an object the assumption is violated: the edge of the back cover of this book is a couple of feet away from you, but if you glance just to its right you might be looking at the moon a quarter of a million miles away. But boundaries make up a small portion of the visual field (you need much less ink to sketch a line drawing than to color it in), and these exceptions can be tolerated. What the assumption rules out is a world made up of dust storms, swarms of gnats, fine wires, deep crevasses between craggy peaks, beds of nails viewed point-on, and so on.

The assumptions sound reasonable in the abstract, but something still has to *find* the matches that satisfy them. Chicken-and-egg problems can sometimes be solved with the technique called constraint satisfaction that we met in Chapter 2 when looking at Necker cubes and accented speech. When the parts of a puzzle cannot be solved one at a time, the puzzle-solver can keep in mind several guesses for each one, compare the guesses for the different parts of the puzzle, and see which ones are mutually consistent. A good analogy is working on a crossword puzzle with a pencil and an eraser. Often a clue for a horizontal word is so vague that several words can be penciled in, and a clue for a vertical word is so vague that several words can be penciled in. But if only one of the vertical guesses shares a letter with any of the horizontal guesses, that pair of words is kept and the others are erased. Imagine doing that for all the clues and squares at once and you have the idea of constraint satisfaction. In the case of solving the correspondence problem in stereo vision, the dots are the clues, the matchups and their depths are the guesses, and the three assumptions about the world are like the rules that say that every letter of every word must sit in a box, every box must have a letter in it, and all the sequences of letters must spell out words.

Constraint satisfaction can sometimes be implemented in a constraint network like the one I presented on page 107. Marr and the theoretical neuroscientist Tomaso Poggio designed one for stereo vision. The input units stand for points, such as the black and white squares of a random-dot stereogram. They feed into an array of units that represent all of the $n \times n$ possible matchups of a point in the left eye with some other point in the right eye. When one of these units turns on, the network is guessing that there is a splotch at a particular depth in the world (relative to where the eyes have converged). Here is a bird's-eye view of one plane of the network, showing a fraction of the units.

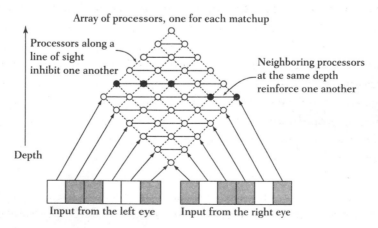

Array of processors, one for each matchup

Processors along a line of sight inhibit one another

Neighboring processors at the same depth reinforce one another

Depth

Input from the left eye   Input from the right eye

The model works as follows. A unit turns on only if it gets the same inputs from the two eyes (black or white), embodying the first assumption (each mark anchored to a surface). Because the units are interconnected, the activation of one unit nudges the activations of its neighbors up or down. Units for different matches lying along the same line of sight inhibit one another, embodying the second assumption (no coincidental markings aligned along a line of sight). Units for neighboring points at nearby depths excite one another, embodying the third assumption (matter is cohesive). The activations reverberate around the network, and it eventually stabilizes, with the activated units tracing out a contour in depth. In the diagram, the filled-in units are showing an edge hovering over its background.

The constraint-satisfaction technique, in which thousands of processors make tentative guesses and hash it out among themselves until a global solution emerges, is consistent with the general idea that the brain works with lots of interconnected processors computing in parallel. It captures some of the psychology, too. When viewing a complicated random-dot stereogram, often you don't see the hidden figure erupt instantaneously. A bit of edge might pop out from the pepper, which then lifts up a sheet, which cleans and straightens a fuzzy border on the other side, and so on until the whole shape coalesces. We experience the solution emerging, but not the struggle of the processors to come up with it. The experience is a good reminder that as we see and think, dozens of iterations of information processing go on beneath the level of consciousness.

The Marr-Poggio model captures the flavor of the brain's computation of stereo vision, but our real circuitry is surely more sophisticated. Experiments have shown that when people are put in artificial worlds that violate assumptions about uniqueness and smoothness, they don't see as badly as the model predicts. The brain must be using additional kinds of information to help solve the matchup problem. For one thing, the world is not made up of random dots. The brain can match up all the little diagonals, T's, zigzags, inkblots, and other jots and tittles in the two eyes' views (which even a random-dot stereogram has in abundance). There are far fewer false matches among jots and tittles than there are among dots, so the number of matches that have to be ruled out is radically shaved.

Another matchmaking trick is to exploit a different geometric consequence of having two eyes, the one noticed by Leonardo: there are parts of an object that one eye can see but that the other eye cannot. Hold a pen vertically in front of you, with the clip facing away at eleven o'clock. When you close each eye in turn, you will notice that only the left eye can see the clip; it is hidden from the right eye by the rest of the pen. Was natural selection as astute as Leonardo when it designed the brain, letting it use this valuable clue to an object's boundary? Or does the brain ignore the clue, grudgingly chalking up each mismatch as an exception to the cohesive-matter assumption? The psychologists Ken Nakayama and Shinsuke Shimojo have shown that natural selection did not ignore the clue. They created a random-dot stereogram whose depth information lay not in shifted dots but in dots that were visible in one eye's view and absent in the other's. Those dots lay at the corners of an imaginary square, with dots at the top and bottom right corners only in the right eye's picture, and dots in the top and bottom left corners only in the left eye's picture. When people view the stereogram, they see a floating square defined by the four points, showing that the brain indeed interprets features visible to only one eye as coming from an edge in space. Nakayama and the psychologist Barton Anderson suggest that there are neurons that detect these occlusions; they would respond to a pair of marks in one eye, one of which can be matched with a mark in the other eye and the other of which cannot be matched. These 3-D boundary detectors would help a stereo network home in on the outlines of the floating patches.

Stereo vision does not come free with the two eyes; the circuitry has to be wired into the brain. We know this because about two percent of the population can see perfectly well out of each eyeball but not with the cyclopean eye; random-dot stereograms remain flat. Another four percent can see stereo only poorly. An even larger minority has more selective deficits. Some can't see stereo depth behind the point of fixation; others can't see it in front. Whitman Richards, who discovered these forms of stereoblindness, hypothesized that the brain has three pools of neurons that detect differences in the position of a spot in the two eyes. One pool is for pairs of spots that coincide exactly or almost exactly, for fine-grained depth perception at the point of focus. Another is for pairs of spots flanking the nose, for farther objects. A third is for pairs of spots approaching the temples, for nearer objects. Neurons with all these properties have since been found in the brains of monkeys and cats. The different kinds of stereoblindness appear to be genetically determined, suggesting that each pool of neurons is installed by a different combination of genes.

Stereo vision is not present at birth, and it can be permanently damaged in children or young animals if one of the eyes is temporarily deprived of input by a cataract or a patch. So far, this sounds like the tiresome lesson that stereo vision, like everything else, is a mixture of nature and nurture. But a better way of thinking about it is that the brain has to be assembled, and the assembly requires project scheduling over an extended timetable. The timetable does not care about when the organism is extruded from the womb; the installation sequence can carry on after birth. The process also requires, at critical junctures, the intake of information that the genes cannot predict.

Stereo vision appears abruptly in infants. When newborns are brought into a lab at regular intervals, for week after week they are unimpressed by stereograms, and then suddenly they are captivated. Close to that epochal week, usually around three or four months of age, the babies converge their eyes properly for the first time (for example, they smoothly track a toy brought up to their nose), and they find rivalrous displays—a different pattern in each eye—annoying, whereas before they had found them interesting.

It is not that babies "learn to see in stereo," whatever that would mean. The psychologist Richard Held has a simpler explanation. When infants are born, every neuron in the receiving layer of the visual cortex *adds up* the inputs from corresponding locations in the two eyes rather than keeping them separate. The brain can't tell which eye a given bit of pattern came from, and simply melts one eye's view on top of the other's in a 2-D overlay. Without information about which eye a squiggle came from, stereo vision, convergence, and rivalry are logically impossible. Around the three-month mark each neuron settles on a favorite eye to respond to. The neurons lying one connection downstream can now know when a mark falls on one spot in one eye and on the same spot, or a slightly shifted-over spot, in the other eye—the grist for stereo vision.

In cats and monkeys, whose brains have been studied directly, this is indeed what happens. As soon as the animal's cortex can tell the eyes apart, the animal sees stereograms in depth. That suggests that when the inputs are first tagged "left eye" or "right eye," the circuitry for stereo computation one layer downstream is already installed and functioning.

In monkeys it's all over in two months: by then each neuron has a favorite eye and the baby monkeys see in depth. Compared with other primates, humans are "altricial": babies are born early and helpless, and complete their development outside the womb. Because human infants are born earlier than monkeys in proportion to the length of their childhood, the installation of their binocular circuitry appears at a later age as measured from the date of birth. More generally, when biologists compare the milestones of the maturation of the visual systems of different animals, some born early and helpless, others born late and seeing, they find that the sequence is pretty much the same whether the later steps take place in the womb or in the world.

The emergence of the crucial left-eye and right-eye neurons can be disrupted by experience. When the neurobiologists David Hubel and Torsten Wiesel raised kittens and baby monkeys with one eye covered, the input neurons of the cortex all tuned themselves to the other eye, making the animal functionally blind in the eye that was covered. The damage was permanent, even with only brief deprivation, if the eye was covered in a critical period in the animal's development. In monkeys, the visual system is especially vulnerable during the first two weeks of life, and the vulnerability tapers off during the first year. Covering the eye of an adult monkey, even for four years, does no harm.

At first this all looked like a case of "use it or lose it," but a surprise was in store. When Hubel and Wiesel covered *both* eyes, the brain did not show twice the damage; half the cells showed no damage at all. The havoc in the single-eyepatch experiment came about not because a neuron destined for the covered eye was starved of input but because the input signals from the *uncovered* eye elbowed the covered eye's inputs out of the way. The eyes compete for real estate in the input layer of the cortex. Each neuron begins with a slight bias for one eye or the other, and the input from that eye exaggerates the bias until the neuron responds to it alone. The inputs do not even have to originate in the world; waves of activation from intermediate way-stations, a kind of internally generated test pattern, can do the trick. The developmental saga, though it is sensitive to changes in the animal's experience, is not exactly "learning," in the sense of registering information from the world. Like an architect who hands a rough sketch to a low-level draftsman to straighten out the lines, the genes build eye-specific neurons crudely and then kick off a process that is guaranteed to sharpen them unless a neurobiologist meddles.

Once the brain has segregated the left eye's image from the right eye's, subsequent layers of neurons can compare them for the minute disparities that signal depth. These circuits, too, can be modified by the animal's experience, though again in surprising ways. If an experimenter makes an animal cross-eyed or wall-eyed by cutting one of the eye muscles, the eyes point in different directions and never see the same thing on the two retinas at the same time. Of course, the eyes don't point 180 degrees apart, so in theory the brain could learn to match the out-of-whack segments that do overlap. But apparently it is not equipped for matches that stretch more than a few degrees across the two eyes; the animal grows up stereoblind, and often functionally blind in one of the two eyes as well, a condition called amblyopia. (Amblyopia is sometimes called "lazy eye," but that is misleading. It is the brain, not the eye, that is insensitive, and the insensitivity is caused by the brain actively suppressing one eye's input in a kind of permanent rivalry, not by the brain lazily ignoring it.)

The same thing can happen in children. If one of the eyes is more far-sighted than the other, the child habitually strains to focus on nearby objects, and the reflex that couples focusing and convergence draws that eye inward. The two eyes point in different directions (a condition called strabismus), and their views don't align closely enough for the brain to use the disparity information in them. The child will grow up amblyopic and stereoblind unless early surgery on the eye muscles lines the eyeballs up. Until Hubel and Wiesel discovered these effects in monkeys and Held found similar ones in children, surgery for strabismus was considered cosmetic and done only on school-aged children. But there is a critical period for the proper alignment of two-eye neurons, a bit longer than the one for one-eye neurons but probably fading out near the age of one or two. Surgery after that point is often too late.

Why is there a critical period, as opposed to rigid hard-wiring or life-long openness to experience? In kittens, monkeys, and human babies, the face keeps growing after birth, and the eyes get pushed farther apart. Their relative vantage points change, and the neurons must keep up by retuning the range of intereye disparities they detect. Genes cannot anticipate the degree of spreading of the vantage points, because it depends on other genes, nutrition, and various accidents. So the neurons track the drifting eyes during the window of growth. When the eyes arrive at their grownup separation in the skull, the need disappears, and that is when the critical period ends. Some animals, like rabbits, have precocious babies whose eyes are set in adult positions within faces that grow very little. (These tend to be prey animals, which don't have the luxury of a long, helpless childhood.) The neurons that receive inputs from the two eyes don't need to retune themselves, and in fact these animals are wired at birth and do without a critical period of sensitivity to the input.

The discoveries about the tunability of binocular vision in different species offers a new way of thinking about learning in general. Learning is often described as indispensable shaper of amorphous brain tissue. Instead it might be an innate adaptation to the project-scheduling demands of a self-assembling animal. The genome builds as much of the animal as it can, and for the parts of the animal that cannot be specified in advance (such as the proper wiring for two eyes that are moving apart at an unpredictable rate), the genome turns on an information-gathering mechanism at the time in development at which it is most needed. In *The Language Instinct* I develop a similar explanation for the critical period for learning language in childhood.

I have led you through magic-eye stereograms not just because it is fun to understand how the magic works. I think stereo vision is one of the glories of nature and a paradigm of how other parts of the mind might work. Stereo vision is information processing that we experience as a particular flavor of consciousness, a connection between mental computation and awareness that is so lawful that computer programmers can manipulate it to enchant millions. It is a module in several senses: it works without the rest of the mind (not needing recognizable objects), the rest of the mind works without it (getting by, if it has to, with other depth analyzers), it imposes particular demands on the wiring of the brain, and it depends on principles specific to its problem (the geometry of binocular parallax). Though stereo vision develops in childhood and is sensitive to experience, it is not insightfully described as "learned" or as "a mixture of nature and nurture"; the development is part of an assembly schedule and the sensitivity to experience is a circumscribed intake of information by a structured system. Stereo vision shows off the engineering acumen of natural selection, exploiting subtle theorems in optics rediscovered millions of years later by the likes of Leonardo da Vinci,

Kepler, Wheatstone, and aerial reconnaissance engineers. It evolved in response to identifiable selection pressures in the ecology of our ancestors. And it solves unsolvable problems by making tacit assumptions about the world that were true when we evolved but are not always true now.

LIGHTING, SHADING, SHAPING

Stereo vision is part of a crucial early stage of vision that figures out the depths and materials of surfaces, but it is not the only part. Seeing in three dimensions doesn't require two eyes. You can get a rich sense of shape and substance from the meagerest hints in a picture. Look at these drawings, designed by the psychologist Edward Adelson.

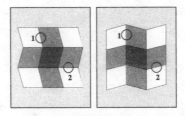

The left one appears to be white cardboard with a gray vertical stripe, folded horizontally and lit from above. The right one appears to be white cardboard with a gray horizontal stripe, folded vertically and lit from the side. (If you stare long enough, either might flip in depth, like a Necker cube; let's ignore that for now.) But the ink on the page (and the projection on your retina) is virtually the same in the two pictures. Each is a zigzag tic-tac-toe box with some of the squares shaded in. In both drawings, the corner squares are white, the top and side squares are light gray, and the middle square is a darker gray. Somehow the combination of shading and zigzagging pops them into the third dimension and colorizes each square, but in different ways. The borders labeled "1" are physically the same in the two drawings. But in the left drawing the border looks like a paint boundary—a white stripe next to a gray one—and in the right drawing it looks like a shape-and-shading boundary—a white stripe falling into a shadow on the other side of a fold. The borders labeled "2" are also identical, but you see them in the opposite way: shadow in the left drawing, paint stripe in the right one. All these differences come from one box zigging where the other one zags!

To see so much world in so little image, you have to undo three laws that make images from the world. Each needs a mental "expert" to do the undoing. Like stereo vision, these experts work to give us an accurate grasp of the world's surfaces, but they run on different kinds of information, solve different kinds of problems, and make different kinds of assumptions about the world.

⌒

The first problem is perspective: a 3-D object gets projected into a 2-D shape on the retina. Unfortunately, any projection could have come from an infinite number of objects, so there is no way to recover a shape from its projection alone (as Ames reminded his viewers). "So," evolution seems to have said, "no one's perfect." Our shape analyzer plays the odds and makes us see the most *probable* state of the world, given the retinal image.

How can a visual system calculate the most probable state of the world from the evidence on the retina? Probability theory offers a simple answer: Bayes' theorem, the most straightforward way of assigning a probability to a hypothesis based on some evidence. Bayes' theorem says that the odds favoring one hypothesis over another can be calculated from just two numbers for each hypothesis. One is the prior probability: how confident are you in the hypothesis before you even look at the evidence? The other is the likelihood: if the hypothesis *were* true, what is the probability that the evidence as you are seeing it now would have appeared? Multiply the prior probability of Hypothesis 1 by the likelihood of the evidence under Hypothesis 1. Multiply the prior probability of Hypothesis 2 by the likelihood of the evidence under Hypothesis 2. Take the ratio of the two numbers. You now have the odds in favor of the first hypothesis.

How does our 3-D line analyzer use Bayes' theorem? It puts its money on the object that has the greatest likelihood of producing those lines if it were really in the scene, and that has a good chance of being in scenes in general. It assumes, as Einstein once said about God, that the world is subtle but not malicious.

So the shape analyzer must be equipped with some probabilities about projection (how objects appear in perspective) and some probabilities about the world (what kinds of objects it has). Some of the probabilities about projection are very good indeed. A penny, theoretically, can project to a thin line, but it does so only when it is viewed edge-on. If there's a penny in the scene, what is the probability that you are viewing it edge-on? Unless someone has choreographed the two of you, not very high. The vast majority of viewpoints will make the penny project an ellipse instead. The shape analyzer assumes the current viewpoint is *generic*—not poised with pinpoint accuracy to line things up, Ames-style—and places its chips accordingly. A matchstick, on the other hand, *will* project to a straight line almost all the time, so if there is a line in an image, a stick is a better guess than a disk, all else being equal.

A *collection* of lines in an image can narrow the odds even further. For example, a set of parallel or near-parallel lines is seldom an accident. Nonparallel lines in the world rarely project near-parallel lines in an image: most pairs of sticks strewn on a floor cross at moderate to sharp angles. But lines that are parallel in the world, such as the edges of a telephone pole, almost always project near-parallel lines. So if there are near-parallel lines in an image, the odds favor parallel edges in the world. There are many other rules of thumb that say what kinds of sculptings of the world can be counted on to give off various markings in an image. Little T's, Y's, angles, arrows, crows' feet, and parallel wiggles are the fingerprints of various straight edges, corners, right angles, and symmetrical shapes. Cartoonists have exploited these rules for millennia, and a wily shape analyzer can run them backwards when betting on what is in the world.

But of course running a likelihood backwards—saying that parallel stuff usually projects near-parallel images, therefore near-parallel images imply parallel stuff—is unsound. It is like hearing hoofbeats outside your window and concluding that they came from a zebra, because zebras often make hoofbeats. The *prior* probability that the world contains some entity—how many zebras are out there, how many parallel edges are out there—has to be multiplied in. For an odds-playing shape analyzer to work, the world had better contain lots of the straight, regular, symmetrical, compact kinds of objects that it likes to guess. Does it? A romantic might think that the natural world is organic and soft, its hard edges bulldozed in by the Army Corps of Engineers. As a literature professor recently declared to his class, "Straight lines on the landscape are put there by man." A skeptical student, Gail Jensen Sanford, published a list of straight lines in nature, recently reprinted in *Harper's* magazine:

> line along the top of a breaking wave; distant edge of a prairie; paths of hard rain and hail; snow-covered fields; patterns in crystals; lines of white quartz in a granite surface; icicles, stalactites, stalagmites; surface of a calm lake; markings on zebras and tigers; bill of a duck; legs of a sand-piper; angle of migrating birds; dive of a raptor; new frond of a fern; spikes of a cactus; trunks of young, fast-growing trees; pine needles; silk strands woven by spiders; cracks in the surface of ice; strata of metamor-phic rock; sides of a volcano; wisp of windblown altocumulus clouds; inside edge of a half-moon.

Some of these are arguable, and others will do a shape guesser more harm than good. (The horizon of a lake or prairie and the edge of a half-moon do *not* come from lines in the world.) But the point is right. Many laws of the world give it nice, analyzable shapes. Motion, tension, and gravity make straight lines. Gravity makes right angles. Cohesion makes smooth contours. Organisms that move evolve to be symmetrical. Nat-ural selection shapes their body parts into tools, duplicating the human engineer's demand for well-machined parts. Large surfaces collect pat-terns with roughly equal sizes, shapes, and spacing: cracks, leaves, peb-bles, sand, ripples, needles. Not only are the seemingly carpentered and wallpapered parts of the world the parts most recoverable by a shape analyzer; they are the parts most worth recovering. They are the telltale signs of potent forces that fill and shape the environment at hand, and are more worthy of attention than heaps of random detritus.

~

Even the best line analyzer is equipped only for a cartoon world. Sur-faces are not just bounded by lines; they are composed of material. Our sense of lightness and color is a way of assaying materials. We avoid bit-ing into a plaster apple because the color tips us off that it is not made of fruit flesh.

Analyzing matter from the light it reflects is a job for a reflectance specialist. Different kinds of matter reflect back different wavelengths of light in different amounts. (To keep things simple, I'll stay in black and white; color is, roughly, the same problem multiplied by three.)

Unfortunately, a given amount of reflected light could have come from an infinite number of combinations of matter and lighting. One hundred units of light could have come from coal reflecting back 10% of the light of 1,000 candles or from snow reflecting back 90% of the light of 111 candles. So there is no foolproof way to deduce an object's material from its reflected light. The lightness analyzer must somehow factor out the level of illumination. This is another ill-posed problem, exactly equivalent to this one: I give you a number, you tell me which two numbers were multiplied to get it. The problem can be solved only by adding in assumptions.

A camera is faced with the same task—how to render the snowball as white whether it is indoors or out. A camera's meter, which controls the amount of light reaching the film, embodies two assumptions. The first is that lighting is uniform: the whole scene is in sun, or in shade, or under a lightbulb. When the assumption is violated, the snapshooter is disappointed. Aunt Mimi is a muddy silhouette against the blue sky because the camera is fooled by her face being in shade while the sky is lit directly by the sun. The second assumption is that the scene is, on average, medium gray. If you throw together a random collection of objects, their many colors and lightnesses will usually average out to a medium shade of gray that reflects back 18% of the light. The camera "assumes" it is looking at an average scene and lets in just enough light to make the middle of the range of lightnesses in the scene come out as medium gray on the film. Patches that are lighter than the middle are rendered pale gray and white; patches that are darker, deep gray and black. But when the assumption is wrong and the scene does not really average out to gray, the camera is fooled. A picture of a black cat on black velvet comes out medium gray, a picture of a polar bear on the snow comes out medium gray, and so on. A skilled photographer analyzes how a scene differs from the average scene and uses various tricks to compensate. A crude but effective one is to carry around a standard medium gray card (which reflects back exactly 18% of the light), lean it on the subject, and aim the meter at the card. The camera's assumption about the world is now satisfied, and its estimate of the ambient illumination level (made by dividing the light reflecting off the card by 18%) is guaranteed to be correct.

Edwin Land, inventor of the polarizing filter and the instant Polaroid Land camera, was challenged by this problem, which is all the more vexing in color photography. Light from lightbulbs is orange; light from fluorescents is olive; light from the sun is yellow; light from the sky is blue. Our brain somehow factors out the color of the illumination, just as it factors out the intensity of the illumination, and sees an object in its correct color in all those lights. Cameras don't. Unless they send out their *own* white light from a flash, they render an indoor scene with a thick rusty cast, a shady scene as pasty blue, and so on. A knowledgeable photographer can buy special film or screw a filter on the lens to compensate, and a good lab technician can correct the color when printing the photograph, but an instant camera obviously cannot. So Land had a practical interest in how to remove the intensity and color of the illumination, a problem called color constancy.

But he was also a self-taught, ingenious perception scientist, curious about how the brain solves the problem. He set up a color perception lab and developed a clever theory of color constancy. His idea, called the Retinex theory, gave the perceiver several assumptions. One is that earthly illumination is a rich mixture of wavelengths. (The exception that proves the rule is the sodium vapor lamp, the energy-saving fixture found in parking lots. It sends out a narrow range of wavelengths which our perception system can't factor out; cars and faces are dyed a ghastly yellow.) The second assumption is that gradual changes in brightness and color across the visual field probably come from the way the scene is illuminated, whereas abrupt transitions probably come from the boundary where one object ends and another begins. To keep things simple, he tested people and his model on artificial worlds composed of 2-D rectangular patches, which he called Mondrians, after the Dutch painter. In a Mondrian lit from the side, a yellow patch at one edge can reflect very different light from the same yellow patch at the other. But people see them both as yellow, and the Retinex model, which removes the lighting gradient from edge to edge, does too.

The Retinex theory was a good start, but it turned out to be too simple. One problem is the assumption that the world is a Mondrian, a big flat plane. Go back to Adelson's drawings on page 242, which are zigzag Mondrians. The Retinex model treats all sharp boundaries alike, interpreting Edge 1 in the left drawing like Edge 1 in the right drawing. But to you, the left one looks like a border between stripes of different colors, and the right one looks like a single stripe that is folded and partly in shade. The difference comes from your interpretation of 3-D shape. Your shape analyzer has bent the Mondrians into striped room dividers, but the Retinex model sees them as the same old checkerboard. Obviously, it is missing something.

~

That something is the effect of slant on shading, the third law that turns a scene into an image. A surface facing a light source head-on reflects back a lot of light, because the light smacks into the surface and rebounds right back. A surface angled almost parallel to the source reflects much less, because most of the light grazes off it and continues on its way. If you are positioned near the light source, your eye picks up more light when the surface faces you than when it faces almost sideways. You may be able to see the difference by rotating a flashlight aimed at a piece of gray cardboard.

How might our shading analyzer run the law backwards and figure out how a surface is slanted based on how much light it reflects? The benefits go beyond estimating the slant of a panel. Many objects, like cubes and gems, are composed of slanted faces, so recovering the slants is a way to ascertain their shape. In fact, any shape can be thought of as a carving made up of millions of tiny facets. Even when the surface is smoothly curved so the "facets" shrink to points, the shading law applies to the light coming off each point. If the law could be run backwards, our shading analyzer could apprehend the shape of a surface by registering the slant of the tangent plane resting on each point.

Unfortunately, a given amount of light reflecting off a patch could have come from a dark surface angled toward the light or from a light surface angled away. So there is no foolproof way to recover a surface's angle from the light it reflects without making additional assumptions.

A first assumption is that surface lightness is uniform: the world is made of plaster. When surfaces are unevenly pigmented, the assumption is violated, and our shading analyzer should be fooled. It is. Paintings and photographs are the most obvious example. A less obvious one is countershading in animal camouflage. The hides of many animals lighten from back to belly in a gradient that cancels out the effects of light on their 3-D shapes. This flattens the animal, making it harder to detect by the assumption-making, shape-from-shading analyzer in the brain of a predator. Makeup is another example. When applied in sub-Tammy Faye Bakker amounts, pigment on the skin can fool the beholder into seeing the flesh and bone as having a more ideal shape. Dark blush on the sides of the nose makes them look as if they are at a shallower angle to the light, which makes the nose appear narrower. White powder on the upper lip works the other way: the lip seems to intercept the light source head-on as if it were fuller, bestowing that desirable pouty look.

The shape-from-shading analyzer has to make other assumptions, too. Surfaces in the world are made of thousands of materials, and light bounces off their slanted surfaces in very different ways. A matte surface like chalk or dull paper follows a simple law, and the brain's shading analyzer often seems to assume that the world is matte. Surfaces with glosses, patinas, fuzz, pits, and prickles do other, stranger things with light, and they can fool the eye.

A famous example is the full moon. It looks like a flat disk, but of course it is a sphere. We have no trouble seeing other spheres from their shading, like ping-pong balls, and any good artist can sketch a sphere with charcoal. The problem with the moon is that it is pockmarked with craters of all sizes, most too small to be discerned from the earth, and they combine into a surface that behaves differently from the matte ideal that our shading analyzer takes for granted. The center of the full moon faces the viewer flat-on, so it should be brightest, but it has little nooks and crannies whose walls are seen edge-on from the viewer's earthly vantage point, making the center of the moon look darker. The surfaces near the perimeter of the moon graze the line of sight and should look darker, but they present their canyon walls face-on and reflect back lots of light, making the perimeter look lighter. Over the whole moon, the angle of its surface and the angles of the facets of its craters cancel out. All portions reflect back the same amount of light, and the eye sees it as a disk.

~

If we had to depend on any one of these analyzers, we would be eating bark and stepping off cliffs. Each analyzer makes assumptions, but those assumptions are often contradicted by other analyzers. Angle, shape, material, lighting—they're all scrambled together, but somehow we unscramble them and see one shape, with one color, at one angle, in one kind of light. What's the trick?

Adelson, together with the psychologist Alex Pentland, used his zigzag illusion in a little parable. You are a designer who must build a stage set that looks just like the right-hand diagram. You go to a workshop where specialists build scenery for dramatic productions. One is a lighting designer. Another is a painter. A third is a sheet-metal worker. You show them the picture and ask them to build a scene that looks like it. In effect, they have to do what the visual system does: given an image, figure out the arrangement of matter and lighting that could have brought it about.

There are many ways the specialists can satisfy you. Each could almost do it alone. The painter could simply paint the arrangement of parallelograms on a flat sheet of metal and ask the lighting designer to illuminate it with a single flood:

The lighting designer could take a plain white sheet and set up nine custom spotlights, each with a special mask and filter, aimed just right to project nine parallelograms onto the sheet (six of the spotlights are shown here):

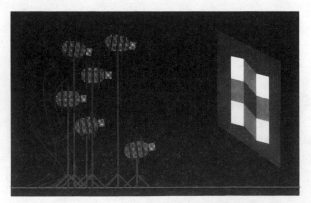

The sheet-metal worker could bend some metal into special shapes that when illuminated and viewed from just the right angle give rise to the image:

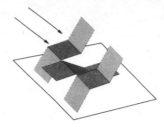

Finally, the figure could be produced by the specialists cooperating. The painter would paint a stripe across the middle of a square sheet of metal, the sheet-metal worker would bend it into a zigzag, and the lighting designer would illuminate the piece with a floodlight. That, of course, is how a human being interprets the image.

Our brain faces the same embarrassment of riches as the set designer in the parable. Once we allow in a mental "expert" that can hypothesize pigmented surfaces out there, it could explain everything in the image as paint: the world would be seen as a masterful trompe l'oeil. Likewise, a lighting expert in the head could tell us that the world is a movie. Since these interpretations are undesirable, the mental specialists should somehow be discouraged from making them. One way would be to force them to stick with their assumptions, come what may (color and lighting are even, shapes are regular and parallel), but that's too extreme. The world is not always a pile of blocks on a sunny day; sometimes it *does* have complicated pigments and lighting, and we see them. We don't want the experts to deny that the world can be complex. We want them to propose exactly as much complexity as there is in the world, and no more. The problem now is how to get them all to do it.

Return now to the parable. Suppose the set design department is on a budget. The specialists charge for their services, using a fee schedule that reflects how difficult and unusual a request is. Simple and common operations are cheap; complex and unusual operations are expensive.

Painter Fees:

Paint a rectangular patch:	$5 each
Paint a regular polygon:	$5 per side

Sheet-Metal Worker Fees:

Right-angle cuts:	$2 each
Odd-angle cuts:	$5 each
Right-angle bends:	$2 each
Odd-angle bends:	$5 each

> **Lighting Designer Fees:**
> Floodlight: $5 each
> Custom spotlight: $30 each

We need one more specialist: a supervisor, who decides how to contract out the job.

> **Supervisor Fees**
> Consultation: $30 per job

The prices for the four solutions will differ. Here are the estimates:

> **Painter's Solution:**
> Paint 9 polygons: $180
> Set up 1 floodlight: $5
> Cut 1 rectangle: $8
> **Total:** **$193**

> **Lighting Designer's Solution:**
> Cut 1 rectangle: $8
> Set up 9 custom spotlights: $270
> **Total:** **$278**

> **Sheet-Metal Worker's Solution:**
> Cut 24 odd angles: $120
> Bend 6 odd angles: $30
> Set up 1 floodlight: $5
> **Total:** **$155**

> **Supervisor's Solution:**
> Cut 1 rectangle: $8
> Bend 2 right angles: $4
> Paint 3 rectangles: $15
> Set up 1 floodlight: $5
> Supervisor's fee: $30
> **Total:** **$62**

The supervisor's solution is the cheapest because it uses each specialist optimally, and the savings more than make up for the supervisor's fee. The moral is that the specialists must be coordinated, not necessarily by a homunculus or demon, but by some arrangement that minimizes the costs, where cheap equals simple equals probable. In the parable, simple operations are easier to perform; in the visual system, simpler descriptions correspond to likelier arrangements in the world.

Adelson and Pentland have brought their parable to life by programming a computer simulation of vision that is designed to interpret scenes with painted polygons much as we do. First, a shape analyzer (a software version of the sheet-metal worker) strives for the most regular shape that duplicates the image. Take the simple shape on the left in this diagram, which people see as a folded sheet, like a book held sideways.

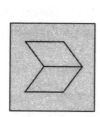 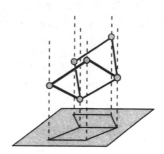

The shape specialist tries to assemble a 3-D model of the input shape, shown on the right. When it begins, all it knows is that the corners and edges in the model have to line up with the dots and lines in the image; it does not know how far away they are in depth. The model's vertices are beads sliding on rods (like rays of projection), and the lines between them are infinitely elastic strings. The specialist slides the beads around until it arrives at a shape with the following desiderata. Each polygon making up the shape should be as regular as possible; that is, a polygon's angles should not be too different. For example, if the polygon has four sides, the specialist strives for a rectangle. The polygon should be as planar as possible, as if the polygon is filled in with a plastic panel that is hard to bend. And the polygons should be as compact as possible, rather than elongated along the line of sight, as if the plastic panel is also hard to stretch.

When the shape specialist is done, it passes on a rigid assembly of white panels to the lighting specialist. The lighting specialist knows the laws that dictate how reflected light depends on the illumination, the lightness of the surface, and the angle of the surface. The specialist is allowed to move a single distant light source around to illuminate the model from various directions. The optimal direction is the one that makes each pair of panels meeting at an edge look as much as possible like their counterparts in the image, requiring as little gray paint as possible to finish the job.

Finally, the reflectance specialist—the painter—gets the model. It is the specialist of last resort, and its task is to take care of any remaining discrepancy between the image and the model. It finishes the job by proposing different shades of pigmentation for the various surfaces.

Does the program work? Adelson and Pentland presented it with a fanfold object and let it rip. The program displays its current guess about the object's shape (first column), its current guess about the direction of the light source (second column), its current guess about where the shadows fall (third column), and its current guess about how the object is painted (fourth column). The program's very first guesses are shown in the top row.

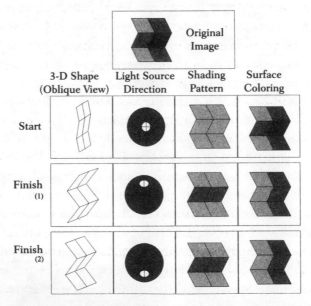

The program initially assumed that the object was flat, like a 2-D painting lying on a table, as in the top of the first column. (It is hard to depict this for you, because your brain insists on seeing a zigzag shape as being folded in depth. The sketch is trying to show some lines sitting flat on the page.) The program assumed the light source was head-on, from the direction of the eye (top of the second column). With this flat lighting, there are no shadows (top of the third column). The reflectance specialist bears all the responsibility for duplicating the image, and it just paints it in. The program thinks it is looking at a painting.

Once the program has a chance to adjust its guesses, it settles into the interpretation shown in the middle row. The shape specialist finds the most regular 3-D shape (shown in side view in the left column): square panels joined at right angles. The lighting specialist finds that by shining the light from above, it can make the play of shadows look something like the image. Finally the reflectance specialist touches up the model with paint. The four columns—zigzag 3-D shape, lighting from above, shadow in the middle, light stripe next to a darker one—correspond to how people interpret the original image.

Does the program do anything else reminiscent of humans? Remember how the fanfold flips in depth like a Necker cube. The outer fold becomes an inner one, and vice versa. The program, in a way, can see the flip, too; the flipped interpretation is shown in the bottom row. The program assigned the same costs to the two interpretations and arrived at one or the other randomly. When people see a 3-D shape flip, they usually see the direction of the light source flip, too: top fold out, light from above; bottom fold out, light from below. The program does the same. Unlike a person, the program does not actually flip between the two interpretations, but if Adelson and Pentland had had the specialists pass around their guesses in a constraint network (like the Necker cube network on p. 107 or the stereo vision model), rather than in an assembly line, it might have done so.

The workshop parable clarifies the idea that the mind is a collection of modules, a system of organs, or a society of experts. Experts are needed because expertise is needed: the mind's problems are too technical and specialized to be solved by a jack-of-all-trades. And most of the information needed by one expert is irrelevant to another and would only interfere with its job. But working in isolation, an expert can consider too many solutions or doggedly pursue an unlikely one; at some point the experts must confer. The many experts are trying to make sense of a single world, and that world is indifferent to their travails, neither offering easy solutions nor going out of its way to befuddle. So a supervisory scheme should aim to keep the experts within a budget in which improbable guesses are more expensive. That forces them to cooperate in assembling the most likely overall guess about the state of the world.

SEEING IN TWO AND A HALF DIMENSIONS

Once the experts have completed their work, what do they post on the blackboard that the rest of the brain accesses? If we could somehow show the visual field from a rest-of-the-brain's-eye view, like the hypothetical camera behind the eye of the Terminator, what would it look like? The very question may sound like a thick-witted little-man-in-the-head fallacy, but it is not. It is about the information in one of the brain's data representations and the form the information takes. Indeed, taking the question seriously sends a bracing shock to our naive intuitions about the mind's eye.

The experts in stereo, motion, contour, and shading have worked hard to recover the third dimension. It would be natural to use the fruits of their labors to build a three-dimensional representation of the world. The retinal mosaic in which the scene is depicted gives way to a mental sandbox in which it is sculpted; the picture becomes a scale model. A 3-D model would correspond to our ultimate understanding of the world. When a child looms up to us and then shrinks away, we know we are not in Wonderland, where one pill makes you larger and one pill makes you small. And unlike the proverbial (and apocryphal) ostrich, we do not think that objects vanish when we look away or cover them up. We negotiate reality because our thought and action are guided by knowledge of a large, stable, solid world. Perhaps vision gives us that knowledge in the form of a scale model.

There is nothing inherently fishy about the scale-model theory. Many computer-aided design programs use software models of solid objects, and CAT-scan and MRI machines use sophisticated algorithms to assemble them. A 3-D model might have a list of the millions of coordinates of the tiny cubes that make up a solid object, called volume elements or "voxels" by analogy to the picture elements or "pixels" making up a picture. Each coordinate-triplet is paired with a piece of information, such as the density of the tissue at that spot in the body. Of course, if the brain stored voxels, they would not have to be arranged in a 3-D cube in the head, any more than voxels are arranged in a 3-D cube inside a computer. All that matters is that each voxel have a consistent set of neurons dedicated to it, so the patterns of firing can register the contents of the voxel.

But now is the time to be vigilant about the homunculus. There is no problem with the idea that some software demon or look-up algorithm or neural network accesses information from a scale model, as long as we are clear that it accesses the information *directly*: coordinates of a voxel in, contents of the voxel out. Just don't think about the look-up algorithm *seeing* the scale model. It's pitch black in there, and the looker-upper doesn't have a lens or a retina or even a vantage point; it is anywhere and everywhere. There is no projection, no perspective, no field of view, no occlusion. Indeed, the whole point of the scale model is to eliminate these nuisances. If you want to think of a homunculus at all, imagine exploring a room-sized scale model of a city in the dark. You can wander through it, coming at a building from any direction, palpating its exterior or sticking fingers through windows and doors to probe its insides. When you grasp a building, its sides are always parallel, whether you are at arm's length or up close. Or think about feeling the shape of a small toy in your hands, or a candy in your mouth.

But vision—even the 3-D, illusion-free vision that the brain works so hard to achieve—is nothing like that! At best, we have an abstract appreciation of the stable structure of the world around us; the immediate, resplendent sense of color and form that fills our awareness when our eyes are open is completely different.

First, vision is not a theater in the round. We vividly experience only what is in front of our eyes; the world beyond the perimeter of the visual field and behind the head is known only in a vague, almost intellectual way. (I *know* there is a bookshelf behind me and a window in front of me, but I *see* only the window, not the bookshelf.) Worse, the eyes flit from spot to spot several times a second, and outside the crosshairs of the fovea the view is surprisingly coarse. (Hold your hand a few inches from your line of sight; it is impossible to count the fingers.) I am not just reviewing the anatomy of the eyeball. One could imagine the brain assembling a collage out of the snapshots taken at each glimpse, like the panoramic cameras that expose a frame of film, pan a precise amount, expose the adjacent stretch of film, and so on, yielding a seamless wide-angle picture. But the brain is not a panoramic camera. Laboratory studies have shown that when people move their eyes or head, they immediately lose the graphic details of what they were looking at.

Second, we don't have x-ray vision. We see surfaces, not volumes. If you watch me put an object inside a box or behind a tree, you know it's there but don't *see* it there and cannot report its details. Once again, this is not just a reminder that you are not Superman. We mortals could have been equipped with a photographic memory that updates a 3-D model by pasting in information from previous views wherever it belongs. But we were not so equipped. When it comes to rich visual detail, out of sight is out of mind.

Third, we see in perspective. When you stand between railroad tracks, they seem to converge toward the horizon. Of course you know they do not *really* converge; if they did, the train would derail. But it's impossible not to *see* them as converging, even though your sense of depth provides plenty of information that your brain could use to cancel the effect. We also are aware that moving objects loom, shrink, and foreshorten. In a genuine scale model, none of this can happen. To be sure, the visual system eliminates perspective to a certain degree. People other than artists have trouble seeing that the near corner of a desk projects an acute angle and the far corner an obtuse angle; both look like the right angles they are in reality. But the railroad tracks show that perspective is not completely eliminated.

Fourth, in a strict geometric sense we see in two dimensions, not three. The mathematician Henri Poincaré came up with an easy way to determine the number of dimensions of some entity. Find an object that can divide the entity into two pieces, then count the dimensions of the divider and add one. A point cannot be divided at all; therefore, it has zero dimensions. A line has one dimension, because it can be severed by a point. A plane has two dimensions, because it can be rent by a line, though not by a point. A sphere has three, because nothing less than a two-dimensional blade can cleave it; a pellet or a needle leaves it whole. What about the visual field? It can be sundered by a line. The horizon, for example, divides the visual field in two. When we stand in front of a taut cable, everything we see is on one side or the other. The perimeter of a round table also partitions the visual field: every point is either within it or outside. Add one to the one-dimensionality of a line, and you get two. By this criterion, the visual field is two-dimensional. Incidentally, this does not mean that the visual field is *flat*. Two-dimensional surfaces can be curved in the third dimension, like a rubber mold or a blister package.

Fifth, we don't immediately see "objects," the movable hunks of matter that we count, classify, and label with nouns. As far as vision is concerned, it's not even clear what an object is. When David Marr considered how to design a computer vision system that finds objects, he was forced to ask:

Is a nose an object? Is a head one? Is it still one if it is attached to a body? What about a man on horseback? These questions show that the difficulties in trying to formulate what should be recovered as a region from an image are so great as to amount almost to philosophical problems. There is really no answer to them—all these things can be an object if you want to think of them that way, or they can be part of a larger object.

A drop of Krazy Glue can turn two objects into one, but the visual system has no way of knowing that.

We have, however, an almost palpable sense of *surfaces* and the *boundaries* between them. The most famous illusions in psychology come from the brain's unflagging struggle to carve the visual field into surfaces and to decide which is in front of the other. One example is the Rubin face-vase, which flips between a goblet and a pair of profiles tête-à-tête. The faces and vase cannot be seen at the same time (even if one imagines two men holding up a goblet between their noses), and whichever shape predominates "owns" the border as its demarcating line, relegating the other patch to an amorphous backdrop.

Another is the Kanisza triangle, a stretch of nothingness that blocks out a shape as real as if it had inscribed it in ink.

The faces, vase, and triangle are familiar objects, but the illusions do not depend on their familiarity; meaningless blobs are just as compelling.

We perceive surfaces involuntarily, impelled by information surging up from our retinas; contrary to popular belief, we do not see what we expect to see.

So what is the product of vision? Marr called it a $2\frac{1}{2}$-D sketch; others call it a visible surface representation. Depth is whimsically downgraded to half a dimension because it does not define the medium in which visual information is held (unlike the left-right and high-low dimensions); it is just a piece of information held in that medium. Think of the toy made of hundreds of sliding pins which you press against a 3-D surface (such as a face), forming a template of the surface in the contour of the pins on the other side. The contour has three dimensions, but they are not created equal. Position from side to side and position from top to bottom are defined by particular pins; position in depth is defined by how far a pin protrudes. For any depth there may be many pins; for any pin there is only one depth.

The $2\frac{1}{2}$-D sketch looks a bit like this:

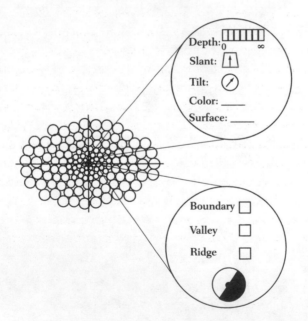

It is a mosaic of cells or pixels, each dedicated to a line of sight from the cyclopean eye's vantage point. It is wider than it is tall because our two eyes sit side by side in our skulls rather than one being above the other. The cells are smaller in the center of the visual field than in the periphery because our resolution is greater in the center. Each cell can represent information about a surface or about an edge, as if it had two kinds of forms with blanks to be filled in. The form for a piece of surface has blanks for depth, for slant (how much the surface leans backward or forward), for tilt (how much it lists left or right), and for color, plus a label for which surface it is seen as belonging to. The form for a piece of edge has boxes to be checked, indicating whether it is at the boundary of an object, a groove, or a ridge, plus a dial for its orientation, which also shows (in the case of an object boundary) which side belongs to the surface that "owns" the boundary and which side is merely the backdrop. Of course, we won't literally find bureaucratic forms in the head. The diagram is a composite that depicts the *kinds* of information in the 2½-D sketch. The brain presumably uses clusters of neurons and their activities to hold the information, and they may be distributed across different patches of cortex as a collection of maps that are accessed in register.

Why do we see in two and a half dimensions? Why not a model in the head? The costs and benefits of storage give part of the answer. Any computer user knows that graphics files are voracious consumers of storage space. Rather than agglomerating the incoming gigabytes into a composite model, which would be obsolete as soon as anything moved, the brain lets the world itself store the information that falls outside a glance. Our heads crane, our eyes flit, and a new, up-to-date sketch is loaded in. As for the second-class status of the third dimension, it is almost inevitable. Unlike the other two dimensions, which announce themselves in the rods and cones that are currently active, depth must be painstakingly wrung out of the data. The stereo, contour, shading, and motion experts that work on computing depth are equipped to send along information about distance, slant, tilt, and occlusion relative to the viewer, not 3-D coordinates in the world. The best they can do is to pool their efforts to give us a two-and-a-half-dimensional acquaintance with the surfaces in front of our eyes. It's up to the rest of the brain to figure out how to use it.

FRAMES OF REFERENCE

The 2½-D sketch is the masterwork of the ingeniously designed, harmoniously running machinery of the visual system. It has only one problem. As delivered, it is useless.

Information in the 2½-D array is specified in a *retinal* frame of reference, a coordinate system centered on the viewer. If a particular cell says, "There's an edge here," what "here" means is the position of that cell on the retina—say, dead straight ahead where you're looking. That would be fine if you were a tree looking at another tree, but as soon as something moves—your eyes, your head, your body, a sighted object—the information lurches to a new resting place in the array. Any part of the brain being guided by information in the array would find that its information is now defunct. If your hand was being guided toward the center of the visual field because that spot had contained an apple, the hand would now be heading toward empty space. If yesterday you memorized an image of your car as you were looking at its door handle, today the image would not match your view of the fender; the two views would barely overlap. You can't even make simple judgments like whether two lines are parallel; remember the converging railroad tracks.

These problems make one long for a scale model in the head, but that isn't what vision delivers. The key to using visual information is not to remold it but to *access* it properly, and that calls for a useful reference frame or coordinate system. Reference frames are inextricable from the very idea of location. How do you answer the question "Where is it?" By naming an object that the asker already knows—the frame of reference—and describing how far and in what direction the "it" is, relative to the frame. A description in words like "next to the fridge," a street address, compass directions, latitude and longitude, Global Positioning System satellite coordinates—they all indicate distance and direction relative to a reference frame. Einstein built his theory of relativity by questioning Newton's fictitious reference frame that was somehow anchored in empty space, independent of anything in it.

The frame of reference packaged with the $2^1/_2$-D sketch is position on the retina. Since the retinas constantly gyrate, it is as useless as directions like "Meet me next to the beige Pontiac that's stopped here at the light." We need a reference frame that stays put as the eyes rock and roll. Suppose there is a circuit that can slide an invisible reference frame over the visual field, like the crosshairs of a rifle sight sliding over a landscape. And suppose that any mechanism that scoops information out of the visual field is locked onto positions defined by the rifle sight (for example, at the hair-crossing, two notches above them, or a notch to the left). Computer displays have a vaguely similar device, the cursor. The commands that read and write information do so relative to a special point that can be positioned at will over the screen, and when the material on the screen scrolls around, the cursor moves with it, glued to its piece of text or graphics. For the brain to use the contents of the 2½-D sketch, it must employ a similar mechanism, indeed, several of them.

The simplest reference frame that moves over the 2½-D sketch is one that stays riveted to the head. Thanks to the laws of optics, when the eyes move right, the image of the apple scoots left. But suppose the neural command to the eye muscles is cc'd to the visual field, and is used to shift the crosshairs over by the same amount in the opposite direction. The crosshairs will stay on the apple, and so will any mental process that funnels information through the crosshairs. The process can happily continue as if nothing had happened, even though the contents of the visual field have slid around.

Here's a simple demonstration of the cc'ing. Move your eyes; the world stands still. Now close one eye and nudge the other one with your finger; the world jumps. In both cases the eye moves, and in both cases the retinal image moves, but only when the eye is moved by a finger do you see the movement. When you move your eyes by deciding to look somewhere, the command to the eye muscles is copied to a mechanism that moves the reference frame together with the sliding images so as to cancel your subjective sense of motion. But when you move your eye by poking it with your finger, the frame-shifter is bypassed, the frame is not shifted, and you interpret the jerking image as coming from a jerking world.

There may also be reference frames that compensate for movements of the head and body. They give each bit of surface in the visual field a fixed address relative to the room or relative to the ground; the address stays the same as the body moves. These frame shifts might be driven by copies of commands to the neck and body muscles, though they may also be driven by circuitry that tracks the slippage of the contents of the visual field.

Another handy overlay would be a trapezoidal mental grid that marked out equal-sized extents in the world. A gridmark near our feet would cover a large stretch of the visual field; a gridmark near the horizon would cover a smaller stretch of the visual field but the same number of inches as measured along the ground. Since the 2½-D sketch contains depth values at every point, the gridmarks would be easy for the brain to calculate. This world-aligned reference frame would allow us to judge the genuine angles and extents of the matter outside our skin. The perceptual psychologist J. J. Gibson argued that we do have this sense of real-world scale superimposed on the retinal projection, and we can mentally flip between not using it and using it. Standing between the railroad tracks, we can assume one frame of mind in which we see the tracks converge, or another in which we see them as parallel. These two attitudes, which Gibson called "the visual field" and "the visual world," come from accessing the same information by either the retinal frame or a world-aligned frame.

Yet another invisible frame is the direction of gravity. The mental plumb bob comes from the vestibular system of the inner ear, a labyrinth of chambers that includes three semicircular canals oriented at right angles to each other. If anyone doubts that natural selection uses principles of engineering rediscovered by humans, let them behold the XYZ Cartesian coordinate axes etched into the bones of the skull! As the head pitches, rolls, and yaws, fluid in the canals sloshes around and triggers neural signals registering the motion. A heavy mass of grit pressing down on other membranes registers linear motion and the direction of gravity. These signals can be used to rotate the mental crosshairs so they are always correctly pointing "up." That is why the world does not seem to list even though people's heads are seldom plumb perpendicular. (The eyes themselves tilt clockwise and counterclockwise in the head, but only enough to undo small head tilts.) Oddly enough, our brains do not compensate for gravity very much. If the compensation were perfect, the world would look normal when we are lying sideways or even standing on our heads. Of course, it does not. It's hard to watch television lying on your side unless you prop your head on your hand, and it's impossible to read unless you hold the book sideways. Perhaps because we are terrestrial creatures, we use the gravity signal mostly to keep our bodies upright rather than to compensate for out-of-kilter visual input when they are not.

The coordination of the retina's frame with the inner ear's frame affects our lives in a surprising way: it causes motion sickness. Ordinarily, when you move about, two signals work in synchrony: the swoops of texture and color in the visual field, and the messages about gravity and inertia sent by the inner ear. But if you are moving inside a container like a car, a boat, or a sedan chair—evolutionarily unprecedented ways to get around—the inner ear says, "You're moving," but the walls and floor say,

"You're staying put." Motion sickness is triggered by this mismatch, and the standard treatments have you eliminate it: don't read; look out the window; stare at the horizon.

Many astronauts are chronically space-sick, because there *is* no gravitational signal, a rather extreme mismatch between gravity and vision. (Space-sickness is measured in *garns*, a unit named after the Republican senator from Utah, Jake Garn, who parlayed his position on the NASA appropriations subcommittee into the ultimate junket, a trip into space. Space Cadet Garn made history as the all-time champion upchucker.) Worse, spacecraft interiors do not give the astronauts a world-aligned frame of reference, because the designers figure that without gravity the concepts "floor," "ceiling," and "walls" are meaningless, so they might as well put instruments on all six surfaces. The astronauts, unfortunately, carry their terrestrial brains with them and literally get lost unless they stop and say to themselves, "I'm going to pretend that thataway is 'up,' thataway is 'forward,'" and so on. It works for a while, but if they look out the window and see terra firma above them, or catch sight of a crewmate floating upside down, a wave of nausea slams them. Space sickness is a concern to NASA, and not only because of the decline in productivity during expensive flight time; you can well imagine the complications of vomiting in zero gravity. It will also affect the burgeoning technology of virtual reality, in which a person wears a wide-field helmet showing a synthetic world whizzing by. *Newsweek*'s assessment: "The most barfogenic invention since the Tilt-a-Whirl. We prefer Budweiser."

Why on earth—or space—should a mismatch between vision and gravity or inertia lead, of all things, to nausea? What does up-and-down have to do with the gut? The psychologist Michel Treisman has come up with a plausible though still unproven explanation. Animals vomit to expel toxins they have eaten before the toxins do further harm. Many naturally occurring toxins act on the nervous system. This raises the problem faced by Ingrid Bergman in *Notorious*: how do you know when you have been poisoned? Your judgment would be addled, but that would affect your judgment about whether your judgment has been addled! More generally, how could a malfunction detector distinguish between the brain's malfunctioning and its accurately registering an unusual situation? (Old bumper sticker: "The world is experiencing technical difficulties. Do not adjust your mind.") Gravity, of course, is the most stable, predictable feature of the world. If two parts of the brain have different opinions about it, chances are that one or both is malfunctioning or that the signals they are

getting have been delayed or garbled. The rule would be: if you think gravity is acting up, you've been poisoned; jettison the rest of the poison, now.

The mental up-down axis is also a powerful organizer of our sense of
shape and form. What do we have here?

Few people recognize that it is an outline of Africa rotated ninety
degrees, even if they tilt their heads counterclockwise. The mental representation of a shape—how our minds "describe" it—does not just
reflect its Euclidean geometry, which remains unchanged as a shape is
turned. It reflects the geometry relative to our up-down reference
frame. Our minds think of Africa as a thing with a fat bit "at the top"
and a skinny bit "at the bottom." Change what's at the top and what's at
the bottom, and it's no longer Africa, even if not a jot of coastline has
been altered.

The psychologist Irvin Rock has found many other examples, including this simple one:

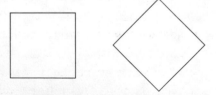

People see the drawings as two different shapes, a square and a diamond. But as far as a geometer is concerned, they are one and the same
shape. They are pegs that fit the same holes; every angle and line is the
same. The only difference is in how they are aligned with respect to
the viewer's up-and-down reference frame, and that difference is enough
to earn them different words in the English language. A square is flat on
top, a diamond is pointy on top; there's no avoiding the "on top." It is
even hard to see that the diamond is made of right angles.

Finally, objects themselves can plot out reference frames:

The shape at the top right flips between looking like a square and looking like a diamond, depending on whether you mentally group it with the three shapes to its left or the eight shapes below. The imaginary lines aligned with the rows of shapes have become Cartesian reference frames—one frame aligned with the retinal up-down, the other tilted diagonally—and a shape looks different when it is mentally described within one or the other.

And in case you are still skeptical about all these colorless, odorless, and tasteless reference frames allegedly overlaying the visual field, I give you a wonderfully simple demonstration from the psychologist Fred Attneave. What is going on in the triangles on the left?

Look at them long enough, and they snap from one appearance to another. They don't move around, they don't reverse in depth, but something changes. People describe the change as "which way they point." What is leaping around the page is not the triangles themselves but a mental frame of reference overlaying the triangles. The frame comes not from the retina, the head, the body, the room, the page, or gravity, but from an axis of symmetry of the triangles. The triangles have three such axes, and they take turns dominating. Each axis has the equivalent of a north and a south pole, which grant the feeling that the triangles are pointing. The triangles flip en masse, as if in a chorus line; the brain likes its reference frames to embrace entire neighborhoods of shapes. The triangles in the right diagram are even more jumpy, hopping among six impressions. They can be interpreted either as obtuse triangles lying flat on the page or as right-angle triangles standing in depth, each with a reference frame that can sit three ways.

ANIMAL CRACKERS

The ability of objects to attract reference frames to themselves helps to solve one of the great problems in vision, the next problem we face as we continue our climb from the retina to abstract thought. How do people recognize shapes? An average adult knows names for about ten thousand things, most of them distinguished by shape. Even a six-year-old knows names for a few thousand, having learned them at a rate of one every few hours for years. Of course, objects can be recognized from many give-aways. Some can be recognized by their sounds and smells, and others, such as shirts in a hamper, can be identified only by their color and material. But most objects can be recognized by their shapes. When we recognize an object's shape, we are acting as pure geometers, surveying the distribution of matter in space and finding the closest match in memory. The mental geometer must be acute indeed, for a three-year-old can look through a box of animal crackers or a pile of garish plastic chips and rattle off the names of exotic fauna from their silhouettes.

The diagram at the bottom of page 9 introduced you to why the problem is so hard. When an object or the viewer moves, the contours in the $2^1/_2$-D sketch change. If your memory for the shape—say, a suitcase—was a copy of the $2^1/_2$-D sketch when you first saw it, the moved version would no longer match. Your memory of a suitcase would be "a rectangular slab and a horizontal handle at twelve o'clock," but the handle you are now looking at is not horizontal and not at twelve o'clock. You would stare blankly, not knowing what it is.

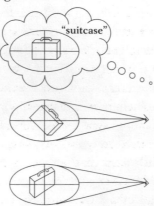

But suppose that instead of using the retinal reference frame, your memory file uses a frame aligned with the object itself. Your memory would be "a rectangular slab with a handle *parallel to the edge of the slab, at the top of the slab.*" The "of the slab" part means that you remember the positions of the parts relative to the object itself, not relative to the visual field. Then, when you see an unidentified object, your visual system would automatically align a 3-D reference frame on it, just as it did with Attneave's chorus line of squares and triangles. Now when you match what you see with what you remember, the two coincide, regardless of how the suitcase is oriented. You recognize your luggage.

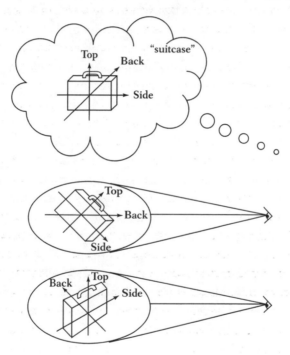

That, in a nutshell, is how Marr explained shape recognition. The key idea is that a shape memory is not a copy of the 2½-D sketch but rather is stored in a format that differs from it in two ways. First, the coordinate system is centered on the object—not, as in the 2½-D sketch, on the viewer. To recognize an object, the brain aligns a reference frame on its axes of elongation and symmetry and measures the positions and angles of the parts in that reference frame. Only then are vision and memory matched. The second difference is that the matcher does not compare vision and memory pixel by pixel, as if placing a jigsaw puzzle piece in a gap. If it did, shapes that ought to match still might not. Real objects have dents and wobbles and come in different styles and models. No two suitcases have identical dimensions, and some have rounded or beefed-up corners and fat or skinny handles. So the representation of the shape about to be matched shouldn't be an exact mold of every hill and valley. It should be couched in forgiving categories like "slab" and "U-shaped thingy." The attachments, too, can't be specified to the millimeter but have to allow for some slop: the handles of different cups are all "on the side," but they can be a bit higher or lower from cup to cup.

The psychologist Irv Biederman has fleshed out Marr's two ideas with an inventory of simple geometric parts that he calls "geons" (by analogy to the protons and electrons making up atoms). Here are five geons along with some combinations:

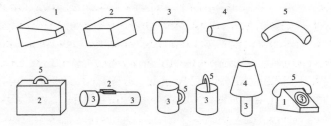

Biederman proposes twenty-four geons altogether, including a cone, a megaphone, a football, a tube, a cube, and a piece of elbow macaroni. (Technically, they are all just different kinds of cones. If an ice cream cone is the surface swept out by an expanding circle as its center is moved along a line, geons are the surfaces swept out by *other* 2-D shapes as they expand or contract while moving along straight or curved lines.) Geons can be assembled into objects with a few attachment relations like "above," "beside," "end to end," "end to off-center," and "parallel." These relations are defined in a frame of reference centered on the object, of course, not the visual field; "above" means "above the main geon," not "above the fovea." So the relations stay the same when the object or viewer moves.

Geons are combinatorial, like grammar. Obviously we don't describe shapes to ourselves in words, but geon assemblies are a kind of internal language, a dialect of mentalese. Elements from a fixed vocabulary are fitted together into larger structures, like words in a phrase or sentence. A sentence is not the sum of its words but depends on their syntactic arrangement; *A man bites a dog* is not the same as *A dog bites a man*. Likewise, an object is not the sum of its geons but depends on their spatial arrangement; a cylinder with an elbow on the side is a cup, while a cylinder with an elbow at the top is a pail. And just as a small number of words and rules combine into an astronomical number of sentences, a small number of geons and attachments combine into an astronomical number of objects. According to Biederman, each of the twenty-four geons comes in fifteen sizes and builds (a bit fatter, a bit skinnier), and there are eighty-one ways to join them. That allows for 10,497,600 objects built out of two geons, and 306 billion objects made of three geons. In theory, that should be more than enough to fit the tens of thousands of shapes we know. In practice, it's easy to build instantly recognizable models of everyday objects out of three, and often only two, geons.

Language and complex shapes even seem to be neighbors in the brain. The left hemisphere is not only the seat of language but also the seat of the ability to recognize and imagine shapes defined by arrangements of parts. A neurological patient who had suffered a stroke to his left hemisphere reported, "When I try to imagine a plant, an animal, an object, I can recall but one part. My inner vision is fleeting, fragmented; if I'm asked to imagine the head of a cow, I know it has ears and horns, but I can't revisualize their places." The right hemisphere, in contrast, is good for measuring whole shapes; it can easily judge whether a rectangle is taller than it is wide or whether a dot lies more or less than an inch from an object.

One advantage of the geon theory is that its demands on the $2^{1}/_{2}$-D sketch are not unreasonable. Carving objects into parts, labeling the parts as geons, and ascertaining their arrangement are not insurmountable problems, and vision researchers have developed models of how the brain might solve them. Another advantage is that a description of an object's anatomy helps the mind to think about objects, not just to blurt out their names. People understand how objects work and what they are for by analyzing the shapes and arrangements of their parts.

The geon theory says that at the highest levels of perception the mind "sees" objects and parts as idealized geometric solids. That would explain a curious and long-noted fact about human visual aesthetics. Anyone who has been to a figure-drawing class or a nude beach quickly learns that real human bodies do not live up to our sweet imaginations. Most of us look better in clothes. In his history of fashion, the art historian Quentin Bell gives an explanation that could have come right out of the geon theory:

If we wrap an object in some kind of envelope, so that the eyes infer rather than see the object that is enclosed, the inferred or imagined form is likely to be more perfect than it would appear if it were uncovered. Thus a square box covered with brown paper will be imagined as a perfect square. Unless the mind is given some very strong clue it is unlikely to visualize holes, dents, cracks, or other accidental qualities. In the same way, if we cast a drapery over a thigh, a leg, an arm or a breast, the imagination supposes a perfectly formed member; it does not and usually cannot envisage the irregularities and the imperfections which experience should lead us to expect.

. . . We know what [a body] is probably like from experience, and yet we are willing to suspend our disbelief in favour of the fictions of [the person's] wardrobe. Indeed I think that we are ready to go further in the way of self-deception. When we slip on our best jacket and see our deplorably unimpressive shoulders artfully magnified and idealised we do, for a moment, rise in our own esteem.

Geons are not good for everything. Many natural objects, such as mountains and trees, have complicated fractal shapes, but geons turn them into pyramids and lollipops. And though geons can be built into a passable generic human face, like a snowman or Mr. Potato Head, it is almost impossible to build a model of a *particular* face—John's face, your grandmother's face—that is different enough from other faces not to confuse them, but stable enough across smiles, frowns, weight gains, and aging to identify that person every time. Many psychologists believe that face recognition is special. In a social species like ours, faces are so important that natural selection gave us a processor that registers the kinds of geometric contours and ratios needed to tell them apart. Babies lock onto facelike patterns, but not onto other complex and symmetrical arrangements, when they are only thirty minutes old, and quickly learn to recognize their mothers, perhaps as early as the second day of life.

Face recognition may even use distinct parts of the brain. An inability to recognize faces is called prosopagnosia. It is not the same as Oliver Sacks' famous man who mistook his wife for a hat: prosopagnosics can tell a face from a hat; they just can't tell whose face it is. But many of them can recognize hats and almost everything else. For example, the patient "LH" was tested in my laboratory by the psychologists Nancy Etcoff and Kyle Cave and the neurologist Roy Freeman. LH is an intelligent, knowledgeable man who suffered head injuries in a car accident twenty years before the tests. Since the accident he has been utterly unable to recognize faces. He cannot recognize his wife and children (except by voice, scent, or gait), his own face in a mirror, or celebrities in photographs (unless they have a visual trademark like Einstein, Hitler, and the Beatles in their moptop days). It was not that he had trouble making out the details of a face; he could match full faces with their profiles, even in arty sidelighting, and assess their age, sex, and beauty. And he was virtually normal at recognizing complicated objects that were not faces, including words, clothing, hairstyles, vehicles, tools, vegetables, musical instruments, office chairs, eyeglasses, dot patterns, and television antenna-like shapes. There were only two kinds of shapes he had trouble with. He was embarrassed that he could not name his children's animal crackers; similarly, in the lab he was below average at naming drawings of animals. And he had some trouble recognizing facial expressions such as frowns, sneers, and fearful looks. But neither animals nor facial expressions were as hard for him as faces, which drew utter blanks.

It's not that faces are the hardest things our brains are ever called upon to recognize, so that if a brain is not running on all eight cylinders, face recognition will be the first thing to suffer. The psychologists Marlene Behrmann, Morris Moscovitch, and Gordon Winocur studied a young man who had been hit on the head by the rear-view mirror of a passing truck. He has trouble recognizing everyday objects but no trouble recognizing faces, even when the faces are disguised with glasses, wigs, or mustaches. His syndrome is the opposite of prosopagnosia, and it proves that face recognition is different from object recognition, not just harder.

So do prosopagnosics have a broken face-recognition module? Some psychologists, noting that LH and other prosopagnosics have *some* trouble with *some* other shapes, would rather say that prosopagnosics have trouble processing the kinds of geometric features that are most useful in recognizing faces, though also useful in recognizing certain other kinds of shapes. I think the distinction between recognizing faces and recognizing objects with the geometry of faces is meaningless. From the brain's point of view, nothing is a face until it has been recognized as a face. The only thing that *can* be special about a perception module is the kind of geometry it pays attention to, such as the distance between symmetrical blobs, or the curvature pattern of 2-D elastic surfaces that are drawn over a 3-D skeleton and filled out by underlying soft pads and connectors. If objects other than faces (animals, facial expressions, or even cars) have some of these geometric features, the module will have no choice but to analyze them, even if they are most useful for faces. To call a module a face-recognizer is not to say it can handle only faces; it is to say that it is optimized for the geometric features that distinguish faces because the organism was selected in its evolutionary history for an ability to recognize them.

The geon theory is lovely, but is it true? Certainly not in its purest form, in which every object would get one description of its 3-D geometry, uncontaminated by the vagaries of vantage point. Most objects are opaque, with some surfaces obscuring others. That makes it literally impossible to arrive at the same description of the object from every vantage point. For example, you can't know what the back of a house looks like when you are standing in front of it. Marr got around the problem by ignoring surfaces altogether and analyzing animals' shapes as if they were built out of pipe cleaners. Biederman's version concedes the problem and gives each object *several* geon models in the mental shape catalogue, one for each view required to reveal all its surfaces.

But this concession opens the door to a completely different way of doing shape recognition. Why not go all the way and give each shape a large number of memory files, one for every vantage point? Then the files wouldn't need a fancy object-centered reference frame; they could use the retinal coordinates available free in the 2½-D sketch, as long as there were enough files to cover all the angles of view. For many years this idea was dismissed out of hand. If the continuum of viewing angles were chopped into one-degree differences, one would need forty thousand files for every object to cover them all (and those are just to cover the viewing angles; they don't embrace the viewing positions at which the object is not dead-center, or the different viewing distances). One cannot skimp by specifying a few views, like an architect's plan and elevation, because in principle any of the views might be crucial. (Simple proof: Imagine a shape consisting of a hollow sphere with a toy glued on the inside and a small hole drilled opposite it. Only by sighting the toy exactly through the hole can the entire shape be seen.) But recently the idea has made a comeback. By choosing views judiciously, and using a pattern-associator neural network to interpolate between them when an object doesn't match a view spot-on, one can get away with storing a manageable number of views per object, forty at most.

It still seems unlikely that people have to see an object from forty different angles to recognize it thereafter, but another trick is available. Remember that people rely on the up-down direction to construe shapes: squares aren't diamonds, sideways Africa goes unrecognized. This introduces another contamination of the pure geon theory: relations like "above" and "top" must come from the retina (with some adjustment from gravity), not from the object. That concession may be inevitable, because there's often no way of pinpointing the "top" of an object before you've recognized it. But the real problem comes from what people *do* with sideways objects they don't recognize at first. If you *tell* people that a shape has been turned sideways, they recognize it quickly, as you surely did when I told you that the Africa drawing was on its side. People can mentally rotate a shape to the upright and then recognize the rotated image. With a mental image-rotator available, the object-centered frame of the geon theory becomes even less necessary. People could store some $2\frac{1}{2}$-D views from a few standard vantage points, like police mug shots, and if an object in front of them didn't match one of the shots, they would mentally rotate it until it did. Some combination of multiple views and a mental rotator would make geon models in object-centered reference frames unnecessary.

With all these options for shape recognition, how can we tell what the mind actually does? The only way is to study real human beings recognizing shapes in the laboratory. One famous set of experiments pointed to mental rotation as a key. The psychologists Lynn Cooper and Roger Shepard showed people letters of the alphabet at different orientations—upright, tilted 45 degrees, sideways, tilted 135 degrees, and upside down. Cooper and Shepard didn't have people blurt out the letter's name because they were worried about shortcuts: a distinctive squiggle like a loop or a tail might be detectable in any orientation and give away the answer. So they forced their subjects to analyze the full geometry of each letter by showing either the letter or its mirror image, and having the subjects press one button if the letter was normal and the other if it was mirror-reversed.

When Cooper and Shepard measured how long it took people to press the button, they observed a clear signature of mental rotation. The farther the letter was misoriented from the upright, the longer people took. That's exactly what you would expect if people gradually dialed an image of the letter to the upright; the more it has to be turned, the longer the turning takes. Maybe, then, people recognize shapes by turning them over in their minds.

But maybe not. People were not just recognizing shapes; they were discriminating them from their mirror images. Mirror images are special. It is fitting that the sequel to *Alice's Adventures in Wonderland* was called *Through the Looking-Glass*. The relation of a shape to its mirror image gives rise to surprises, even paradoxes, in many branches of science. (They are explored in fascinating books by Martin Gardner and by Michael Corballis and Ivan Beale.) Consider the detached right and left hands of a mannequin. In one sense they are identical: each has four fingers and a thumb attached to a palm and a wrist. In another sense they are utterly different; one shape cannot be superimposed on the other. The difference lies only in how the parts are aligned with respect to a frame of reference in which all three axes are labeled with directions: up-down, frontward-backward, left-right. When a right hand is pointing fingers-up palm-frontward (as in a "halt" gesture), its thumb points left; when a left hand is pointing fingers-up palm-frontward, its thumb points right. That's the only difference, but it is real. The molecules of life have a handedness; their mirror images often do not exist in nature and would not work in bodies.

A fundamental discovery of twentieth-century physics is that the universe has a handedness, too. At first that sounds absurd. For any object and event in the cosmos, you have no way of knowing whether you are seeing the actual event or its reflection in a mirror. You may protest that organic molecules and human-made objects like letters of the alphabet are an exception. The standard versions are all over the place and familiar; the mirror images are rare and can easily be recognized. But for a physicist, they don't count, because their handedness is a historical accident, not something ruled out by the laws of physics. On another planet, or on this one if we could rewind the tape of evolution and let it happen again, they could just as easily go the other way. Physicists used to think that this was true for everything in the universe. Wolfgang Pauli wrote, "I do *not* believe that the Lord is a weak left-hander," and Richard Feynman bet fifty dollars to one (he was unwilling to bet a hundred) that no experiment would ever reveal a law of nature that looked different through the looking glass. He lost. The cobalt 60 nucleus is said to spin counterclockwise if you look down on its north pole, but that description by itself is circular because "north pole" is simply what we call the end of the axis from which a rotation looks counterclockwise. The logical circle would be broken if *something else* differentiated the so-called north pole from the so-called south pole. Here is the something else: when the atom decays, electrons are more likely to be flung out of the end we call south. "North" versus "south" and "clockwise" versus "counterclockwise" are no longer arbitrary labels but can be distinguished relative to the electron spurt. The decay, hence the universe, would look different in the mirror. God is not ambidextrous after all.

So right- and left-handed versions of things, from subatomic particles to the raw material of life to the spin of the earth, are fundamentally different. But the mind usually treats them as if they were the same:

> Pooh looked at his two paws. He knew that one of them was the right, and he knew that when you had decided which one of them was the right, then the other one was the left, but he never could remember how to begin.

None of us is good at remembering how to begin. Left and right shoes look so alike that children must be taught tricks to distinguish them, like placing the shoes side by side and sizing up the gap. Which way is Abraham Lincoln facing on the American one-cent piece? There is only a fifty percent chance you will get the answer right, the same as if you had answered by flipping the penny. What about Whistler's famous painting, *Arrangement in Black and Gray: The Artist's Mother*? Even the English language likes to collapse left and right: *beside* and *next to* denote side-by-side without specifying who's on the left, but there is no word like *bebove* or *aneath* that denotes up-and-down without specifying who's on top. Our obliviousness to left-and-right stands in stark contrast to our hypersensitivity to up-and-down and front-and-back. Apparently the human mind does not have a preexisting label for the third dimension of its object-centered reference frame. When it sees a hand, it can align the wrist-fingertip axis with "down-up," and the back-palm axis with "backward-forward," but the direction of the pinkie-thumb axis is up for grabs. The mind calls it, say, "thumbward," and the left and right hands become mental synonyms. Our indecisiveness about left and right needs an explanation, because a geometer would say they are no different from up and down or front and back.

The explanation is that mirror-image confusions come naturally to a bilaterally symmetrical animal. A perfectly symmetrical creature is logically incapable of telling left from right (unless it could react to the decay of cobalt 60!). Natural selection had little incentive to build animals asymmetrically so that they could mentally represent shapes differently from their reflections. Actually, this puts it backwards: natural selection had every incentive to build animals symmetrically so that they would *not* represent shapes differently from their reflections. In the intermediate-sized world in which animals spend their days (bigger than subatomic particles and organic molecules, smaller than a weather front), left and right make no difference. Objects from dandelions to mountains have tops that differ conspicuously from their bottoms, and most things that move have fronts that differ conspicuously from their behinds. But no natural object has a left side that differs nonrandomly from its right, making its mirror-image version behave differently. If a predator comes from the right, next time it might come from the left. Anything learned from the first encounter should generalize to the mirror-image version. Another way of putting it is that if you took a photographic slide of any natural scene, it would be obvious if someone had turned it upside down, but you wouldn't notice if someone had flipped it left-to-right, unless the scene contained a human-made object like a car or writing.

And that brings us back to letters and mental rotation. In a few human activities, like driving and writing, left and right do make a difference, and we learn to tell them apart. How? The human brain and body are *slightly* asymmetrical. One hand is dominant, owing to the asymmetry of the brain, and we can feel the difference. (Older dictionaries used to *define* "right" as the side of the body with the stronger hand, based on the assumption that people are righties. More recent dictionaries, perhaps out of respect for an oppressed minority, use a different asymmetrical object, the earth, and define "right" as east when you are facing north.) The usual way that people tell an object from its mirror image is by turning it so it faces up and forward and looking at which side of their body—the side with the dominant hand or the side with the nondominant hand—the distinctive part is pointing to. The person's body is used as the asymmetrical frame of reference that makes the distinction between a shape and its mirror image logically possible.

Now, Cooper and Shepard's subjects may have been doing the same thing, except that they were rotating the shape *in their minds* instead of in the world. To decide whether they were seeing a normal or a backwards *R*, they mentally rotated an image of the shape until it was upright, and then judged whether the imaginary loop was on their right side or their left side.

So Cooper and Shepard have demonstrated that the mind *can* rotate objects, and they have demonstrated that *one* aspect of an object's intrinsic shape—its handedness—is not stored in a 3-D geon model. But for all its fascination, handedness is such a peculiar feature of the universe that we cannot conclude much about shape recognition in general from the experiments on mental rotation. For all we know, the mind *could* overlay objects with a 3-D reference frame (for geon matching), specified up to, but not including, which way to put the arrow on the side-to-side axis. As they say, more research is needed.

The psychologist Michael Tarr and I did some more research. We created our own little world of shapes and despotically controlled people's exposure to them, aiming at clean tests of the three hypotheses on the table.

The shapes were similar enough that people could not use shortcuts like a telltale squiggle. None was a mirror image of any other, so we would not get sidetracked by the peculiarities of the world in the looking glass. Each shape had a giveaway little foot, so people would never have a problem finding the top and the bottom. We gave each person three shapes to learn, and then asked them to identify the shapes by pressing one of three buttons whenever a shape flashed on a computer screen. Each shape appeared at a few orientations over and over. For example, Shape 3 might appear with its top at four o'clock hundreds of times, and with its top at seven o'clock hundreds of times. (All the shapes and tilts were mixed up in a random order.) People thus had the opportunity to learn what each shape looked like in a few views.

Finally, we hit them with a flurry of new trials in which every shape appeared at twenty-four evenly spaced orientations (again randomly ordered). We wanted to see how people dealt with the old shapes at the new orientations. Every button-press was timed to the thousandth of a second.

According to the multiple-view theory, people should create a separate memory file for every orientation in which an object commonly appeared. For example, they would set up a file showing what Shape 3 looks like right-side up (which is how they learned it), and then a second file for what it looks like at four o'clock and a third for seven o'clock. The people should soon recognize Shape 3 at these orientations very quickly. When we then surprised them with the same shapes at new orientations, however, they should take much longer, because they would have to interpolate a new view between the familiar ones to accommodate it. The new orientations should all take an extra increment of time.

According to the mental-rotation theory, people should be quick to recognize the shape when it is upright, and slower and slower the farther it has been misoriented. An upside-down shape should take the longest, because it needs a full 180-degree turn; the four o'clock shape should be quicker, for it needs only 120 degrees, and so on.

According to the geon theory, orientation shouldn't matter at all. People would learn the objects by mentally describing the various arms and crosses in a coordinate system centered on the object. Then, when a test shape flashed on the screen, it should make no difference if it was sideways, tilted, or upside down. Overlaying a frame should be quick and foolproof, and the shape's description relative to the frame would match the memory model every time.

The envelope, please. And the winner is . . .

All of the above. People definitely stored several views: when a shape appeared in one of its habitual orientations, people were very quick to identify it.

And people definitely rotate shapes in their minds. When a shape appeared at a new, unfamiliar orientation, the farther it would have to be rotated to be aligned with the nearest familiar view, the more time people took.

And at least for some shapes, people use an object-centered reference frame, as in the geon theory. Tarr and I ran a variant of the experiment in which the shapes had simpler geometries:

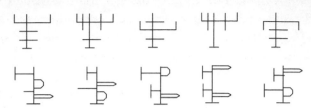

The shapes were symmetrical or nearly symmetrical, or always had the same kinds of frills on each side, so people would never have to describe the parts' up-down and side-to-side arrangements in the same reference frame. With these shapes, people were uniformly quick at identifying them in *all* their orientations; upside down was no slower than right-side up.

So people use all the tricks. If a shape's sides are not too different, they store it as a 3-D geon model centered on the object's own axes. If the shape is more complicated, they store a copy of what it looks like at each orientation they see it in. When the shape appears at an unfamiliar orientation, they mentally rotate it into the nearest familiar one. Perhaps we shouldn't be surprised. Shape recognition is such a hard problem that a single, general-purpose algorithm may not work for every shape under every viewing condition.

Let me finish the story with my happiest moment as an experimenter. You may be skeptical about the mental turntable. All we know is that tilted shapes are recognized more slowly. I've glibly written that people rotate an image, but maybe tilted shapes are just harder to analyze for other reasons. Is there any evidence that people actually simulate a physical rotation in real time, degree by degree? Does their behavior show some signature of the geometry of rotation that could convince us that they play a movie in their minds?

Tarr and I had been baffled by one of our findings. In a different experiment, we had tested people both on the shapes they had studied and on their mirror images, at a variety of orientations:

What people were taught

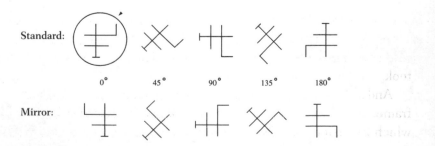

It wasn't a mirror-image test, like the Cooper and Shepard experiments; people were told to treat the two versions the same, just as they use the same word for a left and a right glove. This, of course, is just people's natural tendency. But somehow our subjects were treating them differently. For the standard versions (top row), people took longer when the shape was tilted farther: every picture in the top row took a bit longer than the one before. But for the reflected versions (bottom row), tilt made no difference: every orientation took the same time. It looked as if people mentally rotated the standard shapes but not their mirror images. Tarr and I glumly wrote up a paper begging the reader to believe that people use a different strategy to recognize mirror images. (In psychology, invoking "strategies" to explain funny data is the last refuge of the clueless.) But just as we were touching up the final draft for publication, an idea hit.

We remembered a theorem of the geometry of motion: a 2-D shape can always be aligned with its mirror image by a rotation of no more than 180 degrees, as long as the rotation can be *in the third dimension* around an optimal axis. In principle, any of our mirror-reversed shapes could be flipped in depth to match the standard upright shape, and the flip would take the same amount of time. The mirror image at 0 degrees would simply swivel around a vertical axis like a revolving door. The upside-down shape at 180 degrees could turn like a chicken on a rotisserie. The sideways shape could pivot around a diagonal axis, like this: look at the back of your right hand, fingertips up; now look at your palm, fingertips left. Different tilted axes could serve as the hinge for the other misoriented shapes; in every case, the rotation would be exactly 180 degrees. It would fit the data perfectly: people may have been mentally rotating *all* the shapes but were optimal rotators, dialing the standard shapes in the picture plane and flipping the mirror-reversed shapes in depth around the best axis.

We could scarcely believe it. Could people have found the optimal axis before even knowing what the shape was? We knew it was mathematically possible: by identifying just three non-collinear landmarks in each of two views of a shape, one can calculate the axis of rotation that would align one with the other. But can people really do this calculation? We convinced ourselves with a bit of computer animation. Roger Shepard once showed that if people see a shape alternating with a tilted copy, they see it rock back and forth. So we showed ourselves the standard upright shape alternating with one of its mirror images, back and forth once a second. The perception of flipping was so obvious that we didn't bother to recruit volunteers to confirm it. When the shape alternated with its upright reflection, it seemed to pivot like a washing machine agitator. When it alternated with its upside-down reflection, it did backflips. When it alternated with its sideways reflection, it swooped back and forth around a diagonal axis, and so on. The brain finds the axis every time. The subjects in our experiment were smarter than we were.

The clincher came from Tarr's thesis. He had replicated our experiments using three-dimensional shapes and their mirror images, rotated in the picture plane (shown below) and in depth:

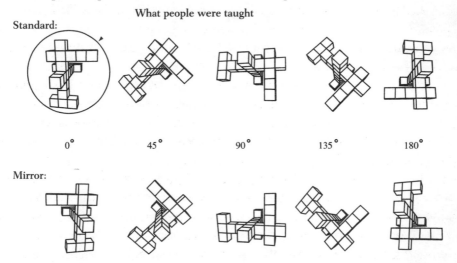

What people were taught

Standard:

0°   45°   90°   135°   180°

Mirror:

Everything came out the same as for the 2-D shapes, except what people did with the mirror images. Just as a misoriented 2-D shape can be matched to the standard orientation by a rotation in the 2-D picture plane, and its mirror image can be rotated to the standard orientation by a 180-degree flip in the third dimension, a misoriented 3-D shape (top row) can be rotated to the standard orientation in 3-D space, and its mirror image (bottom row) can be rotated to the standard by a 180-degree flip in the *fourth* dimension. (In H. G. Wells' "The Plattner Story," an explosion blows the hero into four-dimensional space. When he returns, his heart is on the right side and he writes backwards with his left hand.) The only difference is that mere mortals should not be able to mentally rotate a shape in the fourth dimension, our mental space being strictly 3-D. All the versions should show an effect of tilt, unlike what we had found for 2-D shapes, where the mirror images did not. That's what happened. The subtle difference between two- and three-dimensional objects sewed up the case: the brain rotates shapes around an optimal axis in three dimensions, but no more than three dimensions. Mental rotation is clearly one of the tricks behind our ability to recognize objects.

Mental rotation is another talent of our gifted visual systems, with a special twist. It does not merely analyze the contours coming in from the world, but creates some its own in the form of a ghostly moving image. This brings us to a final topic in the psychology of vision.

IMAGINE THAT!

What shape are a beagle's ears? How many windows are in your living room? What's darker, a Christmas tree or a frozen pea? What's larger, a guinea pig or a gerbil? Does a lobster have a mouth? When a person stands up straight, is her navel above her wrist? If the letter *D* is turned on its back and put on top of a *J*, what does the combination remind you of?

Most people say that they answer these questions using a "mental image." They visualize the shape, which feels like conjuring up a picture available for inspection in the mind's eye. The feeling is quite unlike the experience of answering abstract questions, such as "What is your mother's maiden name?" or "What is more important, civil liberties or a lower rate of crime?"

Mental imagery is the engine that drives our thinking about objects in space. To load a car with suitcases or rearrange the furniture, we imagine the different spatial arrangements before we try them. The anthropologist Napoleon Chagnon described an ingenious use of mental imagery by the Yanomamö Indians of the Amazon rainforest. They had blown smoke down the opening of an armadillo hole to asphyxiate the animal, and then had to figure out where to dig to extract it from its tunnel, which could run underground for hundreds of feet. One of the Yanomamö men hit on the idea of threading a long vine with a knot at the end down the hole as far as it would go. The other men kept their ears to the ground listening for the knot bumping the sides of the burrow so they could get a sense of the direction in which the burrow ran. The first man broke off the vine, pulled it out, laid it along the ground, and began to dig where the end of the vine lay. A few feet down they struck armadillo. Without an ability to visualize the tunnel and the vine and armadillo inside it, the men would not have connected a sequence of threading, listening, yanking, breaking, measuring, and digging actions to an expectation of finding an animal corpse. In a joke we used to tell as children, two carpenters are hammering nails into the side of a house, and one asks the other why he is examining each nail as he takes it out of the box and throwing half of them away. "They're defective," replies the second carpenter, holding one up. "The pointy end is facing the wrong way." "You fool!" shouts the first carpenter. "Those are for the other side of the house!"

But people do not use imagery just to rearrange the furniture or dig up armadillos. The eminent psychologist D. O. Hebb once wrote, "You can hardly turn around in psychology without bumping into the image." Give people a list of nouns to memorize, and they will imagine them interacting in bizarre images. Give them factual questions like "Does a flea have a mouth?" and they will visualize the flea and "look for" the mouth. And, of course, give them a complex shape at an unfamiliar orientation, and they will rotate its image to a familiar one.

Many creative people claim to "see" the solution to a problem in an image. Faraday and Maxwell visualized electromagnetic fields as tiny tubes filled with fluid. Kekulé saw the benzene ring in a reverie of snakes biting their tails. Watson and Crick mentally rotated models of what was to become the double helix. Einstein imagined what it would be like to ride on a beam of light or drop a penny in a plummeting elevator. He once wrote, "My particular ability does not lie in mathematical calculation, but rather in visualizing effects, possibilities, and consequences." Painters and sculptors try out ideas in their minds, and even novelists visualize scenes and plots in their mind's eye before putting pen to paper.

Images drive the emotions as well as the intellect. Hemingway wrote, "Cowardice, as distinguished from panic, is almost always simply a lack of ability to suspend the functioning of the imagination." Ambition, anxiety, sexual arousal, and jealous rage can all be triggered by images of what isn't there. In one experiment, volunteers were hooked up to electrodes and asked to imagine their mates being unfaithful. The authors report, "Their skin conductance increased 1.5 microSiemens, the corrugator muscle in their brow showed 7.75 microvolts units of contraction, and their heart rates accelerated by five beats per minute, equivalent to drinking three cups of coffee at one sitting." Of course, the imagination revives many experiences at a time, not just seeing, but the visual image makes a mental simulation especially vivid.

Imagery is an industry. Courses on How to Improve Your Memory teach age-old tricks like imagining items in the rooms of your house and then mentally walking through it, or finding a visual allusion in a person's name and linking it to his face (if you were introduced to me, you would imagine me in a cerise leisure suit). Phobias are often treated by a kind of mental Pavlovian conditioning where an image substitutes for the bell. The patient relaxes deeply and then imagines the snake or spider, until the image—and, by extension, the real thing—is associated with the relaxation. Highly paid "sports psychologists" have athletes relax in a comfy chair and visualize the perfect swing. Many of these techniques work, though some are downright flaky. I am skeptical of cancer therapies in which patients visualize their antibodies munching the tumor, even more so when it is the patient's support group that does the visualizing. (A woman once called to ask if I thought it would work over the Internet.)

But what *is* a mental image? Many philosophers with behaviorist leanings think the whole idea is a terrible blunder. An image is supposed to be a picture in the head, but then you would need a little man et cetera, et cetera, et cetera. In fact, the computational theory of mind makes the notion perfectly straightforward. We already know that the visual system uses a $2^{1}/_{2}$-D sketch which is picturelike in several respects. It is a mosaic of elements that stand for points in the visual field. The elements are arranged in two dimensions so that neighboring elements in the array stand for neighboring points in the visual field. Shapes are represented by filling in some of the elements in a pattern that matches the shape's projected contours. Shape-analysis mechanisms—not little men—process information in the sketch by imposing reference frames, finding geons, and so on. A mental image is simply a pattern in the $2^{1}/_{2}$-D sketch that is loaded from long-term memory rather than from the eyes. A number of artificial intelligence programs for reasoning about space are designed in exactly this way.

A depiction like the $2^{1}/_{2}$-D sketch contrasts starkly with a description in a language-like representation like a geon model, a semantic network, a sentence in English, or a proposition in mentalese. In the proposition *A symmetrical triangle is above a circle*, the words do not stand for points in the visual field, and they are not arranged so that nearby words represent nearby points. Words like *symmetrical* and *above* can't be pinned to any piece of the visual field; they denote complicated relationships among the filled-in pieces.

One can even make an educated guess about the anatomy of mental imagery. The incarnation of a 2½-D sketch in neurons is called a topographically organized cortical map: a patch of cortex in which each neuron responds to contours in one part of the visual field, and in which neighboring neurons respond to neighboring parts. The primate brain has at least fifteen of these maps, and in a very real sense they are pictures in the head. Neuroscientists can inject a monkey with a radioactive isotope of glucose while it stares at a bull's-eye. The glucose is taken up by the active neurons, and one can literally *develop the monkey's brain* as if it were a piece of film. It comes out of the darkroom with a distorted bull's-eye laid out over the visual cortex. Of course, nothing "looks at" the cortex from above; connectivity is all that matters, and the activity pattern is interpreted by networks of neurons plugged into each cortical map. Presumably space in the world is represented by space on the cortex because neurons are connected to their neighbors, and it is handy for nearby bits of the world to be analyzed together. For example, edges are not scattered across the visual field like rice but snake along a line, and most surfaces are not archipelagos but cohesive masses. In a cortical map, lines and surfaces can be handled by neurons that are highly interconnected.

The brain is also ready for the second computational demand of an imagery system, information flowing down from memory instead of up from the eyes. The fiber pathways to the visual areas of the brain are two-way. They carry as much information down from the higher, conceptual levels as up from the lower, sensory levels. No one knows what these top-down connections are for, but they could be there to download memory images into visual maps.

So mental images *could be* pictures in the head. Are they? There are two ways to find out. One is to see if thinking in images engages the visual parts of the brain. The other is to see if thinking in images works more like computing with graphics or more like computing with a database of propositions.

In the first act of *Richard II*, the exiled Bolingbroke pines for his native England. He is not consoled by a friend's suggestion to fantasize that he is in more idyllic surroundings:

O, who can hold a fire in his hand
By thinking on the frosty Caucasus?
Or cloy the hungry edge of appetite
By bare imagination of a feast?
Or wallow naked in December snow
By thinking on fantastic summer's heat?

Clearly an image is different from an experience of the real thing. William James said that images are "devoid of pungency and tang." But in a 1910 Ph.D. thesis, the psychologist Cheves W. Perky tried to show that images were like *very faint* experiences. She asked her subjects to form a mental image, say of a banana, on a blank wall. The wall was actually a rear-projection screen, and Perky surreptitiously projected a *real* but dim slide on it. Anyone coming into the room at that point would have seen the slide, but none of the subjects noticed it. Perky claimed that they had incorporated the slide into their mental image, and indeed, the subjects reported details in their image that could only have come from the slide, such as the banana's standing on end. It was not a great experiment by modern standards, but state-of-the-art methods have borne out the crux of the finding, now called the Perky effect: holding a mental image interferes with seeing faint and fine visual details.

Imagery can affect perception in gross ways, too. When people answer questions about shapes from memory, like counting off the right angles in a block letter, their visual-motor coordination suffers. (Since learning about these experiments I try not to get too caught up in a hockey game on the radio while I am driving.) Mental images of lines can affect perception just as real lines do: they make it easier to judge alignment and can even induce visual illusions. When people see some shapes and imagine others, later they sometimes have trouble remembering which was which.

So do imagery and vision share space in the brain? The neuropsychologists Edoardo Bisiach and Claudio Luzzatti studied two Milanese patients with damage to their right parietal lobes that left them with visual neglect syndrome. Their eyes register the whole visual field, but they attend only to the right half: they ignore the cutlery to the left of the plate, draw a face with no left eye or nostril, and when describing a room, ignore large details—like a piano—on their left. Bisiach and Luzzatti asked the patients to imagine standing in the Piazza del Duomo in Milan facing the cathedral and to name the buildings in the piazza. The patients named only the buildings that would be visible on the right—neglecting the left half of *imaginary* space! Then the patients were asked to mentally walk across the square and stand on the cathedral steps facing the piazza and describe what was in it. They mentioned the buildings that they had left out the first time, and left out the buildings that they had mentioned. Each mental image depicted the scene from one vantage point, and the patients' lopsided window of attention examined the image exactly as it examined real visual inputs.

These discoveries implicate the visual brain as the seat of imagery, and recently there has been a positive identification. The psychologist Stephen Kosslyn and his colleagues used Positron Emission Tomography (PET scanning) to see which parts of the brain are most active when people have mental images. Each subject lay with his head in a ring of detectors, closed his eyes, and answered questions about uppercase letters of the alphabet, such as whether *B* has any curves. The occipital lobe or visual cortex, the first gray matter that processes visual input, lit up. The visual cortex is topographically mapped—it forms a picture, if you will. In some runs, the subjects visualized large letters, in others, small letters. Pondering large letters activated the parts of the cortex representing the periphery of the visual field; pondering small letters activated the parts representing the fovea. Images really do seem to be laid across the cortical surface.

Could the activation be just a spillover of activity from other parts of the brain, where the real computation is being done? The psychologist Martha Farah showed that it isn't. She tested a woman's ability to form mental images before and after surgery that removed her visual cortex in one hemisphere. After the surgery, her mental images shrank to half their normal width. Mental images live in the visual cortex; indeed, parts of images take up parts of cortex, just as parts of scenes take up parts of pictures.

Still, an image is not an instant replay. It lacks that pungency and tang, though not because it has been bleached or watered down: imagining red is not like seeing pink. And curiously, in the PET studies the mental image sometimes caused *more* activation of the visual cortex than a real display, not less. Visual images, though they share brain areas with perception, are somehow different, and perhaps that is not surprising. Donald Symons notes that reactivating a visual experience may well have benefits, but it also has costs: the risk of confusing imagination with reality. Within moments of awakening from a dream, our memory for its plot is wiped out, presumably to avoid contaminating autobiographical memory with bizarre confabulations. Similarly, our voluntary, waking mental images might be hobbled to keep them from becoming hallucinations or false memories.

~

Knowing *where* mental images are says little about what they are or how they work. Are mental images really patterns of pixels in a 2¹/₂-D array (or patterns of active neurons in a cortical map)? If they are, how do we *think* with them, and what would make imagery different from any other form of thought?

Let's compare an array or sketch to its rival as a model of imagery, symbolic propositions in mentalese (similar to geon models and to semantic networks). The array is on the left, the propositional model on the right. The diagram collapses many propositions, like "A bear has a head" and "The bear has the size XL," into a single network.

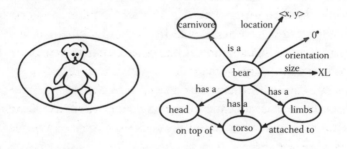

The array is straightforward. Each pixel represents a small piece of surface or boundary, period; anything more global or abstract is only implicit in the pattern of filled pixels. The propositional representation is quite different. First, it is schematic, filled with qualitative relations like "attached to"; not every detail of the geometry is represented. Second, the spatial properties are factored apart and listed *explicitly*. Shape (the arrangement of an object's parts or geons), size, location, and orientation get their own symbols, and each can be looked up independently of the others. Third, propositions mix spatial information, like parts and their positions, with conceptual information, like bearhood and membership in the carnivore class.

Of the two data structures, it is the pictorial array that best captures the flavor of imagery. First, images are thumpingly concrete. Consider this request: Visualize a lemon and a banana next to each other, but don't imagine the lemon either to the right or to the left, just next to the banana. You will protest that the request is impossible; if the lemon and banana are next to each other in an image, one or the other *has* to be on the left. The contrast between a proposition and an array is stark. Propositions can represent cats without grins, grins without cats, or any other disembodied abstraction: squares of no particular size, symmetry with no particular shape, attachment with no particular place, and so on. That is the beauty of a proposition: it is an austere statement of some abstract fact, uncluttered with irrelevant details. Spatial arrays, because they consist only of filled and unfilled patches, commit one to a concrete arrangement of matter in space. And so do mental images: forming an image of "symmetry," without imagining a something or other that is symmetrical, can't be done.

The concreteness of mental images allows them to be co-opted as a handy analogue computer. Amy is richer than Abigail; Alicia is not as rich as Abigail; who's the richest? Many people solve these syllogisms by lining up the characters in a mental image from least rich to richest. Why should this work? The medium underlying imagery comes with cells dedicated to each location, fixed in a two-dimensional arrangement. That supplies many truths of geometry for free. For example, left-to-right arrangement in space is transitive: if A is to the left of B, and B is to the left of C, then A is to the left of C. Any lookup mechanism that finds the locations of shapes in the array will automatically respect transitivity; the architecture of the medium leaves it no choice.

Suppose the reasoning centers of the brain can get their hands on the mechanisms that plop shapes into the array and that read their locations out of it. Those reasoning demons can exploit the geometry of the array as a surrogate for keeping certain logical constraints in mind. Wealth, like location on a line, is transitive: if A is richer than B, and B is richer than C, then A is richer than C. By using location in an image to symbolize wealth, the thinker takes advantage of the transitivity of location built into the array, and does not have to enter it into a chain of deductive steps. The problem becomes a matter of plop down and look up. It is a fine example of how the form of a mental representation determines what is easy or hard to think.

Mental images also resemble arrays in shmooshing together size, shape, location, and orientation into one pattern of contours, rather than neatly factoring them into separate assertions. Mental rotation is a good example. In assessing an object's shape, a person cannot ignore its orientation—which would be a simple matter if orientation were sequestered in its own statement. Instead, the person must nudge the orientation gradually and watch as the shape changes. The orientation is not re-computed in one step like a matrix multiplication in a digital computer; the farther a shape is dialed, the longer the dialing takes. There must be a rotator network overlaid on the array that shifts the contents of cells a few degrees around its center. Larger rotations require iterating the rotator, bucket-brigade style. Experiments on how people solve spatial problems have uncovered a well-stocked mental toolbox of graphic operations, such as zooming, shrinking, panning, scanning, tracing, and coloring. Visual thinking, such as judging whether two objects lie along the same line or whether two blobs of different sizes have the same shape, strings these operations into mental animation sequences.

Finally, images capture the geometry of an object, not just its meaning. The surefire way of getting people to experience imagery is to ask them about obscure details of an object's shape or coloring—the beagle's ears, the curves in the B, the shade of frozen peas. When a feature is noteworthy—cats have claws, bees have stingers—we file it away as an explicit statement in our conceptual database, available later for instant lookup. But when it is not, we call up a memory of the appearance of the object and run our shape analyzers over the image. Checking for previously unnoticed geometric properties of absent objects is one of the main functions of imagery, and Kosslyn has shown that this mental process differs from dredging up explicit facts. When he asked people questions about well-rehearsed facts, like whether a cat has claws or a lobster has a tail, the speed of the answer depended on how strongly the object and its part were associated in memory. People must have retrieved the answer from a mental database. But when the questions were more unusual, like whether a cat has a head or a lobster has a mouth, and people consulted a mental image, the speed of the answer depended on the *size* of the part; smaller parts were slower to verify. Since size and shape are mixed together in an image, smaller shape details are harder to resolve.

For decades, philosophers have suggested that the perfect test of whether mental images are depictions or descriptions was whether people can reinterpret ambiguous shapes, like the duck-rabbit:

If the mind stores only descriptions, then a person who sees the duck-rabbit as a rabbit should tuck away only the label "rabbit." Nothing in the label captures anything about ducks, so later on, the rabbit-seers should be at a loss when asked whether some other animal lurked in the shape; the ambiguous geometric information has been sloughed off. But if the mind stores images, the geometry is still available, and people should be able to call back the image and inspect it for new interpretations. The duck-rabbit itself turns out to be a hard case, because people store shapes with a front-back frame of reference attached, and reinterpreting the duck-rabbit requires reversing the frame. But with some gentle nudging (such as encouraging people to concentrate on the curve at the back of the head), many people do see the duck in the rabbit image or vice versa. Almost everyone can flip simpler ambiguous images. The psychologist Ronald Finke, Martha Farah, and I got people to reinterpret images from verbal descriptions alone, which we read aloud while their eyes were closed. What object can you "see" in each of these descriptions?

> Imagine the letter *D*. Rotate it 90 degrees to the right. Put the number 4 above it. Now remove the horizontal segment of the 4 to the right of the vertical line.

> Imagine the letter *B*. Rotate it 90 degrees to the left. Put a triangle directly below it having the same width and pointing down. Remove the horizontal line.

> Imagine the letter *K*. Place a square next to it on the left side. Put a circle inside the square. Now rotate the figure 90 degrees to the left.

Most people had no trouble reporting the sailboat, the valentine, and the television set that were implicit in the verbiage.

Imagery is a wonderful faculty, but we must not get carried away with the idea of pictures in the head.

For one thing, people cannot reconstruct an image of an entire visual scene. Images are fragmentary. We recall glimpses of parts, arrange them in a mental tableau, and then do a juggling act to refresh each part as it fades. Worse, each glimpse records only the surfaces visible from one vantage point, distorted by perspective. (A simple demonstration is the railroad track paradox—most people see the tracks converge in their mental image, not just in real life.) To remember an object, we turn it over or walk around it, and that means our memory for it is an album of separate views. An image of the whole object is a slide show or pastiche.

That explains why perspective in art took so long to be invented, even though everyone sees in perspective. Paintings without Renaissance craftsmanship look unrealistic, but not because they lack perspective outright. (Even Cro-Magnon cave paintings have a measure of accurate perspective.) Usually, distant objects are smaller, opaque objects hide their backgrounds and take bites out of objects behind them, and many tilted surfaces are foreshortened. The problem is that different parts of the painting are shown as they would appear from *different* vantage points, rather than from the fixed viewing reticle behind Leonardo's window. No incarnate perceiver, chained to one place at one time, can experience a scene from several vantage points at once, so the painting does not correspond to anything a person ever sees. The imagination, of course, is not chained to one place at one time, and paintings without true perspective may, strangely enough, be evocative renditions of our mental imagery. Cubist and surrealist painters, who were avid consumers of psychology, used multiple perspectives in a painting deliberately, perhaps to awaken photograph-jaded viewers to the evanescence of the mind's eye.

A second limitation is that images are slaves to the organization of memory. Our knowledge of the world could not possibly fit into one big picture or map. There are too many scales, from mountains to fleas, to fit into one medium with a fixed grain size. And our visual memory could not very well be a shoebox stuffed with photographs, either. There would be no way to find the one you need without examining each one to recognize what's in it. (Photo and video archives face a similar problem.)

Memory images must be labeled and organized within a propositional superstructure, perhaps a bit like hypermedia, where graphics files are linked to attachment points within a large text or database.

Visual thinking is often driven more strongly by the conceptual knowledge we use to organize our images than by the contents of the images themselves. Chess masters are known for their remarkable memory for the pieces on a chessboard. But it's not because people with photographic memories become chess masters. The masters are no better than beginners when remembering a board of randomly arranged pieces. Their memory captures meaningful relations among the pieces, such as threats and defenses, not just their distribution in space.

Another example comes from a wonderfully low-tech experiment by the psychologists Raymond Nickerson and Marilyn Adams. They asked people to draw both sides of a penny, which everyone has seen thousands of times, from memory. (Try it before you read on.) The results are sobering. An American penny has eight features: Abraham Lincoln's profile, IN GOD WE TRUST, a year, and LIBERTY on one side, and the Lincoln Memorial, UNITED STATES OF AMERICA, E PLURIBUS UNUM, and ONE CENT on the other. Only five percent of the subjects drew all eight. The median number remembered was three, and half were in the wrong place. Intruding into the drawings were ONE PENNY, laurel wreaths, sheaves of wheat, the Washington monument, and Lincoln sitting in a chair. People did better when asked to tick off the features in a penny from a list. (Thankfully, no one selected MADE IN TAIWAN.) But when they were shown fifteen drawings of possible pennies, fewer than half the people picked out the correct one. Obviously, visual memories are not accurate pictures of whole objects.

And if you did get the penny right, try this quiz. Which of these statements are true?

Madrid is farther north than Washington, D.C.
Seattle is farther north than Montreal.
Portland, Oregon, is farther north than Toronto.
Reno is farther west than San Diego.
The Atlantic entrance to the Panama Canal is farther west than the
    Pacific entrance.

They are all true. Almost everyone gets them wrong, reasoning along these lines: Nevada is east of California; San Diego is in California; Reno is in Nevada; therefore Reno is east of San Diego. Of course, this kind of syllogism is invalid whenever regions don't form a checkerboard. Our geographic knowledge is not a big mental map but a set of smaller maps, organized by assertions about how they are related.

Finally, images cannot serve as our concepts, nor can they serve as the meanings of words in the mental dictionary. A long tradition in empiricist philosophy and psychology tried to argue that they could, since it fit the dogma that there is nothing in the intellect that was not previously in the senses. Images were supposed to be degraded or super-imposed copies of visual sensations, the sharp edges sanded off and the colors blended together so that they could stand for entire categories rather than individual objects. As long as you don't think too hard about what these composite images look like, the idea has a ring of plausibility. But then, how would one represent abstract ideas, even something as simple as the concept of a triangle? A triangle is any three-sided polygon. But any *image* of a triangle must be isosceles, scalene, or equilateral. John Locke made the enigmatic claim that our image of a triangle is "all and none of these at once." Berkeley called him on it, challenging his readers to form a mental image of a triangle that was isosceles, scalene, equilateral, and none of the above, all at the same time. But rather than abandoning the theory that abstract ideas are images, Berkeley con-cluded that we don't have abstract ideas!

Early in the twentieth century, Edward Titchener, one of America's first experimental psychologists, rose to the challenge. By carefully intro-specting on his own images, he argued that they could represent any idea, no matter how abstract:

> I can quite well get Locke's picture, the triangle that is no triangle and all triangles at one and the same time. It is a flashy thing, come and gone from moment to moment; it hints two or three red angles, with the red lines deepening into black, seen on a dark green ground. It is not there long enough for me to say whether the angles join to form the complete figure, or even whether all three of the necessary angles are given.
>
> Horse is, to me, a double curve and a rampant posture with a touch of mane about it; cow is a longish rectangle with a certain facial expres-sion, a sort of exaggerated pout.
>
> I have been ideating meanings all my life. And not only meanings, but meaning also. Meaning in general is represented in my conscious-ness by another of these impressionistic pictures. I see meaning as the blue-grey tip of a kind of scoop, which has a bit of yellow above it (proba-bly a part of the handle), and which is just digging into a dark mass of

what appears to be plastic material. I was educated on classical lines; and it is conceivable that this picture is an echo of the oft-repeated admonition to "dig out the meaning" of some passage of Greek or Latin.

Exaggerated pout indeed! Titchener's Cheshire Cow, his triangle with red angles that don't even join, and his meaning shovel could not possibly be the concepts underlying his thoughts. Surely he did not *believe* that cows are rectangular or that triangles can do just fine without one of their angles. Something else in his head, not an image, must have embodied that knowledge.

And that is the problem with other claims that all thoughts are images. Suppose I try to represent the concept "man" by an image of a prototypical man—say, Fred MacMurray. The problem is, what makes the image serve as the concept "man" as opposed to, say, the concept "Fred MacMurray"? Or the concept "tall man," "adult," "human," "American," or "actor who plays an insurance salesman seduced into murder by Barbara Stanwyck"? *You* have no trouble distinguishing among a particular man, men in general, Americans in general, vamp-victims in general, and so on, so you must have more than a picture of a prototypical man in your head.

And how could a concrete image represent an abstract concept, like "freedom"? The Statue of Liberty is already taken; presumably it is representing the concept "the Statue of Liberty." What would you use for negative concepts, like "not a giraffe"? An image of a giraffe with a red diagonal line through it? Then what would represent the concept "a giraffe with a red diagonal line through it"? How about disjunctive concepts, like "either a cat or a bird," or propositions, like "All men are mortal"?

Pictures are ambiguous, but thoughts, virtually by definition, cannot be ambiguous. Your common sense makes distinctions that pictures by themselves do not; therefore your common sense is not just a collection of pictures. If a mental picture is used to represent a thought, it needs to be accompanied by a caption, a set of instructions for how to interpret the picture—what to pay attention to and what to ignore. The captions cannot themselves be pictures, or we would be back where we started. When vision leaves off and thought begins, there's no getting around the need for abstract symbols and propositions that pick out *aspects* of an object for the mind to manipulate.

Incidentally, the ambiguity of pictures has been lost on the designers of graphical computer interfaces and other icon-encrusted consumer products. My computer screen is festooned with little cartoons that do various things when selected by a click of the mouse. For the life of me I can't remember what the tiny binoculars, eyedropper, and silver platter are supposed to do. A picture is worth a thousand words, but that is not always such a good thing. At some point between gazing and thinking, images must give way to ideas.

5

GOOD IDEAS

"I hope you have not murdered too completely your own and my child." So wrote Darwin to Alfred Russel Wallace, the biologist who had independently discovered natural selection. What prompted the purple prose? Darwin and Wallace were mutual admirers, so like-minded that they had been inspired by the same author (Malthus) to forge the same theory in almost the same words. What divided these comrades was the human mind. Darwin had coyly predicted that "psychology will be placed on a new foundation," and in his notebooks was positively grandiose about how evolutionary theory would revolutionize the study of mind:

> Origin of man now proved.—Metaphysics must flourish.—He who understand baboon would do more toward metaphysics than Locke.

> Plato says . . . that our "imaginary ideas" arise from the preexistence of the soul, are not derivable from experience—read monkeys for preexistence.

He went on to write two books on the evolution of human thoughts and feelings, *The Descent of Man* and *The Expression of the Emotions in Man and Animals*.

But Wallace reached the opposite conclusion. The mind, he said, is overdesigned for the needs of evolving humans and cannot be explained by natural selection. Instead, "a superior intelligence has guided the

299

development of man in a definite direction, and for a special purpose."
Et tu!

Wallace became a creationist when he noted that foragers—"savages," in nineteenth-century parlance—were biologically equal to modern Europeans. Their brains were the same size, and they could easily adapt to the intellectual demands of modern life. But in the foragers' way of life, which was also the life of our evolutionary ancestors, that level of intelligence was not needed, and there was no occasion to show it off. How, then, could it have evolved in response to the needs of a foraging lifestyle? Wallace wrote:

> Our law, our government, and our science continually require us to reason through a variety of complicated phenomena to the expected result. Even our games, such as chess, compel us to exercise all these faculties in a remarkable degree. Compare this with the savage languages, which contain no words for abstract conceptions; the utter want of foresight of the savage man beyond his simplest necessities; his inability to combine, or to compare, or to reason on any general subject that does not immediately appeal to his senses. . . .
>
> . . . A brain one-half larger than that of the gorilla would . . . fully have sufficed for the limited mental development of the savage; and we must therefore admit that the large brain he actually possesses could never have been solely developed by any of those laws of evolution, whose essence is, that they lead to a degree of organization exactly proportionate to the wants of each species, never beyond those wants. . . . Natural selection could only have endowed savage man with a brain a few degrees superior to that of an ape, whereas he actually possesses one very little inferior to that of a philosopher.

Wallace's paradox, the apparent evolutionary uselessness of human intelligence, is a central problem of psychology, biology, and the scientific worldview. Even today, scientists such as the astronomer Paul Davies think that the "overkill" of human intelligence refutes Darwinism and calls for some other agent of a "progressive evolutionary trend," perhaps a self-organizing process that will be explained someday by complexity theory. Unfortunately this is barely more satisfying than Wallace's idea of a superior intelligence guiding the development of man in a definite direction. Much of this book, and this chapter in particular, is aimed at demoting Wallace's paradox from a foundation-shaking mystery to a challenging but otherwise ordinary research problem in the human sciences.

Stephen Jay Gould, in an illuminating essay on Darwin and Wallace, sees Wallace as an extreme adaptationist who ignores the possibility of exaptations: adaptive structures that are "fortuitously suited to other roles if elaborated" (such as jaw bones becoming middle-ear bones) and "features that arise without functions . . . but remain available for later co-optation" (such as the panda's thumb, which is really a jury-rigged wristbone).

> Objects designed for definite purposes can, as a result of their structural complexity, perform many other tasks as well. A factory may install a computer only to issue the monthly pay checks, but such a machine can also analyze the election returns or whip anyone's ass (or at least perpetually tie them) in tic-tac-toe.

I agree with Gould that the brain has been exapted for novelties like calculus or chess, but this is just an avowal of faith by people like us who believe in natural selection; it can hardly fail to be true. It raises the question of who or what is doing the elaborating and co-opting, and why the original structures were suited to being co-opted. The factory analogy is not helpful. A computer that issues paychecks *cannot* also analyze election returns or play tic-tac-toe, unless someone has reprogrammed it first.

Wallace went off the tracks not because he was too much of an adaptationist but because he was a lousy linguist, psychologist, and anthropologist (to judge him, unfairly, by modern standards). He saw a chasm between the simple, concrete, here-and-now thinking of foraging peoples and the abstract rationality exercised in modern pursuits like science, mathematics, and chess. But there is no chasm. Wallace, to give him his due, was ahead of his time in realizing that foragers were not on the lower rungs of some biological ladder. But he was wrong about their language, thought, and lifestyle. Prospering as a forager is a more difficult problem than doing calculus or playing chess. As we saw in Chapter 3, people in all societies have words for abstract conceptions, have foresight beyond simple necessities, and combine, compare, and reason on general subjects that do not immediately appeal to their senses. And people everywhere put these abilities to good use in outwitting the defenses of the local flora and fauna. We will soon see that all people, right from the cradle, engage in a *kind* of scientific thinking. We are all intuitive physicists, biologists, engineers, psychologists, and mathematicians. Thanks to these inborn talents, we outperform robots and have wreaked havoc on the planet.

On the other hand, our intuitive science is different from what the people in white coats do. Though most of us would not agree with Lucy in *Peanuts* that fir trees give us fur, sparrows grow into eagles that we eat on Thanksgiving, and you can tell a tree's age by counting its leaves, our beliefs are sometimes just as daffy. Children insist that a piece of styrofoam weighs nothing and that people know the outcome of events they did not witness or hear about. They grow into adults who think that a ball flying out of a spiral tube will continue in a spiral path and that a string of heads makes a coin more likely to land tails.

This chapter is about human reasoning: how people make sense of their world. To reverse-engineer our faculties of reasoning, we must begin with Wallace's paradox. To dissolve it, we have to distinguish the intuitive science and mathematics that is part of the human birthright from the modern, institutionalized version that most people find so hard. Then we can explore how our intuitions work, where they came from, and how they are elaborated and polished to give the virtuoso performances of modern civilization.

ECOLOGICAL INTELLIGENCE

Ever since the Swiss psychologist Jean Piaget likened children to little scientists, psychologists have compared the person in the street, young and old, to the person in the lab. The analogy is reasonable up to a point. Both scientists and children have to make sense of the world, and children are curious investigators striving to turn their observations into valid generalizations. Once I had family and friends staying over, and a three-year-old boy accompanied my sister as she bathed my infant niece. After staring quietly for several minutes he announced, "Babies don't have penises." The boy deserves our admiration, if not for the accuracy of his conclusion, then for the keenness of his scientific spirit.

Natural selection, however, did not shape us to earn good grades in science class or to publish in refereed journals. It shaped us to master the local environment, and that led to discrepancies between how we naturally think and what is demanded in the academy.

For many years the psychologist Michael Cole and his colleagues studied a Liberian people called the Kpelle. They are an articulate group, enjoying argument and debate. Most are illiterate and unschooled, and they do poorly on tests that seem easy to us. This dialogue shows why:

Experimenter: Flumo and Yakpalo always drink cane juice [rum] together. Flumo is drinking cane juice. Is Yakpalo drinking cane juice?

Subject: Flumo and Yakpalo drink cane juice together, but the time Flumo was drinking the first one Yakpalo was not there on that day.

Experimenter: But I told you that Flumo and Yakpalo always drink cane juice together. One day Flumo was drinking cane juice. Was Yakpalo drinking cane juice?

Subject: The day Flumo was drinking the cane juice Yakpalo was not there on that day.

Experimenter: What is the reason?

Subject: The reason is that Yakpalo went to his farm on that day and Flumo remained in town on that day.

The example is not atypical; Cole's subjects often say things like "Yakpalo isn't here at the moment; why don't you go and ask him about the matter?" The psychologist Ulric Neisser, who excerpted this dialogue, notes that these answers are by no means stupid. They are just not answers to the experimenter's question.

A ground rule when you solve a problem at school is to base your reasoning on the premises mentioned in a question, ignoring everything else you know. The attitude is important in modern schooling. In the few thousand years since the emergence of civilizations, a division of labor has allowed a class of knowledge professionals to develop methods of inference that are widely applicable and can be disseminated by writing and formal instruction. These methods literally have no content. Long division can calculate miles per gallon, or it can calculate income per capita. Logic can tell you that Socrates is mortal, or, in the examples in Lewis Carroll's logic textbook, that no lamb is accustomed to smoking cigars, all pale people are phlegmatic, and a lame puppy would not say "thank you" if you offered to lend it a skipping rope. The statistical tools of experimental psychology were borrowed from agronomy, where they were invented to gauge the effects of different fertilizers on crop yields. The tools work just fine in psychology, even though, as one psychological statistician wrote, "we do not deal in manure, at least not knowingly." The power of these tools is that they can be applied to any problem—how color vision works, how to put a man on the moon, whether mitochondrial Eve was an African—no matter how ignorant one is at the outset. To master the techniques, students must feign the ignorance they will later be saddled with when solving problems in their professional lives. A high school student doing Euclidean geometry gets no credit for pulling out a ruler and measuring the triangle, even though that guarantees a correct answer. The point of the lesson is to inculcate a method that later can be used to calculate the unmeasurable, such as the distance to the moon.

But outside of school, of course, it never makes sense to ignore what you know. A Kpelle could be forgiven for asking, Look, do you want to know whether Yakpalo is drinking cane juice, or don't you? That is true for both the knowledge acquired by an individual and the knowledge acquired by the species. No organism needs content-free algorithms applicable to any problem no matter how esoteric. Our ancestors encountered certain problems for hundreds of thousands or millions of years—recognizing objects, making tools, learning the local language, finding a mate, predicting an animal's movement, finding their way—and encountered certain other problems never—putting a man on the moon, growing better popcorn, proving Fermat's last theorem. The knowledge that solves a familiar kind of problem is often irrelevant to any other one. The effect of slant on luminance is useful in calculating shape but not in assessing the fidelity of a potential mate. The effects of lying on tone of voice help with fidelity but not with shape. Natural selection does not care about the ideals of a liberal education and should have no qualms about building parochial inference modules that exploit eons-old regularities in their own subject matters. Tooby and Cosmides call the subject-specific intelligence of our species "ecological rationality."

A second reason we did not evolve into true scientists is the cost of knowledge. Science is expensive, and not just the superconducting supercollider, but the elementary analysis of cause and effect in John Stuart Mill's canons of induction. Recently I was dissatisfied with the bread I had been baking because it was too dry and fluffy. So I increased the water, decreased the yeast, and lowered the temperature. To this day I don't know which of these manipulations made the difference. The scientist in me knew that the proper procedure would have been to try out all eight logical combinations in a factorial design: more water, same yeast, same temperature; more water, more yeast, same temperature; more water, same yeast, lower temperature; and so on. But the experiment would have taken eight days (twenty-seven if I wanted to test two increments of each factor, sixty-four if I wanted to test three) and required a notebook and a calculator. I wanted tasty bread, not a contribution to the archives of human knowledge, so my multiply-confounded one-shot was enough. In a large society with writing and institutionalized science, the cost of an exponential number of tests is repaid by the benefit of the resulting laws to a large number of people. That is why taxpayers are willing to fund scientific research. But for the provincial interests of a single individual or even a small band, good science isn't worth the trouble.

A third reason we are so-so scientists is that our brains were shaped for fitness, not for truth. Sometimes the truth is adaptive, but sometimes it is not. Conflicts of interest are inherent to the human condition (see Chapters 6 and 7), and we are apt to want *our version* of the truth, rather than the truth itself, to prevail.

For example, in all societies, expertise is distributed unevenly. Our mental apparatus for understanding the world, even for understanding the meanings of simple words, is designed to work in a society in which we can consult an expert when we have to. The philosopher Hilary Putnam confesses that, like most people, he has no idea how an elm differs from a beech. But the words aren't synonyms for him or for us; we all know that they refer to different kinds of trees, and that there are experts out there who could tell us which is which if we ever had to know. Experts are invaluable and are usually rewarded in esteem and wealth. But our reliance on experts puts temptation in their path. The experts can allude to a world of wonders—occult forces, angry gods, magical potions—that is inscrutable to mere mortals but reachable through their services. Tribal shamans are flimflam artists who supplement their considerable practical knowledge with stage magic, drug-induced trances, and other cheap tricks. Like the Wizard of Oz, they have to keep their beseechers from looking at the man behind the curtain, and that conflicts with the disinterested search for the truth.

In a complex society, a dependence on experts leaves us even more vulnerable to quacks, from carnival snake-oil salesman to the mandarins who advise governments to adopt programs implemented by mandarins. Modern scientific practices like peer review, competitive funding, and open mutual criticism are meant to minimize scientists' conflicts of interest in principle, and sometimes do so in practice. The stultification of good science by nervous authorities in closed societies is a familiar theme in history, from Catholic southern Europe after Galileo to the Soviet Union in the twentieth century.

It is not only science that can suffer under the thumb of those in power. The anthropologist Donald Brown was puzzled to learn that over the millennia the Hindus of India produced virtually no histories, while the neighboring Chinese had produced libraries full. He suspected that the potentates of a hereditary caste society realized that no good could come from a scholar nosing around in records of the past where he might stumble upon evidence undermining their claims to have descended from heroes and gods. Brown looked at twenty-five civilizations and compared the ones organized by hereditary castes with the others. None of the caste societies had developed a tradition of writing accurate depictions of the past; instead of history they had myth and legend. The caste societies were also distinguished by an absence of political science, social science, natural science, biography, realistic portraiture, and uniform education.

Good science is pedantic, expensive, and subversive. It was an unlikely selection pressure within illiterate foraging bands like our ancestors', and we should expect people's native "scientific" abilities to differ from the genuine article.

LITTLE BOXES

The humorist Robert Benchley said that there are two classes of people in the world: those who divide the people of the world into two classes, and those who do not. In Chapter 2, when I asked why the mind keeps track of individuals, I took it for granted that the mind forms categories. But the habit of categorizing deserves scrutiny as well. People put things and other people into mental boxes, give each box a name, and thereafter treat the contents of a box the same. But if our fellow humans are as unique as their fingerprints and no two snowflakes are alike, why the urge to classify?

Psychology textbooks typically give two explanations, neither of which makes sense. One is that memory cannot hold all the events that bombard our senses; by storing only their categories, we cut down on the load. But the brain, with its trillion synapses, hardly seems short of storage space. It's reasonable to say that entities cannot fit in memory when the entities are combinatorial—English sentences, chess games, all shapes in all colors and sizes at all locations—because the numbers from combinatorial explosions can exceed the number of particles in the universe and overwhelm even the most generous reckoning of the brain's capacity. But people live for a paltry two billion seconds, and there is no known reason why the brain could not record every object and event we experience if it had to. Also, we often remember both a category *and* its members, such as months, family members, continents, and baseball teams, so the category *adds* to the memory load.

The other putative reason is that the brain is compelled to organize; without categories, mental life would be chaos. But organization for its own sake is useless. I have a compulsive friend whose wife tells callers that he cannot come to the phone because he is alphabetizing his shirts. Occasionally I receive lengthy manuscripts from theoreticians who have discovered that everything in the universe falls into classes of three: the Father, the Son, and the Holy Ghost; protons, neutrons, and electrons; masculine, feminine, and neuter; Huey, Dewey, and Louie; and so on, for page after page. Jorge Luis Borges writes of a Chinese encyclopedia that divided animals into: (a) those that belong to the Emperor, (b) embalmed ones, (c) those that are trained, (d) suckling pigs, (e) mermaids, (f) fabulous ones, (g) stray dogs, (h) those that are included in this classification, (i) those that tremble as if they were mad, (j) innumerable ones, (k) those drawn with a very fine camel's hair brush, (l) others, (m) those that have just broken a flower vase, (n) those that resemble flies from a distance.

No, the mind has to get something out of forming categories, and that something is *inference*. Obviously we can't know everything about every object. But we can observe some of its properties, assign it to a category, and from the category predict properties that we have *not* observed. If Mopsy has long ears, he is a rabbit; if he is a rabbit, he should eat carrots, go hippety-hop, and breed like, well, a rabbit. The smaller the category, the better the prediction. Knowing that Peter is a cottontail, we can predict that he grows, breathes, moves, was suckled, inhabits open country or woodland clearings, spreads tularemia, and can contract myxomatosis. If we knew only that he was a mammal, the list would include only growing, breathing, moving, and being suckled. If we knew only that he was an animal, it would shrink to growing, breathing, and moving.

On the other hand, it's much harder to tag Peter as a cottontail than as a mammal or an animal. To tag him as a mammal we need only notice that he is furry and moving, but to tag him as a cottontail we have to notice that he has long ears, a short tail, long hind legs, and white on the underside of his tail. To identify *very* specific categories we have to examine so many properties that there would be few left to predict. Most of our everyday categories are somewhere in the middle: "rabbit," not mammal or cottontail; "car," not vehicle or Ford Tempo; "chair," not furniture or Barcalounger. They represent a compromise between how hard it is to identify the category and how much good the category does you. The psychologist Eleanor Rosch called them basic-level categories. They are the first words children learn for objects and generally the first mental label we assign when seeing them.

What makes a category like "mammal" or "rabbit" better than a category like "shirts made by companies beginning with *H*" or "animals drawn with a very fine camel's hair brush"? Many anthropologists and philosophers believe that categories are arbitrary conventions that we learn along with the other cultural accidents standardized in our language. Deconstructionism, poststructuralism, and postmodernism in the humanities take this view to an extreme. But categories would be useful only if they meshed with the way the world works. Fortunately for us, the world's objects are not evenly sprinkled throughout the rows and columns of the inventory list defined by the properties we notice. The world's inventory is lumpy. Creatures with cotton tails tend have long ears and live in woodland clearings; creatures with fins tend to have scales and live in the water. Other than in the children's books with split pages for assembling do-it-yourself chimeras, there are no finned cottontails or floppy-eared fish. Mental boxes work because things come in clusters that fit the boxes.

What makes the birds of a feather cluster together? The world is sculpted and sorted by laws that science and mathematics aim to discover. The laws of physics dictate that objects denser than water are found on the bottom of a lake, not its surface. Laws of natural selection and physics dictate that objects that move swiftly through fluids have streamlined shapes. The laws of genetics make offspring resemble their parents. Laws of anatomy, physics, and human intentions force chairs to have shapes and materials that make them stable supports.

People form two kinds of categories, as we saw in Chapter 2. We treat games and vegetables as categories that have stereotypes, fuzzy boundaries, and family-like resemblances. That kind of category falls naturally out of pattern-associator neural networks. We treat odd numbers and females as categories that have definitions, in-or-out boundaries, and common threads running through the members. That kind of category is naturally computed by systems of rules. We put some things into both kinds of mental categories—we think of "a grandmother" as a gray-haired muffin dispenser; we also think of "a grandmother" as the female parent of a parent.

Now we can explain what these two ways of thinking are for. Fuzzy categories come from examining objects and uninsightfully recording the correlations among their features. Their predictive power comes from similarity: if A shares some features with B, it probably shares others. They work by recording the clusters in reality. Well-defined categories, in contrast, work by ferreting out the laws that put the clusters there. They fall out of the intuitive theories that capture people's best guess about what makes the world tick. Their predictive power comes from deduction: if A implies B, and A is true, then B is true.

Real science is famous for transcending fuzzy feelings of similarity and getting at underlying laws. Whales are not fish; people are apes; solid matter is mostly empty space. Though ordinary people don't think exactly like scientists, they too let their theories override similarity when they reason about how the world works. Which two out of three belong together: white hair, gray hair, black hair? How about white cloud, gray cloud, black cloud? Most people say that black is the odd hair out, because aging hair turns gray and then white, but that white is the odd cloud out, because gray and black clouds give rain. Say I tell you I have a three-inch disk. Which is it more similar to, a quarter or a pizza? Which is it more likely to be, a quarter or a pizza? Most people say it is more *similar* to a quarter but more *likely to be* a pizza. They reason that quarters have to be standardized but pizzas can vary. On a trip to an unexplored forest, you discover a centipede, a caterpillar that looks like it, and a butterfly that the caterpillar turns into. How many kinds of animals have you found, and which belong together? Most people feel, along with biologists, that the caterpillar and the butterfly are the same animal, but the caterpillar and the centipede are not, despite appearances to the contrary. During your first basketball game, you see blond players with green jerseys run toward the east basket with the ball, and black players with yellow jerseys run toward the west basket with the ball. The whistle blows and a black player with a green jersey enters. Which basket will he run to? Everyone knows it is the east basket.

These similarity-defying guesses come from intuitive theories about aging, weather, economic exchange, biology, and social coalitions. They belong to larger systems of tacit assumptions about kinds of things and the laws governing them. The laws can be played out combinatorially in the mind to get predictions and inferences about events unseen. People everywhere have homespun ideas about physics, to predict how objects roll and bounce; psychology, to predict what other people think and do;

logic, to derive some truths from others; arithmetic, to predict the effects of aggregating; biology, to reason about living things and their powers; kinship, to reason about relatedness and inheritance; and a variety of social and legal rule systems. The bulk of this chapter explores those intuitive theories. But first we must ask: when does the world *allow* theories (scientific or intuitive) to work, and when does it force us all to fall back on fuzzy categories defined by similarity and stereotypes?

Where do our fuzzy similarity clusters come from? Are they just the parts of the world that we understand so poorly that the underlying laws escape us? Or does the world really *have* fuzzy categories even in our best scientific understanding? The answer depends on what part of the world we look at. Mathematics, physics, and chemistry trade in crisp categories that obey theorems and laws, such as triangles and electrons. But in any realm in which history plays a role, such as biology, members drift in and out of lawful categories over time, leaving their boundaries ragged. Some of the categories are definable, but others really are fuzzy.

Most biologists consider species to be lawful categories: they are populations that have become reproductively isolated and adapted to their local environment. Adaptation to a niche and inbreeding homogenize the population, so a species at a given time is a real category in the world that taxonomists can identify using well-defined criteria. But a higher taxonomic category, representing the descendants of an ancestral species, is not as well behaved. When the ancestral organisms dispersed and their descendants lost touch and adopted new homelands, the original pretty picture became a palimpsest. Robins, penguins, and ostriches share some features, like feathers, because they are great-great-grandchildren of a single population adapted to flight. They differ because ostriches are African and adapted to running and penguins are Antarctic and adapted to swimming. Flying, once a badge of all the birds, is now merely part of their stereotype.

For birds, at least, there *is* a kind of crisp biological category into which they can be fitted: a clade, exactly one branch of the genealogical tree of organisms. The branch represents the descendants of a single ancestral population. But not all of our familiar animal categories can be pegged onto one branch. Sometimes the descendants of a species diverge so unevenly that some of their scions are almost unrecognizable. Those branchlets have to be hacked off to keep the category as we know it, and the main branch is disfigured by jagged stumps. It turns into a fuzzy category whose boundaries are defined by similarity, without a crisp scientific definition.

Fish, for example, do not occupy one branch in the tree of life. One of their kind, a lungfish, begot the amphibians, whose descendants embrace the reptiles, whose descendants embrace the birds and the mammals. There is no definition that picks out all and only the fish, no branch of the tree of life that includes salmon and lungfish but excludes lizards and cows. Taxonomists fiercely debate what to do with categories like fish that are obvious to any child but have no scientific definition because they are neither species nor clades. Some insist that there is no such thing as a fish; it is merely a layperson's stereotype. Others try to rehabilitate everyday categories like fish using computer algorithms that sort creatures into clusters sharing properties. Still others wonder what the fuss is about; they see categories like families and orders as matters of convenience and taste—which similarities are important for the discussion at hand.

Classification is particularly fuzzy at the stump where a branch was hacked off, that is, the extinct species that became the inauspicious ancestor of a new group. The fossil *Archaeopteryx*, thought to be the ancestor of the birds, has been described by one paleontologist as "a piss-poor reptile, and not very much of a bird." The anachronistic shoehorning of extinct animals into the modern categories they spawned was a bad habit of early paleontologists, dramatically recounted in Gould's *Wonderful Life*.

So the world sometimes presents us with fuzzy categories, and registering their similarities is the best we can do. Now we may turn the question around. Does the world ever present us with *crisp* categories?

In his book *Women, Fire, and Dangerous Things*, named after a fuzzy grammatical category in an Australian language, the linguist George Lakoff argues that pristine categories are fictions. They are artifacts of the bad habit of seeking definitions, a habit that we inherited from Aristotle and now must shake off. He defies his readers to find a sharp-edged category in the world. Crank up the microscope, and the boundaries turn to fuzz. Take a textbook example, "mother," a category with the seemingly straightforward definition "female parent." Oh, yeah? What about surrogate mothers? Adoptive mothers? Foster mothers? Egg donors? Or take species. A species, unlike the controversial larger categories like "fish," is supposed to have a clear definition: usually, a population of organisms whose members can mate to form fertile offspring. But even that vaporizes under scrutiny. There are widely dispersed, gradually varying species in which an animal from the western edge of the range can mate with an animal from the center, and an animal from the center can mate with an animal from the east, but an animal from the west cannot mate with an animal from the east.

The observations are interesting, but I think they miss an important point. Systems of rules are *idealizations* that abstract away from complicating aspects of reality. They are never visible in pure form, but are no less real for all that. No one has ever actually sighted a triangle without thickness, a frictionless plane, a point mass, an ideal gas, or an infinite, randomly interbreeding population. That is not because they are useless figments but because they are masked by the complexity and finiteness of the world and by many layers of noise. The concept of "mother" is perfectly well defined within a number of idealized theories. In mammalian genetics, a mother is the source of the sex cell that always carries an X chromosome. In evolutionary biology, she is the producer of the larger gamete. In mammalian physiology, she is the site of prenatal growth and birth; in genealogy, the immediate female ancestor; in some legal contexts, the guardian of the child and the spouse of the child's father. The omnibus concept "mother" depends on an idealization of the idealizations in which all the systems pick out the same entities: the contributor of the egg nurtures the embryo, bears the offspring, raises it, and marries the sperm donor. Just as friction does not refute Newton, exotic disruptions of the idealized alignment of genetics, physiology, and law do not make "mother" any fuzzier *within* each of these systems. Our theories, both folk and scientific, can idealize away from the messiness of the world and lay bare its underlying causal forces.

It's hard to read about the human mind's tendency to put things in boxes organized around a stereotype without pondering the tragedy of racism.

If people form stereotypes even about rabbits and fish, does racism come naturally to us? And if racism is both natural and irrational, does that make the love of stereotypes a bug in our cognitive software? Many social and cognitive psychologists would answer yes. They link ethnic stereotypes to an overeagerness to form categories and to an insensitivity to the laws of statistics that would show the stereotypes to be false. An Internet discussion group for neural-network modelers once debated what kinds of learning algorithms would best model Archie Bunker. The discussants assumed that people are racists when their neural networks perform poorly or are deprived of good training examples. If only our networks could use a proper learning rule and take in enough data, they would transcend false stereotypes and correctly register the facts of human equality.

Some ethnic stereotypes are indeed based on bad statistics or none at all; they are a product of a coalitional psychology that automatically denigrates outsiders (see Chapter 7). Others may be based on good statistics about nonexistent people, the virtual characters we meet every day on the big and small screens: Italian goodfellas, Arab terrorists, black drug dealers, Asian kung fu masters, British spies, and so on.

But sadly, some stereotypes may be based on good statistics about real people. In the United States at present, there are real and large differences among ethnic and racial groups in their average performance in school and in their rates of committing violent crimes. (The statistics, of course, say nothing about heredity or any other putative cause.) Ordinary people's estimates of these differences are fairly accurate, and in some cases, people with more contact with a minority group, such as social workers, have more pessimistic, and unfortunately more accurate, estimates of the frequency of negative traits such as illegitimacy and welfare dependency. A good statistical category-maker could develop racial stereotypes and use them to make actuarially sound but morally repugnant decisions about individual cases. This behavior is racist not because it is irrational (in the sense of statistically inaccurate) but because it flouts the moral principle that it is wrong to judge an *individual* using the statistics of a racial or ethnic *group*. The argument against bigotry, then, does not come from the design specs for a rational statistical categorizer. It comes from a rule system, in this case a rule of ethics, that tells us when to turn our statistical categorizers off.

CORE CURRICULUM

You have channel-surfed to a rerun of *L.A. Law*, and you want to know why the harpy lawyer Rosalind Shays is weeping on the witness stand. If someone began to explain that the fluid in her tear ducts had increased in volume until the pressure exceeded the surface tension by such and such an amount, you would squelch the lecture. What you want to find out is that she hopes to win a lawsuit against her former employers and is shedding crocodile tears to convince the jury that when the firm fired her she was devastated. But if you saw the next episode and wanted to know why she plummeted to the bottom of an elevator shaft after she accidentally stepped through the open door, her motives would be irrelevant to anyone but a Freudian gone mad. The explanation is that matter in free fall, Rosalind Shays included, accelerates at a rate of 9.8 meters per second per second.

There are many ways to explain an event, and some are better than others. Even if neuroscientists someday decode the entire wiring diagram of the brain, human behavior makes the most sense when it is explained in terms of beliefs and desires, not in terms of volts and grams. Physics provides no insight into the machinations of a crafty lawyer, and even fails to enlighten us about many simpler acts of living things. As Richard Dawkins observed, "If you throw a dead bird into the air it will describe a graceful parabola, exactly as physics books say it should, then come to rest on the ground and stay there. It behaves as a solid body of a particular mass and wind resistance ought to behave. But if you throw a live bird in the air it will not describe a parabola and come to rest on the ground. It will fly away, and may not touch land this side of the county boundary." We understand birds and plants in terms of their innards. To know why they move and grow, we cut them open and put bits under a microscope. We need yet another kind of explanation for artifacts like a chair and a crowbar: a statement of the function the object is intended to perform. It would be silly to try to understand why chairs have a stable horizontal surface by cutting them open and putting bits of them under a microscope. The explanation is that someone designed the chair to hold up a human behind.

Many cognitive scientists believe that the mind is equipped with innate intuitive theories or modules for the major ways of making sense of the world. There are modules for objects and forces, for animate beings, for artifacts, for minds, and for natural kinds like animals, plants, and minerals. Don't take the "theory" idiom literally; as we have seen, people don't really work like scientists. Don't take the "module" metaphor too seriously, either; people can mix and match their ways of knowing. A concept like "throwing," for example, welds an intention (intuitive psychology) to a motion (intuitive physics). And we often apply modes of thinking to subject matters they were not designed for, such as in slapstick humor (person as object), animistic religion (tree or mountain as having a mind), and anthropomorphic animal stories (animals with human minds). As I have mentioned, I prefer to think of the ways of knowing in anatomical terms, as mental systems, organs, and tissues, like the immune system, blood, or skin. They accomplish specialized functions, thanks to their specialized structures, but don't necessarily come in encapsulated packages. I would also add that the list of intuitive theories or modules or ways of knowing is surely too short. Cognitive scientists think of people as Mr. Spock without the funny ears. A more realistic inventory would include modes of thought and feeling for danger, contamination, status, dominance, fairness, love, friendship, sexuality, children, relatives, and the self. They are explored in later chapters.

Saying that the different ways of knowing are innate is different from saying that knowledge is innate. Obviously we have to learn about Frisbees, butterflies, and lawyers. Talking about innate modules is not meant to minimize learning but to explain it. Learning involves more than recording experience; learning requires couching the records of experience so that they generalize in useful ways. A VCR is excellent at recording, but no one would look to this modern version of the blank slate as a paradigm of intelligence. When we watch lawyers in action, we draw conclusions about their goals and values, not their tongue and limb trajectories. Goals and values are one of the vocabularies in which we mentally couch our experiences. They cannot be built out of simpler concepts from our physical knowledge the way "momentum" can be built out of mass and velocity or "power" can be built out of energy and time. They are primitive or irreducible, and higher-level concepts are defined in terms of them. To understand learning in other domains, we have to find their vocabularies, too.

Because a combinatorial system like a vocabulary can generate a vast number of combinations, one might wonder whether human thoughts can be generated by a single system, a general-purpose Esperanto of the mind. But even a very powerful combinatorial system has its limits. A calculator can add and multiply a vast number of vast numbers, but it will never spell a sentence. A dedicated word processor can type Borges' infinite library of books with all combinations of characters, but it can never add the numbers it spells out. Modern digital computers can do a lot with a little, but that "little" still includes distinct, hard-wired vocabularies for text, graphics, logic, and several kinds of numbers. When the computers are programmed into artificial intelligence reasoning systems, they have to be innately endowed with an understanding of the basic categories of the world: objects, which can't be in two places at once, animals, which live for a single interval of time, people, who don't like pain, and so on. That is no less true of the human mind. Even a dozen innate mental vocabularies—a wild and crazy idea, according to critics—would be a small number with which to spell the entirety of human thought and feeling, from the meanings of the 500,000 words in the *Oxford English Dictionary* to the plots of Scheherazade's 1,001 tales.

~

We live in the material world, and one of the first things in life we must figure out is how objects bump into each other and fall down elevator shafts. Until recently, everyone thought that the infant's world was a kaleidoscope of sensations, a "blooming, buzzing confusion," in William James' memorable words. Piaget claimed that infants were sensorimotor creatures, unaware that objects cohere and persist and that the world works by external laws rather than the infants' actions. Infants would be like the man in the famous limerick about Berkeley's idealist philosophy:

> There once was a man who said, "God
> Must think it exceedingly odd
> If he finds that this tree
> Continues to be
> When there's no one about in the Quad."

Philosophers are fond of pointing out that the belief that the world is a hallucination or that objects do not exist when you aren't looking at them is not refutable by any observation. A baby could experience the blooming and buzzing all its life unless it was equipped with a mental mechanism that interpreted the blooms and buzzes as the outward signs of persisting objects that follow mechanical laws. We should expect infants to show *some* appreciation of physics from the start.

Only careful laboratory studies can tell us what it is like—rather, what it *was* like—to be a baby. Unfortunately, infants are difficult experimental subjects, worse than rats and sophomores. They can't easily be conditioned, and they don't talk. But an ingenious technique, refined by the psychologists Elizabeth Spelke and Renée Baillargeon, capitalizes on one feat that infants are good at: getting bored. When infants see the same old thing again and again, they signal their boredom by looking away. If a new thing appears, they perk up and stare. Now, "old thing" and "new thing" are in the mind of the beholder. By seeing what revives babies' interest and what prolongs their ennui, we can guess at what things they see as the same and what things they see as different—that is, how they categorize experience. It's especially informative when a screen first blocks part of the infant's view and then falls away, for we can try to tell what the babies were thinking about the invisible part of their world. If the baby's eyes are only momentarily attracted and then wander off, we can infer that the scene was in the baby's mind's eye all along. If the baby stares longer, we can infer that the scene came as a surprise.

Three- to four-month-old infants are usually the youngest tested, both because they are better behaved than younger babies and because their stereo vision, motion perception, visual attention, and acuity have just matured. The tests cannot, by themselves, establish what is and is not innate. Three-month-olds were not born yesterday, so anything they know they could, in theory, have learned. And three-month-olds still have a lot of maturing to do, so anything they come to know later could emerge without learning, just as teeth and pubic hair do. But by telling us what babies know at what age, the findings narrow the options.

Spelke and Philip Kelman wanted to see what infants treated as an object. Remember from Chapter 4 that it is not easy, even for an adult, to say what an "object" is. An object can be defined as a stretch of the visual field with a smooth silhouette, a stretch with a homogeneous color and texture, or a collection of patches with a common motion. Often these definitions pick out the same pieces, but when they don't, it is common motion that wins the day. When pieces move together, we see them as a single object; when pieces go their separate ways, we see them as separate objects. The concept of an object is useful because bits of matter that are attached to one another usually move together. Bicycles and grapevines and snails may be jagged agglomerations of different materials, but if you pick up one end, the other end comes along for the ride.

Kelman and Spelke bored babies with two sticks poking out from behind the top and bottom edges of a wide screen. The question was whether the babies would see the sticks as part of a single object. When the screen was removed, the babies saw either one long stick or two short ones with a gap between them. If the babies had *visualized* a single object, then *seeing* a single object would be a bore, and two would come as a surprise. If they had thought of each piece as its own object, then seeing a single object would be a surprise, and two a bore. Control experiments measured how long infants looked at one versus two objects without having seen anything beforehand; these baseline times were subtracted out.

Infants might have been expected to see the two pieces as two pieces, or, if they had mentally united them at all, to have used all the correlations among the features of an object as criteria: smooth silhouettes, common colors, common textures, and common motions. But apparently infants have an idea of objecthood early in life, and it is the core of the adult concept: parts moving together. When two sticks peeking out from behind the screen moved back and forth in tandem, babies saw them as a single object and were surprised if the raised screen revealed two. When they didn't move, babies did not expect them to be a single object, even though the visible pieces had the same color and texture. When a stick peeked out from behind the top edge and a red jagged polygon peeked out from behind the bottom edge, and they moved back and forth in tandem, babies expected *them* to be connected, even though they had nothing in common but motion.

The child is parent to the adult in other principles of intuitive physics. One is that an object cannot pass through another object like a ghost. Renée Baillargeon has shown that four-month-old infants are surprised when a panel just in front of a cube somehow manages to fall back flat to the ground, right through the space that the cube should be occupying. Spelke and company have shown that infants don't expect an object to pass through a barrier or through a gap that is narrower than the object is.

A second principle is that objects move along continuous trajectories: they cannot disappear from one place and materialize in another, as in the transporter room of the *Enterprise*. When an infant sees an object pass behind the left edge of a left screen and then seem to reappear from behind the right edge of a right screen without moving through the gap between the screens, she assumes she is seeing two objects. When she sees an object pass behind the left screen, reappear at the other edge of the screen, cross the gap, and then pass behind the right screen, she assumes she is seeing one object.

A third principle is that objects are cohesive. Infants are surprised when a hand picks up what looks like an object but part of the object stays behind.

A fourth principle is that objects move each other by contact only—no action at a distance. After repeatedly seeing an object pass behind a screen and another object pop out, babies expect to see one launching the other like billiard balls. They are surprised when the screen reveals one ball stopping short and the second just up and leaving.

So three- to four-month-old infants see objects, remember them, and expect them to obey the laws of continuity, cohesion, and contact as they move. Babies are not as stoned as James, Piaget, Freud, and others thought. As the psychologist David Geary has said, James' "blooming, buzzing confusion" is a good description of the *parents'* life, not the infant's. The discovery also overturns the suggestion that babies stop their world from spinning by manipulating objects, walking around them, talking about them, or hearing them talked about. Three-month-olds can barely orient, see, touch, and reach, let alone manipulate, walk, talk, and understand. They could not have learned anything by the standard techniques of interaction, feedback, and language. Nonetheless, they are sagely understanding a stable and lawful world.

Proud parents should not call MIT admissions just yet. Small babies have an uncertain grasp, at best, of gravity. They *are* surprised when a hand pushes a box off a table and it remains hovering in midair, but the slightest contact with the edge of the table or a fingertip is enough for them to act as if nothing were amiss. And they are not fazed when a screen rises to reveal a falling object that has defied gravity by coming to rest in midair. Nor are they nonplussed when a ball rolls right over a large hole in a table without falling through. Infants don't quite have inertia down, either. For example, they don't care when a ball rolls toward one corner of a covered box and then is shown to have ended up in the other corner.

But then, adults' grasp of gravity and inertia is not so firm, either. The psychologists Michael McCloskey, Alfonso Caramazza, and Bert Green asked college students what would happen when a ball shot out of a curved tube or when a whirling tetherball was cut loose. A depressingly large minority, including many who had taken physics, guessed that it would continue in a curving path. (Newton's first law states that a moving object continues to move in a straight line unless a force acts on it.) The students explained that the object acquires a "force" or "momentum" (some students, remembering the lingo but not the concept, called it "angular momentum"), which propels it along the curve until the momentum gets used up and the path straightens out. Their beliefs come right out of the medieval theory in which an object is impressed with an "impetus" that maintains the object's motion and gradually dissipates.

These howlers come from conscious theorizing; they are not what people are prepared to see. When people view their paper-and-pencil answer as a computer animation, they burst out laughing as if watching Wile E. Coyote chasing the Road Runner over a cliff and stopping in midair before plunging straight down. But the cognitive misconceptions do run deep. I toss a ball straight up. After it leaves my hand, which forces act on it on the way up, at the apogee, and on the way down? It's almost impossible not to think that momentum carries the ball up against gravity, the forces equal out, and then gravity is stronger and pushes it back down. The correct answer is that gravity is the only force and that it applies the whole time. The linguist Leonard Talmy points out that the impetus theory infuses our language. When we say *The ball kept rolling because the wind blew on it*, we are construing the ball as having an inherent tendency toward rest. When we say *The ridge kept the pencil on the table*, we are imbuing the pencil with a tendency toward motion, not to mention flouting Newton's third law (action equals reaction) by imputing a greater force to the ridge. Talmy, like most cognitive scientists, believes that the conceptions drive the language, not the other way around.

When it comes to more complicated motions, even perception fails us. The psychologists Dennis Proffitt and David Gilden have asked people simple questions about spinning tops, wheels rolling down ramps, colliding balls, and Archimedes-in-the-bathtub displacements. Even physics professors guess the wrong outcome if they are not allowed to fiddle with equations on paper. (If they are, they spend a quarter of an hour working it out and then announce that the problem is "trivial.") When it comes to these motions, video animations of impossible events look quite natural. Indeed, possible events look unnatural: a spinning top, which leans without falling, is an object of wonder to all of us, even physicists.

It is not surprising to find that the mind is non-Newtonian. The idealized motions of classical mechanics are visible only in perfectly elastic point masses moving in vacuums on frictionless planes. In the real world, Newton's laws are masked by friction from the air, the ground, and the objects' own molecules. With friction slowing everything that moves and keeping stationary objects in place, it's natural to conceive of objects as having an inherent tendency toward rest. As historians of science have noted, it would be hard to convince a medieval European struggling to free an oxcart from the mud that an object in motion continues at a constant speed along a straight line unless acted upon by an external force. Complicated motions like spinning tops and rolling wheels have a double disadvantage. They depend on evolutionarily unprecedented machines with negligible friction, and their motions are governed by complex equations that relate many variables at once; our perceptual system can handle only one at a time even in the best of circumstances.

Even the brainiest baby has a lot to learn. Children grow up in a world of sand, Velcro, glue, Nerf balls, rubbed balloons, dandelion seeds, boomerangs, television remote controls, objects suspended by near-invisible fishing line, and countless other objects whose idiosyncratic properties overwhelm the generic predictions of Newton's laws. The precociousness that infants show in the lab does not absolve them of learning about objects; it makes the learning possible. If children did not carve the world into objects, or if they were prepared to believe that objects could magically disappear and reappear anywhere, they would have no pegs on which to hang their discoveries of stickiness, fluffiness, squishiness, and so on. Nor could they develop the intuitions captured in Aristotle's theory, the impetus theory, Newton's theory, or Wile E. Coyote's theory. An intuitive physics relevant to our middle-sized world has to refer to enduring matter and its lawful motions, and infants see the world in those terms from the beginning.

Here is the plot of a movie. A protagonist strives to attain a goal. An antagonist interferes. Thanks to a helper, the protagonist finally succeeds. This movie does not feature a swashbuckling hero aided by a romantic interest to foil a dastardly villain. Its stars are three dots. One dot moves some distance up an inclined line, back down, and up again, until it is almost at the top. Another abruptly collides with it, and it moves back down. A third gently touches it and moves together with it to the top of the incline. It is impossible not to see the first dot as *trying* to get up the hill, the second as *hindering* it, and the third as *helping* it reach its goal.

The social psychologists Fritz Heider and M. Simmel were the film-makers. Together with many developmental psychologists, they conclude that people interpret certain motions not as special cases in their intuitive physics (perhaps as weird springy objects) but as a different kind of entity altogether. People construe certain objects as animate agents. Agents are recognized by their ability to violate intuitive physics by starting, stopping, swerving, or speeding up without an external nudge, especially when they persistently approach or avoid some other object. The agents are thought to have an internal and renewable source of energy, force, impetus, or oomph, which they use to propel themselves, usually in service of a goal.

These agents are animals, of course, including humans. Science tells us that they follow physical laws, just like everything else in the universe; it's just that the matter in motion consists of tiny little molecules in muscles and brains. But outside the neurophysiology lab ordinary thinkers have to assign them to a different category of uncaused causers.

Infants divide the world into the animate and the inert early in life. Three-month-olds are upset by a face that suddenly goes still but not by an object that suddenly stops moving. They try to bring objects toward them by pushing things, but try to bring people toward them by making noise. By six or seven months, babies distinguish between how hands act upon objects and how other objects act upon objects. They have opposite expectations about what makes people move and what makes objects move: objects launch each other by collisions; people start and stop on their own. By twelve months, babies interpret cartoons of moving dots as if the dots were seeking goals. For example, the babies are not surprised when a dot that hops over a barrier on its way to another dot makes a beeline after the barrier is removed. Three-year-olds describe dot cartoons much as we do, and have no trouble distinguishing things that move on their own, like animals, from things that don't, like dolls, statues, and lifelike animal figurines.

Intuitions about self-propelled agents overlap with three other major ways of knowing. Most agents are animals, and animals, like plants and minerals, are categories that we sense are given by nature. Some self-propelled things, like cars and windup dolls, are artifacts. And many agents do not merely approach and avoid goals but act out of beliefs and desires; that is, they have minds. Let's look at each of them.

~

People everywhere are fine amateur biologists. They enjoy looking at animals and plants, classify them into groups that biologists recognize, predict their movements and life cycles, and use their juices as medicines, poisons, food additives, and recreational drugs. These talents, which have adapted us to the cognitive niche, come from a mode of understanding the world called folk biology, though "folk natural history" may be a more apt term. People have certain intuitions about natural kinds—roughly, the sorts of things found in a museum of natural history, such as animals, plants, and minerals—that they don't apply to artifacts, such as coffeepots, or to kinds stipulated directly by rules, such as triangles and prime ministers.

What is the definition of *lion*? You might say "a large, ferocious cat that lives in Africa." But suppose you learned that a decade ago lions were hunted to extinction in Africa and survive only in American zoos. Suppose scientists discovered that lions weren't innately ferocious; they get that way in a dysfunctional family but otherwise grow up like Bert Lahr in the Wizard of Oz. Suppose it turned out that they were not even cats. I had a teacher who insisted that lions really belonged in the dog family, and though she was wrong, she *could* have been right, just as whales turned out to be mammals, not fish. But if this thought experiment turned out to be true, you would probably feel that these gentle American dogs were still really lions, even if not a word of the definition survived. Lions just don't *have* definitions. They are not even picked out by the picture of a lion in the dictionary next to the definition of the word. A lifelike mechanical lion wouldn't count as the real thing, and one can imagine breeding a striped lion that looked more like a tiger but would still count as a lion.

Philosophers say that the meaning of a natural-kind term comes from an intuition of a hidden trait or essence that the members share with one another and with the first examples dubbed with the term. People don't need to know what the essence is, just that there is one. Some people probably think that lionhood is in the blood; others might mumble something about DNA; still others would have no idea but would sense that lions all have it, whatever it is, and pass it to their offspring. Even when an essence is known, it is not a definition. Physicists tell us that gold is matter with atomic number 79, as good an essence as we can hope for. But if they had miscalculated and it turned out that gold was 78 and platinum 79, we would not think that the word *gold* now refers to platinum or experience much of a change in the way we think about gold. Compare these intuitions with our feelings about artifacts like coffeepots. Coffeepots are pots for making coffee. The possibility that all coffeepots have an essence, that scientists might someday discover it, or that we might have been wrong about coffeepots all along and that they are really pots for making tea are worthy of Monty Python's Flying Circus.

If the driving intuition behind folk physics is the continuous solid object, and the driving intuition behind animacy is an internal and renewable source of oomph, then the driving intuition behind natural kinds is a hidden essence. Folk biology is said to be essentialistic. The essence has something in common with the oomph that powers animals' motions, but it also is sensed to give the animal its form, to drive its growth, and to orchestrate its vegetative processes like breathing and digestion. Of course, today we know that this *élan vital* is really just a tiny data tape and chemical factory inside every cell.

Intuitions about essences can be found long ago and far away. Even before Darwin, the Linnaean classification system used by professional biologists was guided by a sense of proper categories based not on similarity but on underlying constitution. Peacocks and peahens were classified as the same animal, as were a caterpillar and the butterfly it turned into. Some similar animals—monarch and viceroy butterflies, mice and shrews—were put into different groups because of subtle differences in their internal structure or embryonic forms. The classification was hierarchical: every living thing belonged to one species, every species belonged to one genus, and so on up through families, classes, orders, and phyla to the plant and animal kingdoms, all in one tree of life. Again, compare this system with the classification of artifacts—say, the tapes in a video store. They can be arranged by genre, such as dramas and musicals, by period, such as new releases and classics, by alphabetical order, by country of origin, or by various cross-classifications such as foreign new releases or classic musicals. There is no single correct tree of video-tapes.

The anthropologists Brent Berlin and Scott Atran have discovered that folk taxonomies all over the world work the same way as the Linnaean tree. People group all the local plants and animals into kinds that correspond to the biologist's "genus." Since there is usually only one species per genus in a locality, their categories usually match the biologist's "species" as well. Every folk genus belongs to a single "life form," such as mammals, birds, mushrooms, herbs, insects, or reptiles. The life forms are in turn either animals or plants. People override appearances when classifying living things; for example, they lump frogs and tadpoles. They use their classes to reason about how animals work, such as who can breed with whom.

One of Darwin's best arguments for evolution was that it explained why living things are hierarchically grouped. The tree of life is a *family* tree. The members of a species seem to share an essence because they are descendants of a common ancestor that passed it on. Species fall into groups within groups because they diverged from even earlier common ancestors. Embryonic and internal features are more sensible criteria than surface appearance because they better reflect degree of related-ness.

Darwin had to fight his contemporaries' intuitive essentialism because, taken to an extreme, it implied that species could not change. A reptile has a reptilian essence and can no more evolve into a bird than the number seven can evolve into an even number. As recently as the 1940s, the philosopher Mortimer Adler argued that just as there can be no three-and-a-half-sided triangle, there can be nothing intermediate between an animal and a human, so humans could not have evolved. Darwin pointed out that species are populations, not ideal types, with members that vary; in the past they could have shaded into in-between forms.

Today we have gone to the other extreme, and in modern academic life "essentialist" is just about the worst thing you can call someone. In the sciences, essentialism is tantamount to creationism. In the humanities, the label implies that the person subscribes to insane beliefs such as that the sexes are not socially constructed, there are universal human emotions, a real world exists, and so on. And in the social sciences, "essentialism" has joined "reductionism," "determinism," and "reification" as a term of abuse hurled at anyone who tries to explain human thought and behavior rather than redescribe it. I think it is unfortunate that "essentialism" has become an epithet, because at heart it is just the ordinary human curiosity to find out what makes natural things work. Essentialism is behind the success of chemistry, physiology, and genetics, and even today biologists routinely embrace the essentialist heresy when they work on the Human Genome Project (but everyone has a different genome!) or open up *Gray's Anatomy* (but bodies vary!).

How deeply rooted is essentialist thinking? The psychologists Frank Keil, Susan Gelman, and Henry Wellman have taken the philosophers' thought experiments about natural kinds and given them to children. Doctors take a tiger, bleach its fur, and sew on a mane. Is it a lion or a tiger? Seven-year-olds say it's still a tiger, but five-year-olds say it's now a lion. This finding, taken at face value, suggests that older children are essentialists about animals but younger ones are not. (At no age are children essentialists about artifacts—if you make a coffeepot look like a birdfeeder, children, like adults, say it just *is* a birdfeeder.)

But with deeper probing, one can find evidence for essentialist intuitions about living things even in preschoolers. Five-year-olds deny that an animal can be made to cross the deeper boundary into plants or artifacts. For example, they say that a porcupine that looks as if it has been turned into a cactus or a hairbrush in fact has not. And preschoolers think that one species can be turned into another only when the transformation affects a permanent part of the animal's constitution, not when it merely alters appearance. For example, they deny that a lion costume turns a tiger into a lion. They claim that if you remove the innards of a dog, the shell that remains, while looking like a dog, is not a dog and can't bark or eat dogfood. But if you remove the outsides of a dog, leaving something that doesn't look like a dog at all, it's still a dog and does doggy things. Preschoolers even have a crude sense of inheritance. Told that a piglet is being raised by cows, they know it will grow up to oink, not moo.

Children do not merely sort animals like baseball cards but use their categories to reason about how animals work. In one experiment, three-year-olds were shown pictures of a flamingo, a blackbird, and a bat that looked a lot like the blackbird. The children were told that flamingos feed their babies mashed-up food but bats feed their babies milk, and were asked what they thought the blackbird feeds its babies. With no further information, children went with appearances and said that blackbirds, like bats, give milk. But if they were told that a flamingo is a bird, the children thought of them as working like blackbirds, despite their different appearance, and guessed that blackbirds provide their babies with mashed-up food, too.

Children also have a sense that a living thing's properties are there to keep it alive and help it function. Three-year-olds say that a rose has thorns because it helps the rose, but not that barbed wire has barbs to help the wire. They say that claws are good for the lobster, but not that jaws are good for the pliers. This sense of fitness or adaptation is not just a confusion between psychological wants and biological functions. The psychologists Giyoo Hatano and Kayoko Inagaki have shown that children have a clear sense that bodily processes are involuntary. They know that a boy can't digest dinner more quickly to make room for dessert, nor can he make himself fat by wishing alone.

Is essentialism learned? Biological processes are too slow and hidden to show to a bored baby, but testing babies is only one way to show knowledge in the absence of experience. Another is to measure the source of the experience itself. Three-year-olds haven't taken biology, and they have few opportunities to experiment with the innards or the heritability of animals. Whatever they have learned about essences has presumably come from their parents. Gelman and her students analyzed more than four thousand sentences from mothers talking to their children about animals and artifacts. The parents virtually never talked about innards, origins, or essences, and the few times they did, it was about the innards of artifacts. Children are essentialists without their parents' help.

~

Artifacts come with being human. We make tools, and as we evolved our tools made us. One-year-old babies are fascinated by what objects can do for them. They tinker obsessively with sticks for pushing, cloth and strings for pulling, and supports for holding things up. As soon as they can be tested on tool use, around eighteen months, children show an understanding that tools have to contact their material and that a tool's rigidity and shape are more important than its color or ornamentation. Some patients with brain damage cannot name natural objects but can name artifacts, or vice versa, suggesting that artifacts and natural kinds might even be stored in different ways in the brain.

What is an artifact? An artifact is an object suitable for attaining some end that a person intends to be used for attaining that end. The mixture of mechanics and psychology makes artifacts a strange category. Artifacts can't be defined by their shape or their constitution, only by what they can do and by what someone, somewhere, wants them to do. A store in my neighborhood sells nothing but chairs, but its inventory is as varied as a department store's. It has stools, high-backed dining chairs, recliners, beanbags, elastics and wires stretched over frames, hammocks, wooden cubes, plastic S's, and foam-rubber cylinders. We call them all chairs because they are designed to hold people up. A stump or an elephant's foot can become a chair if someone decides to use it as one. Probably somewhere in the forests of the world there is a knot of branches that uncannily resembles a chair. But like the proverbial falling tree that makes no sound, it is not a chair until someone decides to treat it as one. Keil's young subjects who happily let coffeepots turn into birdfeeders get the idea.

An extraterrestrial physicist or geometer, unless it had our psychology, would be baffled by some of the things we think exist in the world when these things are artifacts. Chomsky points out that we can say that the book John is writing will weigh five pounds when it is published: "the book" is both a stream of ideas in John's head and an object with mass. We talk about a house burning down to nothing and being rebuilt; somehow, it's the same house. Consider what kind of object "a city" must be, given that we can say *London is so unhappy, ugly, and polluted that it should be destroyed and rebuilt a hundred miles away.*

When Atran claimed that folk biology mirrors professional biology, he was criticized because folk categories like "vegetable" and "pet" match no Linnaean taxon. He replies that they are artifacts. Not only are they defined by the needs they serve (savory, succulent food; tractable companions), but they are, quite literally, human products. Millennia of selective breeding have created corn out of a grass and carrots out of a root. One has only to imagine packs of French poodles roaming the primeval forests to realize that most pets are human creations, too.

Daniel Dennett proposes that the mind adopts a "design stance" when dealing with artifacts, complementing its "physical stance" for objects like rocks and its "intentional stance" for minds. In the design stance, one imputes an intention to a real or hypothetical designer. Some objects are so suited to accomplishing an improbable outcome that the attribution is easy. As Dennett writes, "There can be little doubt what an axe is, or what a telephone is for; we hardly need to consult Alexander

Graham Bell's biography for clues about what he had in mind." Others are notoriously open to rival interpretations, like paintings and sculpture, which are sometimes *designed* to have an inscrutable design. Still others, like Stonehenge or an assembly of gears found in a shipwreck, probably have a function, though we don't know what it is. Artifacts, because they depend on human intentions, are subject to interpretation and criticism just as if they were works of art, an activity Dennett calls "artifact hermeneutics."

And now we come to the mind's way of knowing other minds. We are all psychologists. We analyze minds not just to follow soap-opera connivings but to understand the simplest human actions.

The psychologist Simon Baron-Cohen makes the point with a story. Mary walked into the bedroom, walked around, and walked out. How do you explain it? Maybe you'd say that Mary was looking for something she wanted to find and thought it was in the bedroom. Maybe you'd say Mary heard something in the bedroom and wanted to know what made the noise. Or maybe you'd say that Mary forgot where she was going; maybe she really intended to go downstairs. But you certainly would not say that Mary just does this every day at this time: she just walks into the bedroom, walks around, and walks out again. It would be unnatural to explain human behavior in the physicist's language of time, distance, and mass, and it would also be wrong; if you came back tomorrow to test the hypothesis, it would surely fail. Our minds explain other people's behavior by their beliefs and desires because other people's behavior is in fact caused by their beliefs and desires. The behaviorists were wrong, and everyone intuitively knows it.

Mental states are invisible and weightless. Philosophers define them as "a relation between a person and a proposition." The relation is an attitude like believes-that, desires-that, hopes-that, pretends-that. The proposition is the content of the belief, something very roughly like the meaning of a sentence—for example, *Mary finds the keys*, or *The keys are in the bedroom*. The content of a belief lives in a different realm from the facts of the world. *There are unicorns grazing in Cambridge Common* is false, but *John thinks there are unicorns grazing in Cambridge Common* could very well be true. To ascribe a belief to someone, we can't just think a thought in the ordinary way, or we wouldn't be able to learn that John believes in unicorns without believing in them ourselves. We have to take a thought, set it aside in mental quotation marks, and think, "That is what John thinks" (or wants, or hopes for, or guesses). Moreover, anything we can think is also something we can think that someone else thinks (Mary knows that John thinks that there are unicorns . . .). These onionlike thoughts-inside-thoughts need a special computational architecture (see Chapter 2) and, when we communicate them to others, the recursive grammar proposed by Chomsky and explained in *The Language Instinct*.

We mortals can't read other people's minds directly. But we make good guesses from what they say, what we read between the lines, what they show in their face and eyes, and what best explains their behavior. It is our species' most remarkable talent. After reading the chapter on vision you might be amazed that people can recognize a dog. Now think about what it takes to recognize the dog in a pantomime of walking one.

But somehow children do it. The skills behind mind reading are first exercised in the crib. Two-month-olds stare at eyes; six-month-olds know when they're staring back; one-year-olds look at what a parent is staring at, and check a parent's eyes when they are uncertain why the parent is doing something. Between eighteen and twenty-four months, children begin to separate the contents of other people's minds from their own beliefs. They show that ability off in a deceptively simple feat: pretending. When a toddler plays along with his mother who tells him the phone is ringing and hands him a banana, he is separating the contents of their pretense (the banana is a telephone) from the contents of his own belief (the banana is a banana). Two-year-olds use mental verbs like *see* and *want*, and three-year-olds use verbs like *think*, *know*, and *remember*. They know that a looker generally wants what he is looking at. And they grasp the idea of "idea." For example, they know that you can't eat the memory of an apple and that a person can tell what's in a box only by looking into it.

By four, children pass a very stringent test of knowledge about other minds: they can attribute to others beliefs they themselves know to be false. In a typical experiment, children open a Smarties box and are surprised to find pencils inside. (Smarties, the British psychologists explain to American audiences, are like M&M's, only better.) Then the children are asked what a person coming into the room expects to find. Though the children know that the box contains pencils, they sequester the knowledge, put themselves in the newcomer's shoes, and say, "Smarties." Three-year-olds have more trouble keeping their knowledge out of the picture; they insist that the newcomer will expect to find pencils in the candy box. But it's unlikely that they lack the very idea of other minds; when the wrong answer is made less alluring or the children are induced to think a bit harder, they attribute false beliefs to others, too. The results come out the same in every country in which children have been tested.

Thinking of other minds comes so naturally that it almost seems like part and parcel of intelligence itself. Can we even imagine what it would be like not to think of other people as having minds? The psychologist Alison Gopnik imagines it would be like this:

> At the top of my field of vision is a blurry edge of nose, in front are waving hands . . . Around me bags of skin are draped over chairs, and stuffed into pieces of cloth; they shift and protrude in unexpected ways. . . . Two dark spots near the top of them swivel restlessly back and forth. A hole beneath the spots fills with food and from it comes a stream of noises. . . . The noisy skin-bags suddenly [move] toward you, and their noises [grow] loud, and you [have] no idea why. . . .

Baron-Cohen, Alan Leslie, and Uta Frith have proposed that there really are people who think like this. They are the people we call autistic.

Autism affects about one in a thousand children. They are said to "draw into a shell and live within themselves." When taken into a room, they disregard people and go for the objects. When someone offers a hand, they play with it like a mechanical toy. Cuddly dolls and stuffed animals hold little interest. They pay little attention to their parents and don't respond when called. In public, they touch, smell, and walk over people as if they were furniture. They don't play with other children. But the intellectual and perceptual abilities of some autistic children are legendary (especially after Dustin Hoffman's performance in *Rain Man*). Some of them learn multiplication tables, put together jigsaw puzzles (even upside down), disassemble and reassemble appliances, read distant license plates, or instantly calculate the day of the week on which any given date in the past or future falls.

Like many psychology undergraduates, I learned about autism from a famous *Scientific American* reprint, "Joey: A Mechanical Boy," by the psychoanalyst Bruno Bettelheim. Bettelheim explained that Joey's autism was caused by emotionally distant parents ("icebox mother" became the favored term) and early, rigid toilet training. He wrote, "It is unlikely that Joey's calamity could befall a child in any time and culture but our own." According to Bettelheim, postwar parents had such an easy time providing their children with creature comforts that they took no pleasure in it, and the children did not develop a feeling of worth from having their basic needs satisfied. Bettelheim claimed to have cured Joey, at first by letting him use a wastebasket instead of the toilet. (He allowed that the therapy "entailed some hardship for his counselors.")

Today we know that autism occurs in every country and social class, lasts a lifetime (though sometimes with improvement), and cannot be blamed on mothers. It almost certainly has neurological and genetic causes, though they have not been pinpointed. Baron-Cohen, Frith, and Leslie suggest that autistic children are mind-blind: their module for attributing minds to others is damaged. Autistic children almost never pretend, can't explain the difference between an apple and a memory of an apple, don't distinguish between someone's looking into a box and someone's touching it, know where a cartoon face is looking but do not guess that it wants what it is looking at, and fail the Smarties (false-belief) task. Remarkably, they pass a test that is logically the same as the false-belief task but not about minds. The experimenter lifts Rubber Ducky out of the bathtub and puts it on the bed, takes a Polaroid snapshot, and then puts it back in the bathtub. Normal three-year-olds believe that the photo will somehow show the duck in the tub. Autistic children know it does not.

Mind-blindness is not caused by real blindness, nor by mental retardation such as Down's syndrome. It is a vivid reminder that the contents of the world are not just there for the knowing but have to be grasped with suitable mental machinery. In a sense, autistic children are right: the universe is nothing but matter in motion. My "normal" mental equipment leaves me chronically dumbfounded at the fact that a microdot and a spoonful of semen can bring about a site of thinking and feeling and that a blood clot or a metal slug can end it. It gives me the delusion that London and chairs and vegetables are on the inventory of the world's objects. Even the objects themselves are a kind of delusion. Buckminster Fuller once wrote: "Everything you've learned . . . as 'obvious' becomes less and less obvious as you begin to study the universe. For example, there are no solids in the universe. There's not even a suggestion of a solid. There are no absolute continuums. There are no surfaces. There are no straight lines."

In another sense, of course, the world *does* have surfaces and chairs and rabbits and minds. They are knots and patterns and vortices of matter and energy that obey their own laws and ripple through the sector of space-time in which we spend our days. They are not social constructions, nor the bits of undigested beef that Scrooge blamed for his vision of Marley's ghost. But to a mind unequipped to find them, they might as well not exist at all. As the psychologist George Miller has put it, "The crowning intellectual accomplishment of the brain is the real world. . . . [A]ll [the] fundamental aspects of the real world of our experience are adaptive interpretations of the really real world of physics."

A TRIVIUM

The medieval curriculum comprised seven liberal arts, divided into the lower-level trivium (grammar, logic, and rhetoric) and the upper-level quadrivium (geometry, astronomy, arithmetic, and music). *Trivium* originally meant three roads, then it meant crossroads, then commonplace (since common people hang around crossroads), and finally trifling or immaterial. The etymology is, in a sense, apt: with the exception of astronomy, none of the liberal arts is *about* anything. They don't explain plants or animals or rocks or people; rather, they are intellectual tools that can be applied in any realm. Like the students who complain that algebra will never help them in the real world, one can wonder whether these abstract tools are useful enough in nature for natural selection to have inculcated them in the brain. Let's look at a modified trivium: logic, arithmetic, and probability.

"Contrariwise," continued Tweedledee, "if it was so, it might be, and if it were so, it would be; but as it isn't, it ain't. That's logic!"

Logic, in the technical sense, refers not to rationality in general but to inferring the truth of one statement from the truth of other statements based only on their form, not their content. I am using logic when I reason as follows. P is true, P implies Q, therefore Q is true. P and Q are true, therefore P is true. P or Q is true, P is false, therefore Q is true. P implies Q, Q is false, therefore P is false. I can derive all these truths not knowing whether P means "There is a unicorn in the garden," "Iowa grows soybeans," or "My car has been eaten by rats."

Does the brain do logic? College students' performance on logic problems is not a pretty sight. There are some archeologists, biologists, and chess players in a room. None of the archeologists are biologists. All of the biologists are chess players. What, if anything, follows? A majority of students conclude that none of the archeologists are chess players, which is not valid. None of them conclude that some of the chess players are not archeologists, which is valid. In fact, a fifth claim that the premises allow *no* valid inferences.

Spock always did say that humans are illogical. But as the psychologist John Macnamara has argued, that idea itself is barely logical. The rules of logic were originally seen as a formalization of the laws of thought. That went a bit overboard; logical truths are true regardless of how people think. But it is hard to imagine a species discovering logic if its brain did not give it a feeling of certitude when it found a logical truth. There is something peculiarly compelling, even irresistible, about P, P implies Q, therefore Q. With enough time and patience, we discover why our own logical errors are erroneous. We come to agree with one another on which truths are necessary. And we teach others not by force of authority but socratically, by causing the pupils to recognize truths by their own standards.

People surely do use some kind of logic. All languages have logical terms like *not, and, same, equivalent,* and *opposite.* Children use *and, not, or,* and *if* appropriately before they turn three, not only in English but in half a dozen other languages that have been studied. Logical inferences are ubiquitous in human thought, particularly when we understand language. Here is a simple example from the psychologist Martin Braine:

> John went in for lunch. The menu showed a soup-and-salad special, with free beer or coffee. Also, with the steak you got a free glass of red wine. John chose the soup-and-salad special with coffee, along with something else to drink.
> (a) Did John get a free beer? (Yes, No, Can't Tell)
> (b) Did John get a free glass of wine? (Yes, No, Can't Tell)

Virtually everyone deduces that the answer to (a) is no. Our knowledge of restaurant menus tells us that the *or* in *free beer or coffee* implies "not both"—you get only one of them free; if you want the other, you have to pay for it. Farther along, we learn that John chose coffee. From the premises "not both free beer and free coffee" and "free coffee," we derive "not free beer" by a logical inference. The answer to question (b) is also no. Our knowledge of restaurants reminds us that food and beverages are not free unless explicitly offered as such by the menu. We therefore add the conditional "if not steak, then no free red wine." John chose the soup and salad, which suggests he did not choose steak; we conclude, using a logical inference, that he did not get a free glass of wine.

Logic is indispensable in inferring true things about the world from piecemeal facts acquired from other people via language or from one's own generalizations. Why, then, do people seem to flout logic in stories about archeologists, biologists, and chess players?

One reason is that logical words in everyday languages like English are ambiguous, often denoting several formal logical concepts. The English word *or* can sometimes mean the logical connective OR (A or B or both) and can sometimes mean the logical connective XOR (exclusive or: A or B but not both). The context often makes it clear which one the speaker intended, but in bare puzzles coming out of the blue, readers can make the wrong guess.

Another reason is that logical inferences cannot be drawn out willy-nilly. Any true statement can spawn an infinite number of true but useless new ones. From "Iowa grows soybeans," we can derive "Iowa grows soybeans or the cow jumped over the moon," "Iowa grows soybeans and either the cow jumped over the moon or it didn't," ad infinitum. (This is an example of the "frame problem" introduced in Chapter 1.) Unless it has all the time in the world, even the best logical inferencer has to guess which implications to explore and which are likely to be blind alleys. Some rules have to be inhibited, so valid inferences will inevitably be missed. The guessing can't itself come from logic; generally it comes from assuming that the speaker is a cooperative conversational partner conveying relevant information and not, say, a hostile lawyer or a tough-grading logic professor trying to trip one up.

Perhaps the most important impediment is that mental logic is not a hand-held calculator ready to accept any A's and B's and C's as input. It is enmeshed with our system of knowledge about the world. A particular step of mental logic, once set into motion, does not depend on world knowledge, but its inputs and outputs are piped directly into that knowledge. In the restaurant story, for example, the links of inference alternate between knowledge of menus and applications of logic.

Some areas of knowledge have their own inference rules that can either reinforce or work at cross-purposes with the rules of logic. A famous example comes from the psychologist Peter Wason. Wason was inspired by the philosopher Karl Popper's ideal of scientific reasoning: a hypothesis is accepted if attempts to falsify it fail. Wason wanted to see how ordinary people do at falsifying hypotheses. He told them that a set of cards had letters on one side and numbers on the other, and asked them to test the rule "If a card has a *D* on one side, it has a 3 on the other," a simple P-implies-Q statement. The subjects were shown four cards and were asked which ones they would have to turn over to see if the rule was true. Try it:

Most people choose either the *D* card or the *D* card and the 3 card. The correct answer is *D* and 7. "P implies Q" is false only if P is true and Q is false. The 3 card is irrelevant; the rule said that *D*'s have 3's, not that 3's have *D*'s. The 7 card is crucial; if it had a *D* on the other side, the rule would be dead. Only about five to ten percent of the people who are given the test select the right cards. Even people who have taken logic courses get it wrong. (Incidentally, it's not that people interpret "If *D* then 3" as "If *D* then 3 and vice versa." If they did interpret it that way but otherwise behaved like logicians, they would turn over *all four* cards.) Dire implications were seen. John Q. Public was irrational, unscientific, prone to confirming his prejudices rather than seeking evidence that could falsify them.

But when the arid numbers and letters are replaced with real-world events, sometimes—though only sometimes—people turn into logicians. You are a bouncer in a bar, and are enforcing the rule "If a person is drinking beer, he must be eighteen or older." You may check what people are drinking or how old they are. Which do you have to check: a beer drinker, a Coke drinker, a twenty-five-year-old, a sixteen-year-old? Most people correctly select the beer drinker and the sixteen-year-old. But mere concreteness is not enough. The rule "If a person eats hot chili peppers, then he drinks cold beer" is no easier to falsify than the *D*'s and 3's.

Leda Cosmides discovered that people get the answer right when the rule is a contract, an exchange of benefits. In those circumstances, showing that the rule is false is equivalent to finding cheaters. A contract is an implication of the form "If you take a benefit, you must meet a requirement"; cheaters take the benefit without meeting the requirement. Beer in a bar is a benefit that one earns by proof of maturity, and cheaters are underage drinkers. Beer after chili peppers is mere cause and effect, so Coke drinking (which logically must be checked) doesn't seem relevant. Cosmides showed that people do the logical thing whenever they construe the P's and Q's as benefits and costs, even when the events are exotic, like eating duiker meat and finding ostrich eggshells. It's not that a logic module is being switched on, but that people are using a different set of rules. These rules, appropriate to detecting cheaters, sometimes coincide with logical rules and sometimes don't. When the cost and benefit terms are flipped, as in "If a person pays $20, he receives a watch," people still choose the cheater card (he receives the watch, he doesn't pay $20)—a choice that is neither logically correct nor the typical error made with meaningless cards. In fact, the very same story can draw out logical or nonlogical choices depending on the reader's interpretation of who, if anyone, is a cheater. "If an employee gets a pension, he has worked for ten years. Who is violating the rule?" If people take the employee's point of view, they seek the twelve-year workers without pensions; if they take the employer's point of view, they seek the eight-year workers who hold them. The basic findings have been replicated among the Shiwiar, a foraging people in Ecuador.

The mind seems to have a cheater-detector with a logic of its own. When standard logic and cheater-detector logic coincide, people act like logicians; when they part company, people still look for cheaters. What gave Cosmides the idea to look for this mental mechanism? It was the evolutionary analysis of altruism (see Chapters 6 and 7). Natural selection does not select public-mindedness; a selfish mutant would quickly outreproduce its altruistic competitors. Any selfless behavior in the natural world needs a special explanation. One explanation is reciprocation: a creature can extend help in return for help expected in the future. But favor-trading is always vulnerable to cheaters. For it to have evolved, it must be accompanied by a cognitive apparatus that remembers who has taken and that ensures that they give in return. The evolutionary biologist Robert Trivers had predicted that humans, the most conspicuous altruists in the animal kingdom, should have evolved a hypertrophied cheater-detector algorithm. Cosmides appears to have found it.

So is the mind logical in the logician's sense? Sometimes yes, sometimes no. A better question is, Is the mind well-designed in the biologist's sense? Here the "yes" can be a bit stronger. Logic by itself can spin off trivial truths and miss consequential ones. The mind does seem to use logical rules, but they are recruited by the processes of language understanding, mixed with world knowledge, and supplemented or superseded by special inference rules appropriate to the content.

Mathematics is part of our birthright. One-week-old babies perk up when a scene changes from two to three items or vice versa. Infants in their first ten months notice how many items (up to four) are in a display, and it doesn't matter whether the items are homogeneous or heterogeneous, bunched together or spread out, dots or household objects, even whether they are objects or sounds. According to recent experiments by the psychologist Karen Wynn, five-month-old infants even do simple arithmetic. They are shown Mickey Mouse, a screen covers him up, and a second Mickey is placed behind it. The babies expect to see two Mickeys when the screen falls and are surprised if it reveals only one. Other babies are shown two Mickeys and one is removed from behind the screen. These babies expect to see one Mickey and are surprised to find two. By eighteen months children know that numbers not only differ but fall into an order; for example, the children can be taught to choose the picture with fewer dots. Some of these abilities are found in, or can be taught to, some kinds of animals.

Can infants and animals really count? The question may sound absurd because these creatures have no words. But registering quantities does not depend on language. Imagine opening a faucet for one second every time you hear a drumbeat. The amount of water in the glass would represent the number of beats. The brain might have a similar mechanism, which would accumulate not water but neural pulses or the number of active neurons. Infants and many animals appear to be equipped with this simple kind of counter. It would have many potential selective advantages, which depend on the animal's niche. They range from estimating the rate of return of foraging in different patches to solving problems such as "Three bears went into the cave; two came out. Should I go in?"

Human adults use several mental representations of quantity. One is analogue—a sense of "how much"—which can be translated into mental images such as an image of a number line. But we also assign number words to quantities and use the words and the concepts to measure, to count more accurately, and to count, add, and subtract larger numbers. All cultures have words for numbers, though sometimes only "one," "two," and "many." Before you snicker, remember that the *concept* of number has nothing to do with the size of a number vocabulary. Whether or not people know words for big numbers (like "four" or "quintillion"), they can know that if two sets are the same, and you add 1 to one of them, that set is now larger. That is true whether the sets have four items or a quintillion items. People know that they can compare the size of two sets by pairing off their members and checking for leftovers; even mathematicians are forced to that technique when they make strange claims about the relative sizes of infinite sets. Cultures without words for big numbers often use tricks like holding up fingers, pointing to parts of the body in sequence, or grabbing or lining up objects in twos and threes.

Children as young as two enjoy counting, lining up sets, and other activities guided by a sense of number. Preschoolers count small sets, even when they have to mix kinds of objects, or have to mix objects, actions, and sounds. Before they really get the hang of counting and measuring, they appreciate much of its logic. For example, they will try to distribute a hot dog equitably by cutting it up and giving everyone two pieces (though the pieces may be of different sizes), and they yell at a counting puppet who misses an item or counts it twice, though their own counting is riddled with the same kinds of errors.

Formal mathematics is an extension of our mathematical intuitions. Arithmetic obviously grew out of our sense of number, and geometry out of our sense of shape and space. The eminent mathematician Saunders Mac Lane speculated that basic human activities were the inspiration for every branch of mathematics:

Counting	→ arithmetic and number theory
Measuring	→ real numbers, calculus, analysis
Shaping	→ geometry, topology
Forming (as in architecture)	→ symmetry, group theory
Estimating	→ probability, measure theory, statistics
Moving	→ mechanics, calculus, dynamics
Calculating	→ algebra, numerical analysis
Proving	→ logic

Puzzling → combinatorics, number theory
Grouping → set theory, combinatorics

Mac Lane suggests that "mathematics starts from a variety of human activities, disentangles from them a number of notions which are generic and not arbitrary, then formalizes these notions and their manifold interrelations." The power of mathematics is that the formal rule systems can then "codify deeper and nonobvious properties of the various originating human activities." Everyone—even a blind toddler—instinctively knows that the path from A straight ahead to B and then right to C is longer than the shortcut from A to C. Everyone also visualizes how a line can define the edge of a square and how shapes can be abutted to form bigger shapes. But it takes a mathematician to show that the square on the hypotenuse is equal to the sum of the squares on the other two sides, so one can calculate the savings of the shortcut without traversing it.

To say that school mathematics comes out of intuitive mathematics is not to say that it comes out *easily*. David Geary has suggested that natural selection gave children some basic mathematical abilities: determining the quantity of small sets, understanding relations like "more than" and "less than" and the ordering of small numbers, adding and subtracting small sets, and using number words for simple counting, measurement, and arithmetic. But that's where it stopped. Children, he suggests, are *not* biologically designed to command large number words, large sets, the base-10 system, fractions, multicolumn addition and subtraction, carrying, borrowing, multiplication, division, radicals, and exponents. These skills develop slowly, unevenly, or not at all.

On evolutionary grounds it would be surprising if children were mentally equipped for school mathematics. These tools were invented recently in history and only in a few cultures, too late and too local to stamp the human genome. The mothers of these inventions were the recording and trading of farming surpluses in the first agricultural civilizations. Thanks to formal schooling and written language (itself a recent, noninstinctive invention), the inventions could accumulate over the millennia, and simple mathematical operations could be assembled into more and more complicated ones. Written symbols could serve as a medium of computation that surmounted the limitations of short-term memory, just as silicon chips do today.

How can people use their Stone Age minds to wield high-tech mathematical instruments? The first way is to set mental modules to work on objects other than the ones they were designed for. Ordinarily, lines and shapes are analyzed by imagery and other components of our spatial sense, and heaps of things are analyzed by our number faculty. But to accomplish Mac Lane's ideal of disentangling the generic from the parochial (for example, disentangling the generic concept of quantity from the parochial concept of the number of rocks in a heap), people might have to apply their sense of number to an entity that, at first, feels like the wrong kind of subject matter. For example, people might have to analyze a line in the sand not by the habitual imagery operations of continuous scanning and shifting, but by counting off imaginary segments from one end to the other.

The second way to get to mathematical competence is similar to the way to get to Carnegie Hall: practice. Mathematical concepts come from snapping together old concepts in a useful new arrangement. But those old concepts are assemblies of still older concepts. Each subassembly hangs together by the mental rivets called chunking and automaticity: with copious practice, concepts adhere into larger concepts, and sequences of steps are compiled into a single step. Just as bicycles are assembled out of frames and wheels, not tubes and spokes, and recipes say how to make sauces, not how to grasp spoons and open jars, mathematics is learned by fitting together overlearned routines. Calculus teachers lament that students find the subject difficult not because derivatives and integrals are abstruse concepts—they're just rate and accumulation—but because you can't do calculus unless algebraic operations are second nature, and most students enter the course without having learned the algebra properly and need to concentrate every drop of mental energy on that. Mathematics is ruthlessly cumulative, all the way back to counting to ten.

Evolutionary psychology has implications for pedagogy which are particularly clear in the teaching of mathematics. American children are among the worst performers in the industrialized world on tests of mathematical achievement. They are not born dunces; the problem is that the educational establishment is ignorant of evolution. The ascendant philosophy of mathematical education in the United States is constructivism, a mixture of Piaget's psychology with counterculture and postmodernist ideology. Children must actively construct mathematical knowledge for themselves in a social enterprise driven by disagreements about the meanings of concepts. The teacher provides the materials and the social milieu but does not lecture or guide the discussion. Drill and practice, the routes to automaticity, are called "mechanistic" and seen as detrimental to understanding. As one pedagogue lucidly explained, "A zone of potential construction of a specific mathematical concept is determined by the modifications of the concept children might make in, or as a result of, interactive communication in the mathematical learning environment." The result, another declared, is that "it is possible for students to construct for themselves the mathematical practices that, historically, took several thousand years to evolve."

As Geary points out, constructivism has merit when it comes to the intuitions of small numbers and simple arithmetic that arise naturally in all children. But it ignores the difference between our factory-installed equipment and the accessories that civilization bolts on afterward. Setting our mental modules to work on material they were not designed for is *hard*. Children do not spontaneously see a string of beads as elements in a set, or points on a line as numbers. If you give them a bunch of blocks and tell them to do something together, they will exercise their intuitive physics and intuitive psychology for all they're worth, but not necessarily their intuitive sense of number. (The better curricula explicitly point out connections across ways of knowing. Children might be told to do every arithmetic problem three different ways: by counting, by drawing diagrams, and by moving segments along a number line.) And without the practice that compiles a halting sequence of steps into a mental reflex, a learner will always be building mathematical structures out of the tiniest nuts and bolts, like the watchmaker who never made subassemblies and had to start from scratch every time he put down a watch to answer the phone.

Mastery of mathematics is deeply satisfying, but it is a reward for hard work that is not itself always pleasurable. Without the esteem for hard-won mathematical skills that is common in other cultures, the mastery is unlikely to blossom. Sadly, the same story is being played out in American reading instruction. In the dominant technique, called "whole language," the insight that language is a naturally developing human instinct has been garbled into the evolutionarily improbable claim that *reading* is a naturally developing human instinct. Old-fashioned practice at connecting letters to sounds is replaced by immersion in a text-rich social environment, and the children don't learn to read. Without an understanding of what the mind was designed to do in the environment in which we evolved, the unnatural activity called formal education is unlikely to succeed.

~

"I shall never believe that God plays dice with the world," Einstein famously said. Whether or not he was right about quantum mechanics and the cosmos, his statement is certainly not true of the games people play in their daily lives. Life is not chess but backgammon, with a throw of the dice at every turn. As a result, it is hard to make predictions, especially about the future (as Yogi Berra allegedly said). But in a universe with any regularities at all, decisions informed by the past are better than decisions made at random. That has always been true, and we would expect organisms, especially informavores such as humans, to have evolved acute intuitions about probability. The founders of probability theory, like the founders of logic, assumed that they were just formalizing common sense.

But then why do people often seem to be "probability-blind," in the words of Massimo Piattelli-Palmarini? Many mathematicians and scientists have bemoaned the innumeracy of ordinary people when they reason about risk. The psychologists Amos Tversky and Daniel Kahneman have amassed ingenious demonstrations of how people's intuitive grasp of chance appears to flout the elementary canons of probability theory. Here are some famous examples.

• People gamble and buy state lottery tickets, sometimes called "the stupidity tax." But since the house must profit, the players, on average, must lose.

• People fear planes more than cars, especially after news of a gory plane crash, though plane travel is statistically far safer. They fear nuclear power, though more people are crippled and killed by coal. Every year a thousand Americans are accidentally electrocuted, but rock stars don't campaign to reduce the household voltage. People clamor for bans on pesticide residues and food additives, though they pose trivial risks of cancer compared to the thousands of natural carcinogens that plants have evolved to deter the bugs that eat them.

• People feel that if a roulette wheel has stopped at black six times in a row, it's due to stop at red, though of course the wheel has no memory and every spin is independent. A large industry of self-anointed seers hallucinate trends in the random walk of the stock market. Hoop fans believe that basketball players get a "hot hand," making baskets in clusters, though their strings of swishes and bricks are indistinguishable from coin flips.

• This problem was given to sixty students and staff members at Harvard Medical School: "If a test to detect a disease whose prevalence is 1/1000 has a false positive rate of 5%, what is the chance that a person found to have a positive result actually has the disease, assuming you know nothing about the person's symptoms or signs?" The most popular answer was .95. The average answer was .56. The correct answer is .02, and only eighteen percent of the experts guessed it. The answer, according to Bayes' theorem, may be calculated as the prevalence or base rate (1/1000) times the test's sensitivity or hit rate (proportion of sick people who test positive, presumably 1), divided by the overall incidence of positive test results (the percentage of the time the test comes out positive, collapsing over sick and healthy people—that is, the sum of the sick people who test positive, 1/1000 × 1, and the healthy people who test positive, 999/1000 × .05). One bugaboo in the problem is that many people misinterpret "false positive rate" as the proportion of positive results that come from healthy people, instead of interpreting it as the proportion of healthy people who test positive. But the biggest problem is that people ignore the base rate (1/1000), which ought to have reminded them that the disease is rare and hence improbable for a given patient even if the test comes out positive. (They apparently commit the fallacy that because zebras make hoofbeats, hoofbeats imply zebras.) Surveys have shown that many doctors needlessly terrify their patients who test positive for a rare disease.

• Try this: "Linda is 31 years old, single, outspoken, and very bright. She majored in philosophy. As a student, she was deeply concerned with issues of discrimination and social justice, and also participated in antinuclear demonstrations. What is the probability that Linda is a bankteller? What is the probability that Linda is a bankteller and is active in the feminist movement?" People sometimes give a higher estimate to the probability that she is a feminist bankteller than to the probability that she is a bankteller. But it's impossible for "A and B" to be more likely than "A" alone.

When I presented these findings in class, a student cried out, "I'm ashamed for my species!" Many others feel the disgrace, if not about themselves, then about the person in the street. Tversky, Kahneman, Gould, Piattelli-Palmarini, and many social psychologists have concluded that the mind is not designed to grasp the laws of probability, even though the laws rule the universe. The brain can process limited amounts of information, so instead of computing theorems it uses crude rules of thumb. One rule is: the more memorable an event, the more likely it is to happen. (I can remember a recent gory plane crash, therefore planes are unsafe.) Another is: the more an individual resembles a stereotype, the more likely he is to belong to that category. (Linda fits my image of a feminist bankteller better than she fits my image of a bankteller, so she's more likely to be a feminist bankteller.) Popular books with lurid titles have spread the bad news: *Irrationality: The Enemy Within*; *Inevitable Illusions: How Mistakes of Reason Rule Our Minds*; *How We Know What Isn't So: The Fallibility of Human Reason in Everyday Life*. The sad history of human folly and prejudice is explained by our ineptness as intuitive statisticians.

Tversky and Kahneman's demonstrations are among the most thought-provoking in psychology, and the research has drawn attention to the depressingly low intellectual quality of our public discourse about societal and personal risk. But in a probabilistic world, could the human mind really be oblivious to probability? The solutions to the problems that people flub can be computed with a few keystrokes on a cheap calculator. Many animals, even bees, compute accurate probabilities as they forage. Could those computations really exceed the information-processing capacity of the trillion-synapse human brain? It is hard to believe, and one does not have to believe it. People's reasoning is not as stupid as it might first appear.

To begin with, many risky choices are just that, choices, and cannot be gainsaid. Take the gamblers, plane phobics, and chemical avoiders. Are they really *irrational*? Some people take pleasure in awaiting the outcomes of events that could radically improve their lives. Some people dislike being strapped in a tube and flooded with reminders of a terrifying way to die. Some people dislike eating foods deliberately laced with poison (just as some people might choose not to eat a hamburger fortified with harmless worm meat). There is nothing irrational in any of these choices, any more than in preferring vanilla over chocolate ice cream.

The psychologist Gerd Gigerenzer, along with Cosmides and Tooby, have noted that even when people's judgments of probability depart from the truth, their reasoning may not be illogical. No mental faculty is omniscient. Color vision is fooled by sodium vapor streetlights, but that does not mean it is badly designed. It is demonstrably well designed, far better than any camera at registering constant colors with changing illumination (see Chapter 4). But it owes its success at this unsolvable problem to tacit assumptions about the world. When the assumptions are violated in an artificial world, color vision fails. The same may be true of our probability-estimators.

Take the notorious "gambler's fallacy": expecting that a run of heads increases the chance of a tail, as if the coin had a memory and a desire to be fair. I remember to my shame an incident during a family vacation when I was a teenager. My father mentioned that we had suffered through several days of rain and were due for good weather, and I corrected him, accusing him of the gambler's fallacy. But long-suffering Dad was right, and his know-it-all son was wrong. Cold fronts aren't raked off the earth at day's end and replaced with new ones the next morning. A cloud cover must have some average size, speed, and direction, and it would not surprise me (now) if a week of clouds really did predict that the trailing edge was near and the sun was about to be unmasked, just as the hundredth railroad car on a passing train portends the caboose with greater likelihood than the third car.

Many events work like that. They have a characteristic life history, a changing probability of occurring over time which statisticians call a hazard function. An astute observer *should* commit the gambler's fallacy and try to predict the next occurrence of an event from its history so far, a kind of statistics called time-series analysis. There is one exception: devices that are *designed* to deliver events independently of their history. What kind of device would do that? We call them gambling machines. Their reason for being is to foil an observer who likes to turn patterns into predictions. If our love of patterns were misbegotten because randomness is everywhere, gambling machines should be easy to build and gamblers easy to fool. In fact, roulette wheels, slot machines, even dice, cards, and coins are precision instruments; they are demanding to manufacture and easy to defeat. Card counters who "commit the gambler's fallacy" in blackjack by remembering the dealt cards and betting they won't turn up again soon are the pests of Las Vegas.

So in any world but a casino, the gambler's fallacy is rarely a fallacy. Indeed, calling our intuitive predictions fallacious because they fail on gambling devices is backwards. A gambling device is, by definition, a machine designed to defeat our intuitive predictions. It is like calling our hands badly designed because they make it hard to get out of handcuffs. The same is true of the hot-hand illusion and other fallacies among sports fans. If basketball shots were easily predictable, we would no longer call basketball a sport. An efficient stock market is another invention designed to defeat human pattern detection. It is set up to let traders quickly capitalize on, hence nullify, deviations from a random walk.

Other so-called fallacies may also be triggered by evolutionary novelties that trick our probability calculators, rather than arising from crippling design defects. "Probability" has many meanings. One is relative frequency in the long run. "The probability that the penny will land heads is .5" would mean that in a hundred coin flips, fifty will be heads. Another meaning is subjective confidence about the outcome of a single event. In this sense, "the probability that the penny will land heads is .5" would mean that on a scale of 0 to 1, your confidence that the next flip will be heads is halfway between certainty that it will happen and certainty that it won't.

Numbers referring to the probability of a single event, which only make sense as estimates of subjective confidence, are commonplace nowadays: there is a thirty percent chance of rain tomorrow; the Canadiens are favored to beat the Mighty Ducks tonight with odds of five to three. But the mind may have evolved to think of probabilities as relative frequencies in the long run, *not* as numbers expressing confidence in a single event. The mathematics of probability was invented only in the seventeenth century, and the use of proportions or percentages to express them arose even later. (Percentages came in after the French Revolution with the rest of the metric system and were initially used for interest and tax rates.) Still more modern is the input to the formulas for probability: data gathered by teams, recorded in writing, checked for errors, accumulated in archives, and tallied and scaled to yield numbers. The closest equivalent for our ancestors would have been hearsay of unknown validity, transmitted with coarse labels like *probably*. Our ancestors' usable probabilities must have come from their own experience, and that means they were frequencies: over the years, five out of the eight people who came down with a purple rash died the following day.

Gigerenzer, Cosmides, Tooby, and the psychologist Klaus Fiedler noticed that the medical decision problem and the Linda problem ask for single-event probabilities: how likely is that *this patient* is sick, how likely is it that *Linda* is a bankteller. A probability instinct that worked in relative frequencies might find the questions beyond its ken. There's only one Linda, and either she is a bankteller or she isn't. "The probability that she is a bankteller" is uncomputable. So they gave people the vexing problems but stated them in terms of frequencies, not single-event probabilities. One out of a thousand Americans has the disease; fifty out of a thousand healthy people test positive; we assembled a thousand Americans; how many who test positive have the disease? A hundred people fit Linda's description; how many are banktellers; how many are feminist banktellers? Now a majority of people—up to ninety-two percent—behave like good statisticians.

This cognitive therapy has enormous implications. Many men who test positive for HIV (the AIDS virus) assume they are doomed. Some have taken extreme measures, including suicide, despite their surely knowing that most men don't have AIDS (especially men who do not fall into a known risk group) and that no test is perfect. But it is hard for doctors and patients to use that knowledge to calibrate the chance of being infected, even when the probabilities are known. For example, in recent years the prevalence of HIV in German men who do not belong to a risk group is 0.01%, the sensitivity (hit rate) of a typical HIV test is 99.99%, and the false positive rate is perhaps 0.01%. The prospects of a patient who has tested positive do not sound very good. But now imagine that a doctor counseled a patient as follows: "Think of 10,000 heterosexual men like you. We expect one to be infected with the virus, and he will almost certainly test positive. Of the 9,999 men who are not infected, one additional man will test positive. Thus we get two who test positive, but only one of them actually has the virus. All we know at this point is that you have tested positive. So the chance that you actually have the virus is about 50–50." Gigerenzer has found that when probabilities are presented in this way (as frequencies), people, including specialists, are vastly more accurate at estimating the probability of a disease following a medical test. The same is true for other judgments under uncertainty, such as guilt in a criminal trial.

～

Gigerenzer argues that people's intuitive equation of probability with frequency not only makes them calculate like statisticians, it makes them think like statisticians about the concept of probability itself—a surprisingly slippery and paradoxical notion. What does the probability of a single event even *mean*? Bookmakers are willing to make up inscrutable numbers such as that the odds that Michael Jackson and LaToya Jackson are the same person are 500 to 1, or that the odds that circles in cornfields emanate from Phobos (one of the moons of Mars) are 1,000 to 1. I once saw a tabloid headline announcing that the chances that Mikhail Gorbachev is the Antichrist are one in eight trillion. Are these statements true? False? Approximately true? How could we tell? A colleague tells me that there is a ninety-five percent chance he will show up at my talk. He doesn't come. Was he lying?

You may be thinking: granted, a single-event probability is just subjective confidence, but isn't it rational to calibrate confidence by relative frequency? If everyday people don't do it that way, wouldn't they be irrational? Ah, but the relative frequency of what? To count frequencies you have to decide on a class of events to count up, and a single event belongs to an infinite number of classes. Richard von Mises, a pioneer of probability theory, gives an example.

In a sample of American women between the ages of 35 and 50, 4 out of 100 develop breast cancer within a year. Does Mrs. Smith, a 49-year-old American woman, therefore have a 4% chance of getting breast cancer in the next year? There is no answer. Suppose that in a sample of women between the ages of 45 and 90—a class to which Mrs. Smith also belongs—11 out of 100 develop breast cancer in a year. Are Mrs. Smith's chances 4%, or are they 11%? Suppose that her mother had breast cancer, and 22 out of 100 women between 45 and 90 whose mothers had the disease will develop it. Are her chances 4%, 11%, or 22%? She also smokes, lives in California, had two children before the age of 25 and one after 40, is of Greek descent . . . What group should we compare her with to figure out the "true" odds? You might think, the more specific the class, the better—but the more specific the class, the smaller its size and the less reliable the frequency. If there were only two people in the world very much like Mrs. Smith, and one developed breast cancer, would anyone say that Mrs. Smith's chances are 50%? In the limit, the only class that is truly comparable with Mrs. Smith in all her details is the class containing Mrs. Smith herself. But in a class of one, "relative frequency" makes no sense.

These philosophical questions about the meaning of probability are not academic; they affect every decision we make. When a smoker rationalizes that his ninety-year-old parents have been puffing a pack a day for decades, so the nationwide odds don't apply to him, he might very well be right. In the 1996 presidential election, the advanced age of the Republican candidate became an issue. *The New Republic* published the following letter:

To the Editors:

In your editorial "Is Dole Too Old?" (April 1) your actuarial information was misleading. The average 72-year-old white man may suffer a 27 percent risk of dying within five years, but more than health and gender must be considered. Those still in the work force, as is Senator Bob Dole, have a much greater longevity. In addition, statistics show that greater wealth correlates to a longer life. Taking these characteristics into consideration, the average 73-year-old (the age that Dole would be if he takes office as president) has a 12.7 percent chance of dying within four years.

Yes, and what about the average seventy-three-year-old wealthy working white male who hails from Kansas, doesn't smoke, and was strong enough to survive an artillery shell? An even more dramatic difference surfaced during the murder trial of O.J. Simpson in 1995. The lawyer Alan Dershowitz, who was consulting for the defense, said on television that among men who batter their wives, only one-tenth of one percent go on to murder them. In a letter to *Nature*, a statistician then pointed out that among men who batter their wives *and whose wives are then murdered by someone*, more than *half* are the murderers.

Many probability theorists conclude that the probability of a single event cannot be computed; the whole business is meaningless. Single-event probabilities are "utter nonsense," said one mathematician. They should be handled "by psychoanalysis, not probability theory," sniffed another. It's not that people can believe anything they want about a single event. The statements that I am more likely to lose a fight against Mike Tyson than to win one, or that I am not likely to be abducted by aliens tonight, are not meaningless. But they are not *mathematical* statements that are precisely true or false, and people who question them have not committed an elementary fallacy. Statements about single events can't be decided by a calculator; they have to be hashed out by weighing the evidence, evaluating the persuasiveness of arguments, recasting the statements to make them easier to evaluate, and all the other fallible processes by which mortal beings make inductive guesses about an unknowable future.

So even the ditziest performance in the *Homo sapiens* hall of shame—saying that Linda is more likely to be a feminist bankteller than a bankteller—is not a fallacy, according to many mathematicians. Since a single-event probability is mathematically meaningless, people are forced to make sense of the question as best they can. Gigerenzer suggests that since frequencies are moot and people don't intuitively

give numbers to single events, they may switch to a third, nonmathe-
matical definition of probability, "degree of belief warranted by the
information just presented." That definition is found in many dictionaries and is used in courts of law, where it corresponds to concepts
such as probable cause, weight of evidence, and reasonable doubt. If
questions about single-event probabilities nudge people into that definition—a natural interpretation for subjects to have made if they
assumed, quite reasonably, that the experimenter had included the
sketch of Linda for some purpose—they would have interpreted the
question as, To what extent does the information given about Linda
warrant the conclusion that she is a bankteller? And a reasonable
answer is, not very much.

A final mind-bending ingredient of the concept of probability is the
belief in a stable world. A probabilistic inference is a prediction today
based on frequencies gathered yesterday. But that was then; this is now.
How do you know that the world hasn't changed in the interim? Philosophers of probability debate whether *any* beliefs in probabilities are truly
rational in a changing world. Actuaries and insurance companies worry
even more—insurance companies go bankrupt when a current event or a
change in lifestyles makes their tables obsolete. Social psychologists
point to the schlemiel who avoids buying a car with excellent repair statistics after hearing that a neighbor's model broke down yesterday.
Gigerenzer offers the comparison of a person who avoids letting his child
play in a river with no previous fatalities after hearing that a neighbor's
child was attacked there by a crocodile that morning. The difference
between the scenarios (aside from the drastic consequences) is that we
judge that the car world is stable, so the old statistics apply, but the river
world has changed, so the old statistics are moot. The person in the
street who gives a recent anecdote greater weight than a ream of statistics is not necessarily being irrational.

Of course, people sometimes reason fallaciously, especially in today's
data deluge. And, of course, everyone should learn probability and statistics. But a species that had no instinct for probability could not learn the
subject, let alone invent it. And when people are given information in a
format that meshes with the way they naturally think about probability,
they can be remarkably accurate. The claim that our species is blind to
chance is, as they say, unlikely to be true.

THE METAPHORICAL MIND

We are almost ready to dissolve Wallace's paradox that a forager's mind is capable of calculus. The human mind, we see, is not equipped with an evolutionarily frivolous faculty for doing Western science, mathematics, chess, or other diversions. It *is* equipped with faculties to master the local environment and outwit its denizens. People form concepts that find the clumps in the correlational texture of the world. They have several ways of knowing, or intuitive theories, adapted to the major kinds of entities in human experience: objects, animate things, natural kinds, artifacts, minds, and the social bonds and forces we will explore in the next two chapters. They wield inferential tools like the elements of logic, arithmetic, and probability. What we now want to know is where these faculties came from and how they can be applied to modern intellectual challenges.

Here is an idea, inspired by a discovery in linguistics. Ray Jackendoff points to sentences like the following:

> The messenger *went from* Paris *to* Istanbul.
> The inheritance finally *went to* Fred.
> The light *went from* green *to* red.
> The meeting *went from* 3:00 *to* 4:00.

The first sentence is straightforward: someone moves from place to place. But in the others, things stay put. Fred could have become a millionaire when the will was read even if no cash changed hands but a bank account was signed over. Traffic signals are set in pavement and don't travel, and meetings aren't even things that *could* travel. We are using space and motion as a metaphor for more abstract ideas. In the Fred sentence, possessions are objects, owners are places, and giving is moving. For the traffic light, a changeable thing is the object, its states (red and green) are places, and changing is moving. For the meeting, time is a line, the present is a moving point, events are journeys, beginnings and ends are origins and destinations.

The spatial metaphor is found not only in talk about changes but in talk about unchanging states. Belonging, being, and scheduling are construed as if they were landmarks situated at a place:

The messenger *is in* Istanbul.
The money *is* Fred's.
The light *is* red.
The meeting *is at* 3:00.

The metaphor also works in sentences about causing something to remain in a state:

The gang *kept* the messenger in Istanbul.
Fred *kept* the money.
The cop *kept* the light red.
Emilio *kept* the meeting on Monday.

Why do we make these analogies? It is not just to co-opt words but to co-opt their inferential machinery. Some deductions that apply to motion and space also apply nicely to possession, circumstances, and time. That allows the deductive machinery for space to be borrowed for reasoning about other subjects. For example, if we know that X went to Y, we can infer that X was not at Y beforehand but is there now. By analogy, if we know that a possession goes to a person, we can infer that the person did not own the possession beforehand but owns it now. The analogy is close, though it is never exact: as a messenger travels he occupies a series of locations between Paris and Istanbul, but as Fred inherits the money it does not gradually come into his possession to varying degrees as the will is being read; the transfer is instantaneous. So the concept of location must not be allowed to merge with the concepts of possession, circumstance, and time, but it can lend them some of its inferential rules. This sharing is what makes the analogies between location and other concepts *good for* something, and not just resemblances that catch our eye.

The mind couches abstract concepts in concrete terms. It is not only words that are borrowed for metaphors, but entire grammatical constructions. The double-object construction—*Minnie sent Mary the marbles*— is dedicated to sentences about giving. But the construction can be co-opted for talking about communication:

Minnie told Mary a story.
Alex asked Annie a question.
Carol wrote Connie a letter.

Ideas are gifts, communication is giving, the speaker is the sender, the audience is the recipient, knowing is having.

Location in space is one of the two fundamental metaphors in language, used for thousands of meanings. The other is force, agency, and causation. Leonard Talmy points out that in each of the following pairs, the two sentences refer to the same event, but the events feel different to us:

> The ball was rolling along the grass.
> The ball kept on rolling along the grass.

> John doesn't go out of the house.
> John can't go out of the house.

> Larry didn't close the door.
> Larry refrained from closing the door.

> Shirley is polite to him.
> Shirley is civil to him.

> Margie's got to go to the park.
> Margie gets to go to the park.

The difference is that the second sentence makes us think of an agent exerting force to overcome resistance or overpower some other force. With the second ball-in-the-grass sentence, the force is literally a physical force. But with John, the force is a *desire*: a desire to go out which has been restrained. Similarly, the second Larry seems to house one psychic force impelling him to close the door and another that overpowers it. For Shirley, those psychodynamics are conveyed by the mere choice of the adjective *civil*. In the first Margie sentence, she is impelled to the park by an external force in spite of an internal resistance. In the second, she is propelled by an internal force that overcomes an external resistance.

The metaphor of force and resistance is even more explicit in this family of sentences:

> Fran forced the door to open.
> Fran forced Sally to go.
> Fran forced herself to go.

The very same word, *force*, is being used literally and metaphorically, with a common thread of meaning that we easily appreciate. Sentences about motion and sentences about desire both allude to a billiard-ball dynamics in which an agonist has an intrinsic tendency to motion or rest, and is opposed by a weaker or stronger antagonist, causing one or both to stop or proceed. It is the impetus theory I discussed earlier in the chapter, the core of people's intuitive theory of physics.

Space and force pervade language. Many cognitive scientists (including me) have concluded from their research on language that a handful of concepts about places, paths, motions, agency, and causation underlie the literal or figurative meanings of tens of thousands of words and constructions, not only in English but in every other language that has been studied. The thought underlying the sentence *Minnie gave the house to Mary* would be something like "Minnie cause [house go-possessionally from Minnie to Mary]." These concepts and relations appear to be the vocabulary and syntax of mentalese, the language of thought. Because the language of thought is combinatorial, these elementary concepts may be combined into more and more complex ideas. The discovery of portions of the vocabulary and syntax of mentalese is a vindication of Leibniz' "remarkable thought": "that a kind of alphabet of human thoughts can be worked out and that everything can be discovered and judged by comparison of the letters of this alphabet and an analysis of the words made from them." And the discovery that the elements of mentalese are based on places and projectiles has implications for both where the language of thought came from and how we put it to use in modern times.

~

Other primates may not think about stories, inheritances, meetings, and traffic lights, but they do think about rocks, sticks, and burrows. Evolutionary change often works by copying body parts and tinkering with the copy. For example, insects' mouth parts are modified legs. A similar process may have given us our language of thought. Suppose ancestral circuits for reasoning about space and force were copied, the copy's connections to the eyes and muscles were severed, and references to the physical world were bleached out. The circuits could serve as a scaffolding whose slots are filled with symbols for more abstract concerns like states, possessions, ideas, and desires. The circuits would retain their computational abilities, continuing to reckon about entities being in one state at a time, shifting from state to state, and overcoming entities with opposite valence. When the new, abstract domain has a logical structure that mirrors objects in motion—a traffic light has one color at a time but flips between them; contested social interactions are determined by the stronger of two wills—the old circuits can do useful inferential work. They divulge their ancestry as space- and force-simulators by the metaphors they invite, a kind of vestigial cognitive organ.

Are there any reasons to believe that this is how our language of thought evolved? A few. Chimpanzees, and presumably their common ancestor with our species, are curious manipulators of objects. When they are trained to use symbols or gestures, they can make them stand for the event of going to a place or putting an object in a location. The psychologist David Premack has shown that chimpanzees can isolate causes. Given a pair of before-and-after pictures, like an apple and a pair of apple halves or a scribbled sheet of paper next to a clean one, they pick out the object that wreaked the change, a knife in the first case and an eraser in the second. So not only do chimpanzees maneuver in the physical world, but they have freestanding thoughts about it. Perhaps the circuitry behind those thoughts was co-opted in our lineage for more abstract kinds of causation.

How do we know that the minds of living human beings really appreciate the parallels between, say, social and physical pressure, or between space and time? How do we know that people aren't just using dead metaphors uncomprehendingly, as when we talk of breakfast without thinking of it as breaking a fast? For one thing, space and force metaphors have been reinvented time and again, in dozens of language families across the globe. Even more suggestive evidence comes from my own main field of research, child language acquisition. The psychologist Melissa Bowerman discovered that preschool children spontaneously coin their *own* metaphors in which space and motion symbolize possession, circumstance, time, and causation:

> You put me just bread and butter.
> Mother takes ball away from boy and puts it to girl.

> I'm taking these cracks bigger [while shelling a peanut].
> I putted part of the sleeve blue so I crossed it out with red [while coloring].

Can I have any reading behind the dinner?
Today we'll be packing because tomorrow there won't be enough space to
   pack.
Friday is covering Saturday and Sunday so I can't have Saturday and
   Sunday if I don't go through Friday.

My dolly is scrunched from someone . . . but not from me.
They had to stop from a red light.

The children could not have inherited the metaphors from earlier speak-
ers; the equation of space with abstract ideas has come naturally to them.

Space and force are so basic to language that they are hardly
metaphors at all, at least not in the sense of the literary devices used in
poetry and prose. There *is* no way to talk about possession, circumstance, and time in ordinary conversation without using words like
going, *keeping*, and *being at*. And the words don't trigger the sense of
incongruity that drives a genuine literary metaphor. We all know when
we are faced with a figure of speech. As Jackendoff points out, it's natural to say, "Of course, the world isn't *really* a stage, but if it were, you
might say that infancy is the first act." But it would be bizarre to say,
"Of course, meetings aren't *really* points in motion, but if they were,
you might say that this one went from 3:00 to 4:00." Models of space
and force don't act like figures of speech intended to convey new
insights; they seem closer to the medium of thought itself. I suspect
that parts of our mental equipment for time, animate beings, minds,
and social relations were copied and modified in the course of our evolution from the module for intuitive physics that we partly share with
chimpanzees.

Metaphors can be built out of metaphors, and we continue to borrow
from concrete thoughts when we stretch our ideas and words to encompass new domains. Somewhere between the basic constructions for
space and time in English and the glories of Shakespeare there is a vast
inventory of everyday metaphors that express the bulk of our experience.
George Lakoff and the linguist Mark Johnson have assembled a list of
the "metaphors we live by"—mental equations that embrace dozens of
expressions:

ARGUMENT IS WAR:
Your claims are *indefensible*.
He *attacked* every *weak* point in my argument.

Her criticisms were *right on target*.
I've never *won* an argument with him.

VIRTUE IS UP:
He is *high*-minded.
She is an *upstanding* citizen.
That was a *low* trick.
Don't be *underhanded*.
I wouldn't *stoop* to that; it is *beneath* me.

LOVE IS A PATIENT:
This is a *sick* relationship.
They have a *healthy* marriage.
This marriage is *dead*—it can't be *revived*.
It's a *tired* affair.

IDEAS ARE FOOD:
What he said *left a bad taste in my mouth*.
All this paper has are *half-baked* ideas and *warmed-over* theories.
I can't *swallow* that claim.
That's *food* for thought.

Once you begin to notice this pedestrian poetry, you find it everywhere. Ideas are not only food but buildings, people, plants, products, commodities, money, tools, and fashions. Love is a force, madness, magic, and war. The visual field is a container, self-esteem is a brittle object, time is money, life is a game of chance.

~

The ubiquity of metaphor brings us closer to a resolution to Wallace's paradox. The answer to the question "Why is the human mind adapted to think about arbitrary abstract entities?" is that it really isn't. Unlike computers and the rules of mathematical logic, we don't think in F's and x's and y's. We have inherited a pad of forms that capture the key features of encounters among objects and forces, and the features of other consequential themes of the human condition such as fighting, food, and health. By erasing the contents and filling in the blanks with new symbols, we can adapt our inherited forms to more abstruse domains. Some of these revisions may have taken place in our evolution, giving us basic mental categories like ownership, time, and will out of forms originally designed for intuitive physics. Other revisions take place as we live our lives and grapple with new realms of knowledge.

Even the most recondite scientific reasoning is an assembly of down-home mental metaphors. We pry our faculties loose from the domains they were designed to work in, and use their machinery to make sense of new domains that abstractly resemble the old ones. The metaphors we think in are lifted not only from basic scenarios like moving and bumping but from entire ways of knowing. To do academic biology, we take our way of understanding artifacts and apply it to organisms. To do chemistry, we treat the essence of a natural kind as a collection of tiny, bouncy, sticky objects. To do psychology, we treat the mind as a natural kind.

Mathematical reasoning both takes from and gives to the other parts of the mind. Thanks to graphs, we primates grasp mathematics with our eyes and our mind's eye. Functions are shapes (linear, flat, steep, crossing, smooth), and operating is doodling in mental imagery (rotating, extrapolating, filling, tracing). In return, mathematical thinking offers new ways to understand the world. Galileo wrote that "the book of nature is written in the language of mathematics; without its help it is impossible to comprehend a single word of it."

Galileo's dictum applies not only to equation-filled blackboards in the physics department but to elementary truths we take for granted. The psychologists Carol Smith and Susan Carey have found that children have odd beliefs about matter. Children know that a heap of rice weighs something but claim that a grain of rice weighs nothing. When asked to imagine cutting a piece of steel in half repeatedly, they say that one will finally arrive at a piece so small that it no longer takes up space or has any steel inside it. They are not of unsound mind. Every physical event has a threshold below which no person or device can detect it. Repeated division of an object results in objects too small to detect; a collection of objects each of which falls below the threshold may be detectable en masse. Smith and Carey note that we find children's beliefs silly because we can construe matter using our concept of number. Only in the realm of mathematics does repeated division of a positive quantity always yield a positive quantity, and repeated addition of zero always yields zero. Our understanding of the physical world is more sophisticated than children's because we have merged our intuitions about objects with our intuitions about number.

So vision was co-opted for mathematical thinking, which helps us see the world. Educated understanding is an enormous contraption of parts within parts. Each part is built out of basic mental models or ways of knowing that are copied, bleached of their original content, connected to other models, and packaged into larger parts, which can be packaged into still larger parts without limit. Because human thoughts are combinatorial (simple parts combine) and recursive (parts can be embedded within parts), breathtaking expanses of knowledge can be explored with a finite inventory of mental tools.

EUREKA!

And what about the genius? How can natural selection explain a Shakespeare, a Mozart, an Einstein, an Abdul-Jabbar? How would Jane Austen, Vincent van Gogh, or Thelonious Monk have earned their keep on the Pleistocene savanna?

All of us are creative. Every time we stick a handy object under the leg of a wobbly table or think up a new way to bribe a child into his pajamas, we have used our faculties to create a novel outcome. But creative geniuses are distinguished not just by their extraordinary works but by their extraordinary way of working; they are not supposed to think like you and me. They burst on the scene as prodigies, *enfants terribles*, young turks. They listen to their muse and defy the conventional wisdom. They work when the inspiration hits, and leap with insight while the rest of us plod in baby steps along well-worn paths. They put a problem aside and let it incubate in the unconscious; then, without warning, a bulb lights up and a fully formed solution presents itself. Aha! The genius leaves us with masterpieces, a legacy of the unrepressed creativity of the unconscious. Woody Allen captures the image in his hypothetical letters from Vincent van Gogh in the story "If the Impressionists Had Been Dentists." Vincent writes to his brother in anguish and despair, "Mrs. Sol Schwimmer is suing me because I made her bridge as I felt it and not to fit her ridiculous mouth! That's right! I can't work to order like a common tradesman! I decided her bridge should be enormous and billowing, with wild, explosive teeth flaring up in every direction like fire! Now she is upset because it won't fit in her mouth! . . . I tried forcing the false plate in but it sticks out like a star burst chandelier. Still, I find it beautiful."

The image came out of the Romantic movement two hundred years ago and is now firmly entrenched. Creativity consultants take millions of dollars from corporations for Dilbertesque workshops on brainstorming, lateral thinking, and flow from the right side of the brain, guaranteed to turn every manager into an Edison. Elaborate theories have been built to explain the uncanny problem-solving power of the dreamy unconscious. Like Alfred Russel Wallace, some have concluded that there can be no natural explanation. Mozart's manuscripts were said to have no corrections. The pieces must have come from the mind of God, who had chosen to express his voice through Mozart.

Unfortunately, creative people are at their most creative when writing their autobiographies. Historians have scrutinized their diaries, notebooks, manuscripts, and correspondence looking for signs of the temperamental seer periodically struck by bolts from the unconscious. Alas, they have found that the creative genius is more Salieri than Amadeus.

Geniuses are wonks. The typical genius pays dues for at least ten years before contributing anything of lasting value. (Mozart composed symphonies at eight, but they weren't very good; his first masterwork came in the twelfth year of his career.) During the apprenticeship, geniuses immerse themselves in their genre. They absorb tens of thousands of problems and solutions, so no challenge is completely new and they can draw on a vast repertoire of motifs and strategies. They keep an eye on the competition and a finger to the wind, and are either discriminating or lucky in their choice of problems. (The unlucky ones, however talented, aren't remembered as geniuses.) They are mindful of the esteem of others and of their place in history. (The physicist Richard Feynman wrote two books describing how brilliant, irreverent, and admired he was and called one of them *What Do You Care What Other People Think?*) They work day and night, and leave us with many works of subgenius. (Wallace spent the end of his career trying to communicate with the dead.) Their interludes away from a problem are helpful not because it ferments in the unconscious but because they are exhausted and need the rest (and possibly so they can forget blind alleys). They do not repress a problem but engage in "creative worrying," and the epiphany is not a masterstroke but a tweaking of an earlier attempt. They revise endlessly, gradually closing in on their ideal.

Geniuses, of course, may also have been dealt a genetic hand with four aces. But they are not freaks with minds utterly unlike ours or unlike anything we can imagine evolving in a species that has always lived by its wits. The genius creates good ideas because we all create good ideas; that is what our combinatorial, adapted minds are for.

6

HOTHEADS

On March 13, 1996, Thomas Hamilton walked into an elementary school in Dunblane, Scotland, carrying two revolvers and two semiautomatic pistols. After wounding staff members who tried to tackle him, he ran to the gymnasium, where a kindergarten class was playing. There he shot twenty-eight children, sixteen fatally, and killed their teacher before turning the gun on himself. "Evil visited us yesterday, and we don't know why," said the school's headmaster the next day. "We don't understand it and I don't think we ever will."

We probably never will understand what made Hamilton commit his vile final acts. But the report of pointless revenge by an embittered loner is disturbingly familiar. Hamilton was a suspected pedophile who had been forced to resign as a Scout leader and then formed his own youth groups so he could continue working with boys. One group held its meetings in the Dunblane school's gymnasium until school officials, responding to parents' complaints about his odd behavior, forced him out. Hamilton was the target of ridicule and gossip, and was known in the area, undoubtedly for good reasons, as "Mr. Creepy." Days before his rampage he had sent letters to the media and to Queen Elizabeth defending his reputation and pleading for reinstatement in the scouting movement.

The Dunblane tragedy was particularly shocking because no one thought it could happen there. Dunblane is an idyllic, close-knit village where serious crime was unknown. It is far from America, land of the wackos, where there are as many guns as people and where murderous rampages by disgruntled postal workers are so common (a dozen incidents in a dozen years) that a slang term for losing one's temper is "going postal." But running amok is not unique to America, to Western nations, or even to modern societies. *Amok* is a Malay word for the homicidal sprees occasionally undertaken by lonely Indochinese men who have suffered a loss of love, a loss of money, or a loss of face. The syndrome has been described in a culture even more remote from the West: the stone-age foragers of Papua New Guinea.

The amok man is patently out of his mind, an automaton oblivious to his surroundings and unreachable by appeals or threats. But his rampage is preceded by lengthy brooding over failure, and is carefully planned as a means of deliverance from an unbearable situation. The amok state is chillingly cognitive. It is triggered not by a stimulus, not by a tumor, not by a random spurt of brain chemicals, but by an idea. The idea is so standard that the following summary of the amok mind-set, composed in 1968 by a psychiatrist who had interviewed seven hospitalized amoks in Papua New Guinea, is an apt description of the thoughts of mass murderers continents and decades away:

> I am not an important or "big man." I possess only my personal sense of dignity. My life has been reduced to nothing by an intolerable insult. Therefore, I have nothing to lose except my life, which is nothing, so I trade my life for yours, as your life is favoured. The exchange is in my favour, so I shall not only kill you, but I shall kill many of you, and at the same time rehabilitate myself in the eyes of the group of which I am a member, even though I might be killed in the process.

The amok syndrome is an extreme instance of the puzzle of the human emotions. Exotic at first glance, upon scrutiny they turn out to be universal; quintessentially irrational, they are tightly interwoven with abstract thought and have a cold logic of their own.

UNIVERSAL PASSION

A familiar tactic for flaunting one's worldliness is to inform listeners that some culture lacks an emotion we have or has an emotion we lack. Allegedly the Utku-Inuit Eskimos have no word for anger and do not feel the emotion. Tahitians supposedly do not recognize guilt, sadness, longing, or loneliness; they describe what we would call grief as fatigue, sickness, or bodily distress. Spartan mothers were said to smile upon hearing that their sons died in combat. In Latin cultures, machismo reigns, whereas the Japanese are driven by a fear of shaming the family. In interviews on language I have been asked, Who but the Jews would have a word, *naches*, for luminous pride in a child's accomplishments? And does it not say something profound about the Teutonic psyche that the German language has the word *Schadenfreude*, pleasure in another's misfortunes?

Cultures surely differ in how often their members express, talk about, and act on various emotions. But that says nothing about what their people feel. The evidence suggests that the emotions of all normal members of our species are played on the same keyboard.

The most accessible signs of emotions are candid facial expressions. In preparing *The Expression of the Emotions in Man and Animals*, Darwin circulated a questionnaire to people who interacted with aboriginal populations on five continents, including populations that had had little contact with Europeans. Urging them to answer in detail and from observation rather than memory, Darwin asked how the natives expressed astonishment, shame, indignation, concentration, grief, good spirits, contempt, obstinacy, disgust, fear, resignation, sulkiness, guilt, slyness, jealousy, and "yes" and "no." For example:

> (5.) When in low spirits, are the corners of the mouth depressed, and the inner corner of the eyebrows raised by that muscle which the French call the "Grief muscle"? The eyebrow in this state becomes slightly oblique, with a little swelling at the inner end; and the forehead is transversely wrinkled in the middle part, but not across the whole breadth, as when the eyebrows are raised in surprise.

Darwin summed up the responses: "The same state of mind is expressed throughout the world with remarkable uniformity; and this fact is in itself interesting as evidence of the close similarity in bodily structure and mental disposition of all the races of mankind."

Though Darwin may have biased his informants with leading questions, contemporary research has borne out his conclusion. When the psychologist Paul Ekman began to study emotions in the 1960s, facial expressions were thought to be arbitrary signs that the infant learns when its random grimaces are rewarded and punished. If expressions appeared universal, it was thought, that was because Western models had become universal; no culture was beyond the reach of John Wayne and Charlie Chaplin. Ekman assembled photographs of people expressing six emotions. He showed them to people from many cultures, including the isolated Fore foragers of Papua New Guinea, and asked them to label the emotion or make up a story about what the person had gone through. Everyone recognized happiness, sadness, anger, fear, disgust, and surprise. For example, a Fore subject said that the American showing fear in the photograph must have just seen a boar. Reversing the procedure, Ekman photographed his Fore informants as they acted out scenarios such as "Your friend has come and you are happy," "Your child has died," "You are angry and about to fight," and "You see a dead pig that has been lying there for a long time." The expressions in the photographs are unmistakable.

When Ekman began to present his findings at a meeting of anthropologists in the late 1960s, he met with outrage. One prominent anthropologist rose from the audience shouting that Ekman should not be allowed to continue to speak because his claims were fascist. On another occasion an African American activist called him a racist for saying that black facial expressions were no different from white ones. Ekman was bewildered because he had thought that if the work had any political moral it was unity and brotherhood. In any case, the conclusions have been replicated and are now widely accepted in some form (though there are controversies over which expressions belong on the universal list, how much context is needed to interpret them, and how reflexively they are tied to each emotion). And another observation by Darwin has been corroborated: children who are blind and deaf from birth display virtually the full gamut of emotions on their faces.

Why, then, do so many people think that emotions differ from culture to culture? Their evidence is much more indirect than Darwin's informants and Ekman's experiments. It comes from two sources that cannot be trusted at all as readouts of people's minds: their language and their opinions.

The common remark that a language does or doesn't have a word for an emotion means little. In *The Language Instinct* I argued that the influence of language on thought has been exaggerated, and that is all the more true for the influence of language on feeling. Whether a language appears to have a word for an emotion depends on the skill of the translator and on quirks of the language's grammar and history. A language accumulates a large vocabulary, including words for emotions, when it has had influential wordsmiths, contact with other languages, rules for forming new words out of old ones, and widespread literacy, which allows new coinages to become epidemic. When a language has not had these stimulants, people describe how they feel with circumlocutions, metaphors, metonyms, and synecdoches. When a Tahitian woman says, "My husband died and I feel sick," her emotional state is hardly mysterious; we can bet she is not complaining about acid indigestion. Even a language with a copious vocabulary has words for only a fraction of emotional experience. The author G. K. Chesterton wrote,

> Man knows that there are in the soul tints more bewildering, more numberless, and more nameless than the colours of an autumn forest; . . . Yet he seriously believes that these things can every one of them, in all their tones and semitones, in all their blends and unions, be accurately represented by an arbitrary system of grunts and squeals. He believes that an ordinary civilized stockbroker can really produce out of his own inside noises which denote all the mysteries of memory and all the agonies of desire.

When English-speakers hear the word *Schadenfreude* for the first time, their reaction is not, "Let me see . . . Pleasure in another's misfortunes . . . What could that possibly be? I cannot grasp the concept; my language and culture have not provided me with such a category." Their reaction is, "You mean there's a *word* for it? Cool!" That is surely what went through the minds of the writers who introduced *Schadenfreude* into written English a century ago. New emotion words catch on quickly, without tortuous definitions; they come from other languages (*ennui, angst, naches, amok*), from subcultures such as those of musicians and drug addicts (*blues, funk, juiced, wasted, rush, high, freaked out*), and from general slang (*pissed, bummed, grossed out, blown away*). I have never heard a foreign emotion word whose meaning was not instantly recognizable.

People's emotions are so alike that it takes a philosopher to craft a genuinely alien one. In an essay called "Mad Pain and Martian Pain," David Lewis defines mad pain as follows:

> There might be a strange man who sometimes feels pain, just as we do, but whose pain differs greatly from ours in its causes and effects. Our pain is typically caused by cuts, burns, pressure, and the like; his is caused by moderate exercise on an empty stomach. Our pain is generally distracting; his turns his mind to mathematics, facilitating concentration on that but distracting him from anything else. Intense pain has no ten-

dency whatever to cause him to groan or writhe, but does cause him to cross his legs and snap his fingers. He is not in the least motivated to prevent pain or to get rid of it.

Have anthropologists discovered a people that feels mad pain or something equally weird? It might seem that way if you look only at stimulus and response. The anthropologist Richard Shweder points out, "It is a trivial exercise for any anthropologist to generate long lists of antecedent events (ingesting cow urine, eating chicken five days after your father dies, kissing the genitals of an infant boy, being complimented about your pregnancy, caning a child, touching someone's foot or shoulder, being addressed by your first name by your wife, ad infinitum) about which the emotional judgments of a Western observer would not correspond to the native's evaluative response." True enough, but if you look a bit deeper and ask how people *categorize* these stimuli, the emotions elicited by the categories make you feel at home. To us, cow urine is a contaminant and cow mammary secretions are a nutrient; in another culture, the categories may be reversed, but we all feel disgust for contaminants. To us, being addressed by your first name by a spouse is not disrespectful, but being addressed by your first name by a stranger might be, and being addressed by your religion by your spouse might be, too. In all the cases, disrespect triggers anger.

But what about the claims of native informants that they just don't have one of our emotions? Do our emotions seem like mad pain to them? Probably not. The Utku-Inuits' claim that they do not feel anger is belied by their behavior: they recognize anger in foreigners, beat their dogs to discipline them, squeeze their children painfully hard, and occasionally get "heated up." Margaret Mead disseminated the incredible claim that Samoans have no passions—no anger between parents and children or between a cuckold and a seducer, no revenge, no lasting love or bereavement, no maternal caring, no tension about sex, no adolescent turmoil. Derek Freeman and other anthropologists found that Samoan society in fact had widespread adolescent resentment and delinquency, a cult of virginity, frequent rape, reprisals by the rape victim's family, frigidity, harsh punishment of children, sexual jealousy, and strong religious feeling.

We should not be surprised at these discrepancies. The anthropologist Renato Rosaldo has noted, "A traditional anthropological description is like a book of etiquette. What you get isn't so much the deep cultural wisdom as the cultural clichés, the wisdom of Polonius, conventions in the trivial rather than the informing sense. It may tell you the official rules, but it won't tell you how life is lived." Emotions, in particular, are often regulated by the official rules, because they are assertions of a person's interests. To me it's a confession of my innermost feelings, but to you it's bitching and moaning, and you may very well tell me to put a lid on it. And to those in power, other people's emotions are even more annoying—they lead to nuisances such as women wanting men as husbands and sons rather than as cannon fodder, men fighting each other when they could be fighting the enemy, and children falling in love with a soulmate instead of accepting a betrothed who cements an important deal. Many societies deal with these nuisances by trying to regulate emotions and spreading the disinformation that they don't exist.

Ekman has shown that cultures differ the most in how the emotions are expressed in public. He secretly filmed the expressions of American and Japanese students as they watched gruesome footage of a primitive puberty rite. (Emotion researchers have extensive collections of gross-out material.) If a white-coated experimenter was in the room interviewing them, the Japanese students smiled politely during scenes that made the Americans recoil in horror. But when the subjects were alone, the Japanese and American faces were equally horrified.

FEELING MACHINES

The Romantic movement in philosophy, literature, and art began about two hundred years ago, and since then the emotions and the intellect have been assigned to different realms. The emotions come from nature and live in the body. They are hot, irrational impulses and intuitions, which follow the imperatives of biology. The intellect comes from civilization and lives in the mind. It is a cool deliberator that follows the interests of self and society by keeping the emotions in check. Romantics believe that the emotions are the source of wisdom, innocence, authenticity, and creativity, and should not be repressed by individuals or society. Often Romantics acknowledge a dark side, the price we must pay for artistic greatness. When the antihero in Anthony Burgess' *A Clockwork Orange* has his violent impulses conditioned out of him, he loses his taste for Beethoven. Romanticism dominates contemporary American popular culture, as in the Dionysian ethos of rock music, the pop psychology imperative to get in touch with your feelings, and the Hollywood formulas about wise simpletons and about uptight yuppies taking a walk on the wild side.

Most scientists tacitly accept the premises of Romanticism even when they disagree with its morals. The irrational emotions and the repressing intellect keep reappearing in scientific guises: the id and the superego, biological drives and cultural norms, the right hemisphere and the left hemisphere, the limbic system and the cerebral cortex, the evolutionary baggage of our animal ancestors and the general intelligence that propelled us to civilization.

In this chapter I present a distinctly unromantic theory of the emotions. It combines the computational theory of mind, which says that the lifeblood of the psyche is information rather than energy, with the modern theory of evolution, which calls for reverse-engineering the complex design of biological systems. I will show that the emotions are adaptations, well-engineered software modules that work in harmony with the intellect and are indispensable to the functioning of the whole mind. The problem with the emotions is not that they are untamed forces or vestiges of our animal past; it is that they were designed to propagate copies of the genes that built them rather than to promote happiness, wisdom, or moral values. We often call an act "emotional" when it is harmful to the social group, damaging to the actor's happiness in the long run, uncontrollable and impervious to persuasion, or a product of self-delusion. Sad to say, these outcomes are not malfunctions but precisely what we would expect from well-engineered emotions.

~

The emotions are another part of the mind that has been prematurely written off as nonadaptive baggage. The neuroscientist Paul MacLean took the Romantic doctrine of the emotions and translated it into a famous but incorrect theory known as the Triune Brain. He described the human cerebrum as an evolutionary palimpsest of three layers. At the bottom are the basal ganglia or Reptilian Brain, the seat of the primitive and selfish emotions driving the "Four Fs": feeding, fighting, fleeing, and sexual behavior. Grafted onto it is the limbic system or Primitive Mammalian Brain, which is dedicated to the kinder, gentler, social emotions, like those behind parenting. Wrapped around that is the Modern Mammalian Brain, the neocortex that grew wild in human evolution and that houses the intellect. The belief that the emotions are animal legacies is also familiar from pop ethology documentaries in which snarling baboons segue into rioting soccer hooligans as the voice-over frets about whether we will rise above our instincts and stave off nuclear doom.

One problem for the triune theory is that the forces of evolution do not just heap layers on an unchanged foundation. Natural selection has to work with what is already around, but it can *modify* what it finds. Most parts of the human body came from ancient mammals and before them ancient reptiles, but the parts were heavily modified to fit features of the human lifestyle, such as upright posture. Though our bodies carry vestiges of the past, they have few parts that were unmodifiable and adapted only to the needs of older species. Even the appendix is currently put to use, by the immune system. The circuitry for the emotions was not left untouched, either.

Admittedly, some traits are so much a part of the architectural plan of an organism that selection is powerless to tinker with them. Might the software for the emotions be burned so deeply into the brain that organisms are condemned to feel as their remote ancestors did? The evidence says no; the emotions are easy to reprogram. Emotional repertoires vary wildly among animals depending on their species, sex, and age. Within the mammals, we find the lion and the lamb. Even within dogs (a single species), a few millennia of selective breeding have given us pit bulls and Saint Bernards. The genus closest to ours embraces common chimpanzees, in which gangs of males massacre rival gangs and females can murder one another's babies, and the pygmy chimpanzees (bonobos), whose philosophy is "Make love not war." Of course, some reactions are widely shared across species—say, panic when one is confined—but the reactions may have been retained because they are adaptive for everyone. Natural selection may not have had complete freedom to reprogram the emotions, but it had a lot.

And the human cerebral cortex does not ride piggyback on an ancient limbic system, or serve as the terminus of a processing stream beginning there. The systems work in tandem, integrated by many two-way connections. The amygdala, an almond-shaped organ buried in each temporal lobe, houses the main circuits that color our experience with emotions. It receives not just simple signals (such as of loud noises) from the lower stations of the brain, but abstract, complex information from the brain's highest centers. The amygdala in turn sends signals to virtually every other part of the brain, including the decision-making circuitry of the frontal lobes.

The anatomy mirrors the psychology. Emotion is not just running away from a bear. It can be set off by the most sophisticated information processing the mind is capable of, such as reading a Dear John letter or coming home to find an ambulance in the driveway. And the emotions help to connive intricate plots for escape, revenge, ambition, and courtship. As Samuel Johnson wrote, "Depend upon it, sir, when a man knows he is to be hanged in a fortnight, it concentrates his mind wonderfully."

The first step in reverse-engineering the emotions is try to imagine what a mind would be like without them. Supposedly Mr. Spock, the Vulcan mastermind, didn't have emotions (except for occasional intrusions from his human side and a seven-year itch that drove him back to Vulcan to spawn). But Spock's emotionlessness really just amounted to his being in control, not losing his head, coolly voicing unpleasant truths, and so on. He must have been driven by *some* motives or goals. Something must have kept Spock from spending his days calculating pi to a quadrillion digits or memorizing the Manhattan telephone directory. Something must have impelled him to explore strange new worlds, to seek out new civilizations, and to boldly go where no man had gone before. Presumably it was intellectual curiosity, a drive to set and solve problems, and solidarity with allies—emotions all. And what would Spock have done when faced with a predator or an invading Klingon? Do a headstand? Prove the four-color map theorem? Presumably a part of his brain quickly mobilized his faculties to scope out how to flee and to take steps to avoid the vulnerable predicament in the future. That is, he had fear. Spock may not have been impulsive or demonstrative, but he must have had drives that impelled him to deploy his intellect in pursuit of certain goals rather than others.

A conventional computer program is a list of instructions that the machine executes until it reaches STOP. But the intelligence of aliens, robots, and animals needs a more flexible method of control. Recall that intelligence is the pursuit of goals in the face of obstacles. Without goals, the very concept of intelligence is meaningless. To get into my locked apartment, I can force open a window, call the landlord, or try to reach the latch through the mail slot. Each of these goals is attained by a chain of subgoals. My fingers won't reach the latch, so the subgoal is to find pliers. But my pliers are inside, so I set up a sub-subgoal of finding a store and buying new pliers. And so on. Most artificial intelligence systems are built around means and ends, like the production system in Chapter 2 with its stack of goal symbols displayed on a bulletin board and the software demons that respond to them.

But where does the topmost goal, the one that the rest of the program tries to attain, come from? For artificial intelligence systems, it comes from the programmer. The programmer designs it to diagnose soybean diseases or predict the next day's Dow Jones Industrial Average. For organisms, it comes from natural selection. The brain strives to put its owner in circumstances like those that caused its ancestors to reproduce. (The brain's goal is not reproduction itself; animals don't know the facts of life, and people who do know them are happy to subvert them, such as when they use contraception.) The goals installed in *Homo sapiens*, that problem-solving, social species, are not just the Four Fs. High on the list are understanding the environment and securing the cooperation of others.

And here is the key to why we have emotions. An animal cannot pursue all its goals at once. If an animal is both hungry and thirsty, it should not stand halfway between a berry bush and a lake, as in the fable about the indecisive ass who starved between two haystacks. Nor should it nibble a berry, walk over and take a sip from the lake, walk back to nibble another berry, and so on. The animal must commit its body to one goal at a time, and the goals have to be matched with the best moments for achieving them. Ecclesiastes says that to every thing there is a season, and a time to every purpose under heaven: a time to weep, and a time to laugh; a time to love, and a time to hate. Different goals are appropriate when a lion has you in its sights, when your child shows up in tears, or when a rival calls you an idiot in public.

The emotions are mechanisms that set the brain's highest-level goals. Once triggered by a propitious moment, an emotion triggers the cascade of subgoals and sub-subgoals that we call thinking and acting. Because the goals and means are woven into a multiply nested control structure of subgoals within subgoals within subgoals, no sharp line divides thinking from feeling, nor does thinking inevitably precede feeling or vice versa (notwithstanding the century of debate within psychology over which comes first). For example, fear is triggered by a signal of impending harm like a predator, a clifftop, or a spoken threat. It lights up the short-term goal of fleeing, subduing, or deflecting the danger, and gives the goal high priority, which we experience as a sense of urgency. It also lights up the longer-term goals of avoiding the hazard in the future and remembering how we got out of it this time, triggered by the state we experience as relief. Most artificial intelligence researchers believe that freely behaving robots (as opposed to the ones bolted to the side of an assembly line) will have to be programmed with something like emotions merely for them to know at every moment what to do next. (Whether the robots would be *sentient* of these emotions is another question, as we saw in Chapter 2.)

Fear also presses a button that readies the body for action, the so-called fight-or-flight response. (The nickname is misleading because the response prepares us for *any* time-sensitive action, such as grabbing a baby who is crawling toward the top of a stairwell.) The heart thumps to send blood to the muscles. Blood is rerouted from the gut and skin, leaving butterflies and clamminess. Rapid breathing takes in oxygen. Adrenaline releases fuel from the liver and helps the blood to clot. And it gives our face that universal deer-in-the-headlights look.

Each human emotion mobilizes the mind and body to meet one of the challenges of living and reproducing in the cognitive niche. Some challenges are posed by physical things, and the emotions that deal with them, like disgust, fear, and appreciation of natural beauty, work in straightforward ways. Others are posed by people. The problem in dealing with people is that people can deal back. The emotions that evolved in response to other people's emotions, like anger, gratitude, shame, and romantic love, are played on a complicated chessboard, and they spawn the passion and intrigue that misleads the Romantic. First let's explore emotions about things, then emotions about people.

THE SUBURBAN SAVANNA

The expression "a fish out of water" reminds us that every animal is adapted to a habitat. Humans are no exception. We tend to think that animals just go where they belong, like heat-seeking missiles, but the animals must experience these drives as emotions not unlike ours. Some places are inviting, calming, or beautiful; others are depressing or scary. The topic in biology called "habitat selection" is, in the case of *Homo sapiens*, the same as the topic in geography and architecture called "environmental aesthetics": what kinds of places we enjoy being in.

Until very recently our ancestors were nomads, leaving a site when they had used up its edible plants and animals. The decision of where to go next was no small matter. Cosmides and Tooby write:

> Imagine that you are on a camping trip that lasts a lifetime. Having to carry water from a stream and firewood from the trees, one quickly learns to appreciate the advantages of some campsites over others. Dealing with exposure on a daily basis quickly gives one an appreciation for sheltered sites, out of the wind, snow, or rain. For hunter-gatherers, there is no escape from this way of life: no opportunities to pick up food at the grocery store, no telephones, no emergency services, no artificial water supplies, no fuel deliveries, no cages, guns, or animal control officers to protect one from the predatory animals. In these circumstances, one's life depends on the operation of mechanisms that cause one to prefer habitats that provide sufficient food, water, shelter, information, and safety to support human life, and that cause one to avoid those that do not.

Homo sapiens is adapted to two habitats. One is the African savanna, in which most of our evolution took place. For an omnivore like our ancestors, the savanna is a hospitable place compared with other ecosystems. Deserts have little biomass because they have little water. Temperate forests lock up much of their biomass in wood. Rainforests—or, as they used to be called, jungles—place it high in the canopy, relegating omnivores on the ground to being scavengers who gather the bits that fall from above. But the savanna—grasslands dotted with clumps of trees—is rich in biomass, much of it in the flesh of large animals, because grass replenishes itself quickly when grazed. And most of the biomass is conveniently placed a meter or two from the ground. Savannas also offer expansive views, so predators, water, and paths can be spotted from afar. Its trees provide shade and an escape from carnivores.

Our second-choice habitat is the rest of the world. Our ancestors, after evolving on the African savannas, wandered into almost every nook and cranny of the planet. Some were pioneers who left the savanna and then other areas in turn, as the population expanded or the climate changed. Others were refugees in search of safety. Foraging tribes can't stand one another. They frequently raid neighboring territories and kill any stranger who blunders into theirs.

We could afford this wanderlust because of our intellect. People explore a new landscape and draw up a mental resource map, rich in details about water, plants, animals, routes, and shelter. And if they can, they make their new homeland into a savanna. Native Americans and Australian aborigines used to burn huge swaths of woodland, opening them up for colonization by grasses. The ersatz savanna attracted grazing animals, which were easy to hunt, and exposed visitors before they got too close.

The biologist George Orians, an expert on the behavioral ecology of birds, recently turned his eye to the behavioral ecology of humans. With Judith Heerwagen, Stephen Kaplan, Rachel Kaplan, and others, he argues that our sense of natural beauty is the mechanism that drove our ancestors into suitable habitats. We innately find savannas beautiful, but we also like a landscape that is easy to explore and remember, and that we have lived in long enough to know its ins and outs.

In experiments on human habitat preference, American children and adults are shown slides of landscapes and asked how much they would like to visit or live in them. The children prefer savannas, even though they have never been to one. The adults like the savannas, too, but they like the deciduous and coniferous forests—which resemble much of the habitable United States—just as much. No one likes the deserts and the rainforests. One interpretation is that the children are revealing our species' default habitat preference, and the adults supplement it with the land with which they have grown familiar.

Of course, people do not have a mystical longing for ancient homelands. They are merely pleased by the landscape features that savannas tend to have. Orians and Heerwagen surveyed the professional wisdom of gardeners, photographers, and painters to learn what kinds of landscapes people find beautiful. They treated it as a second kind of data on human tastes in habitats, supplementing the experiments on people's reactions to slides. The landscapes thought to be the loveliest, they found, are dead ringers for an optimal savanna: semi-open space (neither completely exposed, which leaves one vulnerable, nor overgrown, which impedes vision and movement), even ground cover, views to the horizon, large trees, water, changes in elevation, and multiple paths leading out. The geographer Jay Appleton succinctly captured what makes a landscape appealing: prospect and refuge, or seeing without being seen. The combination allows us to learn the lay of the land safely.

The land itself must be legible, too. Anyone who has lost a trail in a dense forest or seen footage of sand dunes or snow drifts in all directions knows the terror of an environment lacking a frame of reference. A landscape is just a very big object, and we recognize complex objects by locating their parts in a reference frame belonging to the object (see Chapter 4). The reference frames in a mental map are big landmarks, like trees, rocks, and ponds, and long paths or boundaries, like rivers and mountain ranges. A vista without these guideposts is unsettling. Kaplan and Kaplan found another key to natural beauty, which they call mystery. Paths bending around hills, meandering streams, gaps in foliage, undulating land, and partly blocked views grab our interest by hinting that the land may have important features that could be discovered by further exploration.

People also love to look at animals and plants, especially flowers. If you are reading this book at home or in other pleasant but artificial surroundings, chances are you can look up and find animal, plant, or flower motifs in the decorations. Our fascination with animals is obvious. We eat them, they eat us. But our love of flowers, which we don't eat except in salads in overpriced restaurants, needs an explanation. We ran into it in Chapters 3 and 5. People are intuitive botanists, and a flower is a rich source of data. Plants blend into a sea of green and often can be identified only by their flowers. Flowers are harbingers of growth, marking the site of future fruit, nuts, or tubers for creatures smart enough to remember them.

Some natural happenings are deeply evocative, like sunsets, thunder, gathering clouds, and fire. Orians and Heerwagen note that they tell of an imminent and consequential change: darkness, a storm, a blaze. The emotions evoked are arresting, forcing one to stop, take notice, and prepare for what's to come.

Environmental aesthetics is a major factor in our lives. Mood depends on surroundings: think of being in a bus terminal waiting room or a lakeside cottage. People's biggest purchase is their home, and the three rules of home buying—location, location, and location—pertain, apart from nearness to amenities, to grassland, trees, bodies of water, and prospect (views). The value of the house itself depends on its refuge (cozy spaces) and mystery (nooks, bends, windows, multiple levels). And people in the unlikeliest of ecosystems strive for a patch of savanna to call their own. In New England, any land that is left alone quickly turns into a scruffy deciduous forest. During my interlude in suburbia, every weekend my fellow burghers and I would drag out our lawn mowers, leaf blowers, weed whackers, limb loppers, branch pruners, stem snippers, hedge clippers, and wood chippers in a Sisyphean effort to hold the forest at bay. Here in Santa Barbara, the land wants to be an arid chaparral, but decades ago the city fathers dammed wilderness creeks and tunneled through mountains to bring water to thirsty lawns. During a recent drought, homeowners were so desperate for verdant vistas that they sprayed their dusty yards with green paint.

FOOD FOR THOUGHT

> Great green gobs of greasy grimy gopher guts,
> Mutilated monkey meat,
> Concentrated chicken feet.
> Jars and jars of petrified porpoise pus,
> And me without my spoon!
> —fondly remembered camp song, sung to the
> tune of "The Old Gray Mare"; lyricist unknown

Disgust is a universal human emotion, signaled with its own facial expression and codified everywhere in food taboos. Like all the emotions, disgust has profound effects on human affairs. During World War II, American pilots in the Pacific went hungry rather than eat the toads and bugs that they had been taught were perfectly safe. Food aversions are tenacious ethnic markers, persisting long after other traditions have been abandoned.

Judged by the standards of modern science, disgust is manifestly irrational. People who are sickened by the thought of eating a disgusting object will say it is unsanitary or harmful. But they find a sterilized cockroach every bit as revolting as one fresh from the cupboard, and if the sterilized roach is briefly dunked into a beverage, they will refuse to drink it. People won't drink juice that has been stored in a brand-new urine collection bottle; hospital kitchens have found this an excellent way to stop pilferage. People won't eat soup if it is served in a brand-new bedpan or if it has been stirred with a new comb or flyswatter. You can't pay most people to eat fudge baked in the shape of dog feces or to hold rubber vomit from a novelty store between their lips. One's own saliva is not disgusting as long as it is in one's mouth, but most people won't eat from a bowl of soup into which they have spat.

Most Westerners cannot stomach the thought of eating insects, worms, toads, maggots, caterpillars, or grubs, but these are all highly nutritious and have been eaten by the majority of peoples throughout history. None of our rationalizations makes sense. You say that insects are contaminated because they touch feces or garbage? But many insects are quite sanitary. Termites, for example, just munch wood, but Westerners feel no better about eating them. Compare them with chickens, the epitome of palatability ("Try it—it tastes like chicken!"), which commonly eat garbage and feces. And we all savor tomatoes made plump and juicy from being fertilized with manure. Insects carry disease? So does all animal flesh. Just do what the rest of the world does—cook them. Insects have indigestible wings and legs? Pull them off, as you do with peel-and-eat shrimp, or stick to grubs and maggots. Insects taste bad? Here is a report from a British entomologist who was studying Laotian foodways and acquired a firsthand knowledge of his subject matter:

> None distasteful, a few quite palatable, notably the giant waterbug. For the most part they were insipid, with a faint vegetable flavour, but would not anyone tasting bread, for instance, for the first time, wonder why we eat such a flavourless food? A toasted dungbeetle or soft-bodied spider has a nice crisp exterior and soft interior of soufflé consistency which is by no means unpleasant. Salt is usually added, sometimes chili or the leaves of scented herbs, and sometimes they are eaten with rice or added to sauces or curry. Flavour is exceptionally hard to define, but lettuce would, I think, best describe the taste of termites, cicadas, and crickets; lettuce and raw potato that of the giant *Nephila* spider, and concentrated Gorgonzola cheese that of the giant waterbug (*Lethocerus indicus*). I suffered no ill effects from the eating of these insects.

The psychologist Paul Rozin has masterfully captured the psychology of disgust. Disgust is a fear of incorporating an offending substance into one's body. Eating is the most direct way to incorporate a substance, and as my camp song shows, it is the most horrific thought that a disgusting substance can arouse. Smelling or touching it is also unappealing. Disgust deters people from eating certain things, or, if it's too late, makes them spit or vomit them out. The facial expression says it all: the nose is wrinkled, constricting the nostrils, and the mouth is opened and the tongue pushed forward as if to squeegee offending material out.

Disgusting things come from animals. They include whole animals, parts of animals (particularly parts of carnivores and scavengers), and body products, especially viscous substances like mucus and pus and, most of all, feces, universally considered disgusting. Decaying animals and their parts are particularly revolting. In contrast, plants are sometimes distasteful, but distaste is different from disgust. When people avoid plant products—say, lima beans or broccoli—it is because they taste bitter or pungent. Unlike disgusting animal products, they are not felt to be unspeakably vile and polluting. Probably the most complicated thought anyone ever had about a disfavored vegetable was Clarence Darrow's: "I don't like spinach, and I'm glad I don't, because if I liked it I'd eat it, and I just hate it." Inorganic and non-nutritive stuff like sand, cloth, and bark are simply avoided, without strong feelings.

Not only are disgusting things always from animals, but things from animals are almost always disgusting. The nondisgusting animal parts are the exception. Of all the parts of all the animals in creation, people eat an infinitesimal fraction, and everything else is untouchable. Many Americans eat only the skeletal muscle of cattle, chickens, swine, and a few fish. Other parts, like guts, brains, kidneys, eyes, and feet, are beyond the pale, and so is any part of any animal not on the list: dogs, pigeons, jellyfish, slugs, toads, insects, and the other millions of animal species. Some Americans are even pickier, and are repulsed by the dark meat of chicken or chicken on the bone. Even adventurous eaters are willing to sample only a small fraction of the animal kingdom. And it is not just pampered Americans who are squeamish about unfamiliar animal parts. Napoleon Chagnon safeguarded his supply of peanut butter and hot dogs from his begging Yanomamö informants by telling them they were the feces and penises of cattle. The Yanomamö, who are hearty eaters of caterpillars and grubs, had no idea what cattle were but lost their appetite and left him to eat in peace.

A disgusting object contaminates everything it touches, no matter how brief the contact or how invisible the effects. The intuition behind not drinking a beverage that has been stirred with a flyswatter or dunked with a sterilized roach is that invisible contaminating bits—children call them cooties—have been left behind. Some objects, such as a new comb or bedpan, are tainted merely because they are designed to touch something disgusting, and others, such as a chocolate dog turd, are tainted by mere resemblance. Rozin observes that the psychology of disgust obeys the two laws of sympathetic magic—voodoo—found in many traditional cultures: the law of contagion (once in contact, always in contact) and the law of similarity (like produces like).

Though disgust is universal, the list of nondisgusting animals differs from culture to culture, and that implies a learning process. As every parent knows, children younger than two put everything in their mouths, and psychoanalysts have had a field day interpreting their lack of revulsion for feces. Rozin and his colleagues studied the development of disgust by offering children various foods that American adults find disgusting. To the horror of their onlooking parents, sixty-two percent of toddlers ate imitation dog feces ("realistically crafted from peanut butter and odorous cheese"), and thirty-one percent ate a grasshopper.

Rozin suggests that disgust is learned in the middle school-age years, perhaps when children are scolded by their parents or they see the look on their parents' faces when they approach a disgusting object. But I find that unlikely. First, all the subjects older than toddlers behaved virtually the same as the adults did. For example, four-year-olds wouldn't eat imitation feces or drink juice with a grasshopper in it; the only difference between them and the adults was that the children were less sensitive to contamination by brief contact. (Not until the age of eight did the children reject juice briefly dipped with a grasshopper or with imitation dog feces.) Second, children above the age of two are notoriously finicky, and their parents struggle to get them to eat new substances, not to avoid old ones. (The anthropologist Elizabeth Cashdan has documented that children's willingness to try new foods plummets after the third birthday.) Third, if children had to learn what to avoid, then all animals would be palatable except for the few that are proscribed. But as Rozin himself points out, all animals are disgusting except for a few that are permitted. No child has to be taught to revile greasy grimy gopher guts or mutilated monkey meat.

Cashdan has a better idea. The first two years, she proposes, are a sensitive period for learning about food. During those years mothers control children's food intake and children eat whatever they are permitted. Then their tastes spontaneously shrink, and they stomach only the foods they were given during the sensitive period. Those distastes can last to adulthood, though adults occasionally overcome them from a variety of motives: to dine with others, to appear macho or sophisticated, to seek thrills, or to avert starvation when familiar fare is scarce.

~

What is disgust for? Rozin points out that the human species faces "the omnivore's dilemma." Unlike, say, koalas, who mainly eat eucalyptus leaves and are vulnerable when those become scarce, omnivores choose from a vast menu of potential foods. The downside is that many are poison. Many fish, amphibians, and invertebrates contain potent neurotoxins. Meats that are ordinarily harmless can house parasites like tapeworms, and when they spoil, meats can be downright deadly, because the microorganisms that cause putrefaction release toxins to deter scavengers and thereby keep the meat for themselves. Even in industrialized countries food contamination is a major danger. Until recently anthrax and trichinosis were serious hazards, and today public health experts recommend draconian sanitary measures so people won't contract salmonella poisoning from their next chicken salad sandwich. In 1996 a world crisis was set off by the discovery that Mad Cow Disease, a pathology found in some British cattle that makes their brains spongy, might do the same to people who eat the cattle.

Rozin ventured that disgust is an adaptation that deterred our ancestors from eating dangerous animal stuff. Feces, carrion, and soft, wet animal parts are home to harmful microorganisms and ought to be kept outside the body. The dynamics of learning about food in childhood fit right in. Which animal parts are safe depends on the local species and their endemic diseases, so particular tastes cannot be innate. Children use their older relatives the way kings used food tasters: if they ate something and lived, it is not poison. Thus very young children are receptive to whatever their parents let them eat, and when they are old enough to forage on their own, they avoid everything else.

But how can one explain the irrational effects of similarity—the revulsion for rubber vomit, chocolate dog turds, and sterilized roaches? The answer is that these items were *crafted* to evoke the same reaction in people that the objects themselves evoke. That is why novelty shops *sell* rubber vomit. The similarity effect merely shows that reassurance by an authority or by one's own beliefs do not disconnect an emotional response. It is no more irrational than other reactions to modern simulacra, such as being engrossed by a movie, aroused by pornography, or terrified on a roller coaster.

What about our feeling that disgusting things contaminate everything they touch? It is a straightforward adaptation to a basic fact about the living world: germs multiply. Microorganisms are fundamentally different from chemical poisons such as those manufactured by plants. The danger of a chemical depends on its dose. Poisonous plants are bitter-tasting because both the plant and the plant-eater have an interest in the plant-eater stopping after the first bite. But there is no safe dose for a microorganism, because they reproduce exponentially. A single, invisible, untastable germ can multiply and quickly saturate a substance of any size. Since germs are, of course, transmittable by contact, it is no surprise that anything that touches a yucky substance is itself forever yucky, even if it looks and tastes the same. Disgust is intuitive microbiology.

Why are insects and other small creatures like worms and toads— what Latin Americans call "animalitos"—so easy to revile? The anthropologist Marvin Harris has shown that cultures avoid animalitos when larger animals are available, and eat them when they are not. The explanation has nothing to do with sanitation, since bugs are safer than meat. It comes from optimal foraging theory, the analysis of how animals ought to—and usually do—allocate their time to maximize the rate of nutrients they consume. Animalitos are small and dispersed, and it takes a lot of catching and preparing to get a pound of protein. A large mammal is hundreds of pounds of meat on the hoof, available all at once. (In 1978 a rumor circulated that McDonald's was extending the meat in Big Macs with earthworms. But if the corporation were as avaricious as the rumor was meant to imply, the rumor could not be true: worm meat is far more expensive than beef.) In most environments it is not only more efficient to eat larger animals, but the small ones should be avoided altogether— the time to gather them would be better spent hunting for a bigger payoff. Animalitos are thus absent from the diets of cultures that have bigger fish to fry, and since, in the minds of eaters, whatever is not permitted is forbidden, those cultures find them disgusting.

~

What about food taboos? Why, for example, are Hindus forbidden to eat beef? Why are Jews forbidden to eat pork and shellfish and to mix meat with milk? For thousands of years, rabbis have offered ingenious justifications of the Jewish dietary laws. Here are a few listed in the *Encyclopedia Judaica*:

From Aristeas, first century BC: "The dietary laws are ethical in intent, since abstention from the consumption of blood tames man's instinct for violence by instilling in him a horror of bloodshed. . . . The injunction against the consumption of birds of prey was intended to demonstrate that man should not prey on others."

From Isaac ben Moses Arama: "The reason behind all the dietary prohibitions is not that any harm may be caused to the body, but that these foods defile and pollute the soul and blunt the intellectual powers, thus leading to confused opinions and a lust for perverse and brutish appetites which lead men to destruction, thus defeating the purpose of creation."

From Maimonides: "All the food which the Torah has forbidden us to eat have some bad and damaging effect on the body. . . . The principal reason why the Law forbids swine's flesh is to be found in the circumstances that its habits and its food are very dirty and loathsome. . . . The fat of the intestines is forbidden because it fattens and destroys the abdomen and creates cold and clammy blood. . . . Meat boiled in milk is undoubtedly gross food, and makes a person feel overfull."

From Abraham ibn Ezra: "I believe it is a matter of cruelty to cook a kid in its mother's milk."

From Nahmanides: "Now the reason for specifying fins and scales is that fish which have fins and scales get nearer to the surface of the water and are found more generally in freshwater areas. . . . Those without fins and scales usually live in the lower muddy strata which are exceedingly moist and where there is no heat. They breed in musty swamps and eating them can be injurious to health."

With all due respect to rabbinical wisdom, these arguments can be demolished by any bright twelve-year-old, and as a former temple Sunday School teacher I can attest that they regularly are. Many Jewish adults still believe that pork was banned as a public health measure, to prevent trichinosis. But as Harris points out, if that were true the law would have been a simple advisory against undercooking pork: "Flesh of swine thou shalt not eat until the pink has been cooked from it."

Harris observes that food taboos often make ecological and economic sense. The Hebrews and the Muslims were desert tribes, and pigs are animals of the forest. They compete with people for water and nutritious foods like nuts, fruits, and vegetables. Kosher animals, in contrast, are ruminants like sheep, cattle, and goats, which can live off scraggly desert plants. In India, cattle are too precious to slaughter because they are used for milk, manure, and pulling plows. Harris' theory is as ingenious as the rabbis' and far more plausible, though he admits that it can't explain everything. Ancient tribes wandering the parched Judaean sands were hardly in danger of squandering their resources by herding shrimp and oysters, and it is unclear why the inhabitants of a Polish shtetl or a Brooklyn neighborhood should obsess over the feeding habits of desert ruminants.

Food taboos are obviously an ethnic marker, but by itself that observation explains nothing. Why do people wear ethnic badges to begin with, let alone a costly one like banning a source of nutrients? The social sciences assume without question that people submerge their interests to the group, but on evolutionary grounds that is unlikely (as we shall see later in the chapter). I take a more cynical view.

In any group, the younger, poorer, and disenfranchised members may be tempted to defect to other groups. The powerful, especially parents, have an interest in keeping them in. People everywhere form alliances by eating together, from potlatches and feasts to business lunches and dates. If I can't eat with you, I can't become your friend. Food taboos often prohibit a favorite food of a neighboring tribe; that is true, for example, of many of the Jewish dietary laws. That suggests that they are weapons to keep potential defectors in. First, they make the merest prelude to cooperation with outsiders—breaking bread together—an unmistakable act of defiance. Even better, they exploit the psychology of disgust. Taboo foods are absent during the sensitive period for learning food preferences, and that is enough to make children grow up to find them disgusting. That deters them from becoming intimate with the enemy ("He invited me over, but what will I do if they serve . . . EEEEU-UUW!!"). Indeed, the tactic is self-perpetuating because children grow up into parents who don't feed the disgusting things to *their* children. The practical effects of food taboos have often been noticed. A familiar theme in novels about the immigrant experience is the protagonist's torment over sampling taboo foods. Crossing the line offers a modicum of integration into the new world but provokes open conflict with parents and community. (In *Portnoy's Complaint*, Alex describes his mother as pronouncing *hamburger* as if it were *Hitler*.) But since the elders have no desire for the community to see the taboos in this light, they cloak them in talmudic sophistry and bafflegab.

THE SMELL OF FEAR

Language-lovers know that there is a word for every fear. Are you afraid of wine? Then you have *oenophobia*. Tremulous about train travel? You suffer from *siderodromophobia*. Having misgivings about your mother-in-law is *pentheraphobia*, and being petrified of peanut butter sticking to the roof of your mouth is *arachibutyrophobia*. And then there's Franklin Delano Roosevelt's affliction, the fear of fear itself, or *phobophobia*.

But just as not having a word for an emotion doesn't mean that it doesn't exist, having a word for an emotion doesn't mean that it does exist. Word-watchers, verbivores, and sesquipedalians love a challenge. Their idea of a good time is to find the shortest word that contains all the vowels in alphabetical order or to write a novel without the letter *e*. Yet another joy of lex is finding names for hypothetical fears. That is where these improbable phobias come from. Real people do not tremble at the referent of every euphonious Greek or Latin root. Fears and phobias fall into a short and universal list.

Snakes and spiders are always scary. They are the most common objects of fear and loathing in studies of college students' phobias, and have been so for a long time in our evolutionary history. D. O. Hebb found that chimpanzees born in captivity scream in terror when they first see a snake, and the primatologist Marc Hauser found that his laboratory-bred cotton-top tamarins (a South American monkey) screamed out alarm calls when they saw a piece of plastic tubing on the floor. The reaction of foraging peoples is succinctly put by Irven DeVore: "Hunter-gatherers will not suffer a snake to live." In cultures that revere snakes, people still treat them with great wariness. Even Indiana Jones was afraid of them!

The other common fears are of heights, storms, large carnivores, darkness, blood, strangers, confinement, deep water, social scrutiny, and leaving home alone. The common thread is obvious. These are the situations that put our evolutionary ancestors in danger. Spiders and snakes are often venomous, especially in Africa, and most of the others are obvious hazards to a forager's health, or, in the case of social scrutiny, status. Fear is the emotion that motivated our ancestors to cope with the dangers they were likely to face.

Fear is probably several emotions. Phobias of physical things, of social scrutiny, and of leaving home respond to different kinds of drugs, suggesting that they are computed by different brain circuits. The psychiatrist Isaac Marks has shown that people react in different ways to different frightening things, each reaction appropriate to the hazard. An animal triggers an urge to flee, but a precipice causes one to freeze. Social threats lead to shyness and gestures of appeasement. People really do faint at the sight of blood, because their blood pressure drops, presumably a response that would minimize the further loss of one's own blood. The best evidence that fears are adaptations and not just bugs in the nervous system is that animals that have evolved on islands without predators lose their fear and are sitting ducks for any invader—hence the expression "dead as a dodo."

Fears in modern city-dwellers protect us from dangers that no longer exist, and fail to protect us from dangers in the world around us. We ought to be afraid of guns, driving fast, driving without a seatbelt, lighter fluid, and hair dryers near bathtubs, not of snakes and spiders. Public safety officials try to strike fear in the hearts of citizens using everything from statistics to shocking photographs, usually to no avail. Parents scream and punish to deter their children from playing with matches or chasing a ball into the street, but when Chicago schoolchildren were asked what they were most afraid of, they cited lions, tigers, and snakes, unlikely hazards in the Windy City.

Of course, fears do change with experience. For decades psychologists thought that animals learn new fears the way Pavlov's dogs learned to salivate to a bell. In a famous experiment, John B. Watson, the founder of behaviorism, came up behind an eleven-month-old boy playing with a tame white rat and suddenly clanged two steel bars together. After a few more clangs, the boy became afraid of the rat and other white furry things, including rabbits, dogs, a sealskin coat, and Santa Claus. The rat, too, can learn to associate danger with a previously neutral stimulus. A rat shocked in a white room will flee it for a black room every time it is dumped there, long after the shocker has been unplugged.

But in fact creatures cannot be conditioned to fear just any old thing. Children are nervous about rats, and rats are nervous about bright rooms, before any conditioning begins, and they easily associate them with danger. Change the white rat to some arbitrary object, like opera glasses, and the child never learns to fear it. Shock the rat in a black room instead of a white one, and that nocturnal creature learns the association more slowly and unlearns it more quickly. The psychologist Martin Seligman suggests that fears can be easily conditioned only when the animal is evolutionarily prepared to make the association.

Few, if any, human phobias are about neutral objects that were once paired with some trauma. People dread snakes without ever having seen one. After a frightening or painful event, people are more prudent around the cause, but they do not fear it; there are no phobias for electrical outlets, hammers, cars, or air-raid shelters. Television clichés notwithstanding, most survivors of a traumatic event do not get the screaming meemies every time they face a reminder of it. Vietnam veterans resent the stereotype in which they hit the dirt whenever someone drops a glass.

A better way to understand the learning of fears is to think through the evolutionary demands. The world is a dangerous place, but our ancestors could not have spent their lives cowering in caves; there was food to gather and mates to win. They had to calibrate their fears of typical dangers against the actual dangers in the local environment (after all, not *all* spiders are poisonous) and against their own ability to neutralize the danger: their know-how, defensive technology, and safety in numbers.

Marks and the psychiatrist Randolph Nesse argue that phobias are innate fears that have never been unlearned. Fears develop spontaneously in children. In their first year, babies fear strangers and separation, as well they should, for infanticide and predation are serious threats to the tiniest hunter-gatherers. (The film *A Cry in the Dark* shows how easily a predator can snatch an unattended baby. It is an excellent answer to every parent's question of why the infant left alone in a dark bedroom is screaming bloody murder.) Between the ages of three and five, children become fearful of all the standard phobic objects—spiders, the dark, deep water, and so on—and then master them one by one. Most adult phobias are childhood fears that never went away. That is why it is city-dwellers who most fear snakes.

As with the learning of safe foods, the best guides to the local dangers are the people who have survived them. Children fear what they see their parents fear, and often unlearn their fears when they see other children coping. Adults are just as impressionable. In wartime, courage and panic are both contagious, and in some therapies, the phobic watches as an aide plays with a boa constrictor or lets a spider crawl up her arm. Even monkeys watch one another to calibrate their fear. Laboratory-raised rhesus macaques are not afraid of snakes when they first see them, but if they watch a film of another monkey being frightened by a snake, they fear it, too. The monkey in the movie does not instill the fear so much as awaken it, for if the film shows the monkey recoiling from a flower or a bunny instead of a snake, the viewer develops no fear.

The ability to conquer fear selectively is an important component of the instinct. People in grave danger, such as pilots in combat or Londoners during the blitz, can be remarkably composed. No one knows why some people can keep their heads when all about them are losing theirs, but the main calming agents are predictability, allies within shouting distance, and a sense of competence and control, which the writer Tom Wolfe called The Right Stuff. In his book by that name about the test pilots who became Mercury astronauts, Wolfe defined the right stuff as "the ability [of a pilot] to go up in a hurtling piece of machinery and put his hide on the line and then have the moxie, the reflexes, the experience, the coolness, to pull it back in the last yawning moment." That sense of control comes from "pushing the outside of the envelope": testing, in small steps, how high, how fast, how far one can go without bringing on disaster. Pushing the envelope is a powerful motive. Recreation, and the emotion called "exhilaration," come from enduring relatively safe events that look and feel like ancestral dangers. These include most noncompetitive sports (diving, climbing, spelunking, and so on) and the genres of books and movies called "thrillers." Winston Churchill once said, "Nothing in life is so exhilarating as to be shot at without result."

THE HAPPINESS TREADMILL

The pursuit of happiness is an inalienable right, says the Declaration of Independence in its list of self-evident truths. The greatest happiness of the greatest number, wrote Jeremy Bentham, is the foundation of morality. To say that everyone wants to be happy sounds trite, almost circular, but it raises a profound question about our makeup. What is this thing that people strive for?

At first happiness might seem like just desserts for biological fitness (more accurately, the states that would have led to fitness in the environment in which we evolved). We are happier when we are healthy, well-fed, comfortable, safe, prosperous, knowledgeable, respected, noncelibate, and loved. Compared to their opposites, these objects of striving are conducive to reproduction. The function of happiness would be to mobilize the mind to seek the keys to Darwinian fitness. When we are unhappy, we work for the things that make us happy; when we are happy, we keep the status quo.

The problem is, how much fitness is worth striving for? Ice Age people would have been wasting their time if they had fretted about their lack of camping stoves, penicillin, and hunting rifles or if they had striven for them instead of better caves and spears. Even among modern foragers, very different standards of living are attainable in different times and places. Lest the perfect be the enemy of the good, the pursuit of happiness ought to be calibrated by what can be attained through reasonable effort in the current environment.

How do we know what can reasonably be attained? A good source of information is what other people have attained. If they can get it, perhaps so can you. Through the ages, observers of the human condition have pointed out the tragedy: people are happy when they feel better off than their neighbors, unhappy when they feel worse off.

> But, O! how bitter a thing it is to look into happiness through another man's eyes!
> —William Shakespeare (*As You Like It*, V, ii).

> **Happiness**, n. An agreeable sensation arising from contemplating the misery of others.
> —Ambrose Bierce

> It is not enough to succeed. Others must fail.
> —Gore Vidal

> Ven frait zich a hoiker? Ven er zet a gresseren hoiker far zich. (When does a hunchback rejoice? When he sees one with a larger hump.)
> —Yiddish saying

Research on the psychology of happiness has borne out the curmudgeons. Kahneman and Tversky give an everyday example. You open your paycheck and are delighted to find you have been given a five percent raise—until you learn that your co-workers have been given a ten percent raise. According to legend, the diva Maria Callas stipulated that any opera house she sang in had to pay her one dollar more than the next highest paid singer in the company.

People today are safer, healthier, better fed, and longer-lived than at any time in history. Yet we don't spend our lives walking on air, and presumably our ancestors were not chronically glum. It is not reactionary to point out that many of the poor in today's Western nations live in conditions that yesterday's aristocrats could not have dreamed of. People in different classes and countries are often content with their lot until they compare themselves to the more affluent. The amount of violence in a society is more closely related to its inequality than to its poverty. In the second half of the twentieth century, the discontent of the Third World, and later the Second, have been attributed to their glimpses through the mass media of the First.

The other major clue to the attainable is how well off you are now. What you have now is attainable, by definition, and chances are you can do at least a little bit better. Evolutionary theory predicts that a man's reach should exceed his grasp, but not by much. Here we have the second tragedy of happiness: people adapt to their circumstances, good or bad, the way their eyes adapt to sun or darkness. From that neutral point, improvement is happiness, loss is misery. Again, the sages said it first. The narrator of E. A. Robinson's poem (and later Simon and Garfunkel's song) envies the factory owner, Richard Cory, who "glittered when he walked."

> So on we worked, and waited for the light,
> And went without the meat, and cursed the bread;
> And Richard Cory, one calm summer night,
> Went home and put a bullet through his head.

The futility of striving has led many dark souls to deny that happiness is possible. For the show-business personality Oscar Levant, "Happiness is not something you experience, it's something you remember." Freud said that the goal of psychotherapy was "to transform hysterical misery into common unhappiness." A colleague, consulting with me by email about a troubled graduate student, wrote, "sometimes i wish i was young then i remember that wasn't so great either."

But here the curmudgeons are only partly right. People do come to feel the same across an astonishing range of good and bad fortunes. But the baseline that people adapt to, on average, is not misery but satisfaction. (The exact baseline differs from person to person and is largely inherited.) The psychologists David Myers and Ed Diener have found that about eighty percent of people in the industrialized world report that they are at least "fairly satisfied with life," and about thirty percent say they are "very happy." (As far as we can tell, the reports are sincere.) The percentages are the same for all ages, for both sexes, for blacks and whites, and over four decades of economic growth. As Myers and Diener remark, "Compared with 1957, Americans have twice as many cars per person—plus microwave ovens, color TVs, VCRs, air conditioners, answering machines, and $12 billion worth of new brand-name athletic shoes a year. So, are Americans happier than they were in 1957? They are not."

Within an industrialized country, money buys only a little happiness: the correlation between wealth and satisfaction is positive but small. Lottery winners, after their jolt of happiness has subsided, return to their former emotional state. On the brighter side, so do people who have suffered terrible losses, such as paraplegics and survivors of the Holocaust.

These findings do not necessarily contradict the singer Sophie Tucker when she said, "I have been poor and I have been rich. Rich is better." In India and Bangladesh, wealth predicts happiness much better than it does in the West. Among twenty-four Western European and American nations, the higher the gross national product per capita, the happier the citizens (though there are many explanations). Myers and Diener point out that wealth is like health: not having it makes you miserable, but having it does not guarantee happiness.

The tragedy of happiness has a third act. There are twice as many negative emotions (fear, grief, anxiety, and so on) as positive ones, and losses are more keenly felt than equivalent gains. The tennis star Jimmy Connors once summed up the human condition: "I hate to lose more than I like to win." The asymmetry has been confirmed in the lab by showing that people will take a bigger gamble to avoid a sure loss than to improve on a sure gain, and by showing that people's mood plummets more when imagining a loss in their lives (for example, in course grades, or in relationships with the opposite sex) than it rises when imagining an equivalent gain. The psychologist Timothy Ketelaar notes that happiness tracks the effects of resources on biological fitness. As things get better, increases in fitness show diminishing returns: more food is better, but only up to a point. But as things get worse, decreases in fitness can take you out of the game: not enough food, and you're dead. There are many ways to become infinitely worse off (from an infection, starvation, getting eaten, a fall, ad infinitum) and not many ways to become vastly better off. That makes prospective losses more worthy of attention than gains; there are more things that make us unhappy than things that make us happy.

Donald Campbell, an early evolutionary psychologist who studied the psychology of pleasure, described humans as being on a "hedonic tread-mill," where gains in well-being leave us no happier in the long run. Indeed, the study of happiness often sounds like a sermon for traditional values. The numbers show that it is not the rich, privileged, robust, or good-looking who are happy; it is those who have spouses, friends, religion, and challenging, meaningful work. The findings can be overstated, because they apply to averages, not individuals, and because cause and effect are hard to tease apart: being married might make you happy, but being happy might help you get and stay married. But Campbell echoed millennia of wise men and women when he summed up the research: "The direct pursuit of happiness is a recipe for an unhappy life."

THE SIRENS' SONG

When we say that someone is led by emotion rather than reason, we often mean that the person sacrifices long-term interests for short-term gratification. Losing one's temper, surrendering to a seducer, blowing one's paycheck, and turning tail at the dentist's door are examples. What makes us so short-sighted?

The ability to defer a reward is called self-control or delay of gratification. Social scientists often treat it as a sign of intelligence, of the ability to anticipate the future and plan accordingly. But discounting the future, as economists call it, is part of the logic of choice for any agent that lives longer than an instant. Going for the quick reward instead of a distant payoff is often the rational strategy.

Which is better, a dollar now or a dollar a year from now? (Assume there is no inflation.) A dollar now, you might say, because you can invest it and have more than a dollar in a year. Unfortunately, the explanation is circular: the reason that interest exists in the first place is to pay people to give up the dollar that they would rather have now than a year from now. But economists point out that even if the explanation is misplaced, the answer is right: now really *is* better. First, a dollar now is available if a pressing need or opportunity arises in less than a year. Second, if you forgo the dollar now, you have no guarantee that you will get it back a year from now. Third, you might die within a year and never get to enjoy it. It is rational, therefore, to discount the future: to consume a resource now unless investing it brings a high enough return. The interest rate you should demand depends on how important the money is to you now, how likely you are to get it back, and how long you expect to live.

The struggle to reproduce is a kind of economy, and all organisms, even plants, must "decide" whether to use resources now or save them for the future. Some of these decisions are made by the body. We grow frail with age because our genes discount the future and build strong young bodies at the expense of weak old ones. The exchange pays off over the generations because an accident may cause the body to die before it gets old, in which case any sacrifice of vigor for longevity would have gone to waste. But most decisions about the future are made by the mind. At every moment we choose, consciously or unconsciously, between good things now and better things later.

Sometimes the rational decision is "now," particularly when, as the sayings go, life is short or there is no tomorrow. The logic is laid bare in firing-squad jokes. The condemned man is offered the ceremonial last cigarette and responds, "No thanks, I'm trying to quit." We laugh because we know it is pointless for him to delay gratification. Another old joke makes it clear why playing it safe is not always called for. Murray and Esther, a middle-aged Jewish couple, are touring South America. One day Murray inadvertently photographs a secret military installation, and soldiers hustle the couple off to prison. For three weeks they are tortured in an effort to get them to name their contacts in the liberation movement. Finally they are hauled in front of a military court, charged with espionage, and sentenced to death by firing squad. The next morning they are lined up in front of the wall and the sergeant asks them if they have any last requests. Esther wants to know if she can call her daughter in Chicago. The sergeant says that's not possible, and turns to Murray. "This is crazy," Murray shouts, "we're not spies!" and he spits in the sergeant's face. "Murray!" Esther cries. "Please! Don't make trouble!"

Most of the time we are pretty sure that we will not die in minutes. But we all die sometime, and we all risk forgoing the opportunity to enjoy something if we defer it too long. In our ancestors' nomadic lifestyle, without an ability to accumulate possessions or to count on long-lived social institutions like depositors' insurance, the payoffs for consumption must have been even higher. But even if they were not, *some* urge to indulge now had to have been built into our emotions. Most likely, we evolved a mechanism to estimate our longevity and the opportunities and risks posed by different choices (eating now or later, setting up camp or pushing on) and to tune the emotions accordingly.

The political scientist James Q. Wilson and the psychologist Richard Herrnstein have pointed out that many criminals act as if they discount the future steeply. A crime is a gamble whose payoff is immediate and whose possible cost comes later. They attributed the discounting to low intelligence. The psychologists Martin Daly and Margo Wilson have a different explanation. In the American inner cities, life expectancy for young males is low, and they know it. (In *Hoop Dreams*, the documentary about aspiring basketball players in a Chicago ghetto, there is an arresting scene in which the mother of one of the boys rejoices that he is alive on his eighteenth birthday.) Moreover, the social order and long-term ownership rights which would guarantee that investments are repaid are tenuous. These are precisely the circumstances in which steeply discounting the future—taking risks, consuming rather than investing—is adaptive.

More puzzling is *myopic* discounting: the tendency in all of us to prefer a large late reward to a small early one, but then to flip our preference as time passes and both rewards draw nearer. A familiar example is deciding before dinner to skip dessert (a small early reward) in order to lose weight (a large late one), but succumbing to temptation when the waiter takes the dessert orders. Myopic discounting is easy to produce in the lab: give people (or pigeons, for that matter) two buttons, one delivering a small reward now, the other delivering a larger reward later, and the subject will flip from choosing the large reward to choosing the small reward as the small one becomes imminent. The weakness of the will is an unsolved problem in economics and psychology alike. The economist Thomas Schelling asks a question about the "rational consumer" that can also be posed of the adapted mind:

> How should we conceptualize this rational consumer whom all of us know and who some of us are, who in self-disgust grinds his cigarettes down the disposal swearing that this time he means never again to risk orphaning his children with lung cancer and is on the street three hours later looking for a store that's still open to buy cigarettes; who eats a high-calorie lunch knowing that he will regret it, does regret it, cannot understand how he lost control, resolves to compensate with a low-calorie dinner, eats a high-calorie dinner knowing he will regret it, and does regret it; who sits glued to the TV knowing that again tomorrow he'll wake early in a cold sweat unprepared for that morning meeting on which so much of his career depends; who spoils the trip to Disneyland by losing his temper when his children do what he knew they were going to do when he resolved not to lose his temper when they did it?

Schelling notes the strange ways in which we defeat our self-defeating behavior: putting the alarm clock across the room so we won't turn it off and fall back to sleep, authorizing our employers to put part of each paycheck away for retirement, placing tempting snacks out of reach, setting our watches five minutes ahead. Odysseus had his crewmates plug their ears with wax and tie him to the mast so he could hear the Sirens' alluring song and not steer the ship toward them and onto the rocks.

Though myopic discounting remains unexplained, Schelling captures something important about its psychology when he roots the paradox of self-control in the modularity of the mind. He observes that "people behave sometimes as if they had two selves, one who wants clean lungs and long life and another who adores tobacco, or one who wants a lean body and another who wants dessert, or one who yearns to improve himself by reading Adam Smith on self-command . . . and another who would rather watch an old movie on television. The two are in continual contest for control." When the spirit is willing but the flesh is weak, such as in pondering a diet-busting dessert, we can feel two very different kinds of motives fighting within us, one responding to sights and smells, the other to doctors' advice. What about when the rewards are of the same kind, like a dollar today versus two dollars tomorrow? Perhaps an imminent reward engages a circuit for dealing with sure things and a distant one a circuit for betting on an uncertain future. One outranks the other, as if the whole person was designed to believe that a bird in the hand is worth two in the bush. In the modern environment, with its reliable knowledge of the future, that often leads to irrational choices. But our ancestors might have done well to distinguish between what is definitely enjoyable now and what is conjectured or rumored to be more enjoyable tomorrow. Even today, the delay of gratification is sometimes punished because of the frailty of human knowledge. Retirement funds go bankrupt, governments break promises, and doctors announce that everything they said was bad for you is good for you and vice versa.

I AND THOU

Our most ardent emotions are evoked not by landscapes, spiders, roaches, or dessert, but by other people. Some emotions, such as anger, make us want to harm people; others, such as love, sympathy, and gratitude, make us want to help them. To understand these emotions, we first have to understand why organisms should be designed to help or to hurt one another.

Having seen nature documentaries, you may believe that wolves weed out the old and weak deer to keep the herd healthy, that lemmings commit suicide to prevent the population from starving, or that stags ram into each other for the right to breed so that the fittest individuals may perpetuate the species. The underlying assumption—that animals act for the good of the ecosystem, the population, or the species—seems to follow from Darwin's theory. If in the past there were ten populations of lemmings, nine with selfish lemmings who ate their groups into starvation and one in which some died so that others might live, the tenth group would survive and today's lemmings should be willing to make the ultimate sacrifice. The belief is widespread. Every psychologist who has written about the function of the social emotions has talked about their benefit to the group.

When people say that animals act for the good of the group, they seem not to realize that the assumption is in fact a radical departure from Darwinism and almost certainly wrong. Darwin wrote, "Natural selection will never produce in a being any structure more injurious than beneficial to that being, for natural selection acts solely by and for the good of each." Natural selection could select groups with selfless members only if each group could enforce a pact guaranteeing that all their members stayed selfless. But without enforcement, nothing could prevent a mutant or immigrant lemming from thinking, in effect, "To heck with this! I'll let everyone *else* jump off the cliff, and then enjoy the food they leave behind." The selfish lemming would reap the rewards of the others' selflessness without paying any costs himself. With that advantage, his descendants would quickly take over the population, even if the population as a whole was worse off. And that is the fate of any tendency toward sacrifice. Natural selection is the cumulative effect of the relative successes of different replicators. That means that it selects for the replicators that replicate best, namely, the selfish ones.

The inescapable fact that adaptations benefit the replicator was first articulated by the biologist George Williams and later amplified by Richard Dawkins in *The Selfish Gene*. Almost all evolutionary biologists now accept the point, though there are debates over other issues. Selection among groups is possible on paper, but most biologists doubt that the special circumstances that let it happen are ever found in the real world. Selection among branches of the tree of life is possible, but that has nothing to do with whether organisms are designed for unselfishness. Animals just don't care what happens to their group, species, or ecosystem. Wolves catch the old and weak deer because they are the easiest to catch. Hungry lemmings set out for better feeding grounds and sometimes fall or drown by accident, not suicide. Stags fight because each wants to breed, and one concedes when defeat is inevitable, or as part of a strategy that works on average against others playing the same strategy. Males who fight are wasteful to the group—indeed, males *in general* are wasteful to the group when they make up half of it, because a few studs could sire the next generation without eating half the food.

Biologists often describe these acts as self-interested behavior, but what causes behavior is the activity of the brain, especially the circuitry for emotions and other feelings. Animals behave selfishly because of how their emotion circuits are wired. My full stomach, my warmth, my orgasms, feel better to me than yours do, and I want mine, and will seek mine, more than yours. Of course, one animal cannot directly feel what's in another one's stomach, but it could feel it indirectly by observing the second animal's behavior. So it is an interesting psychological fact that animals usually don't experience other animals' observable well-being as their own pleasure. It is an even more interesting fact that they sometimes do.

~

Earlier I said that natural selection selects selfish replicators. If organisms were replicators, all organisms should be selfish. But organisms do not replicate. Your parents did not replicate when they had you, because you are not identical to either of them. The blueprint that made you—your set of genes—is not the same as the blueprint that made them. Their genes were shuffled, randomly sampled to make sperm and eggs, and combined with each other's during fertilization to create a new combination of genes and a new organism unlike them. The only things that actually replicated were the genes and fragments of genes whose copies made it into you, some of which you will in turn pass down to your children, and so on. In fact, even if your mother had cloned herself, she would not have replicated; only her genes would have. That is because any changes she underwent in her lifetime—losing a finger, acquiring a tattoo, having her nose pierced—were not passed on to you. The only change you could have inherited was a mutation of one of the genes in the egg that was to become you. Genes, not bodies, replicate, and that means that genes, not bodies, should be selfish.

DNA, of course, has no feelings; "selfish" means "acting in ways that make one's own replication more likely." The way for a gene to do that in an animal with a brain is to wire the brain so that the animal's pleasures and pains cause it to act in ways that lead to more copies of the gene. Often that means causing an animal to enjoy the states that make it survive and reproduce. A full belly is satisfying because full bellies keep animals alive and moving and reproducing, leading to more copies of the genes that build brains that make full bellies feel satisfying.

By building a brain that makes eating fun, a gene helps to spread copies of itself lying in the animal's gonads. The actual DNA that helps build a brain, of course, doesn't itself get passed into the egg or sperm; only the copies of the gene inside the gonads do. But here is an important twist. The genes in an animal's gonads are not the *only* extant copies of the brain-building genes; they are merely the most convenient ones for the brain-building gene to help replicate. *Any* copy capable of replicating, anywhere in the world, is a legitimate target, if it can be identified and if steps can be taken to help it replicate. A gene that worked to replicate copies of itself inside some *other* animal's gonads could do as well as a gene that worked to replicate copies of itself inside *its own* animal's gonads. As far as the gene is concerned, a copy is a copy; which animal houses it is irrelevant. To a brain-building gene, the only thing special about that animal's gonads is the *certainty* that copies of the gene will be found in those gonads (the certainty comes from the fact that the cells in an animal's body are genetic clones). That is why the brain-building genes make animals enjoy their own well-being so much. If a gene could build a brain that could tell when copies of itself were sitting in *another* animal's gonads, it would make the brain enjoy the *other* animal's well-being, and make it act in ways that increased that other animal's well-being.

When does a copy of a gene in one animal also sit inside another? When the animals are related. In most animals there is a one-in-two chance that any gene in a parent will have a copy lying inside its offspring, because offspring get half their genes from each parent. There is also a one-in-two chance that a copy is lying inside a full sibling, because full siblings inherit their genes from the same pair of parents. There is a one-in-eight chance that a copy is lying inside a first cousin, and so on. A gene that built a brain that made its owner help its relatives would indirectly help to replicate itself. The biologist William Hamilton noted that if the benefit to the relative, multiplied by the probability that a gene is shared, exceeds the cost to the animal, that gene would spread in the population. Hamilton developed and formalized an idea that had been entertained by several other biologists as well, most famously in a wisecrack by the biologist J. B. S. Haldane when he was asked if he would lay down his life for his brother. "No," he said, "but for two brothers or eight cousins."

When an animal behaves to benefit another animal at a cost to itself, biologists call it altruism. When altruism evolves because the altruist is related to the beneficiary so the altruism-causing gene benefits itself, they call it kin selection. But when we look into the psychology of the animal doing the behaving, we can give the phenomenon another name: love.

The essence of love is feeling pleasure in another's well-being and pain in its harm. These feelings motivate acts that benefit the loved one, like nurturing, feeding, and protecting. We now understand why many animals, including humans, love their children, parents, grandparents, grandchildren, siblings, aunts, uncles, nephews, nieces, and cousins: people helping relatives equals genes helping themselves. The sacrifices made for love are modulated by the degree of relatedness: people make more sacrifices for their children than for their nephews and nieces. They are modulated by the expected reproductive life of the beneficiary: parents sacrifice more for children, who have a longer life ahead of them, than children sacrifice for parents. And they are modulated by the beneficiary's own feelings of love. People love their grandmothers not because their grandmothers are expected to reproduce, but because their grandmothers love *them*, and love the rest of their family. That is, you help people who enjoy helping you and helping your relatives. That is also why men and women fall in love. The other parent of my child has as much of a genetic stake in the child as I do, so what is good for her is good for me.

Many people think that the theory of the selfish gene says that "animals try to spread their genes." That misstates the facts and it misstates the theory. Animals, including most people, know nothing about genetics and care even less. People love their children not because they want to spread their genes (consciously or unconsciously) but because they can't help it. That love makes them try to keep their children warm, fed, and safe. What is selfish is not the real motives of the person but the metaphorical motives of the genes that built the person. Genes "try" to spread *themselves* by wiring animals' brains so the animals love their kin and try to keep warm, fed, and safe.

The confusion comes from thinking of people's genes as their true self, and the motives of their genes as their deepest, truest, unconscious motives. From there it's easy to draw the cynical and incorrect moral that all love is hypocritical. That confuses the real motives of the person with the metaphorical motives of the genes. Genes are not puppetmasters; they acted as the recipe for making the brain and body and then they got out of the way. They live in a parallel universe, scattered among bodies, with their own agendas.

Most discussions of the biology of altruism are really not about the biology of altruism. It's easy to see why nature documentaries, with their laudable conservationist ethic, disseminate the agitprop that animals act in the interests of the group. One subtext is, Don't hate the wolf that just ate Bambi; he's acting for the greater good. The other is, Protecting the environment is nature's way; we humans had better shape up. The opposing theory of the selfish gene has been bitterly attacked out of the fear that it vindicates the philosophy of Gordon Gekko in *Wall Street*: greed is good, greed works. Then there are those who believe in selfish genes but urge us to face up to the sad truth: at heart, Mother Teresa is really selfish.

I think moralistic science is bad for morals and bad for science. Surely paving Yosemite is unwise, Gordon Gekko is bad, and Mother Teresa is good regardless of what came out in the latest biology journals. But I suppose it is only human to feel a *frisson* when learning about what made us what we are. So I offer a more hopeful way of reflecting on the selfish gene.

The body is the ultimate barrier to empathy. Your toothache simply does not hurt me the way it hurts you. But genes are not imprisoned in bodies; the same gene lives in the bodies of many family members at once. The dispersed copies of a gene call to one another by endowing bodies with emotions. Love, compassion, and empathy are invisible fibers that connect genes in different bodies. They are the closest we will ever come to feeling someone else's toothache. When a parent wishes she could take the place of a child about to undergo surgery, it is not the species or the group or her body that wants her to have that most unselfish emotion; it is her selfish genes.

～

Animals are nice not just to their relatives. The biologist Robert Trivers developed a suggestion from George Williams on how another kind of altruism could evolve (where altruism, again, is defined as behavior that benefits another organism at a cost to the behaver). Dawkins explains it with a hypothetical example. Imagine a species of bird that suffers from a disease-carrying tick and must spend a good deal of time removing them with its beak. It can reach every part of its body but the top of its head. Every bird would benefit if some other bird groomed its head. If the birds in a group all responded to the sight of a head presented to them by grooming it, the group would prosper. But what would happen if a mutant presented its head for grooming but never groomed anyone else? These freeloaders would be parasite-free, *and* could use the time they saved not grooming others to look for food. With that advantage they would eventually dominate the population, even if it made the group more vulnerable to extinction. The psychologist Roger Brown explains, "One can imagine a pathetic final act in which all birds on stage present to one another heads that none will groom."

But say a different, grudge-bearing mutant arose. This mutant groomed strangers, groomed birds that in the past had groomed it, but refused to groom birds that had refused to groom it. Once a few of them had gained a toehold, these grudgers could prosper, because they would groom one another and not pay the costs of grooming the cheaters. And once they were established, neither indiscriminate groomers nor cheaters could drive them out, though in some circumstances cheaters could lurk as a minority.

The example is hypothetical, illustrating how altruism among non-kin—what Trivers called reciprocal altruism—can evolve. It is easy to confuse the thought experiment with a real observation; Brown remarks, "When I have used the example in teaching, it has sometimes come back to me on exams as a real bird, often as 'Skinner's pigeons,' sometimes the black-headed gull, and once the robin." Some species do practice recip-rocal altruism, but not many, because it evolves only under special conditions. An animal must be able to grant a large benefit to another at a small cost to itself, and the roles must commonly reverse. The animals must devote part of their brains to recognizing each other as individuals (see Chapter 2), and, if repayment comes long after the favor, to remembering who helped them and who refused, and to deciding how to grant and withhold favors accordingly.

Humans are, of course, a brainy species, and are zoologically unusual in how often they help unrelated individuals (Chapter 3). Our lifestyles and our minds are particularly adapted to the demands of reciprocal altruism. People have food, tools, help, and information to trade. With language, information is an ideal trade good because its cost to the giver—a few seconds of breath—is minuscule compared with the benefit to the recipient. Humans are obsessed with individuals; remember the Blick twins from Chapter 2, one of whom bit a police officer but neither of whom could be punished because each benefited from reasonable doubt that he and not his twin did the deed. And the human mind is equipped with goal-setting demons that regulate the doling out of favors; as with kin-directed altruism, reciprocal altruism is behaviorist shorthand for a set of thoughts and emotions. Trivers and the biologist Richard Alexander have shown how the demands of reciprocal altruism are probably the source of many human emotions. Collectively they make up a large part of the moral sense.

The minimal equipment is a cheater-detector and a tit-for-tat strategy that begrudges a gross cheater further help. A gross cheater is one who refuses to reciprocate at all, or who returns so little that the altruist gets back less than the cost of the initial favor. Recall from Chapter 5 that Cosmides has shown that people do reason unusually well about cheaters. But the real intrigue begins with Trivers' observation that there is a more subtle way to cheat. A subtle cheater reciprocates enough to make it worth the altruist's while, but returns less than he is capable of giving, or less than the altruist would give if the situation were reversed. That puts the altruist in an awkward position. In one sense she is being ripped off. But if she insists on equity, the subtle cheater could break off the relationship altogether. Since half a loaf is better than none, the altruist is trapped. She does have one kind of leverage, though. If there are *other* trading partners in the group who don't cheat at all, or who cheat subtly but less stingily, she can give them her business instead.

The game has become more complicated. Selection favors cheating when the altruist will not find out or when she will not break off her altruism if she does find out. That leads to better cheater-detectors, which leads to more subtle cheating, which leads to detectors for more subtle cheating, which leads to tactics to get away with subtle cheating without being detected by the subtle-cheater-detectors, and so on. Each detector must trigger an emotion demon that sets up the appropriate goal—continuing to reciprocate, breaking off the relationship, and so on.

Here is how Trivers reverse-engineered the moralistic emotions as strategies in the reciprocity game. (His assumptions about the causes and consequences of each emotion are well supported by the literature in experimental social psychology and by studies of other cultures, though they are hardly necessary, as real-life examples no doubt will flood into mind.)

Liking is the emotion that initiates and maintains an altruistic partnership. It is, roughly, a willingness to offer someone a favor, and is directed to those who appear willing to offer favors back. We like people who are nice to us, and we are nice to people whom we like.

Anger protects a person whose niceness has left her vulnerable to being cheated. When the exploitation is discovered, the person classifies the offending act as unjust and experiences indignation and a desire to respond with moralistic aggression: punishing the cheater by severing the relationship and sometimes by hurting him. Many psychologists have remarked that anger has moral overtones; almost all anger is righteous anger. Furious people feel they are aggrieved and must redress an injustice.

Gratitude calibrates the desire to reciprocate according to the costs and benefits of the original act. We are grateful to people when their favor helps us a lot and has cost them a lot.

Sympathy, the desire to help those in need, may be an emotion for earning gratitude. If people are most grateful when they most need the favor, a person in need is an opportunity to make an altruistic act go farthest.

Guilt can rack a cheater who is in danger of being found out. H. L. Mencken defined *conscience* as "the inner voice which warns us that someone might be looking." If the victim responds by cutting off all future aid, the cheater will have paid dearly. He has an interest in preventing the rupture by making up for the misdeed and keeping it from happening again. People feel guilty about private transgressions because they may become public; confessing a sin before it is discovered is evidence of sincerity and gives the victim better grounds to maintain the relationship. *Shame*, the reaction to a transgression after it has been discovered, evokes a public display of contrition, no doubt for the same reason.

Lily Tomlin said, "I try to be cynical, but it's hard to keep up." Trivers notes that once these emotions evolved, people had an incentive to mimic them to take advantage of other people's reactions to the real thing. Sham generosity and friendship may induce genuine altruism in return. Sham moral anger when no real cheating took place may nonetheless win reparations. Sham guilt may convince a wronged party that the cheater has reformed his ways, even if cheating is about to resume. Feigning dire straits may evoke genuine sympathy. Sham sympathy which gives the appearance of helping may elicit real gratitude. Sham gratitude may mislead an altruist into expecting a favor to be reciprocated. Trivers notes that none of this hypocrisy need be conscious; indeed, as we shall see, it is most effective when it is not.

The next round in this evolutionary contest is, of course, developing an ability to discriminate between real emotions and sham emotions. We get the evolution of *trust* and *distrust*. When we see someone going through the motions of generosity, guilt, sympathy, or gratitude rather than showing signs of the genuine emotion, we lose the desire to cooperate. For example, if a cheater makes amends in a calculating manner rather than out of credible guilt, he may cheat again when circumstances allow him to get away with it. The search for signs of trustworthiness makes us into mind readers, alert for any twitch or inconsistency that betrays a sham emotion. Since hypocrisy is easiest to expose when people compare notes, the search for trustworthiness makes us avid consumers of gossip. In turn, our reputation becomes our most valuable possession, and we are motivated to protect (and inflate) it with conspicuous displays of generosity, sympathy, and integrity and to take umbrage when it is impugned.

Are you keeping up? The ability to guard against sham emotions can in turn be used as a weapon against real emotions. One can protect one's own cheating by imputing false motives to someone else—by saying that a person really isn't aggrieved, friendly, grateful, guilty, and so on, when she really is. No wonder Trivers was the first to propose that the expansion of the human brain was driven by a cognitive arms race, fueled by the emotions needed to regulate reciprocal altruism.

Like kin selection, reciprocal altruism has been condemned as painting, even condoning, a bleak picture of human motives. Is sympathy nothing but a cheap way to buy gratitude? Is niceness just a business tactic? Not at all. Go ahead and think the worst about the sham emotions. But the reason the real ones are felt is not that they are hoped to help the feeler; it is that they in fact helped the feeler's ancestors. And it's not just that you shouldn't visit the iniquities of the fathers upon the children; the fathers may never have been iniquitous to begin with. The first mutants who felt sympathy and gratitude may have prospered not by their own calculation but because the feelings made it worth their neighbors' while to cooperate with them. The emotions themselves may have been kind and heartfelt in every generation; indeed, once sham-emotion-detectors evolved, they would be most effective when they *are* kind and heartfelt. Of course, the genes are metaphorically selfish in endowing people with beneficent emotions, but who cares about the moral worth of deoxyribonucleic acid?

Many people still resist the idea that the moral emotions are designed by natural selection to further the long-term interests of individuals and ultimately their genes. Wouldn't it be better for everyone if we were built to enjoy what was best for the group? Companies wouldn't pollute, public service unions wouldn't strike, citizens would recycle bottles and take the bus, and those teenagers would stop ruining a quiet Sunday afternoon with their jet-skis.

Once again I think it is unwise to confuse how the mind works with how it would be nice for the mind to work. But perhaps some comfort may be taken in a different way of looking at things. Perhaps we should *rejoice* that people's emotions aren't designed for the good of the group. Often the best way to benefit one's group is to displace, subjugate, or annihilate the group next door. Ants in a colony are closely related, and each is a paragon of unselfishness. That's why ants are one of the few kinds of animal that wage war and take slaves. When human leaders have manipulated or coerced people into submerging their interests into the group's, the outcomes are some of history's worst atrocities. In *Love and Death*, Woody Allen's pacifist character is urged to defend the czar and Mother Russia with the dubious call to duty that under French rule he would have to eat croissants and rich food with heavy sauces. People's desire for a comfortable life for themselves, their family, and their friends may have braked the ambitions of many an emperor.

THE DOOMSDAY MACHINE

It is 1962, and you are the president of the United States. You have just learned that the Soviet Union has dropped an atomic bomb on New York. You know they will not attack again. In front of you is the phone to the Pentagon, the proverbial button, with which you can retaliate by bombing Moscow.

You are about to press the button. The nation's policy is to retaliate in kind against a nuclear attack. The policy was designed to deter attackers; if you don't follow through, the deterrent would have been a sham.

On the other hand, you are thinking, the damage has been done. Killing millions of Russians will not bring millions of dead Americans back to life. The bomb will add radioactive fallout to the atmosphere, harming your own citizens. And you will go down in history as one of the worst mass murderers of all time. Retaliation now would be sheer spite.

But then, it is precisely this line of thinking that emboldened the Soviets to attack. They *knew* that once the bomb fell you would have nothing to gain and much to lose by retaliating. They thought they were calling your bluff. So you had better retaliate to show them it wasn't a bluff.

But then again, what's the point of proving *now* that you weren't bluffing *then*? The present cannot affect the past. The fact remains that if you push the button, you will snuff out millions of lives for no reason.

But wait—the Soviets knew you would think it is pointless to prove you weren't bluffing after they tried to call your bluff. That's why they called your bluff. The very fact that you are thinking this way brought on the catastrophe—so you shouldn't think this way.

But not thinking this way *now* is too late . . .

You curse your freedom. Your predicament is that you have the choice to retaliate, and since retaliating is not in your interests, you may decide not to do it, exactly as the Soviets anticipated. If only you didn't *have* the choice! If only your missiles had been wired to a reliable nuclear-fireball-detector and went off automatically. The Soviets would not have dared to attack, because they would have known retaliation was certain.

This train of reasoning was taken to its logical conclusion in the novel and film *Dr. Strangelove*. A deranged American officer has ordered a nuclear bomber to attack the Soviet Union, and it cannot be recalled. The president and his advisors meet in the war room with the Soviet ambassador to persuade him, and by telephone the Soviet leader, that the imminent attack is an accident and that the Soviets should not retaliate. They learn it is too late. The Soviets had installed the Doomsday Machine: a network of underground nuclear bombs that is set off automatically if the country is attacked or if anyone tries to disarm it. The fallout will destroy all human and animal life on earth. They installed the machine because it was cheaper than pinpoint missiles and bombers, and because they feared the United States might be building one and wanted to prevent a Doomsday gap. President Muffley (played by Peter Sellers) confers with the country's top nuclear strategist, the brilliant Dr. Strangelove (played by Peter Sellers):

"But," Muffley said, "is it really possible for it to be triggered automatically and at the same time impossible to untrigger?"

. . . Doctor Strangelove said quickly, "But precisely. Mister President, it is not only possible, it is essential. That is the whole idea of this machine. Deterrence is the art of producing in the enemy the fear to attack. And so because of the automated and irrevocable decision-making process which rules out human meddling, the Doomsday Machine is terrifying, simple to understand, and completely credible and convincing." . . .

President Muffley said, "But this is fantastic, Doctor Strangelove. How can it be triggered automatically?"

Strangelove said, "Sir, it is remarkably simple to do that. When you merely wish to bury bombs there is no limit to the size. . . . After they are buried they are connected to a gigantic complex of computers. A specific and closely defined set of circumstances under which the bombs are to be exploded is programmed into the tape memory banks. . . ." Strangelove turned so he looked directly at [the Soviet Ambassador]. "There is only one thing I don't understand, Mister Ambassador. The whole point of the Doomsday Machine is lost if you keep it a secret. Why didn't you tell the world?"

[The ambassador] turned away. He said quietly but distinctly, "It was to be announced at the Party Congress on Monday. As you know, the Premier loves surprises."

The German-accented, leather-gloved, wheelchair-bound Dr. Strangelove, with his disconcerting tic of giving the Nazi salute, is one of cinema's all-time eeriest characters. He was meant to symbolize a kind of intellectual who until recently was prominent in the public's imagination: the nuclear strategist, paid to think the unthinkable. These men, who included Henry Kissinger (on whom Sellers based his portrayal), Herman Kahn, John von Neumann, and Edward Teller, were stereotyped as amoral nerds who cheerfully filled blackboards with equations about megadeaths and mutual assured destruction. Perhaps the scariest thing about them was their paradoxical conclusions—for example, that safety in the nuclear age comes from exposing one's cities and protecting one's missiles.

But the unsettling paradoxes of nuclear strategy apply to *any* conflict between parties whose interests are partly competing and partly shared. Common sense says that victory goes to the side with the most intelligence, self-interest, coolness, options, power, and clear lines of communication. Common sense is wrong. Each of these assets can be a liability in contests of strategy (as opposed to contests of chance, skill, or strength), where behavior is calculated by predicting what the other guy will do in response. Thomas Schelling has shown that the paradoxes are ubiquitous in social life. We shall see that they offer great insight into the emotions, particularly the headstrong passions that convinced the Romantics that emotion and reason were opposites. But first let's put the emotions aside and just examine the logic of conflicts of strategy.

Take bargaining. When two people haggle over a car or a house, a bargain is struck when one side makes the final concession. Why does he concede? Because he is sure she will not. The reason she won't concede is that she thinks he will concede. She thinks he will because she thinks he thinks she thinks he will. And so on. There always is a range of prices that the buyer and seller would both accept. Even if a particular price within that range is not the best price for one party, it is preferable to canceling the deal outright. Each side is vulnerable to being forced to settle for the worst acceptable price because the other side realizes that he or she would have no choice if the alternative was to reach no agreement at all. But when both parties can guess the range, *any* price within the range is a point from which at least one party would have been willing to back off, and the other party knows it.

Schelling points out that the trick to coming out ahead is "a voluntary but irreversible sacrifice of freedom of choice." How do you persuade someone that you will not pay more than $16,000 for a car that is really worth $20,000 to you? You can make a public, enforceable $5,000 bet with a third party that you won't pay more than $16,000. As long as $16,000 gives the dealer a profit, he has no choice but to accept. Persuasion would be futile; it's against your interests to compromise. By tying your own hands, you improve your bargaining position. The example is fanciful, but real ones abound. The dealer appoints a salesperson who is not authorized to sell at less than a certain price even if he says he wants to. A homebuyer cannot get a mortgage if the bank's appraiser says he paid too much. The homebuyer exploits that powerlessness to get a better price from the seller.

Not only can power be a liability in conflicts of strategy, communication can be, too. When you are haggling from a pay phone with a friend about where to meet for dinner, you can simply announce that you will be at Ming's at six-thirty and hang up. The friend has to accede if she wants to meet you at all.

Paradoxical tactics also enter into the logic of promises. A promise can secure a favor only when the beneficiary of the promise has good reason to believe it will be carried out. The promiser is thus in a *better* position when the beneficiary knows that the promiser is *bound* by his promise. The law gives companies the right to sue and the right to be sued. The right to be sued? What kind of "right" is that? It is a right that confers the power to make a promise: to enter into contracts, borrow money, and engage in business with someone who might be harmed as a result. Similarly, the law that empowers banks to foreclose on a mortgage makes it worth the bank's while to grant the mortgage, and so, paradoxically, benefits the *borrower*. In some societies, Schelling notes, eunuchs got the best jobs because of what they could not do. How does a hostage persuade his kidnapper not to kill him to prevent him from identifying the kidnapper in court? One option is to deliberately blind himself. A better one is to confess to a shameful secret that the kidnapper can use as blackmail. If he has no shameful secret, he can create one by having the kidnapper photograph him in some unspeakably degrading act.

Threats, and defenses against threats, are the arena in which Dr. Strangelove really comes into his own. There are boring threats, in which the threatener has an interest in carrying out the threat—for example, when a homeowner threatens a burglar that she will call the police. The fun begins when carrying out the threat is costly to the threatener, so its value is only as a deterrent. Again, freedom is costly; the threat is credible only when the threatener has no choice but to carry it out and the target knows it. Otherwise, the target can threaten the threatener right back by refusing to comply. The Doomsday Machine is an obvious example, though the secrecy defeated its purpose. A hijacker who threatens to blow up a plane if anyone tries to disarm him will have a better chance of seeing Cuba if he wears explosives that go off with the slightest jostling. A good way to win the teenagers' game of chicken, in which two cars approach each other at high speed and the first driver to swerve loses face, is to conspicuously remove your own steering wheel and throw it away.

With threats, as with promises, communication can be a liability. The kidnapper remains incommunicado after making the ransom demand so he cannot be persuaded to give up the hostage for a smaller ransom or a safe escape. Rationality is also a liability. Schelling points out that "if a man knocks at the back door and says that he will stab himself unless you give him $10, he is more likely to get the $10 if his eyes are bloodshot." Terrorists, kidnappers, hijackers, and dictators of small countries have an interest in appearing mentally unbalanced. An absence of self-interest is also an advantage. Suicide bombers are almost impossible to stop.

To defend yourself *against* threats, make it impossible for the threatener to make you an offer you can't refuse. Again, freedom, information, and rationality are handicaps. "Driver does not know combination to safe," says the sticker on the delivery truck. A man who is worried that his daughter may be kidnapped can give away his fortune, leave town and remain incommunicado, lobby for a law that makes it a crime to pay ransom, or break the hand with which he signs checks. An invading army may burn bridges behind it to make retreat impossible. A college president tells protesters he has no influence on the town police, and genuinely wants no influence. A racketeer cannot sell protection if the customer makes sure he is not at home when the racketeer comes around.

Because an expensive threat works both ways, it can lead to a cycle of self-incapacitation. Protesters attempt to block the construction of a nuclear power plant by lying down on the railroad tracks leading to the site. The engineer, being reasonable, has no choice but to stop the train. The railroad company counters by telling the engineer to set the throttle so that the train moves very slowly and then to jump out of the train and walk beside it. The protesters must scramble. Next time the protesters handcuff themselves to the tracks; the engineer does not dare leave the train. But the protesters must be certain the engineer sees them in enough time to stop. The company assigns the next train to a nearsighted engineer.

~

In these examples, many of them from Schelling, the paradoxical power comes from a physical constraint like handcuffs or an institutional constraint like the police. But strong passions can do the same thing. Say a bargainer publicly announces that he will not pay more than $16,000 for the car, and everyone knows he could not tolerate the shame of going back on his word. The unavoidable shame is as effective as the enforceable bet, and he will get the car at his price. If Mother Teresa offered to sell you her car, you would not insist on a guarantee because presumably she is constitutionally incapable of cheating you. The hothead who can figuratively explode at any moment enjoys the same tactical advantage as the hijacker who can literally explode at any moment. In *The Maltese Falcon*, Sam Spade (Humphrey Bogart) dares the henchmen of Kasper Gutman (Sidney Greenstreet) to kill him, knowing that they need him to retrieve the falcon. Gutman replies, "That's an attitude, sir, that calls for the most delicate judgment on both sides, because as you know, sir, in the heat of action men are likely to forget where their best interests lie, and let their emotions carry them away." In *The Godfather*, Vito Corleone tells the heads of the other crime families, "I'm a superstitious man. And if some unlucky accident should befall my son, if my son is struck by a bolt of lightning, I will blame some of the people here."

Dr. Strangelove meets The Godfather. Is passion a doomsday machine? People consumed by pride, love, or rage have lost control. They may be irrational. They may act against their interests. They may be deaf to appeals. (The man running amok calls to mind a doomsday machine that has been set off.) But though this be madness, yet there is method in it. Precisely these sacrifices of will and reason are effective tactics in the countless bargains, promises, and threats that make up our social relations.

The theory stands the Romantic model on its head. The passions are no vestige of an animal past, no wellspring of creativity, no enemy of the intellect. The intellect is designed to relinquish control to the passions so that they may serve as guarantors of its offers, promises, and threats against suspicions that they are lowballs, double-crosses, and bluffs. The apparent firewall between passion and reason is not an ineluctable part of the architecture of the brain; it has been programmed in deliberately, because only if the passions are in control can they be credible guarantors.

The doomsday-machine theory has been proposed independently by Schelling, Trivers, Daly and Wilson, the economist Jack Hirshleifer, and the economist Robert Frank. Righteous anger, and the attendant thirst for redress or vengeance, is a credible deterrent if it is uncontrollable and unresponsive to the deterrer's costs. Such compulsions, though useful in the long run, can drive people to fight far out of proportion to the stakes. In 1982 Argentina annexed the British colony of the Falklands, desolate islands with virtually no economic or strategic importance. In earlier decades it might have made sense for Britain to defend them as an immediate deterrent to anyone with designs on the rest of its empire, but at that point there was no empire left to defend. Frank points out that for what they spent to reclaim the islands, Britain could have given each Falklander a Scottish castle and a lifetime pension. But most Britons were proud that they stood up to the Argentinians. The same sense of fairness makes us sue expensively for small amounts or seek a refund for a defective product despite red tape that costs us more in lost wages than the product was worth.

The lust for revenge is a particularly terrifying emotion. All over the world, relatives of the slain fantasize day and night about the bittersweet moment when they might avenge a life with a life and find peace at last. The emotion strikes us as primitive and dreadful because we have contracted the government to settle our scores for us. But in many societies an irresistible thirst for vengeance is one's only protection against deadly raids. Individuals may differ in the resolve with which they will suffer costs to carry out vengeance. Since that resolve is an effective deterrent only if it is advertised, it is accompanied by the emotion traditionally referred to as honor: the desire to publicly avenge even minor trespasses and insults. The hair-trigger of honor and revenge can be tuned to the degree of threat in the environment. Honor and vengeance are raised to godly virtues in societies that lie beyond the reach of law enforcement, such as remote horticulturalists and herders, the pioneers of the Wild West, street gangs, organized crime families, and entire nation-states when dealing with one another (in which case the emotion is called "patriotism"). But even within a modern state society where it serves no purpose, the emotion of vengeance cannot easily be turned off. Most legal theories, even from the highest-minded philosophers, acknowledge that retribution is one of the legitimate goals of criminal punishment, over and above the goals of deterring potential criminals and incapacitating, deterring, and rehabilitating the offender. Enraged crime victims, long disenfranchised from the American legal system, have recently pressed for a say in plea-bargaining and sentencing decisions.

~

As Strangelove explained, the whole point of a doomsday machine is lost if you keep it a secret. That principle may explain one of the longest-standing puzzles of the emotions: why we advertise them on our face.

Darwin himself never argued that facial expressions were naturally selected adaptations. In fact, his theory was downright Lamarckian. Animals have to move their faces for practical reasons: they bare the teeth to bite, widen the eyes for a panoramic view, and pull back the ears to protect them in a fight. These measures turned into habits that the animal performed when it merely anticipated an event. The habits were then passed to their offspring. It may seem strange that Darwin was no Darwinian in one of his most famous books, but remember that Darwin was fighting on two fronts. He had to explain adaptations to satisfy his fellow biologists, but he also made much of pointless features and animal vestiges in humans to combat creationists, who argued that functional design was a sign of God's handiwork. If God had really designed humans from scratch, Darwin asked, why would he have installed features that are useless to us but similar to features that are useful to animals?

Many psychologists still can't understand why broadcasting one's emotional state might be beneficial. Wouldn't the proverbial smell of fear just egg on one's enemies? One psychologist has tried to revive an old idea that facial muscles are tourniquets that send more blood to the parts of the brain that have to cope with the current challenge. Aside from being hydraulically improbable, the theory cannot explain why we are more expressive when there are other people around.

But if the passionate emotions are guarantors of threats and promises, advertising is their reason for being. But here a problem arises. Remember that real emotions create a niche for sham emotions. Why whip yourself into a rage when you can *simulate* a rage, deter your enemies, and not pay the price of pursuing dangerous vengeance if it fails? Let *others* be doomsday machines, and you can reap the benefits of the terror they sow. Of course, when counterfeit facial expressions begin to drive out the real ones, people call each other's bluffs, and the facial expressions, real and fake, become worthless.

Facial expressions are useful only if they are hard to fake. As a matter of fact, they *are* hard to fake. People don't really believe that the grinning flight attendant is happy to see them. That is because a social smile is formed with a different configuration of muscles from the genuine smile of pleasure. A social smile is executed by circuits in the cerebral cortex that are under voluntary control; a smile of pleasure is executed by circuits in the limbic system and other brain systems and is involuntary. Anger, fear, and sadness, too, recruit muscles that can't be controlled voluntarily, and the genuine expressions are hard to fake, though we can pantomime an approximation. Actors must simulate facial expressions for a living, but many cannot avoid a mannered look. Some great actors, like Laurence Olivier, are highly coordinated athletes who have doggedly learned to control every muscle. Others learn method acting, inspired by Konstantin Stanislavsky, in which actors make themselves *feel* an emotion by remembering or imagining a charged experience, and the expression pops on the face reflexively.

The explanation is incomplete, because it raises another question: *why* did we never evolve the ability to control our expressions? You can't just say that it would hurt everyone if counterfeit expressions were circulated. True enough, but in a world of honest emoters the faker would prosper, so fakers should always drive out emoters. I don't know the answer, but there are obvious places to look. Zoologists worry about the same problem: how can honest animal signals, like cries, gestures, and advertisements of health, evolve in a world of would-be fakers? One answer is that honest signals can evolve if they are too expensive to fake. For example, only a healthy peacock can afford a splendiferous tail, so healthy peacocks bear the burden of a cumbersome tail as a display of conspicuous consumption that only they can afford. When the healthiest peacocks display, the less healthy ones have no choice but to follow, because if they hide their health altogether the peahens will assume the worst, namely that they are at death's door.

Is there anything about emotional expressions that would make it inherently costly to put them under voluntary control? Here is a guess. In designing the rest of the human, natural selection had good engineering reasons to segregate the voluntary, cognitive systems from the systems that control housekeeping and physical-plant functions such as the regulation of heartbeat, breathing rate, blood circulation, sweat, tears, and saliva. None of your conscious beliefs are pertinent to how fast your heart ought to beat, so there's no point in letting you control it. In fact, it would be downright dangerous, since you might forget to pump when you got distracted, or you might try out your own harebrained ideas on what the best pulse rate should be.

Now, say selection handcuffed each emotion to a physiological control circuit, and the activity of the circuit was visible to an observer as flushing, blushing, blanching, sweating, trembling, quavering, croaking, weeping, and the facial reflexes Darwin discussed. An observer would have good reason to believe that the emotion was genuine, since a person could not fake it unless he had voluntary control of his heart and other organs. Just as the Soviets would have wanted to show everyone the wiring of the Doomsday Machine to *prove* that it was automatic and irreversible and their description of it no bluff, people might have an interest in showing everyone that an emotion is holding their body hostage and their angry words are no bluff. If so, it would explain why emotions are so intimately tied to the body, a fact that puzzled William James and a century of psychologists after him.

The handcuffing may have been easy for natural selection, because the major human emotions seem to have grown out of evolutionary precursors (anger from fighting, fear from fleeing, and so on), each of which engaged a suite of involuntary physiological responses. (This might be the grain of truth in the Romantic and triune-brain theories: modern emotions may *exploit* the involuntariness of older reflexes, even if they did not inherit it by default.) And once the handcuffs were in place for honest emoters, everyone else would have had little choice but to don them too, like the unhealthy peacocks forced to muster tails. A chronic poker face would suggest the worst: that the emotions a person declares in word and deed are shams.

This theory is unproven, but no one can deny the phenomenon. People are vigilant for sham emotions and put the most faith in involuntary physiological giveaways. That underlies an irony of the telecommunications age. Long-distance phone service, electronic mail, faxes, and videoconferencing should have made the face-to-face business meeting obsolete. But meetings continue to be a major expense for corporations and support entire industries like hotels, airlines, and rental cars. Why do we insist on doing business in the flesh? Because we do not trust someone until we see what makes him sweat.

FOOLS FOR LOVE

Why does romantic love leave us bewitched, bothered, and bewildered? Could it be another paradoxical tactic like handcuffing oneself to railroad tracks? Quite possibly. Offering to spend your life and raise children with someone is the most important promise you'll ever make, and a promise is most credible when the promiser can't back out. Here is how the economist Robert Frank has reverse-engineered mad love.

Unsentimental social scientists and veterans of the singles scene agree that dating is a marketplace. People differ in their value as potential marriage partners. Almost everyone agrees that Mr. or Ms. Right should be good-looking, smart, kind, stable, funny, and rich. People shop for the most desirable person who will accept them, and that is why most marriages pair a bride and a groom of approximately equal desirability. Mate-shopping, however, is only part of the psychology of romance; it explains the statistics of mate choice, but not the final pick.

Somewhere in this world of five billion people there lives the best-looking, richest, smartest, funniest, kindest person who would settle for you. But your dreamboat is a needle in a haystack, and you may die single if you insist on waiting for him or her to show up. Staying single has costs, such as loneliness, childlessness, and playing the dating game with all its awkward drinks and dinners (and sometimes breakfasts). At some point it pays to set up house with the best person you have found so far.

But that calculation leaves your partner vulnerable. The laws of probability say that someday you will meet a more desirable person, and if you are always going for the best you can get, on that day you will dump your partner. But your partner has invested money, time, childrearing, and forgone opportunities in the relationship. If your partner was the most desirable person in the world, he or she would have nothing to worry about, because you would never want to desert. But failing that, the partner would have been foolish to enter the relationship.

Frank compares the marriage market with the rental market. Landlords desire the best of all tenants but settle for the best they can find, and renters want the best of all apartments but settle for the best they can find. Each invests in the apartment (the landlord may paint it the tenant's favorite color; the tenant may install permanent decorations), so each would be harmed if the other suddenly terminated the agreement. If the tenant could leave for a better flat, the landlord would have to bear the costs of an unrented unit and the search for a new tenant; he would have to charge a high rent to cover that risk, and would be loath to paint. If the landlord could evict the tenant for a better one, the tenant would have to search for a new home; she would be willing to pay only a low rent, and would not bother to keep the apartment in good shape, if she had to expose herself to that risk. If the best tenant were renting the best apartment, the worries would be moot; neither would want to end the arrangement. But since both have to compromise, they protect themselves by signing a lease that is expensive for either to break. By agreeing to restrict his own freedom to evict, the landlord can charge a higher rent. By agreeing to restrict her own freedom to leave, the tenant can demand a lower rent. Lack of choice works to each one's advantage.

Marriage laws work a bit like leases, but our ancestors had to find some way to commit themselves before the laws existed. How can you be sure that a prospective partner won't leave the minute it is rational to do so—say, when a 10-out-of-10 moves in next door? One answer is, don't accept a partner who wanted you for rational reasons to begin with; look for a partner who is committed to staying with you because you are you. Committed by what? Committed by an emotion. An emotion that the person did not decide to have, and so cannot decide not to have. An emotion that was not triggered by your objective mate-value and so will not be alienated by someone with greater mate-value. An emotion that is guaranteed not to be a sham because it has physiological costs like tachycardia, insomnia, and anorexia. An emotion like romantic love.

"People who are sensible about love are incapable of it," wrote Douglas Yates. Even when courted by the perfect suitor, people are unable to will themselves to fall in love, often to the bewilderment of the matchmaker, the suitor, and the person himself or herself. Instead it is a glance, a laugh, a manner that steals the heart. Remember from Chapter 2 that spouses of one twin are not attracted to the other; we fall in love with the individual, not with the individual's qualities. The upside is that when Cupid does strike, the lovestruck one is all the more credible in the eyes of the object of desire. Murmuring that your lover's looks, earning power, and IQ meet your minimal standards would probably kill the romantic mood, even though the statement is statistically true. The way to a person's heart is to declare the opposite—that you're in love because you can't help it. Tipper Gore's Parents' Music Resource Center notwithstanding, the sneering, body-pierced, guitar-smashing rock musician is typically not singing about drugs, sex, or Satan. He is singing about love. He is courting a woman by calling attention to the irrationality, uncontrollability, and physiological costs of his desire. I want you so bad, it's driving me mad, Can't eat, can't sleep, Heart beats like a big bass drum, You're the only one, Don't know why I love you like I do, You drive me crazy, Can't stop lovin' you, Ain't nobody can do it to me the way you can, I like the way you walk, I like the way you talk, et cetera, et cetera.

Of course, one can well imagine a woman not being swept off her feet by these proclamations. (Or a man, if it is a woman doing the declaring.) They set off a warning light in the other component of courtship, smart shopping. Groucho Marx said that he would not belong to any club that would have him as a member. Usually people do not want any suitor who wants them too badly too early, because it shows that the suitor is desperate (so they should wait for someone better), and because it shows that the suitor's ardor is too easily triggered (hence too easily triggerable by someone else). The contradiction of courtship—flaunt your desire while playing hard to get—comes from the two parts of romantic love: setting a minimal standard for candidates in the mate market, and capriciously committing body and soul to one of them.

THE SOCIETY OF FEELINGS

Mental life often feels like a parliament within. Thoughts and feelings vie for control as if each were an agent with strategies for taking over the whole person, you. Might our mental agents use paradoxical tactics with one another—handcuffs, doomsday machines, unbreakable contracts with third parties? The analogy is imperfect because natural selection designs people to compete but does not design organs, including mental agents, to compete; the interests of the whole person are paramount. But the whole person has many goals, like food, sex, and safety, and that requires a division of labor among mental agents with different priorities and kinds of expertise. The agents are bound by an entente that benefits the whole person over a lifetime, but over the short term the agents may outwit one another with devious tactics.

Self-control is unmistakably a tactical battle between parts of the mind. Schelling observes that the tactics people use to control themselves are interchangeable with the tactics they use to control others. How do you prevent your child from scratching his hives in his sleep? Put mittens on him. How do you prevent yourself from scratching your hives in your sleep? Put mittens on yourself. If Odysseus had not plugged his shipmates' ears, they would have done it on their own. The self that wants a trim body outwits the self that wants dessert by throwing out the brownies at the opportune moment when it is in control.

So we do seem to use paradoxical tactics against ourselves. The agent in control at one time makes a voluntary but irreversible sacrifice of freedom of choice for the whole body, and gets its way in the long run. That is the bright spot in this whole depressing discussion of selfish genes and doomsday machines. Social life is not always the equivalent of global thermonuclear war because the part of us with the longest view of the future, when in control of the body, can voluntarily sacrifice freedom of choice for the body at other times. We sign contracts, submit to laws, and hitch our reputations to public declarations of loyalty to friends and mates. These are not tactics to defeat someone else, but tactics to defeat the darker parts of ourselves.

One more speculation on the battle inside the head. No one knows what, if anything, grief is for. Obviously the loss of a loved one is unpleasant, but why should it be devastating? Why the debilitating pain that stops people from eating, sleeping, resisting diseases, and getting on with life? Jane Goodall describes a young chimp, Flint, who after the death of his beloved mother became depressed and died himself as if of a broken heart.

Some have suggested that grief is an enforced interlude for reassessment. Life will never be the same, so one must take time to plan how to cope with a world that has been turned upside down. Perhaps grief also gives people time to contemplate how a lapse of theirs may have allowed the death and how they might be more careful in the future. There may be an element of truth to the suggestion. Bereaved people find that they ache all over again every time they discover another habit to unlearn, like setting out an extra plate or buying groceries for two. And blaming oneself is a common symptom. But the pain of grief makes planning harder, not easier, and is too extreme and long-lasting to be useful as a strategy session.

William James wrote, "It takes a mind debauched by learning to carry the process of making the natural seem strange so far as to ask for the 'why' of any instinctive human act." Though legitimate to a scientist, the question "Why do we grieve?" is preposterous to common sense. If you didn't grieve when someone died, could you really have loved him when he was alive? It's logically possible but seems psychologically impossible; grief is the other side of love. And there may lie the answer. Perhaps grief is an internal doomsday machine, pointless once it goes off, useful only as a deterrent. What parents have not lain awake contemplating the horror of losing a child? Or worried themselves sick with awful images when a child is late or lost? These thoughts are powerful reminders to protect and cherish a loved one in the face of myriad other demands on one's time and thoughts. Like all deterrents, grief would be effective only if it is certain and terrible.

KIDDING OURSELVES

The playwright Jerome K. Jerome once said, "It is always the best policy to tell the truth, unless, of course, you are an exceptionally good liar." It's hard to be a good liar, even when it comes to your own intentions, which only you can verify. Intentions come from emotions, and emotions have evolved displays on the face and body. Unless you are a master of the Stanislavsky method, you will have trouble faking them; in fact, they probably evolved *because* they were hard to fake. Worse, lying is stressful, and anxiety has its own telltale markers. They are the rationale for polygraphs, the so-called lie detectors, and humans evolved to be lie detectors, too. Then there is the annoying fact that some propositions logically entail others. Since *some* of the things you say will be true, you are always in danger of exposing your own lies. As the Yiddish saying goes, a liar must have a good memory.

Trivers, pursuing his theory of the emotions to its logical conclusion, notes that in a world of walking lie detectors the best strategy is to believe your own lies. You can't leak your hidden intentions if you don't think that they *are* your intentions. According to his theory of self-deception, the conscious mind sometimes hides the truth from itself the better to hide it from others. But the truth is useful, so it should be registered somewhere in the mind, walled off from the parts that interact with other people. There is an obvious similarity to Freud's theory of the unconscious and the defense mechanisms of the ego (such as repression, projection, denial, and rationalization), though the explanation is completely different. George Orwell stated it in *1984:* "The secret of rulership is to combine a belief in one's own infallibility with a power to learn from past mistakes."

The neuroscientist Michael Gazzaniga has shown that the brain blithely weaves false explanations about its motives. Split-brain patients have had their cerebral hemispheres surgically disconnected as a treatment for epilepsy. Language circuitry is in the left hemisphere, and the left half of the visual field is registered in the isolated right hemisphere, so the part of the split-brain person that can talk is unaware of the left half of his world. The right hemisphere is still active, though, and can carry out simple commands presented in the left visual field, like "Walk" or "Laugh." When the patient (actually, the patient's left hemisphere) is asked why he walked out (which we know was a response to the command presented to the right hemisphere), he ingenuously replies, "To get a Coke." When asked why he is laughing, he says, "You guys come up and test us every month. What a way to make a living!"

Our confabulations, not coincidentally, present us in the best light. Literally hundreds of experiments in social psychology say so. The humorist Garrison Keillor describes the fictitious community of Lake Wobegon, "where the women are strong, the men are good-looking, and all the children are above average." Indeed, most people claim they are above average in any positive trait you name: leadership, sophistication, athletic prowess, managerial ability, even driving skill. They rationalize the boast by searching for an *aspect* of the trait that they might in fact be good at. The slow drivers say they are above average in safety, the fast ones that they are above average in reflexes.

More generally, we delude ourselves about how benevolent and how effective we are, a combination that social psychologists call beneffectance. When subjects play games that are rigged by the experimenter, they attribute their successes to their own skill and their failures to the luck of the draw. When they are fooled in a fake experiment into thinking they have delivered shocks to another subject, they derogate the victim, implying that he deserved the punishment. Everyone has heard of "reducing cognitive dissonance," in which people invent a new opinion to resolve a contradiction in their minds. For example, a person will recall enjoying a boring task if he had agreed to recommend it to others for paltry pay. (If the person had been enticed to recommend the task for generous pay, he accurately recalls that the task was boring.) As originally conceived of by the psychologist Leon Festinger, cognitive dissonance is an unsettled feeling that arises from an inconsistency in one's beliefs.

But that's not right: there is no contradiction between the proposition "The task is boring" and the proposition "I was pressured into lying that the task was fun." Another social psychologist, Eliot Aronson, nailed it down: people doctor their beliefs only to eliminate a contradiction with the proposition "I am nice and in control." Cognitive dissonance is always triggered by blatant evidence that you are not as beneficent and effective as you would like people to think. The urge to reduce it is the urge to get your self-serving story straight.

Sometimes we have glimpses of our own self-deception. When does a negative remark sting, cut deep, hit a nerve? When some part of us knows it is true. If every part knew it was true, the remark would not sting; it would be old news. If no part thought it was true, the remark would roll off; we could dismiss it as false. Trivers recounts an experience that is all too familiar (at least to me). One of his papers drew a published critique, which struck him at the time as vicious and unprincipled, full of innuendo and slander. Rereading the article years later, he was surprised to find that the wording was gentler, the doubts more reasonable, the attitude less biased than he had remembered. Many others have made such discoveries; they are almost the definition of "wisdom."

> If there were a verb meaning "to believe falsely," it would not have any significant first person, present indicative.
> —Ludwig Wittgenstein

> There's one way to find out if a man is honest: ask him; if he says yes, you know he's crooked.
> —Mark Twain

> Our enemies' opinion of us comes closer to the truth than our own.
> —François La Rochefoucauld

> Oh wad some power the giftie gie us
> To see oursels as ithers see us!
> —Robert Burns

No one can examine the emotions without seeing in them the source of much human tragedy. I don't think we should blame the animals; it's clear enough how natural selection engineered our instincts to suit our needs. We shouldn't blame selfish genes, either. They endow us with selfish motives, but they just as surely endow us with the capacity for love and a sense of justice. What we should appreciate and fear is the cunning designs of the emotions themselves. Many of their specs are not for gladness and understanding: think of the happiness treadmill, the Sirens' song, the sham emotions, the doomsday machines, the caprice of romance, the pointless punishment of grief. But self-deception is perhaps the cruelest motive of all, for it makes us feel right when we are wrong and emboldens us to fight when we ought to surrender. Trivers writes,

> Consider an argument between two closely bound people, say, husband and wife. Both parties believe that one is an altruist—of long standing, relatively pure in motive, and much abused—while the other is characterized by a pattern of selfishness spread over hundreds of incidents. They only disagree over who is altruistic and who selfish. It is noteworthy that the argument may appear to burst forth spontaneously, with little or no preview, yet as it rolls along, two whole landscapes of information processing appear to lie already organized, waiting only for the lightning of anger to show themselves.

In cartoons and movies, the villains are mustache-twirling degenerates, cackling with glee at their badness. In real life, villains are convinced of their rectitude. Many biographers of evil men start out assuming that their subjects are cynical opportunists and reluctantly discover that they are ideologues and moralists. If Hitler was an actor, concluded one, he was an actor who believed in the part.

Still, thanks to the complexity of our minds, we need not be perpetual dupes of our own chicanery. The mind has many parts, some designed for virtue, some designed for reason, some clever enough to outwit the parts that are neither. One self may deceive another, but every now and then a third self sees the truth.

7

FAMILY VALUES

Come on, people now, smile on your brother! Everybody get together, try to love one another right now. This is the dawning of the Age of Aquarius: harmony and understanding, sympathy and trust abounding; no more falsehoods or derisions, golden living dreams of visions, mystic crystal revelation, and the mind's true liberation. Imagine no possessions; I wonder if you can. No need for greed or hunger, a brotherhood of man. Imagine all the people sharing all the world. You may say I'm a dreamer, but I'm not the only one. I hope someday you'll join us, and the world will be as one.

Incredible as it may seem, many of us used to believe this treacle. A leading idea of the 1960s and 70s was that mistrust, jealousy, competitiveness, greed, and manipulation were social institutions due for reform. Some people thought they were unnecessary evils, like slavery or the denial of the vote to women. Others thought they were hidebound traditions whose inefficiency had gone unnoticed, as with the genius who figured out that toll bridges could charge a dollar to the traffic going one way instead of fifty cents to the traffic going both ways.

These sentiments came not just from rock musicians but from America's distinguished social critics. In his 1970 book *The Greening of America*, the Yale law professor Charles Reich heralded a nonviolent revolution being led by the college-age generation. The youth of America had evolved a new consciousness, he said. It was less guilty and anxious, nonjudgmental, noncompetitive, nonmaterialistic, affectionate, honest, unmanipulative, unaggressive, communal, and unconcerned with status and careers. The new consciousness, emerging like flowers through the pavement, was expressed in their music, communes, hitchhiking, drugs, moon-gazing, peace salute, and even their clothing. Bell-bottoms, he said, "give the ankles a special freedom as if to invite dancing right on the street." The new consciousness promised "a higher reason, a more human community, and a new and liberated individual. Its ultimate creation will be a new and enduring wholeness and beauty—a renewed relationship of man to himself, to other men, to society, to nature, and to the land."

Greening sold a million copies in a few months. It was serialized in the *New Yorker* and discussed in a dozen articles in the *New York Times* and in a volume of essays by the leading intellectuals of the day. John Kenneth Galbraith gave it a positive review (though with a caveat expressed in his title: "Who's Minding the Store?"). The book recently came out in a twenty-fifth anniversary edition.

Reich wrote his book in the Yale dining halls, and based it on his conversations with the students there. Those students, of course, were among the most privileged individuals in the history of humanity. With Mom and Dad paying the bills, everyone around them coming from the upper classes, and Ivy League credentials about to launch them into the expanding economy of the 1960s, it was easy to believe that all you need is love. After graduation day, Reich's generation became the Gucci-wearing, Beemer-driving, condo-owning, gourmet-baby-breeding urban professionals of the 1980s and 90s. Universal harmony was a style as ephemeral as the bell-bottoms, a status symbol that distanced them from rednecks, jocks, and the less hip preppies. As the post-60s rock musician Elvis Costello asked, "Was it a millionaire who said 'Imagine no possessions'?"

The Woodstock Nation was not the first utopian dream to be shattered. The free-love communes of nineteenth-century America collapsed from sexual jealousy and the resentment of both sexes over the leaders' habit of accumulating young mistresses. The socialist utopias of the twentieth century became repressive empires led by men who collected Cadillacs and concubines. In anthropology, one South Sea island paradise after another has turned out to be nasty and brutish. Margaret Mead said that nonchalant sex made the Samoans satisfied and free of crime; it turned out that the boys tutored one another in rape techniques. She called the Arapesh "gentle"; they were headhunters. She said that the Tshambuli reversed our sex roles, the men wearing curls and makeup. In fact the men beat their wives, exterminated neighboring tribes, and treated homicide as a milestone in a young man's life which entitled him to wear the face paint that Mead thought was so effeminate.

In *Human Universals*, the anthropologist Donald Brown has assembled the traits that as far as we know are found in all human cultures. They include prestige and status, inequality of power and wealth, property, inheritance, reciprocity, punishment, sexual modesty, sexual regulations, sexual jealousy, a male preference for young women as sexual partners, a division of labor by sex (including more child care by women and greater public political dominance by men), hostility to other groups, and conflict within the group, including violence, rape, and murder. The list should come as no surprise to anyone familiar with history, current events, or literature. There are a small number of plots in the world's fiction and drama, and the scholar Georges Polti claims to have listed them all. More than eighty percent are defined by adversaries (often murderous), by tragedies of kinship or love, or both. In the real world, our life stories are largely stories of conflict: the hurts, guilts, and rivalries inflicted by parents, siblings, children, spouses, lovers, friends, and competitors.

This chapter is about the psychology of social relations. The Age of Aquarius notwithstanding, that means it is largely about inborn motives that put us into conflict with one another. Given that our brains were shaped by natural selection, it could hardly be otherwise. Natural selection is driven by the competition among genes to be represented in the next generation. Reproduction leads to a geometric increase in descendants, and on a finite planet not every organism alive in one generation can have descendants several generations hence. Therefore organisms reproduce, to some extent, at one another's expense. If one organism eats a fish, that fish is no longer available to be eaten by another organism. If one organism mates with a second one, it denies an opportunity at parenthood to a third. Everyone alive today is a descendant of millions of generations of ancestors who lived under these constraints but reproduced nonetheless. That means that all people today owe their existence to having winners as ancestors, and everyone today is designed, at least in some circumstances, to compete.

That does *not* mean that people (or any other animals) house an aggressive urge that must be discharged, an unconscious death wish, a rapacious sex drive, a territorial imperative, a thirst for blood, or the other ruthless instincts that are often mistakenly equated with Darwinism. In *The Godfather*, Sollozzo says to Tom Hagen, "I don't like violence, Tom. I'm a businessman. Blood is a big expense." Even in the harshest competition, an intelligent organism must be a strategist, assessing whether its goals might best be served by retreat, conciliation, or living and letting live. As I explained in Chapter 5, it is genes, not organisms, that must compete or die; sometimes the genes' best strategy is to design organisms that cooperate, and yes, even smile on their brother and love one another. Natural selection does not forbid cooperation and generosity; it just makes them difficult engineering problems, like stereoscopic vision. The difficulty of building an organism to see in stereo has not prevented natural selection from installing stereo vision in humans, but we would never have come to understand stereo if we thought it just came free with having two eyes and failed to look for the sophisticated neural programs that accomplish it. Similarly, the difficulty of building an organism to cooperate and be generous has not prevented natural selection from installing cooperation and generosity in humans, but we will never understand these capacities if we think they just come free with living in groups. The on-board computers of social organisms, especially of humans, should run sophisticated programs that assess the opportunities and risks at hand and compete or cooperate accordingly.

The conflict of interest among the members of a species also does not call for a conservative political agenda, as journalists and social scientists often fear. Some worry that if our motives put us into conflict with others, exploitation and violence would be morally correct; since they are deplorable, conflict had better not be part of our nature. The reasoning, of course, is fallacious: nothing says that nature has to be nice, and what people want to do is not necessarily what they ought to do. Others worry that if conflicting motives are inevitable, it would be futile to try to reduce violence and exploitation; our current social arrangements would be the best one can hope for. But that does not follow either. Among modern Western societies, homicide rates vary from 0.5 per million persons per year in Iceland in the first half of the twentieth century, to 10 in most European countries at present, to 25 in Canada, to 100 in the United States and Brazil. There is plenty of room for practical measures that could reduce the murder rate before we are faced with the academic question of whether it can ever be reduced to zero. Moreover, there are ways to reduce conflict other than to dream of a golden future of indiscriminate love. People in all societies not only perpetrate violence but deplore it. And people everywhere take steps to reduce violent conflict, such as sanctions, redress, censure, mediation, ostracism, and law.

I hope this discussion strikes you as trite, so I can get on with the content of the chapter. My goal is not to convince you that people don't always want the best for one another, but to try to explain when and why that should be true. But sometimes the trite has to be stated. The observation that conflict is part of the human condition, banal though it is, contradicts fashionable beliefs. One is expressed in the gluey metaphor of social relations as attachment, bonding, and cohesion. Another is the assumption that we unthinkingly play out the roles society assigns to us, and that social reform is a matter of rewriting the roles. I suspect that if you pressed many academics and social critics you would find views no less utopian than those of Charles Reich.

If the mind is an organ of computation engineered by natural selection, our social motives should be strategies that are tailored to the tournaments we play in. People should have distinct kinds of thoughts and feelings about kin and non-kin, and about parents, children, siblings, dates, spouses, acquaintances, friends, rivals, allies, and enemies. Let's explore them in turn.

KITH AND KIN

Smile on your brother, sang the Youngbloods; a brotherhood of man, sang John Lennon. When we talk of beneficence, we use the language of kinship. Our father who art in heaven; the fatherhood of God; church fathers; Father Christmas; father figure; patriotism. The mother country; the mother church; Mother Superior; motherhood and apple pie; maternal. Blood brothers; black brothers; brothers-in-arms; brotherly love; temple brotherhoods; brethren; fraternities; Brother, can you spare a dime? Sisterhood is powerful; sister cities; soul sisters; sisters of mercy; sororities. The family of man; crime families; one big happy family.

The kinship metaphors have a simple message: treat certain people as kindly as you treat your blood relatives. We all understand the presupposition. The love of kin comes naturally; the love of non-kin does not. That is the fundamental fact of the social world, steering everything from how we grow up to the rise and fall of empires and religions. The explanation is straightforward. Relatives share genes to a greater extent than nonrelatives, so if a gene makes an organism benefit a relative (say, by feeding or protecting it), it has a good chance of benefiting a copy of itself. With that advantage, genes for helping relatives will increase in a population over the generations. The vast majority of altruistic acts in the animal kingdom benefit the actor's kin. The most extreme examples of kin-directed altruism are found among social insects like ants and bees, in which the workers give their all to the colony. They are permanently sterile and defend the colony with kamikaze tactics like blowing up to spray noxious chemicals on an invader or stinging it with a barbed stinger that pulls the insect's body apart when dislodged. Such dedication comes largely from an unusual genetic system which makes them more closely related to their sisters than they would be to their offspring. By defending the colony they help their mothers make sisters instead of making offspring of their own.

Genes can't call to one another or pull the strings of behavior directly. In humans, "kin altruism" and "benefiting one's genes" are shorthand for two collections of psychological machinery, one cognitive, one emotional.

Humans are equipped with a desire and an ability to learn their family tree. Genealogy is a special kind of knowledge. First, the relationships are digital. You're either someone's mother or you aren't. You might be eighty percent sure that Bill is John's father, but that is not the same as thinking that Bill is eighty percent of a father to John. We speak of half-brothers, but everyone knows the expression is shorthand for having the same mother and different fathers or vice versa. Second, kinship is a relation. No one is a father or a sister, period; they have to be the father or the sister *of* someone. Third, kinship is topological. Everyone is a node in a web whose links are defined by parenthood, generation, and gender. Kinship terms are logical expressions that are read off the geometry and labeling of the web: a "parallel cousin," for example, is one's father's brother's child or one's mother's sister's child. Fourth, kinship is self-contained. Age, place of birth, acquaintanceship, status, occupation, zodiac sign, and all the other categories in which we place people lie in a different plane from the categories of kinship and need not be consulted when we calculate kinship.

Homo sapiens is obsessed with kinship. All over the world, when people are asked to talk about themselves, they begin with their parentage and family ties, and in many societies, especially foraging groups, people rattle off endless genealogies. For adoptees, childhood refugees, or descendants of slaves, curiosity about biological kin can drive a lifelong quest. (Entrepreneurs hope to exploit this motive when they send out those computer-generated postcards that offer to trace **Steven Pinker**'s ancestors and find the **Pinker** family seal and coat of arms.) Of course, people ordinarily do not test each other's DNA; they assess kinship by indirect means. Many animals do it by smell. Humans do it with several kinds of information: who grows up together, who resembles whom, how people interact, what reliable sources say, and what can be logically deduced from other kin relationships.

Once we know how we are related to other people, the other component of the psychology of kinship kicks in. We feel a measure of solidarity, sympathy, tolerance, and trust toward our relatives, added on to whatever other feelings we may have for them. ("Home," according to the poem by Robert Frost, is "something you somehow haven't to deserve.") The added good will one feels toward kin is doled out according to a feeling that reflects the probability that the kind act will help a relative propagate copies of one's genes. That in turn depends on the nearness of the relative to oneself in the family tree, the confidence one has in that nearness, and the impact of the kindness on the relative's prospects of reproducing (which depends on age and need). So parents love their children above all others, cousins love each other but not as much as siblings do, and so on. Of course, no one crunches genetic and actuarial data and then decides how much to love. Rather, the mental programs for familial love were calibrated in the course of evolution so that love *correlated with* the probability in the ancestral environment that a loving act would benefit copies of genes for loving acts.

You might think this is just the banal observation that blood is thicker than water. But in today's intellectual climate, the observation is a shocking, radical thesis. A Martian who wanted to learn about human interactions from a textbook in social psychology would have no inkling that humans behave any differently to their relatives than to strangers. Some anthropologists have argued that our sense of kinship has nothing to do with biological relatedness. The conventional wisdom of Marxists, academic feminists, and café intellectuals embraces some astonishing claims: that the nuclear family of husband, wife, and children is a historical aberration unknown in centuries past and in the non-Western world; that in primitive tribes marriage is uncommon and people are indiscriminately promiscuous and free of jealousy; that throughout history the bride and groom had no say in their marriage; that romantic love was invented by the troubadours of medieval Provence and consisted of the adulterous love of a knight for a married lady; that children used to be thought of as miniature adults; that in olden times children died so often that mothers were unaffected by the loss; that concern for one's children is a recent invention. These beliefs are false. Blood really *is* thicker than water, and no aspect of human existence is untouched by that part of our psychology.

Families are important in all societies, and their core is a mother and her biological children. All societies have marriage. A man and a woman enter a publicly acknowledged alliance whose primary goal is children; the man has a "right" of exclusive sexual access to the woman; and they both are obligated to invest in their children. The details vary, often according to the patterns of blood relationships in the society. Generally, when men can be confident that they are the fathers of their wives' children, nuclear families form, usually near the husband's extended kin. In the smaller number of societies where men are not so confident (for example, when they are away for long stretches of military service or farm labor), families live near the mother's kin, and children's principal male benefactors are their closest blood relatives, their maternal uncles. Even then, biological fatherhood is recognized and valued. Both sides of the extended family take an interest in the marriage and the children, and the children feel solidarity with both sides, even when the official rules of descent recognize only one side (as in our own surnames, which are reckoned according to the father's family).

Women fare better when they stay near their relatives and the men move around, because they are surrounded by fathers, brothers, and uncles who can come to their aid in disputes with their husbands. The dynamic was vividly enacted in *The Godfather* when the son of Marlon Brando's character, Sonny Corleone, nearly murdered his sister's husband when he found out that the husband had battered her. Life imitated art two decades later when the real-life son of Brando, Christian Brando, did murder his sister's boyfriend when he found out that the boyfriend had battered her. When a woman has to leave home to live near her husband's family, he can brutalize her with impunity. In many societies, marriages between cousins are encouraged, and the marriages are relatively harmonious because the usual bickering between husband and wife is mitigated by their sympathy for each other as blood relatives.

These days it's impolite to talk about parental love having anything to do with biological relatedness because it sounds like a slur on the many parents with adopted children and stepchildren. Of course couples love their adopted children; if they weren't unusually committed to simulating a natural family experience they would not have adopted to start with. But stepfamilies are different. The stepparent has shopped for a spouse, not a child; the child is a cost that comes as part of the deal. Stepparents have a poor reputation; even Webster's unabridged dictionary defines *stepmother*, in one of its two definitions, as "one that fails to give proper care or attention." The psychologists Martin Daly and Margo Wilson comment:

> The negative characterization of stepparents is by no means peculiar to our culture. The folklorist who consults Stith Thompson's massive *Motif-Index of Folk Literature* will encounter such pithy synopses as "Evil stepmother orders stepdaughter to be killed" (Irish myth), and "Evil step-mother works stepdaughter to death in absence of merchant husband" (India). For convenience, Thompson divided stepfather tales into two categories: "cruel stepfathers" and "lustful stepfathers." From Eskimos to Indonesians, through dozens of tales, the stepparent is a villain in every piece.

Daly and Wilson note that many social scientists assume that the difficulties plaguing step relationships are *caused* by "the myth of the cruel stepparent." But why, they ask, should stepparents in so many cultures be targets of the same slander? Their own explanation is more direct.

> The ubiquity of Cinderella stories . . . is surely a reflection of certain basic, recurring tensions in human society. Women must often have been forsaken with dependent children throughout human history, and both fathers and mothers were often prematurely widowed. If the survivor wished to forge a new marital career, then the fate of the children became problematic. [Among the Tikopia and the Yanomamö, the husband] demands the death of his new wife's prior children. Other solutions have included leaving the children with postmenopausal matrilineal relatives, and the *levirate*, a widespread custom by which a widow and her children are inherited by the dead man's brother or other near relative. In the absence of such arrangements, children were obliged to tag along as stepchildren under the care of nonrelatives with no particular

benevolent interest in their welfare. They surely had genuine cause for alarm.

In one study of emotionally healthy middle-class families in the United States, only half of the stepfathers and a quarter of the stepmothers claimed to have "parental feeling" toward their stepchildren, and fewer still claimed to "love" them. The enormous pop-psychology literature on reconstituted families is dominated by one theme: coping with antagonisms. Many professionals now advise warring families to give up the ideal of duplicating a biological family. Daly and Wilson found that stepparenthood is the strongest risk factor for child abuse ever identified. In the case of the worst abuse, homicide, a stepparent is forty to a hundred times more likely than a biological parent to kill a young child, even when confounding factors—poverty, the mother's age, the traits of people who tend to remarry—are taken into account.

Stepparents are surely no more cruel than anyone else. Parenthood is unique among human relationships in its one-sidedness. Parents give; children take. For obvious evolutionary reasons, people are wired to want to make these sacrifices for their own children but not for anyone else. Worse, as we shall see, children are wired to demand these sacrifices of the adults charged with their care, and that can make them downright annoying to people other than their parents and close kin. The writer Nancy Mitford said, "I love children, especially when they cry, for then someone takes them away." But if you are married to the children's parent, no one ever takes them away. The indifference, even antagonism, of stepparents to stepchildren is simply the standard reaction of a human to another human. It is the endless patience and generosity of a biological parent that is special. This point should not diminish our appreciation of the many benevolent stepparents; if anything, it should enhance it, for they are especially kind and self-sacrificing people.

~

It is often said that you are more likely to be killed by a relative in the home than by a mugger in the street. That sounds suspicious to anyone who knows about evolutionary theory, and it turns out to be false.

Homicide statistics are an important kind of evidence for theories of human relationships. As Daly and Wilson explain, "Killing one's antagonist is the ultimate conflict resolution technique, and our ancestors discovered it long before they were people." Homicides cannot be written off as the product of a diseased mind or a sick society. In most cases a killing is unplanned and undesired; it is the disastrous climax of an escalating battle in which brinkmanship has been carried too far For every killing there must be countless arguments that cool down and countless threats that are not carried out. That makes homicide an excellent assay for conflict and its causes. Unlike lesser conflicts which can only be discovered through reports that the participants can fudge, a homicide leaves a missing person or a dead body, which are hard to ignore, and homicides are meticulously investigated and documented.

People sometimes do murder their relatives. There are infanticides, filicides, parricides, matricides, fratricides, siblicides, uxoricides, familicides, and several unnamed kinds of kin-killing. In a typical data set from an American city, a quarter of the homicides are committed by strangers, a half by acquaintances, and a quarter by "relatives." But most of the relatives are not blood kin. They are spouses, in-laws, and step relations. Only two to six percent of homicide victims are done in by their blood relatives. In fact, that is surely an overestimate. People see their blood relatives more often than they see other people, so relatives are more often within striking distance. When one focuses on people who live together, so that the opportunities for interacting are held constant, one finds that the risk of being killed by a nonrelative is at least eleven times greater than the risk of being killed by a blood relative, and probably much higher than that.

The de-escalation of conflicts among blood relatives is part of a larger pattern of kin solidarity called nepotism. In everyday usage the word refers to bestowing favors on relatives (literally, "nephews") as a perquisite of a job or social rank. Institutional nepotism is officially illicit in our society, though it is widely practiced, and in most societies people are surprised to hear that we consider it a vice. In many countries a newly appointed official openly fires all the civil servants under him and replaces them with relatives. Relatives are natural allies, and before the invention of agriculture and cities, societies were organized around clans of them. One of the fundamental questions of anthropology is how foraging people divide themselves into bands or villages, typically with about fifty members though varying with the time and place. Napoleon Chagnon amassed meticulous genealogies that link thousands of members of the Yanomamö, the foraging and horticultural people of the

Amazon rainforest whom he has studied for thirty years. He showed how kinship is the cement that keeps villages together. Close kin fight each other less often and come to each other's aid in fights more often. A village fissions when its population grows, the villagers become less related to one another, and they increasingly get on each other's nerves. A fight erupts, loyalties divide along blood lines, and one party storms off with his closer kin to form a new village.

A spouse is the most familiar example of *fictive* kin: genetically unrelated people who are called kin and claim the emotions ordinarily directed at kin. The biologist Richard Alexander has pointed out that if spouses are faithful, if each acts on behalf of the union's children rather than other blood relatives, and if the marriage lasts the lifetime of both, the genetic interests of a couple are identical. Their genes are tied up in the same package, their children, and what is good for one spouse is good for the other. Under these idealized conditions, marital love should be stronger than any other kind.

In reality, people's blood kin do claim some of their loyalties, and no one can ever be certain that a spouse is one hundred percent faithful, much less that the spouse will never desert or die. In a simpleminded species, the strength of spousal love might be set at some optimum medium level reflecting the overall probability of nepotism, infidelity, desertion, and widowhood. But humans are sensitive to the particulars of their marriages and fine-tune their emotions accordingly. It is no surprise to a biologist that in-laws, infidelity, and stepchildren are the major causes of marital strife.

Because a couple's genes are in the same boat, and each spouse shares genes with his or her kin, the kin have an interest—in both senses of the word—in their marriage. If your son marries my daughter, our genetic fortunes are partly linked in our common grandchildren, and to that extent what is good for you is good for me. Marriages make in-laws into natural allies, and that is one reason why in all cultures marriages are alliances between clans, not just between spouses. The other reason is that when parents have power over their adult children, as they had in all cultures until recently, the children are excellent trade goods. Since my children don't want to marry each other, you have something I need:

a spouse for my child. Thus dowries and bride-prices are ubiquitous in human cultures, though goods like status and allegiance in conflicts with third parties are also factored into the deal. Like all business transactions, the successful sale or trade of an offspring proves the good faith of the parties and makes them more likely to trust each other in the future. So in-laws are both genetic partners and business partners.

For future-minded parents, in-laws should be chosen carefully. Not only should parents assess the assets and trustworthiness of prospective in-laws, but they should size up whether the dollop of good will that comes free with a common genetic interest in the grandchildren would be put to the best use. It might be wasted on an already secure ally or an implacable foe, but could make all the difference for a clan whose sympathies are somewhere in between. Strategic matchmaking is one outcome of the psychology of kinship; another is rules about who can marry whom. In many cultures people are encouraged to marry their cross cousins and forbidden to marry their parallel cousins. A cross cousin is the child of your mother's brother or of your father's sister; a parallel cousin is a child of your mother's sister or of your father's brother. Why the distinction? Consider the most common arrangement, in which daughters are traded among clans of related males, and imagine yourself contemplating marriage with various cousins (it doesn't matter whether you are male or female). If you marry your cross cousin, you are consummating an exchange with a proven trading partner: a clan with which your own family (presided over by your paternal grandfather) has traded a bride in the past (your mother or your aunt). If you marry your parallel cousin, either you are marrying within the clan (if your father and the father of your betrothed are brothers) and bringing in no external goods, or you are marrying someone from a clan of strangers (if your mother and the mother of your betrothed are sisters).

These intrigues have spawned two of the modern myths of kinship: that in traditional societies, people have no voice in whom they marry, and that kinship has nothing to do with genetic relatedness. The grain of truth in the first myth is that parents everywhere wield as much power as they can to influence whom their children marry. Children do not, however, passively accept their parents' choice. People everywhere have powerful emotions about whom they want to marry—that is, romantic love—and engagements are often fierce battles of wills between parents and children. Even when parents have the final say, the children lobby day and night to make their feelings known, and the feelings almost always enter into the decision. The plot of Sholem Aleichem's *Tevye's Daughters* (adapted into the musical *Fiddler on the Roof*) unfolds on this battlefield, and similar plots are found across the world. When children elope, it is a catastrophe for their parents. The business deal or strategic opportunity of a lifetime may have just been frittered away. Worse, if the parents had pledged the child years before—which often happens, because children are born at different times and the second half of an exchange must wait until a child reaches marriageable age—the parents are now in default and at the mercy of the loan sharks. Or the parents may have mortgaged themselves to the eyeballs to buy a spouse for the departed child. Defaults on marriage agreements are a leading cause of feuding and warfare in traditional societies. With the stakes so high, it is no wonder that the parents' generation always teaches that romantic love is frivolous or does not exist at all. The intellectuals who conclude that romantic love is a recent invention of medieval troubadours or of Hollywood scriptwriters have taken this establishment propaganda at face value.

Those who take fictive kin as evidence that kinship has nothing to do with biology have also bought an official doctrine. A big problem with marriage rules, like the one mandating marriage between cross cousins, is that the age and sex mixture of a group fluctuates, so sometimes there will be no eligible partners for a child. As with all rules, the challenge is to work around them without making them a farce. An obvious solution is to redefine who is related to whom. An eligible bachelor might be called a cross cousin even if the genealogical diagram says otherwise, saving a daughter from spinsterhood without setting the precedent that other children can marry whom they please. But deep down no one is fooled by these face-saving measures. A similar hypocrisy applies to other fictive kin. With kin emotions being so powerful, manipulators try to tap them for solidarity among non-kin by *calling* the non-kin kin. The tactic has been rediscovered again and again, from tribal chiefs to modern preachers and sappy rock musicians. But even in tribes where fictive kin labels are publicly treated with the utmost seriousness, if you press someone in private he will acknowledge that so-and-so is not *really* his brother or cousin. And when people show their true colors in a dispute, the colors go with blood relatives, not fictive ones. Many modern parents tell their children to address family friends as Uncle and Aunt. When I was a child, my friends and I used to refer to them as our fake uncles and fake aunts. Children are even more adamant in resisting the ubiquitous pressure to call their new stepparents Mom and Dad.

For millennia, kin emotions have shaped even the largest societies. The reach of parental love can extend over generations via gifts and inheritance. Parental love causes the fundamental paradox of politics: no society can be simultaneously fair, free, and equal. If it is fair, people who work harder can accumulate more. If it is free, people will give their wealth to their children. But then it cannot be equal, for some people will inherit wealth they did not earn. Ever since Plato called attention to these tradeoffs in *The Republic*, most political ideologies can be defined by the stance they take on which of these ideals should yield.

Another surprising consequence of kin solidarity is that the family is a subversive organization. That conclusion flies in the face of the right-wing view that the church and state have always been steadfast upholders of the family and of the left-wing view that the family is a bourgeois, patriarchal institution designed to suppress women, weaken class solidarity, and manufacture docile consumers. The journalist Ferdinand Mount has documented how every political and religious movement in history has sought to undermine the family. The reasons are obvious. Not only is the family a rival coalition competing for a person's loyalties, but it is a rival with an unfair advantage: relatives innately care for one another more than comrades do. They bestow nepotistic benefits, forgive the daily frictions that strain other organizations, and stop at nothing to avenge wrongs against a member. Leninism, Nazism, and other totalitarian ideologies always demand a new loyalty "higher" than, and contrary to, family ties. So have religions from early Christianity to the Moonies ("*We're* your family now!"). In Matthew 10:34–37, Jesus says:

> Think not that I am come to send peace on earth: I came not to send peace, but a sword. For I am come to set a man at variance against his father, and the daughter against her mother, and the daughter in law against her mother in law. And a man's foes shall be they of his own household. He that loveth father or mother more than me is not worthy of me: and he that loveth son or daughter more than me is not worthy of me.

When Jesus said "Suffer the little children to come unto me," he was saying that they should not go unto their parents.

Successful religions and states eventually realize they have to coexist with families, but they do what they can to contain them, particularly the most threatening ones. The anthropologist Nancy Thornhill has found that the incest laws of most cultures are not created to deal with the problem of brother-sister marriages; brothers and sisters don't want to marry to begin with. Although brother-sister incest may be included in the prohibition and may help to legitimize it, the real targets of the laws are marriages that threaten the interests of the lawmakers. The rules ban marriages among more distant relatives like cousins, and are promulgated by the rulers of stratified societies to prevent wealth and power from accumulating in families, which could be future rivals. The anthropologist Laura Betzig has shown that the medieval church's rules on sex and marriage were also weapons against familial dynasties. In feudal Europe, parents did not bequeath their estates in equal parts to all of their children. Plots of land could not be subdivided every generation or they would become uselessly small, and a title can fall on only one heir. The custom of primogeniture arose, in which everything went to the oldest son and the other sons hit the road to seek their fortunes, often joining armies or the church. The church filled up with disinherited younger sons, who then manipulated marriage rules to make it harder for owners and title-holders to bear legitimate heirs. If they died without sons, the properties and titles passed back to the disinherited brothers or the church they served. According to their laws, a man could not divorce a childless wife, remarry while she was alive, adopt an heir, bear an heir with a woman closer than a seventh cousin, or have sex on various special days that added up to more than half the year. The story of Henry VIII reminds us that much of European history revolves around battles between powerful individuals trying to leverage family feelings for political gain—marrying strategically, striving for heirs—and other powerful individuals trying to foil them.

PARENTS AND CHILDREN

For an organism designed by natural selection, leaving descendants is the reason for being and the goal of all toil and struggle. The love of a parent for a child should be vast, and so it is. But it should not be boundless. Robert Trivers discovered a subtle but profound implication of genetics for the psychology of the family.

In most sexual species, parents bequeath fifty percent of their genes to each offspring. One strategy for maximizing the number of genes in the next generation is to pump out as many babies as possible as quickly as possible. That is what most organisms do. Baby organisms, however, are more vulnerable than adults because they are smaller and less experienced, and in most species the majority never make it to adulthood. All organisms therefore face a "choice" of allocating their time, calories, and risk to caring for an existing offspring and upping its odds of survival, or cranking out new offspring and letting them all fend for themselves. Depending on details of the species' ecosystem and body plan, either strategy can be genetically profitable. Birds and mammals have opted to care for their offspring, mammals by the extreme step of evolving organs that siphon nutrients from their own bodies and package them for their offspring as milk. Birds and mammals invest calories, time, risk, and bodily wear and tear on their offspring, and are repaid in increases in the offspring's life expectancy.

In theory, a parent could go to the other extreme and care for its first-born all its life—say, by suckling it until the parent died of old age. But that would make little sense because at some point the calories being turned into milk could better be invested in bearing and suckling a new offspring. As the first-born grows, each additional pint of milk is less and less crucial to its survival, and it becomes better and better equipped to find its own food. A younger offspring becomes a better investment, and the parent should wean the older one.

A parent should transfer investment from an older child to a younger one when the benefit to the younger exceeds the cost to the older. The reckoning is based on the fact that the two children are equally related to the parent. But these calculations are from the parent's point of view; the first child sees it differently. He shares fifty percent of his genes with his younger sibling, but he shares *one hundred percent* of his genes with *himself*. As far as he is concerned, the parent should continue to invest in him until the benefit to a younger sibling is greater than *twice* the cost to him. The genetic interests of the parent and the child diverge. Each child should want more parental care than the parent is willing to give, because parents want to invest in all of their offspring equally (relative to their needs), whereas each child wants more of the investment for himself. The tension is called parent-offspring conflict. In essence it is sibling rivalry: siblings compete among themselves for their parents' investment, whereas the parents would be happiest if each accepted a share proportional to his or her needs. But sibling rivalry can be played out with parents, too. In evolutionary terms, the only reason a parent withholds investment from one offspring is to save it for future ones. An offspring's conflict with its parents is really a rivalry with unborn siblings.

A tangible example is weaning conflict. The calories a mother converts to milk are not available to grow a new offspring, so nursing suppresses ovulation. At some point mammalian mothers wean their young so their bodies can prepare for bearing a subsequent offspring. When they do, the young mammal puts up a holy stink, hounding the mother for access to the teat for weeks or months before acquiescing.

When I mentioned the theory of parent-offspring conflict to console a colleague whose two-year-old son had become a pest after the birth of a younger brother, he snapped, "All you're saying is that people are selfish!" Sleepless for weeks, he could be forgiven for missing the point. Clearly, parents aren't selfish; parents are the least selfish entities in the known universe. But they aren't infinitely selfless either, or every whine and tantrum would be music to their ears. And the theory predicts that children aren't completely selfish, either. If they were, they would murder each newborn sibling to free up all the parents' investment for themselves and would demand to be breast-fed all their lives. The reason they don't is that they are *partly* related to their present and future siblings. A gene that made a child murder his newborn sister would have a fifty percent chance of destroying a copy of itself, and in most species that cost outweighs the benefit of having one's mother's milk all to oneself. (In some species, like spotted hyenas and some birds of prey, the costs don't outweigh the benefits, and siblings do murder one another.) A gene that made a fifteen-year-old want to nurse would foreclose an opportunity for his mother to manufacture new copies of that gene inside viable siblings. Either cost would exceed twice the benefit, so most organisms have their siblings' interests at heart, though discounted relative to their own. The point of the theory is not that children want to take or that parents don't want to give; it's that children want to take *more* than what their parents want to give.

Parent-offspring conflict begins in the womb. A woman with an unborn child seems like a vision of harmony and nurturance, but beneath the glow a mighty battle goes on inside her. The fetus tries to mine the mother's body for nutrients at the expense of her ability to bear future children. The mother is a conservationist, trying to keep her body in reserve for posterity. The human placenta is a tissue of the fetus that invades the mother's body and taps into her bloodstream. Through it the fetus secretes a hormone that ties up maternal insulin, increasing the levels of blood sugar which it can then skim off. But the resulting diabetes compromises the mother's health, and over evolutionary time she has fought back by secreting more insulin, which prompted the fetus to secrete more of the hormone that ties up insulin, and so on, until the hormones reached a thousand times their usual concentration. The biologist David Haig, who first noticed prenatal parent-offspring conflict, remarks that the raised hormone levels are like raised voices: a sign of conflict. In a similar tug-of-war, the fetus increases the mother's blood pressure, forcing more nutrients its way at the expense of her health.

The battle continues once the baby is born. The first decision of motherhood is whether to let the newborn die. Infanticide has been practiced in all the world's cultures. In ours, "killing babies" is a synonym for depravity, one of the most shocking crimes imaginable. One might think it is a form of Darwinian suicide and proof that other cultures' values are incommensurable with ours. Daly and Wilson show that it is neither.

Parents of all species face the choice of whether to continue to invest in a newborn. Parental investment is a precious resource, and if a newborn is likely to die there is no point in throwing good money after bad by fledging or suckling it. The time and calories would be better spent on its littermates or clutchmates, in starting over with new offspring, or in waiting until the circumstances are better. Thus most animals let their runtish or sickly offspring die. Similar calculations enter into human infanticide. In foraging peoples, women have their first child in their late teens, nurse them on demand for four infertile years, and see many die before adulthood. If a woman is lucky, she might raise two or three children to maturity. (The large broods of our grandparents are historical aberrations resulting from agriculture, which provided substitutes for mother's milk.) To raise even a small number of children to adulthood, a woman has to make hard choices. Women in the world's cultures let infants die in circumstances in which the odds of survival are low: when the infant is deformed, a twin, fatherless, or fathered by a man who isn't the woman's husband, and when the mother is young (and so has opportunities to try again), lacks social support, had the infant soon after another child, is overburdened with older offspring, or is otherwise in desperate straits, such as from a famine. Infanticide in the modern West is similar. The statistics show that the mothers who let their infants die are young, poor, and unwed. There are many explanations, but the parallel with the rest of the world is unlikely to be a coincidence.

Infanticidal mothers are not heartless, and even when infant mortality is common, people never treat young life casually. Mothers experience infanticide as an unavoidable tragedy. They grieve for the child and remember it with pain all their lives. In many cultures people try to distance their emotions from a newborn until they are assured it will survive. They may not touch, name, or grant legal personhood to a baby until a danger period is over, much like our own customs of the christening and the bris (the circumcision of eight-day-old Jewish boys).

The emotions of new mothers, which would drive the decision to keep a baby or let it die, may have been shaped by these actuarial facts. Postpartum depression has been written off as a hormonal delirium, but as with all explanations of complex emotions, one must ask *why* the brain is wired so as to let hormones have their effects. In most of human evolutionary history, a new mother had good reason to pause and take stock. She faced a decision between a definite tragedy now and a chance of an even greater tragedy years hence, and the choice was not to be taken lightly. Even today, the typical rumination of a depressed new mother— how will I cope with this burden?—is a genuine issue. The depression is most severe in the circumstances that lead mothers elsewhere in the world to commit infanticide, such as poverty, marital conflict, and single motherhood.

The emotional response called "bonding" is also surely more sophisticated than the stereotype in which a woman is smitten with a lifelong attachment to her baby if she interacts with it in a critical window after birth, like the victims of Puck in *A Midsummer Night's Dream* who became infatuated with the first person they saw upon awakening. Mothers appear to proceed from a cool assessment of the infant and their current prospects, to an appreciation of the infant as a uniquely wonderful individual after about a week, to a gradual deepening of love over the next few years.

The infant is an interested party, and fights for its interests with the only weapon at its disposal: cuteness. Newborns are precociously responsive to their mothers; they smile, make eye contact, perk up to her speech, even mimic her facial expressions. These advertisements of a functioning nervous system could melt a mother's heart and tip the balance in a close decision of whether to keep the baby. The ethologist Konrad Lorenz pointed out that the geometry of babies—a large head, a bulbous cranium, large eyes low in the face, pudgy cheeks, and short limbs—elicits tenderness and affection. The geometry comes from the baby-assembly process. The head end grows fastest in the womb, and the other end catches up after birth; babies grow into their brain and their eyes. Lorenz showed that animals with that geometry, such as ducks and rabbits, strike people as cute. In his essay "A Biological Homage to Mickey Mouse," Stephen Jay Gould showed that cartoonists exploit the geometry to make their characters more appealing. It's conceivable that the genes exploit it too, exaggerating the juvenile features of a newborn, particularly those that signal good health, to make it look cuter to its mother.

Once a child is allowed to live, the battle between the generations continues. How could an offspring hold its own in the battle? As Trivers notes, babies cannot fling their mothers to the ground and nurse at will; they have to use psychological tactics. A baby has to manipulate its parents' genuine concern for its welfare to induce them to give more than they would otherwise be willing to give. Since parents can learn to ignore cries of "wolf," the tactics have to be more insidious. An infant knows its own condition better than a parent does, because the infant's brain is connected to sensors throughout its body. Both the parent and the infant have an interest in the parent's responding to the infant's needs, such as by feeding it when it is hungry and cuddling it when it is cold. That gives the infant an opening to elicit more care than the parent wants to give. The baby can cry when it is not so cold or hungry, or withhold a smile until it gets its way. The baby need not literally be faking. Since parents should evolve to recognize sham crying, the baby's most effective tactic might be to feel genuinely miserable, even when there is no biological need. Self-deception may begin early.

The child can also resort to extortion by howling at night or throwing a tantrum in public, situations in which the parents are averse to letting the noise continue and are apt to capitulate. Worse, the parents' interest in their children's welfare allows the children to hold themselves hostage, say, by thrashing about in a violent tantrum or refusing to do something both parties know the child would enjoy. Thomas Schelling notes that children are in an excellent position to use paradoxical tactics (Chapter 6). They can cover their ears, scream, avoid their parents' gaze, or regress, all of which prevent them from registering or understanding their parents' threats. We get the evolution of the brat.

The theory of parent-offspring conflict is an alternative to two popular ideas. One is Freud's Oedipal complex, the hypothesis that boys have an unconscious wish to have sex with their mothers and kill their fathers, and therefore fear that their fathers will castrate them. (Similarly, in the Electra complex, little girls want to have sex with their fathers.) There is indeed a fact to be explained. In all cultures, young children are sometimes possessive of their mothers and cool to the mother's consort. Parent-offspring conflict offers a straightforward explanation. Daddy's interest in Mommy takes her attention away from me—and, even worse, threatens to create a baby brother or sister. Children may well have evolved tactics for delaying that sad day by diminishing their mothers' interest in sex and keeping their fathers away from her. It would be a straightforward extension of weaning conflict. The theory explains why so-called Oedipal feelings are as common in girls as in boys, and avoids the preposterous idea that little boys want to copulate with their mothers.

Daly and Wilson, who proposed the alternative, believe that Freud's mistake was to run together two different kinds of parent-offspring conflict. Young children are in conflict with their father over access to their mother, but it is not a sexual rivalry. And older children may have a sexual conflict with their parents, especially their fathers, but it is not a rivalry over the mother. In many societies fathers compete with their sons for sexual partners, explicitly or implicitly. In polygynous societies, where a man can have several wives, they might literally compete for the same women. And in most societies, polygynous or monogamous, a father must subsidize his son's quest for a wife at the expense of his other children or his own aspirations. The son may be impatient for the father to begin diverting resources to him; a still-robust father is a roadblock to his career. Filicides and parricides in most of the world are touched off by such competition.

Parents also arrange marriages, which is a polite way of saying that they sell or trade their children. Here again interests can conflict. Parents may hammer out a package deal in which one child gets a catch and another gets a loser. In polygynous societies a father may trade his daughters for wives for himself. Whether a daughter is traded for a daughter-in-law or for a wife, her value can hinge on her virginity: men don't want to marry a woman who might be carrying another man's child.

(Effective birth control is recent and still far from universal.) Therefore fathers take an interest in their daughters' sexuality, a mimic of the Electra complex but without either party desiring the other. In many societies men take horrifying measures to guarantee a daughter's "purity." They may lock her up, cloak her from head to toe, and extirpate her interest in sex by the horrible custom known by the euphemism "female circumcision" (it is a circumcision in the same sense that Lorena Bobbitt performed a bris). When the measures fail, they may execute an unchaste daughter to preserve what they call, ironically, the family's "honor." (In 1977 a Saudi princess was publicly stoned to death for bringing dishonor to her grandfather, the brother of the king, by having an indiscreet affair in London.) Parent-daughter conflict is a special case of conflict over the "ownership" of women's sexuality, a topic to which we will return.

~

The other popular theory subverted by parent-offspring conflict is the biology-culture distinction, in which babies are a bundle of uncivilized instincts and parents socialize them into competent, well-adjusted members of society. Personality, in this conventional wisdom, is shaped in the formative years by the parenting process. Parents and children both want the children to prosper in the social milieu, and since children are in no position to shape themselves, socialization represents a confluence of their interests.

Trivers reasoned that, according to the theory of parent-offspring conflict, parents should *not* necessarily have their children's interests at heart when they try to socialize them. Just as parents often act against a child's interests, they may try to *train* the child to act against its *own* interests. Parents want each child to act more altruistically to its siblings than the child wants to. That is because it pays the parents for a child to be altruistic when the benefit to a sibling exceeds the cost to the child, but it pays the *child* to be altruistic only when the benefit exceeds *twice* the cost. For more distant kin such as half-siblings and cousins, the difference between the parents' interests and the child's interests is even greater, because the parent is more closely related to the half-sibling or cousin than the child is. Similarly, parents may try to persuade children that staying home to help at the nest, allowing themselves to be sold in marriage, and other outcomes that are good for the parent (and hence the child's unborn siblings) are in fact good for the child. As in all arenas of conflict, parents may resort to deception and, since children are no fools, self-deception. So even if children acquiesce to a parent's rewards, punishments, examples, and exhortations for the time being because they are smaller and have no choice, they should not, according to the theory, allow their personalities to be shaped by these tactics.

Trivers went out on a limb with that prediction. The idea that parents shape their children is so ingrained that most people don't even realize it is a testable hypothesis and not a self-evident truth. The hypothesis has now been tested, and the outcome is one of the most surprising in the history of psychology.

Personalities differ in at least five major ways: whether a person is sociable or retiring (extroversion-introversion), whether a person worries constantly or is calm and self-satisfied (neuroticism-stability), whether the person is courteous and trusting or rude and suspicious (agreeableness-antagonism), whether a person is careful or careless (conscientiousness-undirectedness), and whether a person is daring or conforming (openness-nonopenness). Where do these traits come from? If they are genetic, identical twins should share them, even if they were separated at birth, and biological siblings should share them more than adoptive siblings do. If they are a product of socialization by parents, adoptive siblings should share them, and twins and biological siblings should share them more when they grow up in the same home than when they grow up in different homes. Dozens of studies have tested these kinds of predictions on thousands of people in many countries. The studies have looked not only at these personality traits but at actual outcomes in life such as divorce and alcoholism. The results are clear and replicable, and they contain two shockers.

One result has become well known. Much of the variation in personality—about fifty percent—has genetic causes. Identical twins separated at birth are alike; biological siblings raised together are more alike than adopted siblings. That means that the other fifty percent must come from the parents and the home, right? Wrong! Being brought up in one home versus another accounts, at most, for *five percent* of the differences among people in personality. Identical twins separated at birth are not only similar; they are virtually as similar as identical twins raised together. Adoptive siblings in the same home are not just different; they are about as different as two children plucked from the population at random. The biggest influence that parents have on their children is at the moment of conception.

(I hasten to add that parents are unimportant only when it comes to *differences* among them and differences among their grown children. Anything that *all* normal parents do that affects all children is not measured in these studies. Young children surely need the love, protection, and tutelage of a sane parent. As the psychologist Judith Harris has put it, the studies imply only that children would turn into the same kinds of adults if you left them in their homes and social milieus but switched all the parents around.)

No one knows where the other forty-five percent of the variation comes from. Perhaps personality is shaped by unique events impinging on the growing brain: how the fetus lay in the womb, how much maternal blood it diverted, how it was squeezed during birth, whether it was dropped on its head or caught certain viruses in the early years. Perhaps personality is shaped by unique experiences, like being chased by a dog or receiving an act of kindness from a teacher. Perhaps the traits of parents and the traits of children interact in complicated ways, so that two children growing up with the same parents really have different environments. One kind of parent may reward a rambunctious child and punish a placid one; another kind of parent may do the opposite. There is no good evidence for these scenarios, and I think two others are more plausible, both of which see personality as an adaptation rooted in the divergence of interests between parents and offspring. One is the child's battle plan for competing with its siblings, which I will discuss in the following section. The other is the child's battle plan for competing in its peer group.

Judith Harris has amassed evidence that children everywhere are socialized by their peer group, not by their parents. At all ages children join various play groups, circles, gangs, packs, cliques, and salons, and they jockey for status within them. Each is a culture that absorbs some customs from the outside and generates many of its own. Children's cultural heritage—the rules of Ringolevio, the melody and lyrics of the nyah-nyah song, the belief that if you kill someone you legally have to pay for his gravestone—is passed from child to child, sometimes for thousands of years. As children grow up they graduate from group to group and eventually join adult groups. Prestige at one level gives one a leg up at the next; most significantly, the leaders of young adolescent cliques are the first to date. At all ages children are driven to figure out what it takes to succeed among their peers and to give these strategies precedence over anything their parents foist on them. Weary parents know they are no match for a child's peers, and rightly obsess over the best neighborhood in which to bring their children up. Many successful people immigrated to this country as children and were not handicapped in the least by culturally inept parents who never learned the language or customs. As a researcher of language development I have always been struck by the way in which children rapidly pick up the language (especially the accent) of their peers, though they spend more time with their parents.

Why aren't children putty in parents' hands? Like Trivers and Harris, I suspect it is because children's genetic interests overlap only partly with their parents'. Children take their calories and protection from their parents, because their parents are the only ones willing to provide them, but they get their information from the best sources they can find and forge their strategies for dealing with life themselves. Their own parents may not be the wisest and most knowledgeable adults around, and worse, the rules at home are often stacked against the children in favor of their born and unborn siblings. And as far as reproduction is concerned, the home is a dead end. The child will have to compete for mates, and before that for the status necessary to find and keep them, in other arenas, which play by different rules. The child had better master them.

~

The conflict of interest between parents and offspring is unacknowledged in our public discourse about children. In most times and places, the advantage has been to the parents, and they have wielded their power as cruel tyrants. This century has seen the tables turn. Child-welfare experts flood the bookstores with parenting manuals and the government with policy advice. All politicians paint themselves as friends of children and their opponents as enemies. Childrearing manuals used to advise mothers on how to make it through the day. With Dr. Spock, the spotlight fell on the child and the mother became a nonperson, there only to create mental health in the child and to take the blame if the child turned out bad.

The child-welfare revolution was one of the great liberation movements of all time, but like all realignments of power, it can go too far. Feminist social critics have argued that mothers' interests have been erased by the child-care gurus. In discussing her book *The Myths of Motherhood*, Shari Thurer notes:

The most pervasive myth is the denial of maternal ambivalence: that mothers really both love and hate their children. There's a real silence about the ambivalent feelings; . . . it's tantamount to being a bad mother. [In my clinical practice], anger and rage are normal. Children are endlessly demanding, and they'll just suck you dry. Women shouldn't have to feel that they are supposed to meet all of the child's needs. But the myth is that mother love is natural and operative at all times.

Even the advocates of mothers' rights often feel they must frame their arguments in terms of the interests of the child (an overburdened mother is a bad mother) rather than in terms of the interests of the mother (an overburdened mother is unhappy).

More conservative social critics have also begun to notice that parents' and children's interests can diverge. Barbara Dafoe Whitehead has reviewed data showing that sex education does not succeed in its advertised function of reducing teenage pregnancies. Today's teens know all about sex and its hazards, but the girls end up pregnant anyway, quite possibly because they don't mind the idea of having babies. If the teens' parents do mind, they may have to enforce their interests by controlling the teenagers (with chaperones and curfews), not just by educating them.

I mention these debates not to take a side but to call attention to the long reach of parent-offspring conflict. Evolutionary thinking is often put down as a "reductionistic approach" that aims to redefine all social and political issues as technical problems of biology. The criticism has it backwards. The evolution-free discourse that has prevailed for decades has treated childrearing as a technological problem of determining which practices grow the best children. Trivers' insight is that decisions about childrearing are inherently about how to allocate a scarce resource—the parents' time and effort—to which several parties have a legitimate claim. As such, childrearing will always be partly a question of ethics and politics, not just of psychology and biology.

BROTHERS AND SISTERS

Ever since Cain slew Abel, siblings have been entangled by many emotions. As people of the same generation who know each other well, they react to each other as individuals: they may like or dislike one another, compete if they are of the same sex, or feel sexual attraction if they are not. As close kin, they feel a big extra dose of affection and solidarity. But though they share fifty percent of their genes with each other, each sibling shares one hundred percent of its genes with itself, so brotherly or sisterly love has its limits. Being offspring of the same parents, siblings are rivals for their parents' investment, from weaning to the reading of the will. And though genetic overlap makes a pair of siblings natural allies, it also makes them unnatural parents, and that genetic alchemy tempers their sexual feelings.

If people gave birth to a single litter of interchangeable n-tuplets, parent-offspring conflict would be a raw struggle among the siblings, each demanding more than its share. But all children are different, if for no other reason than that they are born at different times. Parents may not want to invest one nth of their energy in each of their n children, but may, like shrewd portfolio managers, try to pick winners and losers and invest accordingly. The investment decisions are not conscious forecasts of the number of grandchildren expected from each child, but emotional responses that were tuned by natural selection to have outcomes that maximized that number in the environment in which we evolved. Though enlightened parents try mightily never to play favorites, they don't always succeed. In one study, fully two-thirds of British and American mothers confessed to loving one of their children more.

How do parents make Sophie's Choice and sacrifice a child when circumstances demand it? Evolutionary theory predicts that the main criterion should be age. Childhood is a minefield, and the older a child gets, the luckier a parent is to have it alive and the more irreplaceable the child is as an expected source of grandchildren, right up until sexual maturity. (From then on, the reproductive years begin to be used up and the child's expected number of offspring declines.) For example, the actuarial tables show that a four-year-old in a foraging society will, on average, give a parent 1.4 times as many grandchildren as a newborn, an eight-year-old 1.5 times as many, and a twelve-year-old 1.7 times as many. So if parents already have a child when an infant arrives and cannot feed them both, they should sacrifice the infant. In no human society do parents sacrifice an older child when a younger one is born. In our society, the chance that a parent will kill a child drops steadily with the child's age, especially during the vulnerable first year. When parents are asked to imagine the loss of a child, they say they would grieve more for older children, up until the teenage years. The rise and fall of anticipated grief correlates almost perfectly with the life expectancies of hunter-gatherer children.

On the other hand, a younger child, being more helpless, has more use for a parent's daily ministrations. Parents report more tender feelings for their younger offspring, even though they seem to value the older ones more. The calculations begin to change when parents get older and a new child is likely to be their last one. There is nothing to save for, and the baby of the family is likely to be indulged. Parents also favor children that one might call, in a cold-hearted way, better investments: more vigorous, better looking, more talented.

Given that parents are apt to play favorites, offspring should be selected to manipulate their parents' investment decisions in their favor. Children are exquisitely sensitive to favoritism, right through adulthood and after the parents' deaths. They should calculate how to make the best of the hand that nature dealt them and of the dynamics of the poker game they were born into. The historian Frank Sulloway has argued that the elusive nongenetic component of personality is a set of strategies to compete with siblings for parental investment, and that is why children in the same family are so different. Each child develops in a different family ecology and forms a different plan for getting out of childhood alive. (The idea is an alternative to Harris' proposal that personality is a strategy for coping in peer groups, though both could be right.)

A first-born child has been spotted several advantages. The first-born, merely by having survived to its present age, is more precious to the parents, and of course is bigger, stronger, and wiser and will be so for as long as the younger one is a child. Having ruled the roost for a year or more, the first-born sees the newcomer as a usurper. Thus he (or she) should identify with his parents, who have aligned their interests with his, and should resist changes to the status quo, which has always served him well. He should also learn how best to wield the power that fate has granted him. In sum, a first-born should be a conservative and a bully. Second-born children have to cope in a world that contains this obsequious martinet. Since they cannot get their way with thuggery and toadyism, they must cultivate the opposite strategies. They should become appeasers and cooperators. And with less at stake in the status quo, they should be receptive to change. (These dynamics depend, too, on the innate components of the personalities of the siblings and on their sex, size, and spacing; your mileage may vary.)

Later-borns have to be flexible for another reason. Parents invest in the children who show the most promise of success in the world. The first-born has staked a claim in whatever personal and technical skills she is best at. There's no point in a later-born competing on that turf; any success would have to come at the expense of the older and more experienced sibling, and he (or she) would be forcing his parents to pick a winner, with daunting odds against him. Instead, he should find a different niche in which to excel. That gives his parents an opportunity to diversify their investments, because he complements his older sibling's skills in competition outside the family. Siblings in a family exaggerate their differences for the same reason that species in an ecosystem evolve into different forms: each niche supports a single occupant.

Family therapists have discussed these dynamics for decades, but is there any hard evidence? Sulloway analyzed data on 120,000 people from 196 adequately controlled studies of birth order and personality. As he predicted, first-borns are less open (more conforming, traditional, and closely identified with parents), more conscientious (more responsible, achievement-oriented, serious, and organized), more antagonistic (less agreeable, approachable, popular, and easygoing), and more neurotic (less well-adjusted, more anxious). They are also more extroverted (more assertive, more leaderly), though the evidence is cloudy because they are more serious, which makes them seem more introverted.

Family politics affects not only what people say in paper-and-pencil tests but how they act in the world when playing for high stakes. Sulloway analyzed biographical data from 3,894 scientists who had voiced opinions on radical scientific revolutions (such as the Copernican revolution and Darwinism), 893 members of the French National Convention during the Terror of 1793–1794, more than seven hundred protagonists in the Protestant Reformation, and the leaders of sixty-two American reform movements such as the abolition of slavery. In each of these shake-ups, later-borns were more likely to support the revolution, first-borns were more likely to be reactionary. The effects are not by-products of family size, family attitudes, social class, or other confounding factors. When evolutionary theory was first proposed and still incendiary, later-borns were *ten times* as likely to support it as first-borns. Other alleged causes of radicalism, such as nationality and social class, have only minor effects. (Darwin himself, for example, was upper-class but later-born.) Later-born scientists are also less specialized, trying their hands in a greater number of scientific fields.

If personality is an adaptation, why should people carry the strategies that served them in the rumpus room right into adulthood? One possibility is that siblings never completely escape the orbit of their parents, but compete all their lives. That is certainly true in traditional societies, including foraging groups. Another is that tactics like assertiveness and conservatism are skills like any other. As a young person invests more and more in honing them, she becomes increasingly loath to retrace the learning curve to cultivate new strategies for dealing with people.

The discovery that children brought up in the same family are no more similar than they would be if they had been brought up on different planets shows how poorly we understand the development of personality. All we know is that cherished ideas about the influence of parents are wrong. The most promising hypotheses, I suspect, will come from recognizing that childhood is a jungle and that the first problem children face in life is how to hold their own among siblings and peers.

The relationship between a brother and a sister has an added twist: one is male, one is female, and those are the ingredients of a sexual relationship. People have sex with and marry those with whom they interact the most—their co-workers, the girl or boy next door—and the people most like themselves—those of the same class, religion, race, and appearance. The forces of sexual attraction should pull siblings together like magnets. Even if familiarity breeds some contempt and only a tiny fraction of siblings hit it off, there should be millions of brothers and sisters wanting to have sex and get married. There are virtually none. Not in our society, not in any well-studied human society, not in most animals in the wild. (Prepubertal children sometimes engage in sexual play; I'm talking about real intercourse between mature siblings.)

Do brothers and sisters avoid copulating because their parents discourage it? Almost certainly not. Parents try to socialize their children to be more affectionate with each other ("Go ahead—kiss your sister!"), not less. And if they did discourage sex, it would be the only case in all of human experience in which a sexual prohibition worked. Teenage brothers and sisters do not sneak off for trysts in parks and the back seats of cars.

The incest taboo—a public prohibition against sex or marriage between close relatives—has been an obsession of anthropology for a century, but it does not explain what keeps siblings apart. Avoiding incest is universal; taboos against incest are not. And most incest taboos are not about sex within the nuclear family. Some are about sex with fictive kin and merely enforce sexual jealousy. For example, polygynous men may pass laws to keep their sons away from their junior wives, officially the sons' "stepmothers." As we have seen, most taboos prohibit marriage (not sex) between more distant kin, such as cousins, and are ploys that rulers use to prevent wealth from accumulating in rival families. Sometimes sex among family members falls under the umbrella of more general codes against incest, but nowhere is it the target.

Brothers and sisters simply don't find each other appealing as sexual partners. That is an understatement: the thought makes them acutely uncomfortable or fills them with disgust. (People who grew up without siblings of the opposite sex do not understand the emotion.) Freud claimed that the strong emotion is itself proof of an unconscious desire, especially when a male claims revulsion at the thought of coitus with his mother. By that reasoning we may conclude that people have an unconscious desire to eat dog feces and to stick needles in their eyes.

Repugnance at sex with a sibling is so robust in humans and other long-lived, mobile vertebrates that it is a good candidate for an adaptation. The function would be to avoid the costs of inbreeding: a reduction in the fitness of offspring. There is a grain of biological truth behind the folklore that incest "thickens the blood" and the stereotypes of defective hillbillies and royal twits. Harmful mutations steadily drip into the gene pool. Some are dominant, cripple their bearers, and are soon selected out. But most are recessive and do no harm until they build up in the population and meet up with copies of themselves when two carriers mate. Since close relatives share genes, if they mate they run a much higher risk that two copies of a harmful recessive gene will match up in their offspring. Since all of us carry the equivalent of one to two lethal recessive genes, when a brother and sister mate they are quite likely to have a compromised offspring, both in theory and in the studies that have measured the risks. The same is true for mother-son and father-daughter matings (and, to a lesser extent, to matings between more distant kin). It stands to reason that humans (and many other animals) have evolved an emotion that makes the thought of sex with a family member a turnoff.

Incest avoidance showcases the complicated software engineering behind our emotions for other people. We feel stronger bonds of affection to family members than to acquaintances or strangers. We clearly perceive the sexual attractiveness of family members, and even take pleasure in looking at them. But the affection and appreciation of beauty don't translate into a desire to copulate, though if the same emotions had been elicited by a nonrelative, the urge might be irresistible. The way a single bit of knowledge can turn lust into horror has been used to great dramatic effect in the dozens of plots that Polti classifies as "Involuntary crimes of love," of which Sophocles' *Oedipus Rex* is the most famous.

Incest avoidance has two twists. One is that different couplings within the family have different genetic costs and benefits, both for the participants and for the bystanders. We might expect sexual repugnance to be adjusted accordingly. For both males and females, the benefit of having a child with an immediate family member is that the child contains seventy-five percent of each parent's genes, instead of the usual fifty percent (the extra twenty-five percent comes from the genes shared by the parents by virtue of their being related which are then passed on to the child). The costs are the risk of a deformed child and the forgone opportunity to have a child with someone else. The forgone opportunities, however, differ for males and females. Also, children are always sure who their mothers are but are not always sure who their fathers are. For both these reasons, incest has to be costed out separately for each of the possible couplings in a family.

Neither a mother nor a son has any advantage in the mother coupling with the son as opposed to with the boy's father that could offset the genetic risks. And since men are generally not attracted to women old enough to be their mothers, the net result is that mother-son incest virtually never happens.

For incest between fathers and daughters and between brothers and sisters, the calculations come out differently depending on whose point of view we take. A hypothetical ancestral girl made pregnant by a brother or father would be precluded from having a child with a nonrelative for the nine months of pregnancy, and were she to keep the baby, for another two to four years of nursing. She wastes a precious opportunity for reproduction on a child that may be deformed. Incest should be thoroughly repugnant. But a male who impregnates his sister or daughter could be adding to the number of offspring he sires, because her pregnancy does not foreclose his impregnating someone else. There is a risk that the child will be deformed, but if it isn't, the child is a sheer bonus (more accurately, the extra dose of his genes in that child are the bonus).

Incest repugnance might be weaker, making him more likely to cross the line. It is a special case of the lower costs of reproduction for males and their less discriminating sexual desire, to which we will return.

A father, moreover, can never be certain that a daughter is his, so the genetic cost to him could be zero. That could weaken the suppression of desire even further compared to her brother, who is certain to be related to his sister because they share a mother. For stepfathers and stepbrothers, there is no genetic cost at all. It is no surprise, then, that between half and three-quarters of all reported incest cases are between stepfathers and stepdaughters, most of them initiated by the stepfather. Most of the rest are between fathers and daughters, and virtually all are coerced by the father. Some are between girls and other older male relatives, also mostly coerced. A mother gets no genetic benefit from a mating between her husband and her daughter (compared with a mating between her daughter and a son-in-law), but suffers the cost of defective grandchildren, so her interests are aligned with her daughter's and she should be a force opposing incest. Incestuous exploitation of girls might be even more common if their mothers were not around. These battles are driven by strong emotions, but the emotions are not an alternative to the genetic analysis; the analysis explains why they exist. And of course, in science as in detective work, to try to figure out the motive for a crime is not to excuse the crime.

People cannot directly sense their genetic overlap with another person; as with the rest of perception, the brain must combine information from the senses with assumptions about the world to make an intelligent guess. Chapter 4 showed that when the world violates the assumptions, we fall prey to an illusion, and that is exactly what happens in the perception of kinship. The nineteenth-century anthropologist Edward Westermarck conjectured that growing up in intimate closeness with a person in the early years is the key information the brain uses to put the person in the category "sibling." Similarly, when an adult raises a child the adult should perceive the child as "son" or "daughter" and the child should perceive the adult as "mother" or "father." The classifications then negate sexual desire.

These algorithms presuppose a world in which children who are raised together are biological siblings and vice versa. That is certainly true of foraging peoples. A mother's children grow up with her and usually with their father, too. When the assumption is false, people should be the victim of a kinship illusion. If they grow up with a person who is not a relative, they should be sexually indifferent or repelled. If they do not grow up with a person who is a relative, they should fail to be repelled. Being told in so many words that a date is really your brother or sister may be enough to kill the romantic mood, but an unconscious imprinting mechanism at work during a critical period in early childhood is surely even more powerful.

Both kinds of illusions have been documented. The Israeli communal villages called kibbutzim were founded early in the twentieth century by utopian planners determined to break down the nuclear family. Boys and girls of the same age shared living quarters from shortly after birth through adolescence and were raised together by nurses and teachers. When they became sexually mature, the children who had grown up together very rarely married or even had sex, though marriages were not discouraged. In some parts of China, brides used to move into their in-laws' homes, giving rise to frictions that you can well imagine. Parents hit on the brilliant idea of adopting a bride for their son when she was still a child, guaranteeing that she would forever be under her mother-in-law's thumb. What they did not realize was that the arrangement mimicked the psychological cues to siblinghood. When the couple grew up, they found each other unsexy, and compared with conventional couples, their marriages were unhappy, unfaithful, unfecund, and short. In parts of Lebanon, paternal parallel cousins grow up together as if they were siblings. Parents pressure the cousins into marrying, but the couples are sexually apathetic, relatively childless, and prone to divorce. Unconventional childrearing arrangements have been found to have the same outcome on all continents, and various alternative explanations can be ruled out.

Conversely, people who *do* commit incest often have not grown up together. A study of sibling incest offenders in Chicago found that the only ones who had contemplated marriage were those who had been raised apart. Fathers who sexually abuse their daughters tend to have spent less time with them when they were small. Stepfathers who have had as much contact with their young stepdaughters as biological fathers do are no more likely to abuse them. There are anecdotes that adoptees who seek out their biological parents and siblings often find themselves sexually attracted to them, though I know of no controlled studies.

The Westermarck effect explains the most famous incest offender of all: Oedipus. Laius, king of Thebes, was warned by an oracle that his son would slay him. When Jocasta, his wife, bore a son, he tied the baby up and left him exposed on a mountain. Oedipus was found and raised by a shepherd and then adopted by the king of Corinth and brought up as his son. On a visit to Delphi, Oedipus learned that he was fated to kill his father and marry his mother, so he left Corinth vowing never to return. On his way toward Thebes, he encountered Laius and killed him in a quarrel. When he then outwitted the Sphinx, his reward was the throne of Thebes and the hand of its widowed queen, Jocasta—the biological mother he did not grow up with. They had four children before he got the bad news.

But the ultimate triumph of the Westermarck theory has been pointed out by John Tooby. The idea that boys want to sleep with their mothers strikes most men as the silliest thing they have ever heard. Obviously it did not seem so to Freud, who wrote that as a boy he once had an erotic reaction to watching his mother dressing. But Freud had a wet-nurse, and may not have experienced the early intimacy that would have tipped off his perceptual system that Mrs. Freud was his mother. The Westermarck theory has out-Freuded Freud.

MEN AND WOMEN

Men and women. Women and men. It will never work.
—ERICA JONG

Sometimes, of course, it does work. A man and a woman can fall in love, and the key ingredient is an expression of commitment, as we saw in Chapter 6. A man and a woman need each other's DNA and hence can enjoy sex. A man and a woman have a common interest in their children, and their enduring love has evolved to protect that interest. And a husband and wife can be each other's best friends, and can enjoy the lifelong dependability and trust that underlies the logic of friendship (more on this later). These emotions are rooted in the fact that if a man and woman are monogamous, together for life, and not nepotistic toward their own families, their genetic interests are identical.

Unfortunately, that is a big "if." Even the happiest couples can fight like cats and dogs, and today fifty percent of marriages in the United States end in divorce. George Bernard Shaw wrote, "When we want to read of the deeds that are done for love, whither do we turn? To the murder column." Conflict between men and women, sometimes deadly, is universal, and it suggests that sex is not a bonding force in human affairs but a divisive one. Once again, that banality must be stated because the conventional wisdom denies it. One of the utopian ideals of the 1960s, reiterated ever since by sex gurus like Dr. Ruth, is the intensely erotic, mutually enjoyable, guilt-free, emotionally open, life-long monogamous pair-bond. The alternative from the counterculture was the intensely erotic, mutually enjoyable, guilt-free, emotionally open, round-robin orgy. Both were attributed to our hominid ancestors, to earlier stages of civilization, or to primitive tribes still out there somewhere. Both are as mythical as the Garden of Eden.

The battle between the sexes is not just a skirmish in the war between unrelated individuals but is fought in a different theater, for reasons first explained by Donald Symons. "With respect to human sexuality," he wrote, "there is a female human nature and a male human nature, and these natures are extraordinarily different. . . . Men and women differ in their sexual natures because throughout the immensely long hunting and gathering phase of human evolutionary history the sexual desires and dispositions that were adaptive for either sex were for the other tickets to reproductive oblivion."

Many people deny that there are any interesting differences between the sexes. At my own institution, students taking Psychology of Gender used to be taught that the only well-established difference between men and women is that men like women and women like men. Symons' two human natures are dismissed as "gender stereotypes," as if that were proof that they are false. The belief that spiders spin webs and pigs don't is also a stereotype, but is no less true for that. As we shall see, some stereotypes about sexual feelings have been verified beyond a reasonable doubt. In fact, researchers in sex differences have found that many gender stereotypes *underestimate* the documented differences between the sexes.

~

Why is there sex to begin with? Lord Chesterfield noted of sex that "the pleasure is momentary, the position ridiculous, and the expense damnable." Biologically speaking, the costs are damnable indeed, so why do almost all complex organisms reproduce sexually? Why don't women give virgin birth to daughters who are clones of themselves instead of wasting half their pregnancies on sons who lack the machinery to make grandchildren and are nothing but sperm donors? Why do people and other organisms swap out half their genes for the genes of another member of the species, generating variety in their offspring for variety's sake? It's not to evolve faster, because organisms are selected for fitness in the present. It's not to adapt to environmental change, because a random change in an already adapted organism is more likely to be for the worse than for the better, there being vastly more ways to be badly adapted than to be well adapted. The best theory, proposed by John Tooby, William Hamilton, and others, and now supported by several kinds of evidence, is that sex is a defense against parasites and pathogens (disease-causing microorganisms).

From a germ's point of view, you are a big yummy mound of cheesecake, there for the eating. Your body takes a different view, and has evolved a battery of defenses, from your skin to your immune system, to keep them out or do them in. An evolutionary arms race goes on between hosts and pathogens, though a better analogy might be an escalating contest between lockpickers and locksmiths. Germs are small, and they evolve diabolical tricks for infiltrating and hijacking the machinery of the cells, for skimming off its raw materials, and for passing themselves off as the body's own tissues to escape the surveillance of the immune system. The body responds with better security systems, but the germs have a built-in advantage: there are more of them and they can breed millions of times faster, which makes them evolve faster. They can evolve substantially within the lifetime of a host. Whatever molecular locks the body has evolved, the pathogens can evolve keys to open them.

Now, if an organism is asexual, once the pathogens crack the safe of its body they also have cracked the safes of its children and siblings. Sexual reproduction is a way of changing the locks once a generation. By swapping half the genes out for a different half, an organism gives its offspring a head start in the race against the local germs. Its molecular locks have a different combination of pins, so the germs have to start evolving new keys from scratch. A malevolent pathogen is the one thing in the world that rewards change for change's sake.

Sex poses a second puzzle. Why do we come in *two* sexes? Why do we make one big egg and lots of little sperm, instead of two equal blobs that coalesce like mercury? It is because the cell that is to become the baby cannot be just a bag of genes; it needs the metabolic machinery of the rest of a cell. Some of that machinery, the mitochondria, has its own genes, the famous mitochondrial DNA which is so useful in dating evolutionary splits. Like all genes, the ones in mitochondria are selected to replicate ruthlessly. And that is why a cell formed by fusing two equal cells faces trouble. The mitochondria of one parent and the mitochondria of the other parent wage a ferocious war for survival inside it. Mitochondria from each parent will murder their counterparts from the other, leaving the fused cell dangerously underpowered. The genes for the rest of the cell (the ones in the nucleus) suffer from the crippling of the cell, so they evolve a way of heading off the internecine warfare. In each pair of parents, one "agrees" to unilateral disarmament. It contributes a cell that provides no metabolic machinery, just naked DNA for the new nucleus. The species reproduces by fusing a big cell that contains a half-set of genes plus all the necessary machinery with a small cell that contains a half-set of genes and nothing else. The big cell is called an egg and the small cell is called a sperm.

Once an organism has taken that first step, the specialization of its sex cells can only escalate. A sperm is small and cheap, so the organism might as well make many of them, and give them outboard motors to get to the egg quickly and an organ to launch them on their way. The egg is big and precious, so the organism had better give it a head start by packing it with food and a protective cover. That makes it more expensive still, so to protect the investment the organism evolves organs that let the fertilized egg grow inside the body and absorb even more food, and that release the new offspring only when it is large enough to survive. These structures are called male and female reproductive organs. A few animals, hermaphrodites, put both kinds of organs in every individual, but most specialize further and divide up into two kinds, each allocating all their reproductive tissue to one kind of organ or the other. They are called males and females.

Trivers has worked out how all the prominent differences between males and females stem from the difference in the minimum size of their investment in offspring. Investment, remember, is anything a parent does that increases the chance of survival of an offspring while decreasing the parent's ability to produce other viable offspring. The investment can be energy, nutrients, time, or risk. The female, by definition, begins with a bigger investment—the larger sex cell—and in most species commits herself to even more. The male contributes a puny package of genes and usually leaves it at that. Since every offspring requires one of each, the female's contribution is the limiting step on how many offspring can be produced: at most, one offspring for each egg she creates and nurtures. Two cascades of consequences flow from this difference.

First, a single male can fertilize several females, which forces other males to go mateless. That sets up a competition among males for access to females. A male may beat up other males to prevent them from getting to a female, or compete for the resources necessary to mate, or court a female to get her to choose him. Males therefore vary in reproductive success. A winner can beget many offspring, a loser will beget none.

Second, the reproductive success of males depends on how many females they mate with, but the reproductive success of females does not depend on how many males they mate with. That makes females more discriminating. Males woo females and mate with any female that lets them. Females scrutinize males and mate only with the best ones: the ones with the best genes, the ones most willing and able to feed and protect her offspring, or the ones that the other females tend to prefer.

Male competition and female choice are ubiquitous in the animal kingdom. Darwin called attention to these two spectacles, which he dubbed sexual selection, but was puzzled as to why it should be males that compete and females that choose rather than the other way around. The theory of parental investment solves the puzzle. The greater-investing sex chooses, the lesser-investing sex competes. Relative investment, then, is the cause of sex differences. Everything else—testosterone, estrogen, penises, vaginas, Y chromosomes, X chromosomes—is secondary. Males compete and females choose only because the slightly bigger investment in an egg that *defines* being female tends to get multiplied by the rest of the animal's reproductive habits. In a few species, the whole animal reverses the initial difference in investment between egg and sperm, and in those cases *females* should compete and *males* should choose. Sure enough, these exceptions prove the rule. In some fishes, the male broods the young in a pouch. In some birds, the male sits on the egg and feeds the young. In those species, the females are aggressive and try to court the males, who select partners carefully.

In a typical mammal, though, the female does almost all the investing. Mammals have opted for a body plan in which the female carries the fetus inside her, nourishes it with her blood, and nurses and protects it after it is born until the offspring has grown big enough to fend for itself. The male contributes a few seconds of copulation and a sperm cell weighing one ten-trillionth of a gram. Not surprisingly, male mammals compete for opportunities to have sex with female mammals. The details depend on the rest of the animal's way of life. Females live alone or in groups, in small groups or large ones, in stable groups or temporary ones, using sensible criteria like where the food is, where it's safest, where they can easily bear and raise young, and whether they need strength in numbers. Males go where the females are. Female elephant seals, for example, congregate on beach strips which a male can easily patrol. A single male can monopolize the group, and males fight bloody battles for this jackpot. Bigger fighters are better fighters, so the males have evolved to be four times the size of the females.

Apes have a wide variety of sexual arrangements. That means, by the way, that there is no such thing as an "ape legacy" that humans are doomed to live by. Gorillas live on the fringes of forests in small groups of one male and several females, and the males fight each other for control over females, the males evolving to be twice the females' size. Gibbon females are solitary and widely dispersed, and the male finds a female's territory and acts as a faithful consort. Since other males are off in other territories, they fight no more than females do and are no bigger. Orangutan females are solitary but close enough together that a male can monopolize two or more of their ranges, and the males are about 1.7 times the size of the females. Chimps live in large, unstable groups that no male could dominate. Groups of males live with the females, and the males compete for dominance, which confers more opportunities to copulate. The males are about 1.3 times as large as the females. With lots of males around, a female has an incentive to mate with many of them so that a male can never be sure that an infant is not his and hence will not murder the infant to make its mother available to bear his own offspring. Bonobo (pygmy chimp) females are almost indiscriminately promiscuous, and the males fight less and are about the same size as females. They compete in a different way: inside the females' bodies.

Sperm can survive in the vagina for several days, so a promiscuous female can have several males' sperm competing inside her for a chance at fertilizing the egg. The more sperm a male produces, the greater the chance that one of his will get there first. That explains why chimpanzees have enormous testicles for their body size. Bigger testes make more sperm, which have a better chance inside promiscuous females. A gorilla is four times the weight of a chimpanzee, but his testicles are four times smaller. The females in his harem have no chance to copulate with any other male, so his sperm do not have to compete. Gibbons, who are monogamous, have small testicles, too.

In almost all primates (indeed, in almost all mammals), the males are deadbeat dads, contributing nothing to their offspring but DNA. Other species are more fatherly. Most birds, many fishes and insects, and social carnivores such as wolves have males that protect or feed their offspring. The evolution of male parental investment is helped along by several things. One is external fertilization, found in most fishes, where the female drops her eggs and the male fertilizes them in the water. The male is guaranteed that the fertilized eggs carry his genes, and since they have been released while the young are undeveloped, he has an opportunity to help. But in most mammals the cards are stacked against doting fatherhood. The egg is tucked away inside the mother, where some other male can fertilize it, so a male is never certain an offspring is his. He faces the danger of wasting his investment on another male's genes. Also, the embryo does most of its growing inside the mother, where the father can't get at it to help directly. And a father can easily desert and try to mate with another female, whereas the female is left holding the bag and cannot get rid of the fetus or offspring without having to go through the long process of nurturing an embryo all over again to get back to where she started. Fatherhood is also promoted when a species' lifestyle makes the benefits exceed the costs: when the offspring would be vulnerable without him, when he can easily provision them with concentrated food like meat, and when the young are easy to defend.

When males become devoted fathers, the rules of the mating game change. A female may choose a mate based on his ability and his willingness to invest in their offspring, insofar as she can judge. Females, not just males, compete for mates, though the prizes are different: males compete for fertile females willing to copulate, females compete for flush males willing to invest. Polygamy is no longer a matter of one male beating up all the others, or the females all wanting to be inseminated by the fiercest or prettiest male. When males invest *more* than females, as we have seen, the species may be polyandrous, with tough females keeping harems of males. (The mammals' body plan has foreclosed that option.) When one male has much more to invest than others (because, say, he controls a better territory), females may be better off sharing him—polygyny—than each having her own mate, because a fraction of a big resource may be better than the entirety of a small one. When males' contributions are more equal, the undivided attention of one becomes valuable, and the species settles on monogamy.

Many birds appear to be monogamous. In *Manhattan*, Woody Allen says to Diane Keaton, "I think people should mate for life, like pigeons or Catholics." The movie came out before ornithologists began to submit birds to DNA testing, which revealed, to their shock, that pigeons are not so faithful either. In some species of birds, a third of the offspring contain the DNA of a male other than the female's consort. The male bird is adulterous because he tries to raise the offspring of one female and mate with others, hoping that her offspring will survive on their own, or best of all, be raised by a cuckolded consort. The female bird is adulterous because she has a chance of getting the best of both worlds: the genes of the fittest male and the investment of the most willing male. The victim of cuckoldry is worse off than if he had failed to breed at all, because he has devoted his worldly efforts to the genes of a competitor. So in species whose males invest, the male's jealousy is directed not only at rival males but at the female. He may guard her, follow her around, copulate repeatedly, and avoid females that show signs of having recently mated.

~

The human mating system is not like any other animal's. But that does not mean it escapes the laws governing mating systems, which have been documented in hundreds of species. Any gene predisposing a male to be cuckolded, or a female to receive less paternal help than her neighbors, would quickly be tossed from the gene pool. Any gene that allowed a male to impregnate all the females, or a female to bear the most indulged offspring of the best male, would quickly take over. These selection pressures are not small. For human sexuality to be "socially constructed" and independent of biology, as the popular academic view has it, not only must it have miraculously escaped these powerful pressures, but it must have withstood equally powerful pressures of a different kind. If a person played out a socially constructed role, other people could shape the role to prosper at his or her expense. Powerful men could brainwash the others to enjoy being celibate or cuckolded, leaving the women for them. Any willingness to accept socially constructed gender roles would be selected out, and genes for resisting the roles would take over.

What kind of animal is *Homo sapiens*? We are mammals, so a woman's minimum parental investment is much larger than a man's. She contributes nine months of pregnancy and (in a natural environment) two to four years of nursing. He contributes a few minutes of sex and a teaspoon of semen. Men are about 1.15 times as large as women, which tells us that they have competed in our evolutionary history, with some men mating with several women and some men mating with none. Unlike gibbons, who are isolated, monogamous, and relatively sexless, and gorillas, who are clustered, harem-forming, and relatively sexless, we are gregarious, with men and women living together in large groups and constantly facing opportunities to couple. Men have smaller testicles for their body size than chimpanzees but bigger ones than gorillas and gibbons, suggesting that ancestral women were not wantonly promiscuous but were not always monogamous either. Children are born helpless and remain dependent on adults for a large chunk of the human lifespan, presumably because knowledge and skills are so important to the human way of life. So children need parental investment, and men, because they get meat from hunting and other resources, have something to invest. Men far exceed the minimum investment that their anatomy would let them get away with: they feed, protect, and teach their children. That should make cuckoldry a concern to men, and a man's willingness and ability to invest in children a concern to women. Because men and women live together in large groups, like chimps, but the males invest in their offspring, like birds, we developed marriage, in which a man and woman form a reproductive alliance that is meant to limit demands from third parties for sexual access and parental investment.

These facts of life have never changed, but others have. Until recently, men hunted and women gathered. Women were married soon after puberty. There was no contraception, no institutionalized adoption by nonrelatives, and no artificial insemination. Sex meant reproduction and vice versa. There was no food from domesticated plants or animals, so there was no baby formula; all children were breast-fed. There was also no paid day care, and no househusbands; babies and toddlers hung around with their mothers and other women. These conditions persisted through ninety-nine percent of our evolutionary history and have shaped our sexuality. Our sexual thoughts and feelings are adapted to a world in which sex led to babies, whether or not we want to make babies now. And they are adapted to a world in which children were a mother's problem more than a father's. When I use terms like "should," "best," and "optimal," they will be a shorthand for the strategies that would have led to reproductive success in that world. I will not be referring to what is morally right, attainable in the modern world, or conducive to happiness, which are different matters altogether.

~

The first question of strategy is how many partners to want. Remember that when the minimum investment in offspring is greater for females, a male can have more offspring if he mates with many females, but a female does not have more offspring if she mates with many males—one per conception is enough. Suppose a foraging man with one wife can expect two to five children with her. A premarital or extramarital liaison that conceives a child would increase his reproductive output by twenty to fifty percent. Of course, if the child starves or is killed because the father isn't around, the father is genetically no better off. The optimal liaison, then, is with a married woman whose husband would bring up the child. In foraging societies, fertile women are almost always married, so sex with a woman is usually sex with a married woman. Even if she is not, more fatherless children live than die, so a liaison with an unmarried partner can increase reproduction, too. None of this math applies to women. A part of the male mind, then, should want a variety of sexual partners for the sheer sake of having a variety of sexual partners.

Do you think that the only difference between men and women is that men like women and women like men? Any bartender or grand-mother you ask would say that men are more likely to have a wandering eye, but perhaps that is just an old-fashioned stereotype. The psychologist David Buss has looked for the stereotype in the people most likely to refute it—men and women in elite liberal American universities a generation after the feminist revolution, in the heyday of politically correct sensibilities. The methods are refreshingly direct.

Confidential questionnaires asked a series of questions. How strongly are you seeking a spouse? The answers were on average identical for men and women. How strongly are you seeking a one-night stand? The women said, Not very strongly; the men said, Pretty strongly. How many sexual partners would you like to have in the next month? In the next two years? In your lifetime? Women said that in the next month eight-tenths of a sexual partner would be just about right. They wanted one in the next two years, and four or five over their lifetimes. Men wanted two sex partners within the month, eight in the next two years, and eighteen over their lifetimes. Would you consider having sex with a desirable partner that you had known for five years? For two years? For a month? For a week? Women said "probably yes" for a man they had known for a year or more, "neutral" for one they had known for six months, and "definitely not" for someone they had known a week or less. Men said "probably yes" as long as they had known the woman for a week. How short a time would a man have to know a woman before he would definitely *not* have sex with her? Buss never found out; his scale did not go down past "one hour." When Buss presented these findings at a university and explained them in terms of parental investment and sexual selection, a young woman raised her hand and said, "Professor Buss, I have a simpler explanation of your data." Yes, he said, what is it? "Men are slime."

Are men really slime, or are they just trying to look like slime? Perhaps in questionnaires men try to exaggerate their studliness but women want to avoid looking easy. The psychologists R. D. Clark and Elaine Hatfield hired attractive men and women to approach strangers of the opposite sex on a college campus and say to them, "I have been noticing you around campus. I find you very attractive," and then ask one of three questions: (a) "Would you go out with me tonight?" (b) "Would you come over to my apartment tonight?" (c) "Would you go to bed with me tonight?" Half the women consented to a date. Half the men consented to a date. Six percent of the women consented to go to the stooge's apartment. Sixty-nine percent of the men consented to go to the stooge's apartment. None of the women consented to sex. Seventy-five percent of the men consented to sex. Of the remaining twenty-five percent, many were apologetic, asking for a rain check or explaining that they couldn't because their fiancée was in town. The results have been replicated in several states. When the studies were conducted, contraception was widely available and safe-sex practices were heavily publicized, so the results cannot be dismissed simply because women might be more cautious about pregnancy and sexually transmitted diseases.

An awakening of male sexual desire by a new partner is known as the Coolidge effect, after a famous anecdote. One day President Calvin Coolidge and his wife were visiting a government farm and were taken on separate tours. When Mrs. Coolidge was shown the chicken pens, she asked whether the rooster copulated more than once a day. "Dozens

of times," replied the guide. "Please tell that to the president," Mrs.
Coolidge requested. When the president was shown the pens and told
about the rooster, he asked, "Same hen every time?" "Oh, no, Mr. President, a different one each time." The president said, "Tell that to Mrs.
Coolidge." Many male mammals are indefatigable when a new willing
female is available after each copulation. They cannot be fooled by the
experimenter cloaking a previous partner or masking her scent. This
shows, incidentally, that male sexual desire is not exactly "undiscriminat-
ing." Males may not care *what kind of* female they mate with, but they
are hypersensitive to *which* female they mate with. It is another example
of the logical distinction between individuals and categories that I
argued was so important when criticizing associationism in Chapter 2.

Men do not have the sexual stamina of roosters, but they show a kind
of Coolidge effect in their desire over longer periods. In many cultures,
including our own, men report that their sexual ardor for their wives
wanes in the first years of marriage. It is the concept of the individual
person, not her appearance or other qualities, that triggers the decline;
the taste for new partners is not just an example of variety being the
spice of life, as in getting bored with strawberry and wanting to try
chocolate ripple. In Isaac Bashevis Singer's story "Schlemiel the First," a
simpleton from the mythical village of Chelm sets out on a trip but loses
his way and inadvertently returns home, thinking he has come across
another village, which by an amazing coincidence looks just like his. He
meets a woman who looks exactly like the wife he has grown tired of, and
finds her ravishing.

~

Another part of the male sexual mind is an ability to be easily aroused
by a possible sex partner—indeed, by the faintest hint of a possible sex
partner. Zoologists have found that the males of many species will court
an enormous range of objects having a vague resemblance to the female:
other males, females of the wrong species, females of the right species
that have been stuffed and nailed to a board, parts of stuffed females
such as a head suspended in midair, even parts of stuffed females with
important features missing like the eyes and the mouth. The male of the
human species is aroused by the sight of a nude woman, not only in the
flesh but in movies, photographs, drawings, postcards, dolls, and

bit-mapped cathode-ray-tube displays. He takes pleasure in this mistaken identity, supporting a worldwide pornography industry which in the United States alone grosses ten billion dollars a year, almost as much as spectator sports and the movies combined. In foraging cultures, young men make charcoal drawings of breasts and vulvas on rock overhangs, carve them on tree trunks, and scratch them in the sand. Pornography is similar the world over and was much the same a century ago as it is today. It depicts in graphic physical detail a succession of anonymous nude females eager for casual, impersonal sex.

It would make no sense for a woman to be easily aroused by the sight of a nude male. A fertile woman never has a shortage of willing sexual partners, and in that buyer's market she can seek the best husband available, the best genes, or other returns on her sexual favors. If she could be aroused by the sight of a naked man, men could induce her to have sex by exposing themselves and her bargaining position would be compromised. The reactions of the sexes to nudity are quite different: men see nude women as a kind of invitation, women see nude men as a kind of threat. In 1992 a Berkeley student known around campus as the Naked Guy chose to jog, attend class, and eat in the dining halls in the nude as a protest against the repressive sexual traditions of Western society. He was expelled when some female students protested that his behavior should be classified as sexual harassment.

Women do not seek the sight of naked male strangers or enactments of anonymous sex, and there is virtually no female market for pornography. (*Playgirl*, the supposed counterexample, is clearly for gay men. It has no ads for any product a woman would buy, and when a woman gets a subscription as a gag gift she finds herself on mailing lists for gay male pornography and sex toys.) In the laboratory, some early experiments claimed that men and women showed identical physiological arousal to a pornographic passage. The men, however, showed a bigger response to the *neutral* passage in the control condition than the women showed to the *pornography*. The so-called neutral passage, which had been chosen by the female investigators, described a man and a woman chatting about the relative merits of an anthropology major over pre-med. The men found it highly erotic! Women can sometimes be aroused when they have agreed to watch portrayals of intercourse, but they do not seek them out. (Symons points out that women are more choosy than men in *consenting* to sex, but once they have consented, there is no reason to believe they are any less responsive to sexual stimulation.) The closest mass-market equivalents to pornography for women are the romance novel and the bodice-ripper, in which the sex is described in the context of emotions and relationships rather than as a succession of bumping bodies.

～

The desire for sexual variety is an unusual adaptation, for it is insatiable. Most commodities of fitness show diminishing returns or an optimal level. People do not seek mass quantities of air, food, and water, and they want to be not too hot and not too cold but just right. But the more women a man has sex with, the more offspring he leaves; too much is never enough. That gives men a limitless appetite for casual sex partners (and perhaps for the commodities that in ancestral environments would have led to multiple partners, such as power and wealth). Everyday life offers most men few opportunities to plumb the bottom of the desire, but occasionally a man is rich, famous, handsome, and amoral enough to try. Georges Simenon and Hugh Hefner claimed to have had thousands of partners; Wilt Chamberlain estimated that he had twenty thousand. Say we liberally adjust for braggadocio and assume that Chamberlain inflated his estimate by a factor of, say, ten. That would still mean that one thousand nine hundred and ninety-nine sex partners were not enough.

Symons notes that homosexual relations offer a clear window on the desires of each sex. Every heterosexual relationship is a compromise between the wants of a man and the wants of a woman, so differences between the sexes tend to be minimized. But homosexuals do not have to compromise, and their sex lives showcase human sexuality in purer form (at least insofar as the rest of their sexual brains are not patterned like those of the opposite sex). In a study of homosexuals in San Francisco before the AIDS epidemic, twenty-eight percent of gay men reported having had more than a thousand sex partners, and seventy-five percent reported having had more than a hundred. No gay woman reported a thousand partners, and only two percent reported as many as a hundred. Other desires of gay men, like pornography, prostitutes, and attractive young partners, also mirror or exaggerate the desires of heterosexual men. (Incidentally, the fact that men's sexual wants are the same whether they are directed at women or directed at other men refutes the theory that they are instruments for oppressing women.) It's not that gay men are oversexed; they are simply men whose male desires bounce off other male desires rather than off female desires. Symons writes, "I am suggesting that heterosexual men would be as likely as homosexual men to have sex most often with strangers, to participate in anonymous orgies in public baths, and to stop off in public restrooms for five minutes of fellatio on the way home from work if women were interested in these activities. But women are not interested."

Among heterosexuals, if men want variety more than women do, Econ 101 tells us what should follow. Copulation should be conceived of as a female service, a favor that women can bestow on or withhold from men. Scores of metaphors treat sex with a woman as a precious commodity, whether they take the woman's perspective (*saving yourself, giving it away, feeling used*) or the man's (*getting any, sexual favors, getting lucky*). And sexual transactions often obey market principles, as cynics of all persuasions have long recognized. The feminist theorist Andrea Dworkin has written, "A man wants what a woman has—sex. He can steal it (rape), persuade her to give it away (seduction), rent it (prostitution), lease it over the long term (marriage in the United States) or own it outright (marriage in most societies)." In all societies, it is mostly or entirely the men who woo, proposition, seduce, use love magic, give gifts in trade for sex, pay bride-prices (rather than collect dowries), hire prostitutes, and rape.

Sexual economics, of course, also depends on the desirability of the individuals, not just the average desires of the sexes. People "pay" for sex—in cash, commitment, or favors—when the partner is more desirable than they are. Since women are more discriminating than men, the average man has to pay for sex with the average woman. An average man can attract a higher-quality wife than casual sex partner (assuming that a marriage commitment is a kind of payment), whereas a woman can attract a higher-quality casual sex partner (who would pay nothing) than husband. The highest-quality men, in theory, should have a large number of women willing to have sex with them. A cartoon by Dan Wasserman shows a couple leaving the theater after having seen *Indecent Proposal*. The husband says, "Would you sleep with Robert Redford for a million dollars?" She replies, "Yes, but they'd have to give me some time to come up with the money."

The cartoonist's wit, though, exploits our sense of surprise. We don't expect real life to work that way. The men most attractive to women do not hire themselves out as prostitutes; they may even hire prostitutes themselves. In 1995, the actor Hugh Grant, arguably the world's hand-somest man, was arrested for having oral sex with a prostitute in the front seat of his car. The simple economic analysis fails here because money and sex are not completely fungible. As we shall see, part of men's attractiveness comes from their wealth, so the most attractive men don't need the money. And the "payment" that most women hope for is not cash but long-term commitment, which is a scarce resource even for the handsomest and wealthiest man. The economics of the Hugh Grant affair are well summed up by an exchange from another movie, based on the story of Heidi Fleiss, the Hollywood madame. A call girl asks her friend why her handsome tricks have to pay for sex. "They're not paying you for the sex," the friend explains. "They're paying you to go away after-wards."

Could it be that men *learn* to want sexual variety? Perhaps it is a means to an end, the end being status in our society. The Don Juan is revered as a dashing stud; the pretty woman on his arm is a trophy. Cer-tainly anything that is desirable and rare can become a status symbol. But that does not mean that all desirable things are pursued *because* they are status symbols. I suspect that if men were given the hypothetical choice between clandestine sex with many attractive women and a *repu-tation* for sex with many attractive women, but without the sex, they would go for the sex. Not only because sex is incentive enough, but because a reputation for having sex is a *dis*incentive. Don Juans do *not* inspire admiration, especially in women, though they may inspire envy in men, a different and not always welcome reaction. Symons remarks,

> Human males appear to be so constituted that they resist learning *not* to desire variety despite impediments such as Christianity and the doctrine of sin; Judaism and the doctrine of mensch; social science and the doc-trines of repressed homosexuality and psychosexual immaturity; evolu-tionary theories of monogamous pair-bonding; cultural and legal traditions that support and glorify monogamy; the fact that the desire for variety is virtually impossible to satisfy; the time and energy, and the innumerable kinds of risk—physical and emotional—that variety-seeking entails; and the obvious potential rewards of learning to be sexually satis-fied with one woman.

A wandering eye, learned or not, is not the only component of a man's mind. Though desire often leads to behavior, it often does not, because other desires are stronger or because tactics of self-control (see Chapter 6) have been put into effect. Men's sexual tastes can be calibrated and overruled depending on the man's attractiveness, the availability of partners, and his assessment of the costs of a dalliance.

HUSBANDS AND WIVES

In evolutionary terms, a man who has a short-term liaison is betting that his illegitimate child will survive without his help or is counting on a cuckolded husband to bring it up as his own. For the man who can afford it, a surer way to maximize progeny is to seek several wives and invest in all their children. Men should want many wives, not just many sex partners. And in fact, men in power have allowed polygyny in more than eighty percent of human cultures. Jews practiced it until Christian times and outlawed it only in the tenth century. Mormons encouraged it until it was outlawed by the U.S. government in the late nineteenth century, and even today there are thought to be tens of thousands of clandestine polygynous marriages in Utah and other western states. Whenever polygyny is allowed, men seek additional wives and the means to attract them. Wealthy and prestigious men have more than one wife; ne'er-do-wells have none. Typically a man who has been married for some time seeks a younger wife. The senior wife remains his confidante and partner and runs the household; the junior one becomes his sexual interest.

In foraging societies wealth cannot accumulate, but a few fierce men, skilled leaders, and good hunters may have two to ten wives. With the invention of agriculture and massive inequality, polygyny can reach ridiculous proportions. Laura Betzig has documented that in civilization after civilization, despotic men have implemented the ultimate male fantasy: a harem of hundreds of nubile women, closely guarded (often by eunuchs) so no other man can touch them. Similar arrangements have popped up in India, China, the Islamic world, sub-Saharan Africa, and the Americas. King Solomon had a thousand concubines. Roman emperors called them slaves, and medieval European kings called them serving maids.

Polyandry, by comparison, is vanishingly rare. Men occasionally share a wife in environments so harsh that a man cannot survive without a woman, but the arrangement collapses when conditions improve. Eskimos have sporadically had polyandrous marriages, but the co-husbands are always jealous and one often murders the other. As always, kinship mitigates enmity, and among Tibetan farmers two or more brothers sometimes marry a woman simultaneously in the hope of putting together a family that can survive in the bleak territory. The junior brother, though, aspires to have a wife of his own.

Marriage arrangements are usually described from the man's point of view, not because the desires of women are irrelevant but because powerful men have usually gotten their way. Men are bigger and stronger because they have been selected to fight one another, and they can form powerful clans because in traditional societies sons stay near their families and daughters move away. The most florid polygynists are always despots, men who could kill without fear of retribution. (According to the *Guinness Book of World Records*, the man with the most recorded children in history—888—was an emperor of Morocco with the evocative name Moulay Ismail The Bloodthirsty.) The hyperpolygynist not only must fend off the hundreds of men he has deprived of wives, but must oppress his harem. Marriages always have at least a bit of reciprocity, and in most polygynous societies a man may forgo additional wives because of their emotional and financial demands. A despot can keep them imprisoned and terrified.

But oddly enough, in a freer society polygyny is not necessarily bad for women. On financial and ultimately on evolutionary grounds, a woman may prefer to share a wealthy husband than to have the undivided attention of a pauper, and may even prefer it on emotional grounds. Laura Betzig summed up the reason: Would you rather be the third wife of John F. Kennedy or the first wife of Bozo the Clown? Co-wives often get along, sharing expertise and child-care duties, though jealousies among the subfamilies often erupt, much as in stepfamilies but with more factions and adult players. If marriage were genuinely a free market, then in a polygamous society men's greater demand for a limited supply of partners and their inflexible sexual jealousy would give the advantage to women. Laws enforcing monogamy would work to women's disadvantage. The economist Steven Landsburg explains the market principle, using labor instead of money in his example:

> Today, when my wife and I argue about who should do the dishes, we start from positions of roughly equal strength. If polygamy were legal, my wife could hint that she's thought about leaving me to marry Alan and Cindy down the block—and I might end up with dishpan hands.

. . . Antipolygamy laws are a textbook example of the theory of cartels. Producers, initially competitive, gather together in a conspiracy against the public or, more specifically, against their customers. They agree that each firm will restrict its output in an attempt to keep prices high. But a high price invites cheating, in the sense that each firm seeks to expand its own output beyond what is allowable under the agreement. Eventually, the cartel crumbles unless it is enforced by legal sanctions, and even then violations are legion.

That story, told in every economics textbook, is also the story of male producers in the romance industry. Initially fiercely competitive, they gather together in a conspiracy against their "customers"—the women to whom they offer their hands in marriage. The conspiracy consists of an agreement under which each man restricts his romantic endeavors in an attempt to increase the bargaining position of men in general. But the improved position of men invites cheating, in the sense that each man tries to court more women than allowed under the agreement. The cartel survives only because it is enforced by legal sanctions, and even so violations are legion.

Legal monogamy historically has been an agreement between more and less powerful men, not between men and women. Its aim is not so much to exploit the customers in the romance industry (women) as to minimize the costs of competition among the producers (men). Under polygyny, men vie for extraordinary Darwinian stakes—many wives versus none—and the competition is literally cutthroat. Many homicides and most tribal wars are directly or indirectly about competition for women. Leaders have outlawed polygyny when they needed less powerful men as allies and when they needed their subjects to fight an enemy instead of fighting one another. Early Christianity appealed to poor men partly because the promise of monogamy kept them in the marriage game, and in societies since, egalitarianism and monogamy go together as naturally as despotism and polygyny.

Even today, inequality has allowed a kind of polygyny to flourish. Wealthy men support a wife and a mistress, or divorce their wives at twenty-year intervals and pay them alimony and child support while marrying younger women. The journalist Robert Wright has speculated that easy divorce and remarriage, like overt polygyny, increases violence. Women of childbearing age are monopolized by well-to-do men, and the shortage of potential wives trickles down to the lower strata, forcing the poorest young men into desperate competition.

All of these intrigues come from a single difference between the sexes, men's greater desire for multiple partners. But men are not completely indiscriminate, and women are not voiceless in any but the most despotic societies. Each sex has criteria for picking partners for liaisons and for marriages. Like other staunch human tastes, they appear to be adaptations.

Both sexes want spouses, and men want liaisons more than women do, but that does not mean women never want liaisons. If they never did want them, the male urge to philander could not have evolved because it would never have been rewarded (unless the philanderer could always trick his conquest into thinking he was courting her as a wife—but even then, a married woman should never philander or be a target of philandering). Men's testicles would not have evolved to their larger-than-gorilla proportions, for their sperm would never be in danger of being outnumbered. And jealous feelings directed at wives would not exist; as we shall see, they do exist. The ethnographic record shows that in all societies, both sexes commit adultery, and the women do not always take arsenic or throw themselves under the 5:02 from St. Petersburg.

What could ancestral women have gained from liaisons that would have allowed the desire to evolve? One reward is resources. If men want sex for its own sake, women can make them pay for it. In foraging societies, women openly demand gifts from their lovers, usually meat. You may be offended at the thought that our foremothers gave themselves away for a steak dinner, but to foraging peoples in lean times when high-quality protein is scarce, meat is an obsession. (In *Pygmalion*, when Doolittle tries to sell his daughter Eliza to Higgins, Pickering shouts, "Have you no morals, man?" Doolittle replies, "Can't afford them, Governor. Neither could you if you was as poor as me.") From a distance it sounds like prostitution, but to the people involved it may feel more like ordinary etiquette, much as a woman in our own society might be offended if a wealthier lover never took her out to dinner or spent money on her, though both parties would deny there is a quid pro quo. In questionnaires, female college students report that an extravagant lifestyle and a willingness to give gifts are important qualities in picking a short-term lover, though not in picking a husband.

And like many birds, a woman could seek genes from the best-quality male and investment from her husband, because they are unlikely to be the same man (especially under monogamy and when she has little say in her marriage). Women report that looks and strength matter more in a lover than in a husband; as we shall see, looks are an indicator of genetic quality. And when women go through with an affair, they generally pick men of higher status than their husbands; the qualities that lead to status are almost certainly heritable (though a taste for prestigious lovers may also help with the first motive, extracting resources). Liaisons with superior men also may allow a woman to test her ability to trade up in the marriage market, either as a prelude to doing so or to improve her bargaining position within the marriage. Symons' summary of the sex difference in adultery is that a woman has an affair because she feels that the man is in some way superior or complementary to her husband, and a man has an affair because the woman is not his wife.

Do men require *anything* in a casual sex partner other than two X chromosomes? Sometimes it would appear that the answer is no. The anthropologist Bronislaw Malinowski reported that some Trobriand Island women were considered so repulsive that they were absolutely debarred from sexual intercourse. These women nevertheless managed to have several offspring, which the Trobrianders interpreted as conclusive proof of virgin birth. But more systematic research has shown that men, at least American college students, do have some preferences in a short-term partner. They rate looks as important; as we shall see, beauty is a signal of fertility and genetic quality. Promiscuity and sexual experience are also rated as assets. As Mae West explained, "Men like women with a past because they hope history will repeat itself." But these assets turn into liabilities when the men are asked about long-term partners. They subscribe to the infamous madonna-whore dichotomy, which divides the female sex into loose women, who may be dismissed as easy conquests, and coy women, who are valued as potential wives. This mentality is often called a symptom of misogyny, but it is the optimal genetic strategy for males of any species that invest in their offspring: mate with any female that will let you, but make sure your consort does not mate with any other male.

What should women look for in a husband? A bumper sticker from the 1970s read, "A woman without a man is like a fish without a bicycle." But at least for women in foraging societies, that would have been an overstatement. When a foraging woman is pregnant, nursing, and bringing up children, she and the children are vulnerable to hunger, protein deficiency, predation, rape, kidnapping, and murder. Any man who fathers her children should be put to good use in feeding and protecting them. From her point of view, he has nothing better to do, though from his point of view, there is an alternative: competing for and wooing other women. Men vary in their ability and willingness to invest in their children, so a woman should choose wisely. She should be impressed by wealth and status, or, in the case of men too young to have them, by portents that they will get them, such as ambition and industriousness. These are all useless unless the man hangs around once the woman becomes pregnant, and men have an interest in saying they will hang around whether or not they intend to. As Shakespeare wrote, "Men's vows are women's traitors." A woman therefore should look for signs of stability and sincerity. An aptitude for bodyguard duty would also come in handy.

What should men look for in a wife? Aside from faithfulness, which guarantees his paternity, she should be able to bear as many children as possible. (As always, that would be how our tastes were engineered; the reasoning does not imply that a man literally wants lots and lots of babies.) She should be fertile, which means she should be healthy and past the age of puberty but before the age of menopause. But a woman's current fertility is more relevant to a one-night stand than to a lifelong marriage. What counts is the number of offspring he can expect over the long term. Since a woman can bear and nurse one child every few years, and her childbearing years are finite, the younger the bride, the bigger the future family. That is true even though the youngest brides, teenagers, are somewhat less fertile than women in their early twenties. Ironically for the men-are-slime theory, an eye for nubile women may have evolved in the service of marriage and fatherhood, not one-night stands. Among chimpanzees, where a father's role ends with copulation, some of the wrinkled and saggy females are the sexiest.

Are the predictions just old-fashioned stereotypes? Buss designed a questionnaire asking about the importance of eighteen qualities of a mate and gave it to ten thousand people in thirty-seven countries on six continents and five islands—monogamous and polygynous, traditional and liberal, communist and capitalist. Men and women everywhere place the highest value of all on intelligence and on kindness and understanding. But in every country men and women differ on the other qualities. Women value earning capacity more than men do; the size of the difference varies from a third more to one and a half times more, but it's always there. In virtually every country, women place a greater value than men on status, ambition, and industriousness. And in most, they value dependability and stability more than men do. In every country, men place a higher value on youth and on looks than women do. On average, men want a bride 2.66 years younger; women want a groom 3.42 years older. The results have been replicated many times.

People's actions tell the same story. According to the contents of personal advertisements, Men Seeking Women seek youth and looks, Women Seeking Men seek financial security, height, and sincerity. The owner of one dating service observed, "Women really read over our profile forms; guys just look at the pictures." Among married couples, the husband is 2.99 years older than the wife, as if they had split the difference between their preferences. In foraging cultures, everyone agrees that some people are sexier than others, and the sexpots are usually young women and prestigious men. Yanomamö men, for example, say that the most desirable women are *moko dudei*, an expression that when applied to fruit means perfectly ripe and when applied to women means between fifteen and seventeen years old. When shown slides, Western observers of both sexes agree with the Yanomamö men that the *moko dudei* women are the most attractive. In our society, the best predictor of a man's wealth is his wife's looks, and the best predictor of a woman's looks is her husband's wealth. Dumpy-looking cabinet secretaries like Henry Kissinger and John Tower are called sex symbols and womanizers. Octogenarian oil barons like J. Paul Getty and J. Howard Marshall marry women young enough to be their great-granddaughters, such as the model Anna Nicole Smith. Not-so-handsome rock stars like Billy Joel, Rod Stewart, Lyle Lovett, Rick Ocasek, Ringo Starr, and Bill Wyman marry gorgeous actresses and supermodels. But former Representative Patricia Schroeder says she has noticed that a middle-aged congresswoman does not radiate the same animal magnetism to the opposite sex that a middle-aged congressman does.

An obvious retort is that women value wealthy and powerful men because it is the men who have the wealth and power. In a sexist society, women have to marry up to get them. That alternative has been tested and refuted. Women with large salaries, postgraduate degrees, prestigious professions, and high self-esteem place a *greater* value on wealth and status in a husband than other women do. So do the leaders of feminist organizations. Poor men place no higher value on wealth or earning power in a wife than other men do. Among the Bakweri in Cameroon, the women are wealthier and more powerful than the men, and they still insist on men with money.

The humorist Fran Lebowitz once said in an interview, "People who get married because they're in love make a ridiculous mistake. It makes much more sense to marry your best friend. You *like* your best friend more than anyone you're ever going to be in love with. You don't choose your best friend because they have a cute nose, but that's all you're doing when you get married; you're saying, 'I will spend the rest of my life with you because of your lower lip.'"

It *is* a puzzle, and the obvious place to look for an answer is the fact that you don't make children with your best friend but you do with your spouse. Perhaps we care about a few millimeters of flesh here or there because it is a perceptual signal of a deeper trait that cannot be measured directly: how well equipped the person's body is to serve as the other parent of your children. Fitness as a dam or stud is like any other feature of the world. It is not written on a tag but has to be inferred from appearances, using assumptions about how the world works.

Could we really be equipped with an innate eye for beauty? What about the natives in *National Geographic* who file their teeth, stretch their necks with stacks of rings, burn scars into their cheeks, and put plates in their lips? What about the fat women in the Rubens paintings and Twiggy in the 60s? Don't they show that standards of beauty are arbitrary and vary capriciously? They do not. Who says that *everything* people do to their bodies is an attempt to look sexy? That is the tacit assumption behind the *National Geographic* argument, but it's obviously false. People decorate their bodies for many reasons: to look rich, to look well connected, to look tough, to look "in," to earn membership in an elite group by enduring a painful initiation. Sexual attractiveness is different. People outside a culture usually agree with the people inside about who is beautiful and who is not, and people everywhere want good-looking partners. Even three-month-old infants prefer to look at a pretty face.

What goes into sexiness? Both sexes want a spouse who has developed normally and is free of infection. Not only is a healthy spouse vigorous, non-contagious, and more fertile, but the spouse's hereditary resistance to the local parasites will be passed on to the children. We haven't evolved stethoscopes and tongue-depressors, but an eye for beauty does some of the same things. Symmetry, an absence of deformities, cleanliness, unblemished skin, clear eyes, and intact teeth are attractive in all cultures. Orthodontists have found that a good-looking face has teeth and jaws in the optimal align-ment for chewing. Luxuriant hair is always pleasing, possibly because it shows not only current health but a record of health in the years before. Malnutrition and disease weaken the hair as it grows from the scalp, leaving a fragile spot in the shaft. Long hair implies a long history of good health.

A subtler sign of good genes is being average. Not average in attrac-tiveness, of course, but average in the size and shape of every part of the face. The average measurement of a trait in a local population is a good estimate of the optimal design favored by natural selection. If people form a composite of the opposite-sex faces around them, they would have an ideal of the fittest mate against which any candidate could be matched. The exact facial geometry of the local race or ethnic group would not need to be built in. In fact, composite faces, whether formed by superimposing negatives in an enlarger or by sophisticated computer-graphics algorithms, are prettier or handsomer than the individual faces that went into them.

Average faces are a good start, but some faces are even more attrac-tive than the average face. When boys reach puberty, testosterone builds up the bone in their jaws, brows, and nasal region. Girls faces grow more evenly. The difference in 3-D geometry allows us to tell a man's head from a woman's even when they are both bald and shaved. If the geome-try of a woman's face is similar to a man's, she is homelier; if it is less similar, she is prettier. Beauty in a woman comes from a short, delicate, smoothly curved jawbone, a small chin, a small nose and upper jaw, and a smooth forehead without brow ridges. The "high cheekbones" of a beautiful woman are not bones at all but soft tissue, and contribute to beauty because the other parts of a beautiful face (the jaws, forehead, and nose) are small by comparison.

Why are masculine-looking women less attractive? If a woman's face is masculinized, she probably has too much testosterone in her blood (a symptom of many diseases); if she has too much testosterone, she is likely to be infertile. Another explanation is that prettiness-detectors are really female-face detectors, designed to pick them out from every other object in the world and tuned to minimize the risk of a false alarm to a male face, which is the object most similar to a female face. The more unmanly the face, the louder the detector beeps. Similar engineering could explain why men with unfeminine faces are more handsome. A man with a large, angular jaw, a strong chin, and a prominent forehead and brow is undoubtedly an adult male with normal male hormones.

By the callous reckoning of natural selection, young women who have not yet had children are the best wives, because they have the longest reproductive career ahead of them and have no children from another man tagging along. Signs of youth and signs of never having been pregnant should make a woman prettier. Teenage women have larger eyes, fuller and redder lips, smoother, moister, and tighter skin, and firmer breasts, all long recognized as ingredients of pulchritude. Aging lengthens and coarsens a woman's facial bones, and so do pregnancies. Therefore a small-jawed, light-boned face is a clue to four reproductive virtues: being female, having the right hormones, being young, not having been pregnant. The equation of youth and beauty is often blamed on America's being obsessed with youth, but by that reasoning every culture is obsessed with youth. If anything, contemporary America is less youth-oriented. The age of *Playboy* models has *increased* over the decades, and in most times and places women in their twenties have been considered over the hill. Men's looks don't decline as quickly when they age, not because of a double standard in our society but because men's fertility doesn't decline as quickly when they age.

At puberty a girl's hips become wider because her pelvis grows and because fat is deposited on her hips, a reserve of calories available to supply the body during pregnancy. The ratio of waist size to hip size decreases in most fertile women to between .67 and .80, whereas the ratio for most men, children, and postmenopausal women is between .80 and .95. Among women, a low waist-to-hip ratio has been found to correlate with youth, health, fertility, not being pregnant, and never having been pregnant. The psychologist Devendra Singh has shown photographs and computer-generated pictures of female bodies of different sizes and shapes to hundreds of people of various ages, sexes, and cultures. Everyone finds a ratio of .70 or lower the most attractive. The ratio captures the old idea of the hourglass figure, the wasp waist, and the 36–24–36 ideal measurements. Singh also measured the ratio in *Playboy* centerfolds and winners of beauty contests over seven decades. Their weight has gone down, but their waist-to-hip ratio has stayed the same. Even most of the Upper Paleolithic Venus figurines, carved tens of thousands of years ago, have the right proportions.

The geometry of beauty once was an indicator of youth, health, and nonpregnancy, but it no longer has to be. Women today have fewer babies, have them later, are less exposed to the elements, and are better nourished and less disease-ridden than their ancestors. They can look like an ancestral teenager well into middle age. Women also have a technology to simulate and exaggerate the clues to youth, femaleness, and health: eye makeup (to enlarge the eyes), lipstick, eyebrow plucking (to reduce the appearance of a masculine brow ridge), makeup (to exploit the shape-from-shading mechanism of Chapter 4), products that increase the luster, thickness, and color of hair, bras and clothing that simulate young breasts, and hundreds of potions alleged to keep the skin looking young. Dieting and exercise can keep the waist thinner and the waist-to-hip ratio lower, and an illusion can be engineered with bodices, corsets, hoops, crinolines, bustles, girdles, pleats, tapering, and wide belts. Women's fashion has never embraced bulky cummerbunds.

Outside the scientific literature, more has been written about women's weight than any other aspect of beauty. In the West, women in pictures have weighed less and less over the past decades. That has been taken as evidence for the arbitrariness of beauty and for the oppression of women, who are expected to conform to these standards no matter how unreasonable. Slender models are commonly blamed for anorexia nervosa in teenage girls, and a recent book was called *Fat Is a Feminist Issue*. But weight may be the least important part of beauty. Singh found that very fat women and very thin women are judged less attractive (and in fact they are less fertile), but there is a range of weights considered attractive, and shape (waist-to-hip ratio) is more important than size. The hoopla about thinness applies more to women who pose for other women than to women who pose for men. Twiggy and Kate Moss are fashion models, not pinups; Marilyn Monroe and Jayne Mansfield were pinups, not fashion models. Weight is a factor mostly in the competition among women for status in an age in which wealthy women are more likely to be slender than poor ones, a reversal of the usual relation.

Still, the women posing for both sexes today are slimmer today than their historical counterparts, and it may be for reasons other than just changes in the signs of status. My own conjecture is that today's slender centerfolds and supermodels would not have had trouble finding a date at any time in history, because they are *not* like the skinny women eschewed in centuries past. Body parts do not vary independently. Tall men tend to have big feet, people with thick waists tend to have double chins, and so on. Undernourished women may tend to have more masculine bodies, and well-nourished ones more feminine bodies, so historically attractive women may have tended to be heavier. Neither kind of woman has the most beautiful shape conceivable—say, Jessica

Rabbit's—because real bodies did not evolve as cartoon sex lures. They are compromises among the demands of attractiveness, running, lifting, childbearing, nursing, and surviving famines. Perhaps modern technology *has* fabricated a sex lure, not with a cartoonist's brush but with artificial selection. In a world of five billion people there are bound to be women with wide feet and small heads, men with big ears and scrawny necks, and any other combination of body parts you want to specify. There may be a few thousand women with freakish combinations of small waists, flat abdomens, large firm breasts, and curved but medium-sized hips—optical illusions that send the needles of people's fertility and childlessness gauges into the red. When word gets around that they can parlay their freaky bodies into fame and fortune, they come out of the woodwork, and enhance their gifts with makeup, exercise, and glamour photography. The bodies in the beer commercials may be unlike anything seen in history.

Beauty is not, as some feminists have claimed, a conspiracy by men to objectify and oppress women. The *really* sexist societies drape women in chadors from head to foot. Throughout history the critics of beauty have been powerful men, religious leaders, sometimes older women, and doctors, who can always be counted on to say that the latest beauty craze is hazardous to women's health. The enthusiasts are women themselves. The explanation is simple economics and politics (though not the orthodox feminist analysis—quite insulting to women, incidentally—in which women are dupes who have been brainwashed into striving for something they don't want). Women in open societies want to look good because it gives them an edge in competing for husbands, status, and the attention of powerful people. Men in closed societies hate beauty because it makes their wives and daughters indiscriminately attractive to other men, giving the women a measure of control over the profits from their own sexuality and taking it away from the men (and, in the case of daughters, away from their mothers). Similar economics make men want to look good, too, but the market forces are weaker or different because men's looks matter less to women than women's looks matter to men.

Though the beauty industry is not a conspiracy against women, it is not innocuous either. We calibrate our eye for beauty against the people we see, including our illusory neighbors in the mass media. A daily diet of freakishly beautiful virtual people may recalibrate the scales and make the real ones, including ourselves, look ugly.

For humans, like birds, life is complicated because of two of their reproductive habits. Males invest in their offspring, but fertilization happens out
of sight inside the female's body, so a male never knows which offspring
are his. A female, in contrast, can be certain that any egg or baby coming
out of her body carries her genes. A cuckolded male is worse than a celibate one in the evolutionary struggle, and male birds have evolved defenses
against it. So have humans. Sexual jealousy is found in all cultures.

Both sexes can feel intense jealousy at the thought of a dallying mate,
but their emotions are different in two ways. Women's jealousy appears
to be under the control of more sophisticated software, and they can
appraise their circumstances and determine whether the man's behavior
poses a threat to their ultimate interests. Men's jealousy is cruder and
more easily triggered. (Once triggered, though, women's jealousy appears
to be as intensely felt as men's.) In most societies, some women readily
share a husband, but in no society do men readily share a wife. A woman
having sex with another man is *always* a threat to the man's genetic interests, because it might fool him into working for a competitor's genes, but
a man having sex with another woman is not necessarily a threat to the
woman's genetic interests, because his illegitimate child is another
woman's problem. It is only a threat if the man diverts investment from
her and her children to the other woman and her children, either temporarily or, in the case of desertion, permanently.

So men and women should be jealous of different things. Men should
squirm at the thought of their wives or girlfriends having sex with
another man; women should squirm at the thought of their husbands or
boyfriends giving time, resources, attention, and affection to another
woman. Of course no one likes to think of their mate offering sex *or*
affection to anyone else, but even then the reasons may differ: men may
be upset about affection because it could lead to sex; women may be
upset about sex because it could lead to affection. Buss found that men
and women are made as jealous by the thought of alienated sex as by the
thought of alienated affection, but when asked to pick their torture, most
men said they were more upset by the thought of their partner being sexually unfaithful than emotionally unfaithful, and most women had the
opposite reaction. (The same differences are found when men and
women imagine their partners being both sexually *and* emotionally

unfaithful and are asked which aspect of the betrayal bothers them more. That shows that the sex difference is not just a matter of men and women having different expectations of their partners' behavior, the men worrying that a woman having sex must also be in love and the women worrying that a man in love must also be having sex.) Buss then pasted electrodes on people and asked them to imagine the two kinds of treachery. The men sweated, frowned, and palpitated more from images of sexual betrayal; the women sweated, frowned, and palpitated more from images of emotional betrayal. (I cited the experiment in Chapter 4 as an illustration of the power of mental images.) Similar results have been found in several countries in Europe and Asia.

It takes two to commit adultery, and men, always the more violent sex, have directed their anger at both parties. The largest cause of spousal abuse and spousal homicide is sexual jealousy, almost always the man's. Men beat and kill their wives and girlfriends to punish them for real or imagined infidelity and to deter them from becoming unfaithful or leaving them. Women beat and kill their husbands in self-defense or after years of abuse. Critics of feminism have made much of the occasional statistic that American men are victims of beating and homicide by their spouses almost as often as the women are. But that's not true in the vast majority of communities, and even in the few where it is, the husband's jealousy and intimidation are almost always the cause. Often a morbidly jealous man will imprison his wife in the house and interpret every incoming phone call as proof that she is unfaithful. Women are most at risk when they threaten to leave or do it. The forsaken man may stalk her, hunt her down, and execute her, always with the same rationale: "If I can't have her, no one can." The crime is pointless, but it is the undesired outcome of a paradoxical tactic, a doomsday machine. For every killing of an estranged wife or girlfriend there must be thousands of threats made credible by signs that the man is crazy enough to carry them out regardless of the cost.

Many pundits blame violence against women on this or that feature of American society, such as circumcision, war toys, James Bond, or football. But it happens worldwide, including in foraging societies. Among the Yanomamö, a man who suspects his wife of infidelity might slash her with a machete, shoot her with an arrow, hold an ember against her, cut off her ears, or kill her. Even among the idyllic !Kung San of the Kalahari Desert in southern Africa, men batter wives they suspect of being unfaithful. Incidentally, none of these points "condone" the violence or imply that "it's not the man's fault," as it is sometimes claimed. Those non sequiturs could be attached to *any* explanation, such as the common feminist theory that men are brainwashed by media images that glorify violence against women.

All over the world, men also beat and kill cuckolds and suspected cuckolds. Recall that rivalry over women is the leading cause of violence, homicide, and warfare among foraging peoples. As it is written in Proverbs 6:34, "For jealousy is the rage of a man: therefore he will not spare in the day of vengeance."

Unlike birds, though, humans plug their sexual jealousy into a baroque cognitive machine. People think in metaphors, and the metaphor that men have always used for wives is property. In their essay "The Man Who Mistook His Wife for a Chattel," Wilson and Daly show that men do not merely aim to control their wives and fend off rivals; they assert an *entitlement* to wives, especially their reproductive capacity, identical to the right of an owner over inanimate property. An owner can sell, exchange, or dispose of his possessions, can modify them without interference, and can demand redress for theft or damage. These rights are recognized by the rest of society and can be enforced by collective reprisals. In culture after culture, men have deployed the full cognitive apparatus of ownership in conceiving of their relationship to their wives, and until recently they have formalized the metaphor in codes of law.

In most societies, marriage is a blatant transfer of ownership of a woman from her father to her husband. In our own marriage ceremony, the father of the bride still "gives her away," but more commonly he sells her. In seventy percent of societies, someone pays when two people get married. In ninety-six percent of these, the groom or his family pays the bride's family, sometimes in cash or a daughter, sometimes in bride-service, whereby the groom works for the bride's father for a fixed period. (In the Bible, Jacob worked for Laban for seven years for the right to marry his daughter Rachel, but Laban substituted his other daughter, Leah, at the wedding, so Jacob had to work *another* seven years to acquire Rachel as his second wife.) Dowries, which are more familiar to us, are not a mirror image of bride-wealth, because they go to the newlyweds, not to the bride's parents. The husband notifies other men of his ownership in customs retained by many modern couples. The woman, not the man, wears an engagement ring, bears her spouse's surname, and is given a new form of address, *Mrs.*, short for "mistress of."

People can control their property, and husbands (and before them, fathers and brothers) have controlled women's sexuality. They have used chaperones, veils, wigs, chadors, segregation by sex, confinement, foot-binding, genital mutilation, and the many ingenious designs for chastity belts. Despots not only kept harems but kept them guarded. In traditional societies, "protecting a woman" was a euphemism for keeping her chaste. (Mae West observed, "Men always say they're protecting you, but they never say from what.") Only fertile women were controlled in these ways; children and postmenopausal women had more freedom.

The word *adultery* is related to the word *adulterate* and refers to making a woman impure by introducing an improper substance. The infamous double standard, in which a married woman's philandering is punished more severely than a married man's, is common in legal and moral codes in all kinds of societies. Its rationale was succinctly captured when James Boswell remarked, "There is a great difference between the offence of infidelity in a man and that of his wife," and Samuel Johnson replied, "The difference is boundless. The man imposes no bastards on his wife." Both the married woman and her lover are commonly punishable (often by death), but the symmetry is illusory, because it is the woman's marital status, not the man's, that makes it a crime, specifically, a crime against her husband. Until recently most of the world's legal systems treated adultery as a property violation or tort. The husband was entitled to damages, a refund of the bride-price, a divorce, or the right to violent revenge. Rape was an offense against the woman's husband, not against the woman. Elopement was considered an abduction of a daughter from her father. Until very recently, the rape of a woman by her husband was not a crime, or even a coherent concept: husbands were entitled to sex with their wives.

Throughout the English-speaking world, the common law recognizes three circumstances that reduce murder to manslaughter: self-defense, the defense of close relatives, and sexual contact with the man's wife. (Wilson and Daly observe that they are the three main threats to Darwinian fitness.) In several American states, including Texas as recently as 1974, a man who discovered his wife in flagrante delicto and killed her lover was not guilty of a crime. Even today, in many places those homicides are not prosecuted or the killer is treated leniently. Jealous rage at the sight of a wife's adultery is cited as one of the ways a "reasonable man" can be expected to behave.

～

I wish I could have discussed the evolutionary psychology of sexuality without the asides about feminist theory, but in today's intellectual climate that is impossible. The Darwinian approach to sex is often attacked as being antifeminist, but that is just wrong. Indeed, the accusation is baffling on the face of it, especially to the many feminist women who have developed and tested the theory. The core of feminism is surely the goal of ending sexual discrimination and exploitation, an ethical and political position that is in no danger of being refuted by any foreseeable scientific theory or discovery. Even the spirit of the research poses no threat to feminist ideals. The sex differences that have been documented are in the psychology of reproduction, not in economic or political worth, and they are invidious with regard to men, not women. The differences should heighten awareness of incest, exploitation, harassment, stalking, battering, rape (including date rape and marital rape), and legal codes that discriminate against women. If they show that men are especially tempted to commit certain crimes against women, the implication is that the deterrents should be surer and more severe, not that the crimes are somehow less odious. Even evolutionary explanations of the traditional division of labor by sex do not imply that it is unchangeable, "natural" in the sense of good, or something that should be forced on individual women or men who don't want it.

What evolutionary psychology challenges is not the goals of feminism, but parts of the modern orthodoxy about the mind that have been taken up by the intellectual establishment of feminism. One idea is that people are designed to carry out the interests of their class and sex, rather than to act out of their own beliefs and desires. A second is that the minds of children are formed by their parents, and the minds of adults are formed by language and by media images. A third is the romantic doctrine that our natural inclinations are good and that ignoble motives come from society.

The unstated premise that nature is nice lies behind many of the objections to the Darwinian theory of human sexuality. Carefree sex is natural and good, it is assumed, so if someone claims that men want it more than women do, it would imply that men are mentally healthy and women neurotic and repressed. That conclusion is unacceptable, so the claim that men want carefree sex more than women do cannot be correct. Similarly, sexual desire is good, so if men rape for sex (rather than to express anger towards women), rape would not be as evil. Rape is evil; therefore the claim that men rape for sex cannot be correct. More generally, what people instinctively like is good, so if people like beauty, beauty would be a sign of worth. Beauty is not a sign of worth, so the claim that people like beauty cannot be correct.

These kinds of arguments combine bad biology (nature is nice), bad psychology (the mind is created by society), and bad ethics (what people like is good). Feminism would lose nothing by giving them up.

RIVALS

People everywhere strive for a ghostly substance called authority, cachet, dignity, dominance, eminence, esteem, face, position, preeminence, prestige, rank, regard, repute, respect, standing, stature, or status. People go hungry, risk their lives, and exhaust their wealth in pursuit of bits of ribbon and metal. The economist Thorstein Veblen noticed that people sacrificed so many necessities of life to impress one another that they appear to be responding to a "higher, spiritual need." Status and virtue are close in people's minds, as we see in words like *chivalrous, classy, courtly, gentlemanly, honorable, noble*, and *princely*, and their opposites *ill-bred, low-class, low-rent, mean, nasty, rude, shabby*, and *shoddy*. When it comes to the trifles of personal appearance, we express our admiration for the tasteful using ethical metaphors such as *right, good, correct*, and *faultless*, and censure the tacky with tones usually reserved for sin—an attitude that the art historian Quentin Bell dubbed "sartorial morality."

Is this any way to build an intelligent organism? Where do these powerful motives come from?

Many animals are moved by pointless decorations and rituals, and the selective causes are no longer mysterious. Here is the key idea. Creatures differ in their ability to hurt and help others. Some are stronger or fiercer or more poisonous; some have better genes or more largesse. These potent creatures want everyone to know they are potent, and the creatures they can impinge on *also* want to know which ones are potent. But it is impossible for every creature to probe every other one's DNA, muscle mass, biochemical composition, ferocity, and so on. So the consequential creatures advertise their worth with a signal. Unfortunately, the inconsequential creatures can counterfeit the signal and reap the benefits, debasing its value to everyone else. The race is on for the consequential creatures to cook up a display that is hard to counterfeit, for the less consequential ones to become better counterfeiters, and for the third parties to sharpen their powers of discrimination. Like paper currency, the signals are inimitably gaudy and intrinsically worthless, but are treated as if they were valuable and *are* valuable because everyone treats them that way.

The precious stuff behind the displays can be divided into dominance—who can hurt you—and status—who can help you. They often go together, because people who can hurt you can also help you by their ability to hurt others. But it's convenient to look at them separately.

Most people have heard of the dominance hierarchies, pecking orders, and alpha males that are widespread in the animal kingdom. Animals of the same species don't fight to the death every time they contest something of value. They have a ritualized fight or a show of arms or a staring match, and one backs down. Konrad Lorenz and other early ethologists thought that gestures of surrender helped preserve the species against internecine bloodshed, and that humans were in peril because we lost the gestures. But that idea comes from the fallacy that animals evolve to benefit the species. It cannot explain why a truculent mutant that never surrendered and that killed surrenderers would not walk over the competition and soon characterize the species. The biologists John Maynard Smith and Geoffrey Parker came up with a better explanation by modeling how the different aggressive strategies that animals might adopt would stack up against each other and against themselves.

Fighting every contest to the bitter end is a poor strategy for an animal, because chances are its adversary has evolved to do the same thing. A fight is costly to the loser, because it will be injured or dead and hence worse off than if it had relinquished the prize from the start. It also can be costly to the victor because he may sustain injuries in the course of victory. Both parties would have done better if they had assessed who was likely to win beforehand and if the underdog simply conceded. So animals size each other up to see who's bigger, or brandish their weapons to see whose are more dangerous, or wrestle until it's clear who's stronger. Though only one animal wins, both walk away. The loser concedes because he can seek his fortunes elsewhere or bide his time until circumstances are more propitious. When animals size each other up, they evolve ways to exaggerate their size: ruffs, balloons, manes, bristling, rearing, and bellowing, whose low pitch shows off the size of the resonating cavity in the animal's body. If a fight is costly and a winner unpredictable, the faceoff may be decided by an arbitrary difference such as who arrived first, in the same way that human rivals may settle a dispute quickly by flipping a coin. If the animals are closely matched and the stakes are high enough (such as a harem), an all-out fight may ensue, sometimes to the death.

If both creatures walk away, they may remember the outcome and thereafter the loser will defer to the winner. When many animals in a group spar or size one another up in a round-robin, the outcome is a pecking order, which correlates with the probability that each animal would win an all-out duel. When the probabilities change—say, when a dominant animal gets old or injured, or an underling gains in strength or experience—the underling may mount a challenge and the rankings may change. In chimpanzees, dominance depends not only on fighting prowess but on political acumen: a pair in cahoots may depose a stronger animal going it alone. Many group-living primates settle into two dominance hierarchies, one for each sex. The females compete for food; the males compete for females. Dominant males mate more often, both because they can shove other males out of the way and because the females prefer to mate with them, if for no other reason than that a high-ranking sex partner will tend to sire high-ranking sons, who will give the female more grandchildren than low-ranking sons.

Humans don't have rigid pecking orders, but in all societies people recognize a kind of dominance hierarchy, particularly among men. High-ranking men are deferred to, have a greater voice in group decisions, usually have a greater share of the group's resources, and always have more wives, more lovers, and more affairs with other men's wives. Men strive for rank, and achieve it in some ways that are familiar from zoology books and other ways that are uniquely human. Better fighters have higher rank, and men who *look* like better fighters have higher rank. Sheer height is surprisingly potent in a species that calls itself the rational animal. The word for "leader" in most foraging societies is "big man," and in fact the leaders usually *are* big men. In the United States, taller men are hired more, are promoted more, earn more ($600 per inch in annual salary), and are elected president more: the taller candidate won twenty of the twenty-four elections between 1904 and 1996. A glance at the personal ads shows that women want taller men. As in other species whose males compete, the human male is bigger than the female, and has evolved ways of appearing bigger still, like a low voice and a beard (which makes the head look bigger and has evolved separately in lions and monkeys). Leonid Brezhnev claimed that he got to the top because of his eyebrows! Men everywhere exaggerate the size of their heads (with hats, helmets, headdresses, and crowns), their shoulders (with pads, boards, epaulettes, and feathers), and, in some societies, their penises (with impressive codpieces and sheaths, sometimes a yard long).

But humans also evolved language and a new way of propagating information about dominance: reputation. Sociologists have long been puzzled that the largest category of motives for homicide in American cities is not robbery, drug deals gone sour, or other tangible incentives. It is a category they call "altercation of relatively trivial origin; insult, curse, jostling, etc." Two young men argue over who gets to use the pool table in a bar. They shove each other and trade insults and obscenities. The loser, humiliated before onlookers, storms off and returns with a gun. The murders are the epitome of "senseless violence," and the men who commit them are often written off as madmen or animals.

Daly and Wilson point out that these men behave as if a great deal more is at stake than the use of a pool table. And a great deal more *is* at stake:

> Men are known by their fellows as "the sort who can be pushed around" and "the sort who won't take any shit," as people whose word means action or people who are full of hot air, as guys whose girlfriends you can chat up with impunity or guys you don't want to mess with.
>
> In most social milieus, a man's reputation depends in part upon the maintenance of a credible threat of violence. Conflicts of interest are endemic to society, and one's interests are likely to be violated by competitors unless those competitors are *deterred*. Effective deterrence is a matter of convincing our rivals that any attempt to advance their interests at our expense will lead to such severe penalties that the competitive gambit will end up a net loss which should never have been undertaken.

The credibility of the deterrent can be devalued by a public challenge that is not taken up, even if nothing tangible is at stake. Moreover, if a challenger knew that his target was a cool calculator of costs and benefits, he could extort him into backing down with the threat of a fight that was dangerous to both. But a hothead who would stop at nothing to preserve his reputation (a doomsday machine) is unextortable.

The ghetto gang member who stabs the guy who dissed him has honorable counterparts in all the world's cultures. The very meaning of the word *honor* in many languages (including one of its senses in English) is a determination to avenge insults, with bloodshed if necessary. In many foraging societies a boy achieves manly status only after he has killed. A man's respect increases with his verified body count, giving rise to charming customs like scalping and headhunting. Dueling between "men of honor" was traditional in the American South, and many men rose to leadership with the help of their success in duels. The man on the ten-dollar bill, Secretary of the Treasury Alexander Hamilton, was killed in a duel by Vice President Aaron Burr, and the man on the twenty, President Andrew Jackson, won two duels and tried to provoke others.

Why don't we see periodontists or college professors dueling over a parking space? First, they live in a world in which the state has a monopoly on the legitimate use of violence. In places beyond the reach of the state, like urban underworlds or rural frontiers, or in times when the state did not exist, like the foraging bands in which we evolved, a credible threat of violence is one's only protection. Second, the assets of periodontists and professors, such as houses and bank accounts, are hard to steal. "Cultures of honor" spring up when a rapid response to a threat is essential because one's wealth can be carried away by others. They develop among herders, whose animals can be stolen, more often than among crop-growers, whose land stays put. And they develop among people whose wealth is in other liquid forms, like cash or drugs. But perhaps the biggest reason is that periodontists and professors are not male, poor, and young.

Maleness is by far the biggest risk factor for violence. Daly and Wilson report thirty-five samples of homicide statistics from fourteen countries, including foraging and preliterate societies and thirteenth-century England. In all of them, men kill men massively more often than women kill women—on average, twenty-six times more often.

Also, the poolhall avengers and their victims are nobodies: uneducated, unmarried, unprosperous, and often unemployed. Among polygynous mammals such as ourselves, reproductive success varies enormously among males, and the fiercest competition can be at the bottom, among males whose prospects teeter between zero and nonzero. Men attract women by their wealth and status, so if a man doesn't have them and has no way of getting them he is on a one-way road to genetic nothingness. As with birds that venture into dangerous territories when they are near starvation, and hockey coaches that pull the goalie for an extra skater when they are a goal down with a minute to play, an unmarried man without a future should be willing to take any risk. As Bob Dylan pointed out, "When you got nothing, you got nothing to lose."

Youth makes matters even worse. The population geneticist Alan Rogers has calculated from actuarial data that young men should discount the future steeply, and so they do. Young men commit crimes, drive too fast, ignore illnesses, and pick dangerous hobbies like drugs, extreme sports, and surfing on the roofs of tram cars and elevators. The combination of maleness, youth, penury, hopelessness, and anarchy makes young men indefinitely reckless in defending their reputation.

And it's not so clear that professors (or people in any competitive profession) *don't* duel over pool tables, figuratively speaking. Academics are known by their fellows as "the sort who can be pushed around" and "the sort who won't take any shit," as people whose word means action or people who are full of hot air, as guys whose work you can criticize with impunity or guys you don't want to mess with. Brandishing a switchblade at a scholarly conference would somehow strike the wrong note, but there is always the stinging question, the devastating riposte, the moralistic outrage, the withering invective, the indignant rebuttal, and means of enforcement in manuscript reviews and grant panels. Scholarly institutions, of course, try to minimize this rutting, but it is hard to eradicate. The goal of argumentation is to make a case so forceful (note the metaphor) that skeptics are *coerced* into believing it—they are powerless to deny it while still claiming to be rational. In principle, it is the ideas themselves that are, as we say, compelling, but their champions are not always averse to helping the ideas along with tactics of verbal dominance, among them intimidation ("Clearly . . ."), threat ("It would be unscientific to . . ."), authority ("As Popper showed . . ."), insult ("This work lacks the necessary rigor for . . ."), and belittling ("Few people today seriously believe that . . ."). Perhaps this is why H. L. Mencken wrote that "college football would be more interesting if the faculty played instead of the students."

~

Status is the public knowledge that you possess assets that would allow you to help others if you wished to. The assets may include beauty, irreplaceable talent or expertise, the ear and trust of powerful people, and especially wealth. Status-worthy assets tend to be fungible. Wealth can bring connections and vice versa. Beauty can be parlayed into wealth (through gifts or marriage), can attract the attention of important people, or can draw more suitors than the beautiful one can handle. Asset-holders, then, are not just seen as holders of their assets. They exude an aura or charisma that makes people want to be in their graces. It's always handy to have people want to be in your graces, so status itself is worth craving. But there are only so many hours in the day, and sycophants must choose whom to fawn over, so status is a limited resource. If A has more, B must have less, and they must compete.

Even in the dog-eat-dog world of tribal leadership, physical dominance is not everything. Chagnon reports that some Yanomamö headmen are flamboyant bullies but others achieve their station by shrewdness and discretion. A man named Kaobawä, though no wimp, earned his authority by leaning on the support of his brothers and cousins and cultivating alliances with the men with whom he had traded wives. He conserved his authority by giving orders only when he was sure everyone would follow them, and magnified it by breaking up fights, disarming machete-wielding maniacs, and bravely scouting the village alone when raiders were in evidence. His quiet leadership was rewarded with six wives and as many affairs. In foraging societies, status also clings to good hunters and knowledgeable naturalists. Assuming that our ancestors, too, practiced occasional meritocracy, human evolution was not always the survival of the fiercest.

Romantic anthropologists used to claim that foraging peoples were unmoved by wealth. But that is because the foragers they studied didn't have any. Twentieth-century hunter-gatherers are unrepresentative of humanity in one respect. They live on land that no one else wants, land that cannot be farmed. They don't necessarily prefer their deserts, rainforests, and tundras, but farming peoples like us have taken the rest. Though foragers cannot achieve the massive inequality that comes from cultivating and storing food, they do have inequality, both of wealth and of prestige.

The Kwakiutl of the Canadian Pacific coast enjoyed annual runs of salmon and abundant sea mammals and berries. They settled in villages run by wealthy chiefs who tried to outdo one another in competitive feasts called potlatches. The guests at a potlatch were encouraged to gorge themselves on salmon and berries, and the chief boastfully showered them with boxes of oil, baskets of berries, and piles of blankets. The humiliated guests slunk back to their village and plotted revenge with an even bigger feast, in which they would not only give away valuables but ostentatiously destroy them. The chief would start a roaring fire in the center of his house and stoke it with fish oil, blankets, furs, canoe paddles, canoes, and sometimes the house itself, a spectacle of consumption the world would not see again until the American bar mitzvah.

Veblen proposed that the psychology of prestige was driven by three "pecuniary canons of taste": conspicuous leisure, conspicuous consumption, and conspicuous waste. Status symbols are flaunted and coveted not necessarily because they are useful or attractive (pebbles, daisies, and pigeons are quite beautiful, as we rediscover when they delight young children), but often because they are so rare, wasteful, or pointless that only the wealthy can afford them. They include clothing that is too delicate, bulky, constricting, or stain-prone to work in, objects too fragile for casual use or made from unobtainable materials, functionless objects made with prodigious labor, decorations that consume energy, and pale skin in lands where the plebeians work in the fields and suntans in lands where they work indoors. The logic is: You can't see all my wealth and earning power (my bank account, my lands, all my allies and flunkeys), but you can see my gold bathroom fixtures. No one could afford them without wealth to spare, therefore you know I am wealthy.

Conspicuous consumption is counterintuitive because squandering wealth can only reduce it, bringing the squanderer down to the level of his or her rivals. But it works when other people's esteem is useful enough to pay for and when not *all* the wealth or earning power is sacrificed. If I have a hundred dollars and you have forty, I can give away fifty, but you can't; I will impress others and *still* be richer than you. The principle has been confirmed from an unlikely source, evolutionary biology. Biologists since Darwin had been puzzled by displays like the peacock's tail, which impresses the peahen but consumes nutrients, hinders movement, and attracts predators. The biologist Amotz Zahavi proposed that the displays evolved *because* they were handicaps. Only the healthiest animals could afford them, and females choose the healthiest birds to mate with. Theoretical biologists were initially skeptical, but one of them, Alan Grafen, later proved that the theory was sound.

Conspicuous consumption works when only the richest can afford luxuries. When the class structure loosens, or sumptuous goods (or good imitations) become widely available, the upper middle class can emulate the upper class, the middle class can emulate the upper middle class, and so on down the ladder. The upper class cannot very well stand by as they begin to resemble the hoi polloi; they must adopt a new look. But then the look is emulated once again by the upper middle class and begins to trickle down again, prompting the upper class to leap to yet a different look, and so on. The result is fashion. The chaotic cycles of style, in which the chic look of one decade becomes dowdy or slutty, nerdy or foppish in the next, has been explained as a conspiracy of clothing makers, an expression of nationalism, a reflection of the economy, and much else. But Quentin Bell, in his classic analysis of fashion, *On Human Finery*, showed that only one explanation works: people follow the rule, "Try to look like the people above you; if you're at the top, try to look different from the people below you."

Once again animals discovered the trick first. The other dandies of the animal kingdom, butterflies, did not evolve their colors to impress the females. Some species evolved to be poisonous or distasteful, and warned their predators with gaudy colors. Other poisonous kinds copied the colors, taking advantage of the fear already sown. But then some *non*poisonous butterflies copied the colors, too, enjoying the protection while avoiding the expense of making themselves distasteful. When the mimics become too plentiful, the colors no longer conveyed information and no longer deterred the predators. The distasteful butterflies evolved new colors, which were then mimicked by the palatable ones, and so on.

Wealth is not the only asset that people flaunt and covet. In a complicated society, people compete in many leagues, not all of them dominated by plutocrats. Bell added a fourth canon to Veblen's list: conspicuous outrage. Most of us depend on the approval of others. We need the favor of bosses, teachers, parents, clients, customers, or prospective in-laws, and that requires a certain measure of respect and unobtrusiveness. Aggressive nonconformity is an advertisement that one is so confident in one's station or abilities that one can jeopardize the good will of others without ending up ostracized and destitute. It says, "I'm so talented, wealthy, popular, or well-connected that I can afford to offend you." The nineteenth century had the baroness George Sand smoking a cigar in trousers and Oscar

Wilde in knee breeches with long hair and a sunflower. In the last half of the twentieth century conspicuous outrage has become the convention, and we have been treated to a tedious parade of rebels, outlaws, wild ones, bohemians, freaks, punks, shock jocks, gender-benders, mau-maus, bad boys, gangstas, sex divas, bitch goddesses, vamps, tramps, and material girls. Hipness has replaced classiness as the motor of fashion, but the status psychology is the same. Trend-setters are members of upper classes who adopt the styles of lower classes to differentiate themselves from middle classes, who wouldn't be caught dead in lower-class styles because they're the ones in danger of being mistaken for them. The style trickles downward, sending the hip off in search of a new form of outrage. As the media and the merchandisers learn to market each new wave more efficiently, the avant-garde merry-go-round goes faster and more furiously. A regular feature of urban newspapers is the favorable notice of an "alternative" band followed by haughty letters advising that they were good when few had heard of them but that they have now sold out. Tom Wolfe's mordant social commentaries (*The Painted Word, From Bauhaus to Our House, Radical Chic*) document how a thirst for status in the form of hipness drives the worlds of art, architecture, and the politics of the cultural elite.

FRIENDS AND ACQUAINTANCES

People bestow favors on one another even when they are unrelated and have no sexual interest. It is easy to understand why even the most selfish organism might want to do so. If favors are traded, both parties profit as long as the value of what they get is greater to them than the value of what they give up. A clear example is a commodity whose benefit shows diminishing returns. If I have two pounds of meat and no fruit, and you have two pounds of fruit and no meat, the second pound of meat is worth less to me than the first (since there's only so much meat I can eat at a sitting), and you feel the same way about your second pound of fruit. We're both better off if we exchange a pound for a pound. Economists call the benefit a gain in trade.

When traders exchange goods simultaneously, cooperation is easy. If the other guy is reneging, you hang on to your meat or grab it back. Most favors, however, cannot be retracted, such as sharing information, saving a drowning person, or helping in a fight. Also, most favors cannot change hands at

the same time. Needs may change; if I help you now in return for protection of my unborn child, I cannot collect until the child is born. And surpluses often are staggered; if you and I have just felled antelopes, there's no
point in trading identical carcasses. Only if you felled one today and I fell
one in a month does it make sense to trade. Money is one solution, but it is
a recent invention and could not have figured in our evolution.

As we saw in Chapter 6, the problem with *delayed* exchanges, or reciprocation, is that it's possible to cheat, to accept a favor now and not
return it later. Obviously everyone would be better off if no one cheated.
But as long as the other guy *might* cheat (which is inevitable when individuals can vary), I may be discouraged from extending him a favor that
in the long run would help us both. The problem has been compressed
into a parable called the Prisoner's Dilemma. Partners in crime are held
in separate cells, and the prosecutor offers each one a deal. If you rat on
your partner and he stays mum, you go free and he gets ten years. If you
both stay mum, you both get six months. If you both rat, you both get
five years. The partners cannot communicate, and neither knows what
the other will do. Each one thinks: If my partner rats and I stay mum, I'll
do ten years; if he rats and I rat, too, I'll do five years. If he stays mum
and I stay mum, I'll do six months; if he stays mum and I rat, I'll go free.
Regardless of what he does, then, I'm better off betraying him. Each is
compelled to turn in his partner, and they both serve five years—far
worse than if each had trusted the other. But neither could take the
chance because of the punishment he would incur if the other didn't.
Social psychologists, mathematicians, economists, moral philosophers,
and nuclear strategists have fretted over the paradox for decades. There
is no solution.

Real life, however, is not a Prisoner's Dilemma in one respect. The
mythical prisoners are placed in their dilemma once. Real people face
each other in dilemmas of cooperation again and again, and can remember past treacheries or good turns and play accordingly. They can feel
sympathetic and extend good will, feel aggrieved and seek revenge, feel
grateful and return a favor, or feel remorseful and make amends. Recall
that Trivers proposed that the emotions making up the moral sense could
evolve when parties interacted repeatedly and could reward cooperation
now with cooperation later and punish defection now with defection
later. Robert Axelrod and William Hamilton confirmed the conjecture in
a round-robin computer tournament that pitted different strategies for
playing a repeated Prisoner's Dilemma game against each other. They

stripped the dilemma to its essentials and awarded points to a strategy for the equivalent of minimizing jail time. A simple strategy called tit-for-tat—cooperate on the first move, and then do what your partner did on the move before—beat sixty-two other strategies. Then they ran an artificial life simulation in which each strategy "reproduced" in proportion to its winnings and a new round-robin took place among the copies of the strategies. They repeated the process for many generations and found that the Tit for Tat strategy took over the population. Cooperativeness can evolve when the parties interact repeatedly, remember each other's behavior, and reciprocate it.

As we saw in Chapters 5 and 6, people are good at detecting cheaters and are fitted with moralistic emotions that prompt them to punish the cheaters and reward the cooperators. Does that mean that tit-for-tat underlies the widespread cooperation we find in the human species? It certainly underlies much of the cooperation we find in our society. Cash-register tapes, punch clocks, train tickets, receipts, accounting ledgers, and the other accoutrements of transactions that do not rely on the "honor system" are mechanical cheater-detectors. The cheaters, such as thieving employees, are sometimes charged with crimes, but more often they are simply cut off from further reciprocation, that is, fired. Similarly, the businesses that cheat their customers soon lose them. Footloose job applicants, fly-by-night businesses, and strangers calling with "investment opportunities" are often discriminated against because they look like they are playing a one-shot rather than an iterated game of cooperation, and so are immune to tit-for-tat. Even moderately good friends privately remember the most recent Christmas gifts and dinner-party invitations and calculate the proper way to reciprocate.

Does all this accounting come from our alienation and bourgeois values in a capitalist society? One of the fondest beliefs of many intellectuals is that there are cultures out there where everyone shares freely. Marx and Engels thought that preliterate peoples represented a first stage in the evolution of civilization called primitive communism, whose maxim was "From each according to his abilities, to each according to his needs." Indeed, people in foraging societies do share food and risk. But in many of them, people interact mainly with their kin, so in the biologist's sense they are sharing with extensions of themselves. Many cultures also have an *ideal* of sharing, but that means little. Of course I will proclaim how great it is for *you* to share; the question is, will *I* share when my turn comes?

Foraging peoples, to be sure, really do share with nonrelatives, but not out of indiscriminate largesse or a commitment to socialist principles. The data from anthropology show that the sharing is driven by cost-benefit analyses and a careful mental ledger for reciprocation. People share when it would be suicidal not to. In general, species are driven to share when the *variance* of success in gathering food is high. Say in some weeks I am lucky and have more food than I can eat, but in other weeks I am unlucky and in danger of starving. How can I store extra food in the fat weeks and draw on it in the lean weeks? Refrigeration is not an option. I could gorge on it now and store it as blubber, but that works only up to a point; I can't eat enough in a day to avoid hunger for a month. But I *can* store it in the bodies and minds of *other* people, in the form of a memory of my generosity they feel obliged to repay when fortunes reverse. When the prospects are risky, it pays to pool the risks.

The theory has been confirmed in nonhuman species, such as vampire bats, and it has also been confirmed in humans in two elegant studies that control for differences among cultures by contrasting the forms of sharing *within* a culture. The Ache of Paraguay hunt game and gather plant foods. Hunting is largely a matter of luck: on any given day an Ache hunter has a forty percent chance of coming home empty-handed. Gathering is largely a matter of effort: the longer you work, the more you bring home, and an empty-handed gatherer is probably lazy rather than unlucky. As predicted, the Ache share plant foods only within the nuclear family but share meat throughout the band.

The !Kung San of the Kalahari Desert are perhaps the closest thing the world has to primitive communists. Sharing is holy; boasting and hoarding are contemptible. They hunt and gather in a harsh, drought-prone ecosystem, and trade food and access to waterholes. The //Gana San, a neighboring branch of the same people, have taken to cultivating melons, which store water, and to herding goats. They do not yo-yo between good times and bad as much as their cousins, and unlike them, they hoard food and have developed inequalities in wealth and status. In both the Ache and the San, high-variance foods are shared, low-variance foods are hoarded.

These people do not pull out calculators and compute the variances. What goes through their minds when they decide to share? Cosmides and Tooby note that the psychology is hardly exotic; it matches our own sense of fairness and compassion. Consider what makes people more or less willing to help the homeless. Those who urge that we all share with the homeless emphasize the random, variance-driven dimension to homelessness. Homeless people are worthy of aid because they are down on their luck. They are the unfortunate victims of circumstances like unemployment, discrimination, or mental illness. Advocates of the homeless urge us to think, "There but for fortune go I." Those who oppose sharing, on the other hand, emphasize the predictability of rewards in our society to anyone willing to put in the work. Homeless people are unworthy of aid because they are able-bodied but lazy, or brought it on themselves by choosing to drink or take drugs. Defenders of the homeless reply that drug use is itself an illness that could happen to anyone.

Even at their most munificent, foraging people do not act out of hearts filled with loving kindness. They enforce the sharing ethic with obsessively detailed memories of who has helped, a clear expectation of payback, and snide gossip about those who don't pitch in. And all this still does not expunge selfish feelings. The anthropologist Melvin Konner, who lived with the !Kung San for years and has written respectfully about their ways, tells his readers:

> Selfishness, arrogance, avarice, cupidity, fury, covetousness, all these forms of gluttony are held in check in their traditional situation in the same way simple alimentary gluttony is: Namely, it doesn't happen because the situation does not allow it. Nor, as some suppose, because the people or their culture are somehow better. I will never forget the time a !Kung man—the father of a family, about forty years of age, well respected in the community, a good and substantial man in every way— asked me to hold on to a leg of antelope he had killed. He had given away most of it, as one had to. But he saw a chance to hide some of it, for later, for himself and his own family. Ordinarily, of course, there would be no place in the entire Kalahari to hide it; it would either be unsafe from scavengers or unsafe from predatory distant relatives. But the presence of foreigners presented an interface with another world, and he wanted to slip the meat, temporarily, through a chink in that interface, into the only conceivable hiding place.

～

When it comes to friendship, reciprocal altruism does not ring true. It would be in questionable taste for a dinner guest to pull out his wallet and offer to pay the hosts for his dinner. Inviting the hosts back the very next night would not be much better. Tit-for-tat does not cement a friendship; it strains it. Nothing can be more awkward for good friends than a business transaction between them, like the sale of a car. The same is true for one's best friend in life, a spouse. The couples who keep close track of what each has done for the other are the couples who are the least happy.

Companionate love, the emotion behind close friendship and the enduring bond of marriage (the love that is neither romantic nor sexual), has a psychology of its own. Friends or spouses feel as if they are in each other's debt, but the debts are not measured and the obligation to repay is not onerous but deeply satisfying. People feel a spontaneous pleasure in helping a friend or a spouse, without anticipating repayment or regretting the favor if repayment never comes. Of course, the favors may be tabulated somewhere in the mind, and if the ledger has become too lopsided, a person might call in the debt or cut off future credit, that is, end the friendship. But the line of credit is long and the terms of repayment forgiving. Companionate love, then, does not literally contradict the theory of reciprocal altruism, but it does embody an elastic version in which the emotional guarantors—liking, sympathy, gratitude, and trust—are stretched to the limit.

The facts of companionate love are clear enough, but why did it evolve? Tooby and Cosmides have tried to reverse-engineer the psychology of friendship by calling attention to an aspect of the logic of exchange they call the Banker's Paradox. Many frustrated borrowers have learned that a bank will lend you exactly as much money as you can prove you don't need. As Robert Frost put it, "A bank is a place where they lend you an umbrella in fair weather and ask for it back when it begins to rain." The banks say they have only so much money to invest and every loan is a gamble. Their portfolio has to return a profit or they would go out of business, so they measure credit risks and weed out the worst.

The same cruel logic applies to altruism among our ancestors. A person mulling over whether to extend a large favor is like a bank. He must worry not only about cheaters (is the beneficiary willing to repay?) but about bad credit risks (is the beneficiary *able* to repay?). If the recipient dies, is disabled, becomes a pariah, or leaves the group, the favor would have been wasted. Unfortunately, it is the bad credit risks—the sick, starving, injured, and ostracized—who most *need* favors. Anyone can suffer a reversal of fortune, especially in the harsh life of a forager. Once abandoned, a stricken forager is not long for that world. What kinds of thoughts and feelings might evolve as a kind of insurance in which other people would extend "credit" to you even if misfortune were to make you a risk?

One strategy is to make yourself irreplaceable. By cultivating expertise that no one in the group can duplicate, like toolmaking, wayfinding, or conflict resolution, you make yourself costly to abandon in times of need: everyone depends upon you too much to risk letting you die. People today do spend a lot of their social lives publicizing their unique and valuable talents or looking for a clique in which their talents would be unique and valuable. The quest for status is in part a motive for making oneself irreplaceable.

Another is to associate with people who benefit from the things that benefit you. Merely by going about your life and pursuing your own interests, you can advance someone else's interests as a side effect. Marriage is the clearest example: the husband and wife share an interest in their children's welfare. Another was pointed out by Mao Tse-tung in his little red book: "The enemy of my enemy is my friend." A third is to possess skills that benefit others at the same time that they benefit you, like being good at finding your way home. Other examples are living with a person who likes the room at the same temperature or who likes the same music. In all the examples, one delivers a benefit to someone without being altruistic in the biologist's sense of incurring a cost and thereby needing a repayment to make the act worthwhile. The challenge of altruism has attracted so much attention that a more direct form of helping in nature has often been downplayed: symbiosis, in which two organisms, such as the algae and fungi making up lichen, associate because the side effects of each one's lifestyle fortuitously benefit the other one. Symbionts give benefits and take them, but neither pays a cost. Roommates with the same taste in music are a kind of symbiotic pair, and each can value the other without an exchange of favors.

Once you have made yourself valuable to someone, the person becomes valuable to you. You value him or her because if you were ever in trouble, they would have a stake—albeit a selfish stake—in getting you out. But now that you value the person, they should value you even more. Not only are you valuable because of your talents or habits, but you are valuable because of your stake in rescuing him or her from hard times. The more you value the person, the more the person values you, and so on. This runaway process is what we call friendship. If you ask people why they are friends, they are likely to say, "We like the same things, and we know we'll always be there for each other."

Friendship, like other kinds of altruism, is vulnerable to cheaters, and we have a special name for them: fair-weather friends. These sham friends reap the benefits of associating with a valuable person and mimic signs of warmth in an effort to become valued themselves. But when a little rain falls, they are nowhere in sight. People have an emotional response that seems designed to weed out fair-weather friends. When we are neediest, an extended hand is deeply affecting. We are moved, never forget the generosity, and feel compelled to tell the friend we will never forget it. Hard times show you who your real friends are. That is because the point of friendship, in evolutionary terms, is to save you in hard times when it's not worth anyone else's trouble.

Tooby and Cosmides go on to speculate that the design of our friendship emotions may explain the alienation and loneliness that so many people feel in modern society. Explicit exchanges and turn-taking reciprocation are the kinds of altruism we fall back on when friendship is absent and trust is low. But in modern market economies we trade favors with strangers at unprecedented rates. It may create the perception that we are not deeply engaged with our fellows and are vulnerable to desertion in difficult times. And ironically, the comfortable environment that makes us physically more secure may make us emotionally less secure, because it minimizes the crises that tell us who our real friends are.

ALLIES AND ENEMIES

No account of human relationships could be complete without a discussion of war. War is not universal, but people in all cultures feel that they are members of a group (a band, tribe, clan, or nation) and feel animosity toward other groups. And warfare itself is a major fact of life for foraging tribes. Many intellectuals believe that primitive warfare is rare, mild, and ritualized, or at least was so until the noble savages were contaminated by contact with Westerners. But this is romantic nonsense. War has always been hell.

Yanomamö villages raid one another endlessly. Seventy percent of all adults over forty have lost a family member to violence. Thirty percent of the men are killed by other men. Forty-four percent of the men have killed someone. The Yanomamö call themselves the Fierce People, but other pristine tribes give similar numbers. The archeologist Lawrence Keeley has documented that New Guineans, Australian aborigines, Pacific Islanders, and Native Americans have been wracked by warfare, especially in the centuries before the Pax Brittanica ended this nuisance to the colonial administrators in much of the world. In primitive warfare, mobilization was more complete, battles were more frequent, casualties higher, prisoners fewer, and weapons more damaging. War is, to put it mildly, a major selection pressure, and since it appears to have been a recurring event in our evolutionary history, it must have shaped parts of the human psyche.

Why would anyone be so stupid as to start a war? Tribal people can fight over anything of value, and the causes of tribal wars are as difficult to disentangle as the causes of World War I. But one motive that is surprising to Westerners appears over and over. In foraging societies, men go to war to get or keep women—not necessarily as a conscious goal of the warriors (though often it is exactly that), but as the ultimate payoff that allowed a willingness to fight to evolve. Access to women is the limiting factor on males' reproductive success. Having two wives can double a man's children, having three wives can triple it, and so on. For a man who is not at death's door, no other resource has as much impact on evolutionary fitness. The most common spoils of tribal warfare are women. Raiders kill the men, abduct the nubile women, gang-rape them, and allocate them as wives. Chagnon discovered that Yanomamö men who had killed an enemy had three times as many wives and three times as many children as those who had not. Most young men who had killed were married; most young men who had never killed were not. The difference is not an accident of other differences between the killers and the non-killers, such as size, strength, or number of kin. Killers are held in esteem in Yanomamö villages; they attract and are ceded more wives.

The Yanomamö sometimes plan raids just to abduct women. More frequently, they plan them to avenge a past killing or abduction, but they always try to abduct women, too. Blood feuds, in which relatives avenge a death with a death, either of the killer or of his relatives, are the major impetus to extended violence everywhere; the motive that drives them has an obvious deterrent function, as we saw in Chapter 6. Blood feuds can extend for decades or longer because each side counts the score differently, so at any time each remembers injustices that must be redressed. (Imagine your feelings toward a neighboring people that has murdered your husband, your brothers, and your sons, or has raped and abducted your wife, your daughters, and your sisters.) But the feuders do not stop at an eye for an eye. If they see an opportunity to get rid of a headache once and for all by massacring their opponents, they may do so, with the women as an extra incentive. The desire for women not only helps to fuel blood feuds; it also helps to spark them in the first place. Usually the first killing was over a woman: a man seduces or abducts someone's wife, or reneges on a deal to trade a daughter.

Modern people have trouble believing that preliterate tribes go to war over women. One anthropologist wrote to Chagnon, "Women? Fighting over women? Gold and diamonds I can understand, but women? Never." The reaction, of course, is biologically topsy-turvy. Other anthropologists argued that the Yanomamö suffered from a protein shortage and were fighting over game. But their protein intake, when measured, turned out to be more than adequate. Across the world the best-fed foraging peoples are the *most* warlike. When Chagnon mentioned the meat-shortage hypothesis to his Yanomamö informants, they laughed incredulously and said, "Even though we like meat, we like women a whole lot more." Chagnon points out that they are not so different from us. "Some Saturday night just visit a hard-hat bar where fights are frequent. What are the fights usually about? Are they about the amount of meat in someone's hamburger? Or study the words of a dozen country-and-western songs. Do any of them say, 'Don't take your cow to town'?"

The similarities run deeper. Warfare among Western peoples is different from primitive warfare in many ways, but it is similar in at least one way: the invaders rape or abduct women. It was codified in the Bible:

> And they warred against the Midianites, as the LORD commanded Moses; and they slew all the males. . . . And the children of Israel took all the women of Midian captives, and their little ones, and took the spoil of all their cattle, and all their flocks, and all their goods. . . . And Moses said unto them, Have ye saved all the women alive? . . . Now therefore kill every male among the little ones, and kill every woman that hath known man by lying with him. But all the women children, that have not known a man by lying with him, keep alive for yourselves. (Numbers 31)

> When thou comest nigh unto a city to fight against it, then proclaim peace unto it. . . . And if it will make no peace with thee, but will make war against thee, then thou shalt besiege it: And when the LORD thy God hath delivered it into thine hands, thou shalt smite every male thereof

with the edge of the sword: But the women, and the little ones, and the cattle, and all that is in the city, even all the spoil thereof, shalt thou take unto thyself. (Deuteronomy 20)

When thou goest forth to war against thine enemies, and the LORD thy God hath delivered them into thine hands, and thou hast taken them captive, And seest among the captives a beautiful woman, and hast a desire unto her, that thou wouldest have her to thy wife; Then thou shalt bring her home to thine house; and she shall shave her head, and pare her nails; And she shall put the raiment of her captivity from off her, and shall remain in thine house, and bewail her father and her mother a full month; and after that thou shalt go in unto her, and be her husband, and she shall be thy wife. (Deuteronomy 21)

According to the *Iliad*, the Trojan War began with the abduction of Helen of Troy. During the First Crusade, Christian soldiers raped their way across Europe to Constantinople. Shakespeare has Henry V threatening a French village during the Hundred Years War that if they do not surrender, it will be their fault that their "pure maidens fall into the hand of hot and forcing violation":

> If not, why, in a moment look to see
> The blind and bloody soldier with foul hand
> Defile the locks of your shrill-shrieking daughters;
> Your fathers taken by the silver beards,
> And their most reverend heads dash'd to the walls,
> Your naked infants spitted upon pikes,
> Whiles the mad mothers with their howls confused
> Do break the clouds, as did the wives of Jewry
> At Herod's bloody-hunting slaughtermen.

The feminist writer Susan Brownmiller has documented that rape was systematically practiced by the English in the Scottish Highlands, the Germans invading Belgium in World War I and eastern Europe in World War II, the Japanese in China, the Pakistanis in Bangladesh, the Cossacks during the pogroms, the Turks persecuting the Armenians, the Ku Klux Klan in the American South, and, to a lesser extent, Russian soldiers marching toward Berlin and American soldiers in Vietnam. Recently the Serbs in Bosnia and the Hutus in Rwanda have added themselves to this list. Prostitution, which in wartime is often hard to distinguish from rape, is a ubiquitous perquisite of soldiers. Leaders may sometimes use rape as a terror tactic to attain other ends, as Henry V obviously did, but the tactic is effective precisely because the soldiers are so eager to implement it, as Henry took pains to remind the Frenchmen. In fact it often backfires by giving the defenders an incalculable incentive to fight on, and probably for that reason, more than out of compassion for enemy women, modern armies have outlawed rape. Even when rape is not a prominent part of our warfare, we invest our war leaders with enormous prestige, just as the Yanomamö do, and by now you know the effects of prestige on a man's sexual attractiveness and, until recently, his reproductive success.

War, or aggression by a coalition of individuals, is rare in the animal kingdom. You would think that the second-, third-, and fourth-strongest elephant seals would gang up, kill the strongest male, and divide his harem among them, but they never do. Aside from the social insects, whose unusual genetic system makes them a special case, only humans, chimpanzees, dolphins, and perhaps bonobos join up in groups of four or more to attack other males. These are some of the largest-brained species, hinting that war may require sophisticated mental machinery. Tooby and Cosmides have worked out the adaptive logic of coalitional aggression and the cognitive mechanisms necessary to support it. (That does not, of course, mean that they think war is unavoidable or "natural" in the sense of "good.")

People often are conscripted into armies, but sometimes they enlist with gusto. Jingoism is alarmingly easy to evoke, even without a scarce resource to fight over. In numerous experiments by Henri Tajfel and other social psychologists, people are divided into two groups, actually at random but ostensibly by some trivial criterion such as whether they underestimate or overestimate the number of dots on a screen or whether they prefer the paintings of Klee or Kandinsky. The people in each group instantly dislike and think worse of the people in the other group, and act to withhold rewards from them even if doing so is costly to their own group. This instant ethnocentrism can be evoked even if the experimenter drops the charade with the dots or paintings and divides people into groups by flipping a coin before their eyes! The behavioral consequences are by no means minor. In a classic experiment, the social psychologist Muzafer Sherif carefully selected a group of well-adjusted, middle-class American boys for a summer camp, and randomly divided them into two groups which then competed in sports and skits. Within days the groups were brutalizing and raiding each other with sticks, bats, and rocks in socks, forcing the experimenters to intervene for the boys' safety.

The enigma of war is why people volunteer for an activity that has an excellent chance of getting them killed. How could a desire to play Russian roulette have evolved? Tooby and Cosmides explain it by the fact that natural selection favors traits that increase fitness *on average*. Every gene contributing to a trait is embodied in many individuals in many generations, so if one individual with the gene dies childless, the success of many others with the gene can make up for it. Imagine a game of Russian roulette where if you don't get killed you have one more offspring. A gene for joining in the game could be selected, because five-sixths of the time it would leave an extra copy in the gene pool and one-sixth of the time it would leave none. On average, that yields .83 more copies than staying out of the game. Joining a coalition of five other men that is certain to capture five women but suffer one fatality is in effect the same choice. The key idea is that the coalition acting together can gain a benefit that its members acting alone cannot, and that spoils are distributed according to the risks undertaken. (There are several complications, but they do not change the point.)

In fact, if the spoils are certain and divided up fairly, the level of danger doesn't matter. Say your coalition has eleven members and can ambush an enemy coalition of five, taking their women. If one member of your coalition is likely to be killed, you have a ten-in-eleven chance of surviving, which would entitle you to a one-in-two chance (five captive women, ten men) of gaining a wife, an expected gain of .45 wives (averaged over many situations with these payoffs). If two members will be killed, you have a smaller chance of surviving (nine in eleven), but if you do survive you have a larger chance of gaining a wife, since your dead allies won't be taking theirs. The average gain ($9/11 \times 5/9$) is the same, .45 wives. Even if *six* members are likely to be killed, so that your survival odds fall to less than even (five in eleven), the spoils are divided fewer ways (five women among five victors), so if you survive you are guaranteed a wife, for an expected gain, once again, of .45 wives.

Tooby and Cosmides' calculations assume that a man's children can do just fine when he is dead, so the loss of fitness with death is zero, not negative. Of course that is not true, but they point out that if the group is relatively prosperous the fatherless children's survival chances may not diminish too much and it still could pay men to raid. They predict that men should be more willing to fight when their group is secure in food than when it is hungry, contrary to the protein-shortage hypothesis. The data bear them out. Another implication is that females should never have an interest in starting a war (even if they had weapons or allies that made up for their smaller size). The reason that females never evolved an appetite to band together and raid neighboring villages for husbands is that a woman's reproductive success is rarely limited by the number of available males, so any risk to her life while pursuing additional mates is a sheer loss in expected fitness. (Foraging women do, however, encourage men to fight in defense of the group and to avenge slain family members.) The theory also explains why in modern warfare most people are unwilling to send women into combat and feel morally outraged when women are casualties, even though no ethical argument makes a woman's life more precious than a man's. It is hard to shake the intuition that war is a game that benefits men (which was true for most of our evolutionary history), so they should bear the risks.

The theory also predicts that men should be willing to fight collectively only if they are confident of victory and none of them knows in advance who will be injured or killed. If defeat is likely, it's pointless to fight on. And if you bear more than your share of the risk—say, if your platoonmates are exposing you to danger by looking out for their own hides—it's also pointless to fight on. These two principles shape the psychology of war.

Among foragers, warring bands are usually factions of the same people and have the same kinds of weaponry, so the predictor of victory in our evolutionary past would have been sheer numbers. The side with more warriors was invincible, and the odds of victory could be estimated from the manpower on each side. The Yanomamö are obsessed with the size of their villages for just that reason, and they often form alliances or rethink secessions because they know that smaller villages are helpless in wars. Even in modern societies, a mob of people on your side is emboldening and a mob on the other side terrifying. Mustering a crowd is a common tactic for whipping up patriotism, and a mass demonstration can incite panic even in a militarily secure ruler. A major principle of battlefield strategy is to surround an enemy unit, making defeat look certain and causing panic and rout.

Just as important is an equitable distribution of risk. A war party faces the problem of altruism par excellence. Every member has an incentive to cheat by keeping himself out of harm's way and exposing the others to greater risk. Just as benevolent cooperation cannot evolve unless the favor-granter detects and punishes cheaters, aggressive cooperation cannot evolve unless the fighters detect and punish cowards or shirkers. Bravery and discipline are the obsessions of fighting men. They affect everything from a soldier's sense of whom he wants in his foxhole to the command structure that coerces soldiers into assuming risk equitably and that rewards bravery and punishes desertion. War is rare in the animal kingdom because animals, like humans, ought to be cowards unless they can enforce a multiparty contract to share the risks. Unlike ancestral humans, they did not have the cognitive machinery from which an enforcement calculator could easily evolve.

Here is another peculiarity of the logic and psychology of war. A man should agree to stay in a coalition for as long as he does not *know* that he is about to die. He may know the odds, but he cannot know whether the spinner of death is slowing down at him. But at *some* point he may see it coming. He may glimpse an archer who has him in his sights, or detect an impending ambush, or notice that he has been sent on a suicide mission. At that point everything changes, and the only rational move is to desert. Of course, if the uncertainty collapses only seconds before death, it's too late. The farther in advance a fighter can predict that he is about to become an unknown soldier, the more easily he can desert, and the more likely the coalition is to unravel. In a coalition of animals attacking another coalition or an individual, an attacker has some warning if he is being picked out for a counterattack, and can flee before they give chase. For that reason a coalition of animals would be especially prone to unraveling. But humans have invented weapons, from spears and arrows to bullets and bombs, that make fate unknowable until the last second. Behind this veil of ignorance, men can be motivated to fight to the last.

Decades before Tooby and Cosmides spelled out this logic, the psychologist Anatol Rapoport illustrated it with a paradox from World War II. (He believed the scenario was true but was unable to verify it.) At a bomber base in the Pacific, a flier had only a twenty-five percent chance of surviving his quota of missions. Someone calculated that if the fliers carried twice as many bombs, a mission could be carried out with half as many flights. But the only way to increase the payload was to reduce the fuel, which meant that the planes would have to fly on one-way missions. If the fliers would be willing to draw lots and take a one-in-two chance of flying off to a certain death instead of hanging on to their three-in-four chance of flying off to an unpredictable death, they would *double* their chance of survival: only half of them would die instead of three-quarters. Needless to say, it was never implemented. Few of us would accept such an offer, though it is completely fair and would save many lives, including, possibly, our own. The paradox is an intriguing demonstration that our mind is equipped to volunteer for a risk of death in a coalition but only if we do not know when death will come.

HUMANITY

So should we all just take poison now and be done with it? Some people think that evolutionary psychology claims to have discovered that human nature is selfish and wicked. But they are flattering the researchers and anyone who would claim to have discovered the opposite. No one needs a scientist to measure whether humans are prone to knavery. The question has been answered in the history books, the newspapers, the ethnographic record, and the letters to Ann Landers. But people treat it like an open question, as if someday science might discover that it's all a bad dream and we will wake up to find that it is human nature to love one another. The task of evolutionary psychology is not to weigh in on human nature, a task better left to others. It is to add the satisfying kind of insight that only science can provide: to connect what we know about human nature with the rest of our knowledge of how the world works, and to explain the largest number of facts with the smallest number of assumptions. Already a large part of our social psychology, well documented in the lab and the field, can be shown to fall out of a few assumptions about kin selection, parental investment, reciprocal altruism, and the computational theory of mind.

So does human nature doom us to a nightmare of exploitation by ruthless fitness-maximizers? Again, it is silly to look to science for the answer. Everyone knows that people are capable of monumental kindness and sacrifice. The mind has many components, and accommodates not only ugly motives but love, friendship, cooperation, a sense of fairness, and an ability to predict the consequences of our actions. The different parts of the mind struggle to engage or disengage the clutch pedal of behavior, so bad thoughts do not always cause bad deeds. Jimmy Carter, in his famous *Playboy* interview, said, "I have looked on a lot of women with lust. I've committed adultery in my heart many times." But the prying American press has found no evidence that he has committed it in real life even once.

And on the larger stage, history has seen terrible blights disappear permanently, sometimes only after years of bloodshed, sometimes as if in a puff of smoke. Slavery, harem-holding despots, colonial conquest, blood feuds, women as property, institutionalized racism and anti-Semitism, child labor, apartheid, fascism, Stalinism, Leninism, and war have vanished from expanses of the world that had suffered them for decades, centuries, or millennia. The homicide rates in the most vicious American urban jungles are twenty times lower than in many foraging societies. Modern Britons are twenty times less likely to be murdered than their medieval ancestors.

If the brain has not changed over the centuries, how can the human condition have improved? Part of the answer, I think, is that literacy, knowledge, and the exchange of ideas have undermined some kinds of exploitation. It's not that people have a well of goodness that moral exhortations can tap. It's that information can be framed in a way that makes exploiters look like hypocrites or fools. One of our baser instincts—claiming authority on a pretext of beneficence and competence—can be cunningly turned on the others. When everyone sees graphic representations of suffering, it is no longer possible to claim that no harm is being done. When a victim gives a first-person account in words the victimizer might use, it's harder to maintain that the victims are a lesser kind of being. When a speaker is shown to be echoing the words of his enemy or of a past speaker whose policies led to disaster, his authority can crumble. When peaceable neighbors are described, it's harder to insist that war is inevitable. When Martin Luther King said, "I have a dream that one day this nation will rise up and live out the true meaning of its creed: 'We hold these truths to be self-evident, that all men are created equal,'" he made it impossible for segregationists to maintain they were patriots without looking like charlatans.

And as I mentioned at the outset, though conflict is a human universal, so are efforts to reduce it. The human mind occasionally catches a glimmering of the brute economic fact that often adversaries can both come out ahead by dividing up the surplus created by their laying down their arms. Even some of the Yanomamö see the futility of their ways and long for a means to break the cycle of vengeance. People throughout history have invented ingenious technologies that turn one part of the mind against another and eke increments of civility from a human nature that was not selected for niceness: rhetoric, exposés, mediation, face-saving measures, contracts, deterrence, equal opportunity, mediation, courts, enforceable laws, monogamy, limits on economic inequality, abjuring vengeance, and many others. Utopian theoreticians ought to be humble in the face of this practical wisdom. It is likely to remain more effective than "cultural" proposals to make over childrearing, language, and the media, and "biological" proposals to scan the brains and genes of gang members for aggression markers and to hand out antiviolence pills in the ghettos.

Tenzin Gyatso, the Dalai Lama of Tibet, was identified at the age of two as the fourteenth reincarnation of the Buddha of Compassion, Holy Lord, Gentle Glory, Eloquent, Compassionate, Learned Defender of the Faith, Ocean of Wisdom. He was taken to Lhasa and brought up by doting monks, who tutored him in philosophy, medicine, and metaphysics. In 1950 he became the spiritual and secular leader in exile of the Tibetan people. Despite not having a power base, he is recognized as a world statesman on the sheer force of his moral authority, and in 1989 was awarded the Nobel Peace Prize. No human being could be more predisposed by his upbringing and by the role he has been thrust into to have pure and noble thoughts.

In 1993 an interviewer for the *New York Times* asked him about himself. He said that as a boy he loved war toys, especially his air rifle. As an adult, he relaxes by looking at battlefield photographs and had just ordered a thirty-volume Time-Life illustrated history of World War II. Like guys everywhere, he enjoys studying pictures of military hardware, like tanks, airplanes, warships, U-boats, submarines, and especially aircraft carriers. He has erotic dreams and finds himself attracted to beautiful women, often having to remind himself, "I'm a monk!" None of this has stood in the way of his being one of history's great pacifists. And despite the oppression of his people, he remains an optimist and predicts that the twenty-first century will be more peaceful than the twentieth. Why? asked the interviewer. "Because I believe," he said, "that in the 20th century, humanity has learned something from many, many experiences. Some positive, and many negative. What misery, what destruction! The greatest number of human beings were killed in the two world wars of this century. But human nature is such that when we face a tremendous critical situation, the human mind can wake up and find some other alternative. That is a human capacity."

8

THE MEANING OF LIFE

Man does not live by bread alone, nor by know-how, safety, children, or sex. People everywhere spend as much time as they can afford on activities that, in the struggle to survive and reproduce, seem pointless. In all cultures, people tell stories and recite poetry. They joke, laugh, and tease. They sing and dance. They decorate surfaces. They perform rituals. They wonder about the causes of fortune and misfortune, and hold beliefs about the supernatural that contradict everything else they know about the world. They concoct theories of the universe and their place within it.

As if that weren't enough of a puzzle, the more biologically frivolous and vain the activity, the more people exalt it. Art, literature, music, wit, religion, and philosophy are thought to be not just pleasurable but noble. They are the mind's best work, what makes life worth living. Why do we pursue the trivial and futile and experience them as sublime? To many educated people the question seems horribly philistine, even immoral. But it is unavoidable for anyone interested in the biological makeup of *Homo sapiens*. Members of our species do mad deeds like taking vows of celibacy, living for their music, selling their blood to buy movie tickets, and going to graduate school. Why? How might we understand the psychology of the arts, humor, religion, and philosophy within the theme of this book, that the mind is a naturally selected neural computer?

Every college has a faculty of arts, which usually dominates the institution in numbers and in the public eye. But the tens of thousands of scholars and millions of pages of scholarship have shed almost no light on the question of why people pursue the arts at all. The function of the arts is almost defiantly obscure, and I think there are several reasons why.

One is that the arts engage not only the psychology of aesthetics but the psychology of status. The very uselessness of art that makes it so incomprehensible to evolutionary biology makes it all too comprehensible to economics and social psychology. What better proof that you have money to spare than your being able to spend it on doodads and stunts that don't fill the belly or keep the rain out but that require precious materials, years of practice, a command of obscure texts, or intimacy with the elite? Thorstein Veblen's and Quentin Bell's analyses of taste and fashion, in which an elite's conspicuous displays of consumption, leisure, and outrage are emulated by the rabble, sending the elite off in search of new inimitable displays, nicely explains the otherwise inexplicable oddities of the arts. The grand styles of one century become tacky in the next, as we see in words that are both period labels and terms of abuse (*gothic, mannerist, baroque, rococo*). The steadfast patrons of the arts are the aristocracy and those who want to join them. Most people would lose their taste for a musical recording if they learned it was being sold at supermarket checkout counters or on late-night television, and even the work of relatively prestigious artists, such as Pierre Auguste Renoir, draws derisive reviews when it is shown in a popular "blockbuster" museum show. The value of art is largely unrelated to aesthetics: a priceless masterpiece becomes worthless if it is found to be a forgery; soup cans and comic strips become high art when the art world says they are, and then command conspicuously wasteful prices. Modern and postmodern works are intended not to give pleasure but to confirm or confound the theories of a guild of critics and analysts, to *épater la bourgeoisie*, or to baffle the rubes in Peoria.

The banality that the psychology of the arts is partly the psychology of status has been repeatedly pointed out, not just by cynics and barbarians but by erudite social commentators such as Quentin Bell and Tom Wolfe. But in the modern university, it is unmentioned, indeed, unmentionable. Academics and intellectuals are culture vultures. In a gathering of today's elite, it is perfectly acceptable to laugh that you barely passed Physics for Poets and Rocks for Jocks and have remained ignorant of science ever since, despite the obvious importance of scientific literacy to informed choices about personal health and public policy. But saying that you have never heard of James Joyce or that you tried listening to

Mozart once but prefer Andrew Lloyd Webber is as shocking as blowing your nose on your sleeve or announcing that you employ children in your sweatshop, despite the obvious *un*importance of your tastes in leisure-time activity to just about anything. The blending in people's minds of art, status, and virtue is an extension of Bell's principle of sartorial morality that we met in Chapter 7: people find dignity in the signs of an honorably futile existence removed from all menial necessities.

I mention these facts not to denigrate the arts but to clarify my topic. I want you to look at the psychology of the arts (and later, humor and religion) with the disinterested eye of an alien biologist trying to make sense of the human species rather than as a member of the species with a stake in how the arts are portrayed. *Of course* we find pleasure and enlightenment in contemplating the products of the arts, and not all of it is a pride in sharing the tastes of the beautiful people. But to understand the psychology of the arts that remains when we subtract out the psychology of status, we must leave at the door our terror of being mistaken for the kind of person who prefers Andrew Lloyd Webber to Mozart. We need to begin with folk songs, pulp fiction, and paintings on black velvet, not Mahler, Eliot, and Kandinsky. And that does *not* mean compensating for our slumming by dressing up the lowly subject matter in highfalutin "theory" (a semiotic analysis of *Peanuts*, a psychoanalytic exegesis of Archie Bunker, a deconstruction of *Vogue*). It means asking a simple question: What is it about the mind that lets people take pleasure in shapes and colors and sounds and jokes and stories and myths?

That question might be answerable, whereas questions about art in general are not. Theories of art carry the seeds of their own destruction. In an age when any Joe can buy CDs, paintings, and novels, artists make their careers by finding ways to avoid the hackneyed, to challenge jaded tastes, to differentiate the cognoscenti from the dilettantes, and to flout the current wisdom about what art is (hence the fruitless attempts over the decades to define art). Any discussion that fails to recognize that dynamic is doomed to sterility. It can never explain why music pleases the ear, because "music" will be defined to encompass atonal jazz, chromatic compositions, and other intellectual exercises. It will never understand the bawdy laughs and convivial banter that are so important in people's lives because it will define humor as the arch wit of an Oscar Wilde. Excellence and the avant-garde are designed for the sophisticated palate, a product of years of immersion in a genre and a familiarity with its conventions and clichés. They rely on one-upmanship and arcane allusions and displays of virtuosity. However fascinating and worthy of our support they are, they tend to obscure the psychology of aesthetics, not to illuminate it.

～

Another reason the psychology of the arts is obscure is that they are not adaptive in the biologist's sense of the word. This book has been about the adaptive design of the major components of the mind, but that does not mean that I believe that everything the mind does is biologically adaptive. The mind is a neural computer, fitted by natural selection with combinatorial algorithms for causal and probabilistic reasoning about plants, animals, objects, and people. It is driven by goal states that served biological fitness in ancestral environments, such as food, sex, safety, parenthood, friendship, status, and knowledge. That toolbox, however, can be used to assemble Sunday afternoon projects of dubious adaptive value.

Some parts of the mind register the attainment of increments of fitness by giving us a sensation of pleasure. Other parts use a knowledge of cause and effect to bring about goals. Put them together and you get a mind that rises to a biologically pointless challenge: figuring out how to get at the pleasure circuits of the brain and deliver little jolts of enjoyment without the inconvenience of wringing bona fide fitness increments from the harsh world. When a rat has access to a lever that sends electrical impulses to an electrode implanted in its medial forebrain bundle, it presses the lever furiously until it drops of exhaustion, forgoing opportunities to eat, drink, and have sex. People don't yet undergo elective neurosurgery to have electrodes implanted in their pleasure centers, but they have found ways to stimulate them by other means. An obvious example is recreational drugs, which seep into the chemical junctions of the pleasure circuits.

Another route to the pleasure circuits is via the senses, which stimulate the circuits when they are in environments that would have led to fitness in past generations. Of course a fitness-promoting environment cannot announce itself directly. It gives off patterns of sounds, sights, smells, tastes, and feels that the senses are designed to register. Now, if the intellectual faculties could identify the pleasure-giving patterns, purify them, and concentrate them, the brain could stimulate itself without the messiness of electrodes or drugs. It could give itself intense artificial doses of the sights and sounds and smells that ordinarily are given off by healthful environments. We enjoy strawberry cheesecake, but not because we evolved a taste for it. We evolved circuits that gave us trickles of enjoyment from the sweet taste of ripe fruit, the creamy mouth feel of fats and oils from nuts and meat, and the coolness of fresh water. Cheesecake packs a sensual wallop unlike anything in the natural world because it is a brew of megadoses of agreeable stimuli which we concocted for the express purpose of pressing our pleasure buttons. Pornography is another pleasure technology. In this chapter I will suggest that the arts are a third.

There is another way that the design of the mind can throw off fascinating but biologically functionless activities. The intellect evolved to crack the defenses of things in the natural and social world. It is made up of modules for reasoning about how objects, artifacts, living things, animals, and other human minds work (Chapter 5). There are problems in the universe other than those: where the universe came from, how physical flesh can give rise to sentient minds, why bad things happen to good people, what happens to our thoughts and feelings when we die. The mind can pose such questions but may not be equipped to answer them, even if the questions have answers. Given that the mind is a product of natural selection, it should not have a miraculous ability to commune with all truths; it should have a mere ability to solve problems that are sufficiently similar to the mundane survival challenges of our ancestors. According to a saying, if you give a boy a hammer, the whole world becomes a nail. If you give a species an elementary grasp of mechanics, biology, and psychology, the whole world becomes a machine, a jungle, and a society. I will suggest that religion and philosophy are in part the application of mental tools to problems they were not designed to solve.

Some readers may be surprised to learn that after seven chapters of reverse-engineering the major parts of the mind, I will conclude by arguing that some of the activities we consider most profound are nonadaptive by-products. But both kinds of argument come from a single standard, the criteria for biological adaptation. For the same reason that it is wrong to write off language, stereo vision, and the emotions as evolutionary accidents—namely, their universal, complex, reliably developing, well-engineered, reproduction-promoting design—it is wrong to invent functions for activities that lack that design merely because we want to ennoble them with the imprimatur of biological adaptiveness. Many writers have said that the "function" of the arts is to bring the community together, to help us see the world in new ways, to give us a sense of harmony with the cosmos, to allow us to experience the sublime, and so on. All these claims are true, but none is about adaptation in the technical sense that has organized this book: a mechanism that brings about effects that would have increased the number of copies of the genes building that mechanism in the environment in which we evolved. Some aspects of the arts, I think, do have functions in this sense, but most do not.

ARTS AND ENTERTAINMENT

The visual arts are a perfect example of a technology designed to defeat the locks that safeguard our pleasure buttons and to press the buttons in various combinations. Recall that vision solves the unsolvable problem of recovering a description of the world from its projection onto the retina by making assumptions about how the world is put together, such as smooth matte shading, cohesive surfaces, and no razor-edge alignment. Optical illusions—not just cereal-box material but the ones that use Leonardo's window, such as paintings, photographs, movies, and television—cunningly violate those assumptions and give off patterns of light that dupe our visual system into seeing scenes that aren't there. That's the lock-picking. The pleasure buttons are the content of the illusions. Everyday photographs and paintings (remember—think "motel room," not "Museum of Modern Art") depict plants, animals, landscapes, and people. In previous chapters we saw how the geometry of beauty is the visible signal of adaptively valuable objects: safe, food-rich, explorable, learnable habitats, and fertile, healthy dates, mates, and babies.

Less obvious is why we take pleasure in abstract art: the zigzags, plaids, tweeds, polka dots, parallels, circles, squares, stars, spirals, and splashes of color with which people decorate their possessions and bodies all over the world. It cannot be a coincidence that exactly these kinds of motifs have been posited by vision researchers as the features of the world that our perceptual analyzers lock onto as they try to make sense of the surfaces and objects out there (see Chapter 4). Straight lines, parallel lines, smooth curves, and right angles are some of the nonaccidental properties that the visual system seeks out because they are giveaways of parts of the world that contain solid objects or that have been shaped by motion, tension, gravity, and cohesion. A swath of the visual field sprinkled with repetitions of a pattern usually comes from a single surface in the world, like a tree trunk, a field, a rock face, or a body of water. A hard boundary between two regions usually comes from one surface occluding another. Bilateral symmetry almost always comes from animals, parts of plants, or human artifacts.

Other patterns that we find pretty help us to recognize objects by their three-dimensional shapes. Frames of reference are fitted onto bounded, elongated shapes, onto symmetrical shapes, and onto shapes with parallel or near-parallel edges. Once fitted, the shapes are mentally carved into geons (cones, cubes, and cylinders) before being matched against memory.

All of the optimal geometric features for visual analysis that I have listed in the last two paragraphs are popular in visual decorations. But how do we explain the overlap? Why is the optimal feedstock for visual processing pretty to look at?

First, we seem to get pleasure out of looking at purified, concentrated versions of the geometric patterns that in dilute form give us pips of microsatisfaction as we orient ourselves toward informative environments and fine-tune our vision to give us a clear picture of them. Think of the annoyance you feel when a movie is out of focus and your relief when the projectionist wakes up and twiddles the lens. The fuzzy picture resembles your own retinal image when you are not properly accommodating the lens of your eye. The dissatisfaction is the impetus to accommodate; the satisfaction tells you when you have succeeded. Bright, crisp, saturated, contrasty images, whether from an expensive television set or from a colorful painting, may exaggerate the click of pleasure we get when we have adjusted our eyes properly.

And it is frustrating, even frightening, to gaze at a scene in poor viewing conditions—far away, at night, or through haze, water, or foliage—and be unable to make head or tail of it, not knowing, for example, whether something is a hole or a bump or where one surface leaves off and another begins. A canvas that is cleanly divided into solid shapes and continuous backgrounds may exaggerate the reduction of anxiety we experience when we find viewing conditions that resolve the visual field into unambiguous surfaces and objects.

Finally, we find some parts of the world snazzy and other parts dreary to the extent that they convey information about improbable, information-rich, consequential objects and forces. Imagine scooping out the entire scene in front of you, putting it in a giant blender set on LIQUEFY, and pouring the detritus back in front of you. The scene no longer contains any object of interest. Any food, predators, shelter, hiding places, vantage points, tools, and raw materials have been ground into sludge. And what does it look like? It has no lines, no shapes, no symmetry, and no repetition. It is brown, just like the color you got when you mixed all your paints together as a child. It has nothing to look at because it has nothing in it. The thought experiment shows that drabness comes from an environment with nothing to offer, and its opposite, visual pizzazz, comes from an environment that contains objects worth paying attention to. Thus we are designed to be dissatisfied by bleak, featureless scenes and attracted to colorful, patterned ones. We push that pleasure button with vivid artificial colors and patterns.

～

Music is an enigma. In *Much Ado About Nothing*, Benedick asks, "Is it not strange that sheep's guts should hale souls out of men's bodies?" In all cultures, certain rhythmic sounds give listeners intense pleasure and heartfelt emotions. What benefit could there be to diverting time and energy to the making of plinking noises, or to feeling sad when no one has died? Many suggestions have been made—music bonds the social group, coordinates action, enhances ritual, releases tension—but they just pass the enigma along rather than explaining it. *Why* do rhythmic sounds bond the group, dissipate tension, and so on? As far as biological cause and effect are concerned, music is useless. It shows no signs of design for attaining a goal such as long life, grandchildren, or accurate perception and prediction of the world. Compared with language, vision, social reasoning, and physical know-how, music could vanish from our species and the rest of our lifestyle would be virtually unchanged. Music appears to be a pure pleasure technology, a cocktail of recreational drugs that we ingest through the ear to stimulate a mass of pleasure circuits at once.

"Music is the universal language," says the cliché, but that is misleading. Anyone who lived through the craze for Indian raga music after George Harrison made it hip in the 1960s appreciates that musical styles vary from culture to culture and that people most enjoy the idiom they grew up with. (During the Concert for Bangladesh, Harrison was mortified when the audience applauded Ravi Shankar for tuning up his sitar.) Musical sophistication also varies across people, cultures, and historical periods in ways that language does not. All neurologically normal children spontaneously speak and understand complex language, and the complexity of spoken vernaculars varies little across cultures and periods. In contrast, while everyone enjoys listening to music, many people cannot carry a tune, fewer can play an instrument, and those who can play need explicit training and extensive practice. Musical idioms vary greatly in complexity across time, cultures, and subcultures. And music communicates nothing but formless emotion. Even a plot as simple as "Boy meets girl, boy loses girl" cannot be narrated by a sequence of tones in any musical idiom. All this suggests that music is quite different from language and that it is a technology, not an adaptation.

But there are *some* parallels. As we shall see, music may borrow some of the mental software for language. And just as the world's languages conform to an abstract Universal Grammar, the world's musical idioms conform to an abstract Universal Musical Grammar. That idea was first broached by the composer and conductor Leonard Bernstein in *The Unanswered Question*, a passionate attempt to apply Noam Chomsky's ideas to music. The richest theory of universal musical grammar has been worked out by Ray Jackendoff in collaboration with the music theorist Fred Lerdahl and incorporating the ideas of many musicologists before them, most prominently Heinrich Schenker. According to the theory, music is built from an inventory of notes and a set of rules. The rules assemble notes into a sequence and organize them into three hierarchical structures, all superimposed on the same string of notes. To understand a musical piece means to assemble these mental structures as we listen to the piece.

The building blocks of a musical idiom are its inventory of notes— roughly, the different sounds that a musical instrument is designed to emit. The notes are played and heard as discrete events with beginnings and ends and a target pitch or coloring. That sets music apart from most other streams of sound, which slide continuously up or down, such as a howling wind, an engine roar, or the intonation of speech. The notes differ in how *stable* they feel to a listener. Some give a feeling of finality or settledness, and are suitable endings of a composition. Others feel unstable, and when they are played the listener feels a tension that is resolved when the piece returns to a more stable note. In some musical idioms, the notes are drumbeats with different timbres (coloring or quality). In others, the notes are pitches that are arrayed from high to low but not placed at precise intervals. But in many idioms the notes are tones of a fixed pitch; we label ours "do, re, mi, . . ." or "C, D, E, . . ." The musical significance of a pitch cannot be defined in absolute terms, but only by an interval between it and a reference pitch, usually the stablest one in the set.

The human sense of pitch is determined by the frequency of vibration of the sound. In most forms of tonal music, the notes in the inventory are related to the frequencies of vibration in a straightforward way. When an object is set into a sustained vibration (a string is plucked, a hollow object is struck, a column of air reverberates), the object vibrates at several frequencies at once. The lowest and often loudest frequency—the fundamental—generally determines the pitch we hear, but the object also vibrates at twice the fundamental frequency (but typically not as intensely), at three times the frequency (even less intensely), at four times (less intensely still), and so on. These vibrations are called harmonics or overtones. They are not perceived as pitches distinct from the fundamental, but when they are all heard together they give a note its richness or timbre.

But now imagine disassembling a complex tone and playing each of its overtones separately and at the same volume. Say the fundamental frequency is 64 vibrations a second, the second C below middle C on the piano. The first overtone is a vibration at 128 cycles a second, twice the frequency of the fundamental. Played by itself, it sounds higher than the fundamental but with the same pitch; on the piano, it corresponds to the next C going up the keyboard, the C below middle C. The interval between the two notes is called an octave, and all people—indeed, all mammals—perceive tones separated by an octave as having the same quality of pitch. The second overtone vibrates at three times the fundamental frequency, 192 times per second, and corresponds to G below middle C; the interval between the pitches is called the perfect fifth. The third overtone, four times the fundamental (256 vibrations per second), is two octaves above it, middle C. The fourth overtone, five times the fundamental (320 vibrations a second), is the E above middle C, separated from it by an interval called the major third.

These three pitches are the heart of the pitch inventory in Western music and many other idioms. The lowest and most stable note, C in our example, is called the tonic, and most melodies tend to return to it and end on it, giving the listener a sense of repose. The perfect fifth or G note is called the dominant, and melodies tend to move toward it and pause there at intermediate points in the melody. The major third or E note, in many (but not all) cases, gives a feeling of brightness, pleasantness, or joy. For example, the opening of Bill Haley's "Rock Around the Clock" begins with the tonic ("One o'clock, two o'clock, three o'clock, rock") proceeds to the major third ("Four o'clock, five o'clock, six o'clock, rock"), goes to the dominant ("Seven o'clock, eight o'clock, nine o'clock, rock") and remains there for several beats before launching into the main verses, each of which ends on the tonic.

More complicated pitch inventories are filled out by adding notes to the tonic and the dominant, often corresponding in pitch to the higher and higher (and softer and softer) overtones of a complex vibration. The seventh overtone of our reference note (448 vibrations a second) is close to middle A (but, for complicated reasons, not exactly at it). The ninth (576 vibrations a second) is the D in the octave above middle C. Put the five pitches together in the same octave and you get the five-tone or pentatonic scale, common in musical systems across the world. (At least, this is a popular explanation of where musical scales come from; not everyone agrees.) Add the pitches of the next two distinct overtones (F and B) and you get the seven-tone or diatonic scale that forms the core of all Western music, from Mozart to folk songs to punk rock to most jazz. With additional overtones you get the chromatic scale, all the white and black keys on the piano. Even the esoteric art music of the twentieth century, incomprehensible to the uninitiated, tends to stick to the notes of the chromatic scale rather than using arbitrary collections of frequencies. Added to the feeling that most notes "want" to return to the tonic (C) are other tensions among the notes. For example, in many musical contexts B wants to go up to C, F wants to be pulled toward E, and A wants to go to G.

Pitch inventories may also contain notes that add an emotional coloring. In the C major scale, if the E is lowered in pitch by half a tone to E-flat, forming an interval with respect to C called the minor third, then in comparison with its major counterpart it tends to evoke a feeling of sadness, pain, or pathos. The minor seventh is another "blue note," which evokes a gentle melancholy or mournfulness. Other intervals give off feelings that have been described as stoic, yearning, needful, dignified, dissonant, triumphant, horrific, flawed, and determined. The feelings are evoked both when the notes are played in succession as part of a melody and when they are played simultaneously as part of a chord or harmony. The emotional connotations of musical intervals are not exactly universal, because people need to be familiar with an idiom to experience them, but they are not arbitrary either. Infants as young as four months old prefer music with consonant intervals such as a major third to music with dissonant intervals such as a minor second. And to learn the more complex emotional colorings of music, people do not have to be conditioned Pavlov-style, say, by hearing intervals paired with joyful or melancholy lyrics or by hearing them while in a joyful or a melancholy mood. A person merely has to listen to melodies in a particular idiom over time, absorbing the patterns and contrasts among the intervals, and the emotional connotations develop automatically.

Those are the pitches; how are they strung into melodies? Jackendoff and Lerdahl show how melodies are formed by sequences of pitches that are organized in three different ways, all at the same time. Each pattern of organization is captured in a mental representation. Take the opening of Woody Guthrie's "This Land Is Your Land":

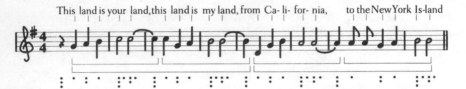

The first representation is called a grouping structure. The listener feels that groups of notes hang together in motifs, which in turn are grouped into phrases, which are grouped into lines or sections, which are grouped into stanzas, movements, and pieces. This hierarchical tree is similar to the phrase structure of a sentence, and when the music has lyrics, the two partly line up. The grouping structure is shown here by the brackets beneath the music. The snatches of melody for "This land is your land" and for "this land is my land" are the smallest-sized chunks. When they are joined together, they form a larger chunk. That larger chunk is joined with the combined chunk "from California to the New York Island" into a still larger chunk, and so on.

The second representation is a metrical structure, the repeating sequence of strong and weak beats that we count off as "ONE-two-THREE-four, ONE-two-THREE-four." The overall pattern is summed up in musical notation as the time signature, such as 4/4, and the major boundaries of the structure itself are demarcated by the vertical lines separating the music into bars. Each bar contains four beats, allocated among the different notes, with the first beat getting the strongest emphasis, the third beat an intermediate emphasis, and the second and fourth beats remaining weak. The metrical structure in this example is illustrated by the columns of dots under the notes. Each column corresponds to one tick of a metronome. The more dots in a column, the stronger the accent on that note.

The third representation is a reductional structure. It dissects the melody into essential parts and ornaments. The ornaments are stripped off and the essential parts further dissected into even *more* essential parts and ornaments on them. The reduction continues until the melody is reduced to a bare skeleton of a few prominent notes. Here is "This Land" boiled down first to half tones, then to four whole tones, then to only two whole tones.

The whole passage is basically a fancy way of getting from C to B. We hear the reductional structure of a melody in the chords of the rhythm guitar line. We also hear it when the band accompanying a tap dancer plays one of the stanzas in stop time, striking a single note in place of an entire line of music so that the tapping is easier to hear. And we sense it when we recognize variations of a piece in classical music or jazz. The skeleton of the melody is conserved while the ornaments differ from variation to variation.

Jackendoff and Lerdahl propose that there are in fact *two* ways that melodies may be dissected into simpler and simpler skeletons. I have shown you the first way, the time-span reduction, which lines up with the grouping and metrical structures and designates some of the groups and beats as ornaments on others. Jackendoff and Lerdahl call the second one a prolongation reduction. It captures the sense of musical flow across phrases, the buildup and release of tension within longer and longer passages over the course of the piece, culminating in a feeling of maximum repose at the end. Tension builds up as the melody departs from the more stable notes to the less stable notes, and is discharged when the melody returns to the stable ones. The contours of tension and release are also defined by changes from dissonant to consonant chords, from nonaccented to accented notes, from higher to lower notes, and from prolonged to nonprolonged notes.

The musicologist Deryck Cooke worked out a theory of the emotional semantics of the prolongation reduction. He showed how music conveys tension and resolution by transitions across unstable and stable intervals, and conveys joy and sorrow by transitions across major and minor intervals. Simple motifs of only four or five notes, he said, convey feelings like "innocent, blessed joy," "demonic horror," "continuous pleasurable longing," and "a burst of anguish." Longer stretches, and passages with motifs within motifs, can convey intricate patterns of feeling. One passage, as Cooke analyzes it, expresses "the feeling of a passionate outburst of painful emotion, which does not protest further, but falls back into acceptance—a flow and ebb of grief. Being neither complete protest nor complete acceptance, it has an effect of restless sorrow." Cooke supports his analyses with lists of examples that have a consensus interpretation, many with lyrics that offer additional corroboration. Some musicologists scoff at theories like Cooke's, finding counterexamples to every claim. But the exceptions tend to come from fine classical music, which uses interleaved, embedded, and ambiguous lines to challenge simple expectations and engage a sophisticated listener. Cooke's particular analyses may be debatable, but his main idea that there are lawful connections between patterns of intervals and patterns of emotion is clearly on the right track.

So that is the basic design of music. But if music confers no survival advantage, where does it come from and why does it work? I suspect that music is auditory cheesecake, an exquisite confection crafted to tickle the sensitive spots of at least six of our mental faculties. A standard piece tickles them all at once, but we can see the ingredients in various kinds of not-quite-music that leave one or more of them out.

1. Language. We can put words to music, and we wince when a lazy lyricist aligns an accented syllable with an unaccented note or vice versa.

That suggests that music borrows some of its mental machinery from language—in particular, from prosody, the contours of sound that span many syllables. The metrical structure of strong and weak beats, the intonation contour of rising and falling pitch, and the hierarchical grouping of phrases within phrases all work in similar ways in language and in music. The parallel may account for the gut feeling that a musical piece conveys a complex message, that it makes assertions by introducing topics and commenting on them, and that it emphasizes some portions and whispers others as asides. Music has been called "heightened speech," and it can literally grade into speech. Some singers slip into "talking on pitch" instead of carrying the melody, like Bob Dylan, Lou Reed, and Rex Harrison in *My Fair Lady*. They sound halfway between animated raconteurs and tone-deaf singers. Rap music, ringing oratory from preachers, and poetry are other intermediate forms.

2. Auditory scene analysis. Just as the eye receives a jumbled mosaic of patches and must segregate surfaces from their backdrops, the ear receives a jumbled cacophony of frequencies and must segregate the streams of sound that come from different sources—the soloist in an orchestra, a voice in a noisy room, an animal call in a chirpy forest, a howling wind among rustling leaves. Auditory perception is inverse acoustics: the input is a sound wave, the output a specification of the soundmakers in the world that gave rise to it. The psychologist Albert Bregman has worked out the principles of auditory scene analysis and has shown how the brain strings together the notes of a melody as if it were a stream of sound coming from a single soundmaker.

One of the brain's tricks as it identifies the soundmakers in the world is to pay attention to harmonic relations. The inner ear dissects a blare into its component frequencies, and the brain glues some of the components back together and perceives them as a complex tone. Components that stand in harmonic relations—a component at one frequency, another component at twice that frequency, yet another component at three times the frequency, and so on—are grouped together and perceived as a single tone rather than as separate tones. Presumably the brain glues them together to make our perception of sound reflect reality. Simultaneous sounds in harmonic relations, the brain guesses, are probably the overtones of a single sound coming from one soundmaker in the world. That is a good guess because many resonators, such as plucked strings, struck hollow bodies, and calling animals, emit sounds composed of many harmonic overtones.

What does this have to do with melody? Tonal melodies are sometimes said to be "serialized overtones." Building a melody is like slicing a complex harmonic sound into its overtones and laying them end to end in a particular order. Perhaps melodies are pleasing to the ear for the same reason that symmetrical, regular, parallel, repetitive doodles are pleasing to the eye. They exaggerate the experience of being in an environment that contains strong, clear, analyzable signals from interesting, potent objects. A visual environment that cannot be seen clearly or that is composed of homogeneous sludge looks like a featureless sea of brown or gray. An auditory environment that cannot be heard clearly or that is composed of homogeneous noise sounds like a featureless stream of radio static. When we hear harmonically related tones, our auditory system is satisfied that it has successfully carved the auditory world into parts that belong to important objects in the world, namely, resonating soundmakers like people, animals, and hollow objects.

Continuing this line of thought, we might observe that the more stable notes in a scale correspond to the lower and typically louder overtones emanating from a single soundmaker, and can confidently be grouped with the soundmaker's fundamental frequency, the reference note. The less stable notes correspond to the higher and typically weaker overtones, and though they *may* have come from the same soundmaker as the reference note, the assignment is less secure. Similarly, notes separated by a major interval are sure to have come from a single resonator, but notes separated by a minor interval might be very high overtones (and hence weak and uncertain ones), *or* they might come from a soundmaker with a complicated shape and material that does not give out a nice clear tone, *or* they might not come from a single soundmaker at all. Perhaps the ambiguity of the source of a minor interval gives the auditory system a sense of unsettledness that is translated as sadness elsewhere in the brain. Wind chimes, church bells, train whistles, claxton horns, and warbling sirens can evoke an emotional response with just two harmonically related tones. Recall that a few jumps among tones are the heart of a melody; all the rest is layer upon layer of ornamentation.

3. Emotional calls. Darwin noticed that the calls of many birds and primates are composed of discrete notes in harmonic relations. He speculated that they evolved because they were easy to reproduce time after time. (Had he lived a century later, he would have said that digital representations are more repeatable than analog ones.) He suggested, not too plausibly, that human music grew out of our ancestors' mating calls. But

his suggestion may make sense if it is broadened to include all emotional calls. Whimpering, whining, crying, weeping, moaning, growling, cooing, laughing, yelping, baying, cheering, and other ejaculations have acoustic signatures. Perhaps melodies evoke strong emotions because their skeletons resemble digitized templates of our species' emotional calls. When people try to describe passages of music in words, they use these emotional calls as metaphors. Soul musicians mix their singing with growls, cries, moans, and whimpers, and singers of torch songs and country-and-western music use catches, cracks, hesitations, and other emotional tics. Ersatz emotion is a common goal of art and recreation; I will discuss the reasons in a following section.

4. Habitat selection. We pay attention to features of the visual world that signal safe, unsafe, or changing habitats, such as distant views, greenery, gathering clouds, and sunsets (see Chapter 6). Perhaps we also pay attention to features of the auditory world that signal safe, unsafe, or changing habitats. Thunder, wind, rushing water, birdsong, growls, footsteps, heartbeats, and snapping twigs all have emotional effects, presumably because they are thrown off by attention-worthy events in the world. Perhaps some of the stripped-down figures and rhythms at the heart of a melody are simplified templates of evocative environmental sounds. In the device called tone painting, composers intentionally try to evoke environmental sounds like thunder or birdsong in a melody.

Perhaps a pure example of the emotional tug of music may be found in cinematic soundtracks. Many movies and television shows literally orchestrate the viewers' emotions from beginning to end with quasi-musical arrangements. They have no real rhythm, melody, or grouping, but can yank the moviegoer from feeling to feeling: the climactic rising scales of silent films, the lugubrious strings in the mushy scenes of old black-and-white movies (the source of the sarcastic violin-bowing gesture that means "You are trying to manipulate my sympathy"), the ominous two-note motif from *Jaws*, the suspenseful cymbal and drumbeats in the *Mission Impossible* television series, the furious cacophony during fights and chase scenes. It's not clear whether this pseudo-music distills the contours of environmental sounds, speech, emotional cries, or some combination, but it is undeniably effective.

5. Motor control. Rhythm is the universal component of music, and in many idioms it is the primary or only component. People dance, nod, shake, swing, stride, clap, and snap to music, and that is a strong hint that music taps into the system of motor control. Repetitive actions like

walking, running, chopping, scraping, and digging have an optimal rhythm (usually an optimal pattern of rhythms within rhythms), which is determined by the impedances of the body and of the tools or surfaces it is working with. A good example is pushing a child on a swing. A constant rhythmic pattern is an optimal way to time these motions, and we get moderate pleasure from being able to stick to it, which athletes call getting in a groove or feeling the flow. Music and dance may be a concentrated dose of that stimulus to pleasure. Muscle control also embraces sequences of tension and release (for example, in leaping or striking), actions carried out with urgency, enthusiasm, or lassitude, and erect or slumping body postures that reflect confidence, submission, or depression. Several psychologically oriented music theorists, including Jackendoff, Manfred Clynes, and David Epstein, believe that music recreates the motivational and emotional components of movement.

6. Something else. Something that explains how the whole is more than the sum of the parts. Something that explains why watching a slide go in and out of focus or dragging a filing cabinet up a flight of stairs does not hale souls out of men's bodies. Perhaps a resonance in the brain between neurons firing in synchrony with a soundwave and a natural oscillation in the emotion circuits? An unused counterpart in the right hemisphere of the speech areas in the left? Some kind of spandrel or crawl space or short-circuit or coupling that came along as an accident of the way that auditory, emotional, language, and motor circuits are packed together in the brain?

This analysis of music is speculative, but it nicely complements the discussions of the mental faculties in the rest of the book. I chose them as topics because they show the clearest signs of being adaptations. I chose music because it shows the clearest signs of not being one.

~

"The fact is I am quite happy in a movie, even a bad movie. Other people, so I have read, treasure memorable moments in their lives." At least the narrator of Walker Percy's novel *The Moviegoer* acknowledges the difference. Television stations get mail from soap-opera viewers with death threats for the evil characters, advice to the lovelorn ones, and booties for the babies. Mexican moviegoers have been known to riddle the screen with bullets. Actors complain that fans confuse them with their roles; Leonard Nimoy wrote a memoir called *I Am Not Spock*, then gave up and wrote another one called *I Am Spock*. These anecdotes appear regularly in the newspapers, usually to insinuate that people today are boobs who cannot distinguish fantasy from reality. I suspect that the people are not literally deluded but are going to extremes to enhance the pleasure we all get from losing ourselves in fiction. Where does this motive, found in all peoples, come from?

Horace wrote that the purpose of literature is "to delight and instruct," a function echoed centuries later by John Dryden when he defined a play as "a just and lively image of human nature, representing its passions and humours, and the changes of fortune to which it is subject; for the delight and instruction of mankind." It's helpful to distinguish the delight, perhaps the product of a useless technology for pressing our pleasure buttons, from the instruction, perhaps a product of a cognitive adaptation.

The technology of fiction delivers a simulation of life that an audience can enter in the comfort of their cave, couch, or theater seat. Words can evoke mental images, which can activate the parts of the brain that register the world when we actually perceive it. Other technologies violate the assumptions of our perceptual apparatus and trick us with illusions that partly duplicate the experience of seeing and hearing real events. They include costumes, makeup, sets, sound effects, cinematography, and animation. Perhaps in the near future we can add virtual reality to the list, and in the more distant future the feelies of *Brave New World*.

When the illusions work, there is no mystery to the question "Why do people enjoy fiction?" It is identical to the question "Why do people enjoy life?" When we are absorbed in a book or a movie, we get to see breathtaking landscapes, hobnob with important people, fall in love with ravishing men and women, protect loved ones, attain impossible goals, and defeat wicked enemies. Not a bad deal for seven dollars and fifty cents!

Of course, not all stories have happy endings. Why would we pay seven dollars and fifty cents for a simulation of life that makes us miserable? Sometimes, as with art films, it is to gain status through cultural machismo. We endure a pummeling of the emotions to differentiate ourselves from the crass philistines who actually go to the movies to enjoy themselves. Sometimes it is the price we pay to satisfy two incompatible desires: stories with happy endings and stories with unpredictable endings, which preserve the illusion of a real world. There have to be some stories in which the murderer does catch up with the heroine in the basement, or we would never feel suspense and relief in the stories in which she escapes. The economist Steven Landsburg observes that happy endings predominate when no director is willing to sacrifice the popularity of his or her film for the greater good of more suspense in the movies in general.

But then how can we explain the tearjerker, aimed at a market of moviegoers who *enjoy* being defrauded into grief? The psychologist Paul Rozin lumps tearjerkers with other examples of benign masochism like smoking, riding on roller coasters, eating hot chili peppers, and sitting in saunas. Benign masochism, recall, is like the drive of Tom Wolfe's test pilots to push the outside of the envelope. It expands the range of options in life by testing, in small increments, how closely one can approach a brink of disaster without falling over it. Of course the theory would be vacuous if it offered a glib explanation for every inexplicable act, and it would be false if it predicted that people would pay to have needles stuck under their fingernails. But the idea is more subtle. Benign masochists must be confident that no serious harm will befall them. They must bring on the pain or fear in measured increments. And they must have an opportunity to control and mitigate the damage. The technology of tearjerkers seems to fit. Moviegoers know the whole time that when they leave the theater they will find their loved ones unharmed. The heroine is done in by a progressive disease, not a heart attack or a piece of hot dog stuck in the throat, so we can prepare our emotions for the tragedy. We only have to accept the abstract premise that the heroine will die; we are excused from witnessing the disagreeable details. (Greta Garbo, Ali MacGraw, and Debra Winger all looked quite lovely as they wasted away from consumption and cancer.) And the viewer must identify with the next of kin, empathize with their struggle to cope, and feel confident that life will go on. Tearjerkers simulate a triumph over tragedy.

Even following the foibles of ordinary virtual people as they live their lives can press a pleasure button, the one labeled "gossip." Gossip is a favorite pastime in all human societies because knowledge is power. Knowing who needs a favor and who is in a position to offer one, who is trustworthy and who is a liar, who is available (or soon to become available) and who is under the protection of a jealous spouse or family—all give obvious strategic advantages in the games of life. That is especially true when the information is not yet widely known and one can be the first to exploit an opportunity, the social equivalent of insider trading. In the small bands in which our minds evolved, everyone knew everyone else, so all gossip was useful. Today, when we peer into the private lives of fictitious characters, we are giving ourselves the same buzz.

Literature, though, not only delights but instructs. The computer scientist Jerry Hobbs has tried to reverse-engineer the fictional narrative in an essay he was tempted to call "Will Robots Ever Have Literature?" Novels, he concluded, work like experiments. The author places a fictitious character in a hypothetical situation in an otherwise real world where ordinary facts and laws hold, and allows the reader to explore the consequences. We can imagine that there was a person in Dublin named Leopold Bloom with the personality, family, and occupation that James Joyce attributed to him, but we would object if we were suddenly to learn that the British sovereign at the time was not Queen Victoria but King Victor. Even in science fiction, we are asked to suspend belief in a few laws of physics, say to get the heroes to the next galaxy, but the events should otherwise unfold according to lawful causes and effects. A surreal story like Kafka's *Metamorphosis* begins with one counterfactual premise—a man can turn into an insect—and plays out the consequences in a world where everything else is the same. The hero retains his human consciousness, and we follow him as he makes his way and people react to him as real people would react to a giant insect. Only in fiction that is *about* logic and reality, such as *Alice's Adventures in Wonderland*, can any strange thing happen.

Once the fictitious world is set up, the protagonist is given a goal and we watch as he or she pursues it in the face of obstacles. It is no coincidence that this standard definition of plot is identical to the definition of intelligence I suggested in Chapter 2. Characters in a fictitious world do exactly what our intelligence allows us to do in the real world. We watch what happens to them and mentally take notes on the outcomes of the strategies and tactics they use in pursuing their goals.

What are those goals? A Darwinian would say that ultimately organisms have only two: to survive and to reproduce. And those are precisely the goals that drive the human organisms in fiction. Most of the thirty-six plots in Georges Polti's catalogue are defined by love or sex or a threat to the safety of the protagonist or his kin (for example, "Mistaken jealousy," "Vengeance taken for kindred upon kindred," and "Discovery of the dishonor of a loved one"). The difference between fiction for children and fiction for adults is commonly summed up in two words: sex and violence. Woody Allen's homage to Russian literature was entitled *Love and Death*. Pauline Kael got the title for one of her books of movie criticism from an Italian movie poster that she said contained "the briefest statement imaginable of the basic appeal of the movies": *Kiss Kiss Bang Bang*.

Sex and violence are not just the obsessions of pulp fiction and trash TV. The language maven Richard Lederer and the computer programmer Michael Gilleland present the following tabloid headlines:

CHICAGO CHAUFFEUR SMOTHERS BOSS'S DAUGHTER,
THEN CUTS HER UP AND STUFFS HER IN FURNACE

DOCTOR'S WIFE AND LOCAL MINISTER EXPOSED FOR CONCEIVING
ILLEGITIMATE DAUGHTER

TEENAGERS COMMIT DOUBLE SUICIDE;
FAMILIES VOW TO END VENDETTA

STUDENT CONFESSES TO AXE MURDER OF
LOCAL PAWNBROKER AND ASSISTANT

GARAGE OWNER STALKS AFFLUENT BUSINESSMAN,
THEN SHOTGUNS HIM IN HIS SWIMMING POOL

MADWOMAN LONG IMPRISONED IN ATTIC SETS HOUSE ON FIRE,
THEN LEAPS TO DEATH

FORMER SCHOOLTEACHER, FOUND TO HAVE BEEN PROSTITUTE,
COMMITTED TO INSANE ASYLUM

PRINCE ACQUITTED OF KILLING MOTHER IN REVENGE
FOR MURDER OF HIS FATHER

Sound familiar? See the endnotes.

Fiction is especially compelling when the obstacles to the protagonist's goals are other people in pursuit of incompatible goals. Life is like chess, and plots are like those books of famous chess games that serious players study so they will be prepared if they ever find themselves in similar straits. The books are handy because chess is combinatorial; at any stage there are too many possible sequences of moves and countermoves for them all to be played out in one's mind. General strategies like "Get your Queen out early" are too vague to be of much use, given the trillions of situations the rules permit. A good training regime is to build up a mental catalogue of tens of thousands of game challenges and the moves that allowed good players to do well in them. In artificial intelligence, it is called case-based reasoning.

Life has even more moves than chess. People are always, to some extent, in conflict, and their moves and countermoves multiply out to an unimaginably vast set of interactions. Partners, like the prisoners in the hypothetical dilemma, can either cooperate or defect, on this move and on subsequent moves. Parents, offspring, and siblings, because of their partial genetic overlap, have both common and competing interests, and any deed that one party directs toward another may be selfless, selfish, or a mixture of the two. When boy meets girl, either or both may see the other as a spouse, as a one-night stand, or neither. Spouses may be faithful or adulterous. Friends may be false friends. Allies may assume less than their fair share of the risk, or may defect as the finger of fate turns toward them. Strangers may be competitors or outright enemies. These games are taken into higher dimensions by the possibility of deception, which allows words and deeds to be either true or false, and self-deception, which allows *sincere* words and deeds to be either true or false. They are expanded into still higher dimensions by rounds of paradoxical tactics and countertactics, in which a person's usual goals—control, reason, and knowledge—are voluntarily surrendered to make the person unthreatenable, trustworthy, or too dangerous to challenge.

The intrigues of people in conflict can multiply out in so many ways that no one could possibly play out the consequences of all courses of action in the mind's eye. Fictional narratives supply us with a mental catalogue of the fatal conundrums we might face someday and the outcomes of strategies we could deploy in them. What are the options if I were to suspect that my uncle killed my father, took his position, and married my mother? If my hapless older brother got no respect in the family, are there circumstances that might lead him to betray me? What's the worst that could happen if I were seduced by a client while my wife and daughter were away for the weekend? What's the worst that could happen if I had an affair to spice up my boring life as the wife of a country doctor? How can I avoid a suicidal confrontation with raiders who want my land today without looking like a coward and thereby ceding it to them tomorrow? The answers are to be found in any bookstore or video shop. The cliché that life imitates art is true because the function of some kinds of art is for life to imitate it.

Can anything be said about the psychology of *good* art? The philosopher Nelson Goodman came up with an insight while examining the difference between art and other symbols. Suppose by coincidence an electrocardiogram and a Hokusai drawing of Mount Fuji both consisted of the same jagged line. Both tracings stand for something, but the only part of the electrocardiogram that matters is the position of each point that the line passes through. Its color and thickness, the size of the tracing, and the color and shading of the paper are irrelevant. If they were changed, the diagram would remain the same. But in the Hokusai drawing, none of the features may be ignored or casually altered; any might have been deliberately crafted by the artist. Goodman calls this property of art "repleteness."

A good artist takes advantage of repleteness and puts every aspect of the medium to good use. She might as well do so. She already has the eye and ear of the audience, and the work, having no practical function, does not have to meet any demanding mechanical specifications; every part is up for grabs. Heathcliff has to show his passion and fury somewhere; why not against the stormy, spooky Yorkshire moors? A scene has to be painted with brushstrokes; why not use jarring swirls to enhance the impact of a starry night, or a smudge of green on a face to give an impression of the dappled reflections that define the mood of a pastoral scene? A song needs a melody and words; in Cole Porter's "Ev'ry Time We Say Goodbye," a line is sung in alternating verses in a major key and a minor key, and the lyrics are:

> When you're here, there's such an air of spring about it.
> I can hear a lark somewhere begin to sing about it.
> There's no love song finer,
> But how strange the change from major to minor,
> Ev'ry time we say goodbye.

The song is about the change from joy to sadness when parting from a lover; the melody changes from joyful to sad; the lyrics say that the mood changes from joy to sadness using the metaphor of a melody that changes from joyful to sad. In the effort to mold a stream of sound to evoke the change, nothing has gone to waste.

A skillful use of repleteness impresses us not only by evoking a pleasurable feeling through several channels at once. Some of the parts are anomalous at first, and in resolving the anomaly we discover for ourselves the clever ways in which the artist shaped the different parts of the medium to do the same thing at the same time. Why, we ask ourselves, did a howling wind suddenly come up? Why does the lady have a green spot on her cheek? Why is a love song talking about musical keys? In solving the puzzles, the audience is led to pay attention to an ordinarily inconspicuous part of the medium, and the desired effect is reinforced. This insight comes from Arthur Koestler's tour de force on creativity, *The Act of Creation*, and underlies his ingenious analysis of that other great enigma of human psychology, humor.

WHAT'S SO FUNNY?

Here is how Koestler introduces the problem of humor:

> What is the survival value of the involuntary, simultaneous contraction of fifteen facial muscles associated with certain noises which are often irrepressible? Laughter is a reflex, but unique in that it serves no apparent biological purpose; one might call it a luxury reflex. Its only utilitarian function, as far as one can see, is to provide temporary relief from utilitarian pressures. On the evolutionary level where laughter arises, an element of frivolity seems to creep into a humourless universe governed by the laws of thermodynamics and the survival of the fittest.
>
> The paradox can be put in a different way. It strikes us as a reasonable arrangement that a sharp light shone into the eye makes the pupil contract, or that a pin stuck into one's foot causes its instant withdrawal—because both the "stimulus" and the "response" are on the same physiological level. But that a complicated mental activity like the reading of a page by Thurber should cause a specific motor response on the reflex level is a lopsided phenomenon which has puzzled philosophers since antiquity.

Let's piece together the clues from Koestler's analysis, from more recent ideas of evolutionary psychology, and from actual studies of humor and laughter.

Laughter, Koestler noted, is involuntary noisemaking. As any schoolteacher knows, it diverts attention from a speaker and makes it difficult to continue. And laughter is contagious. The psychologist Robert Provine, who has documented the ethology of laughter in humans, found that people laugh thirty times more often when they are with other people than when they are alone. Even when people laugh alone, they are often imagining they are with others: they are reading others' words, hearing their voices on the radio, or watching them on television. People laugh when they hear laughter; that is why television comedies use laugh tracks to compensate for the absence of a live audience. (The rim shot or drumbeat that punctuated the jokes of vaudeville comedians was a precursor.)

All this suggests two things. First, laughter is noisy not because it releases pent-up psychic energy but so that others may hear it; it is a form of communication. Second, laughter is involuntary for the same reason that other emotional displays are involuntary (Chapter 6). The brain broadcasts an honest, unfakable, expensive advertisement of a mental state by transferring control from the computational systems underlying voluntary action to the low-level drivers of the body's physical plant. As with displays of anger, sympathy, shame, and fear, the brain is going to some effort to convince an audience that an internal state is heartfelt rather than a sham.

Laughter appears to have homologues in other primate species. The human ethologist Irenaus Eibl-Eibesfeldt hears the rhythmic noise of laughter in the mobbing call that monkeys give when they gang up to threaten or attack a common enemy. Chimpanzees make a different noise that primatologists describe as laughter. It is a breathy pant made both when exhaling and when inhaling, and it sounds more like sawing wood than like the exhaled ha-ha-ha of human laughter. (There may be other kinds of chimpanzee laughter as well.) Chimps "laugh" when they tickle each other, just as children do. Tickling consists of touching vulnerable parts of the body during a mock attack. Many primates, and children in all societies, engage in rough-and-tumble play as practice for fighting. Play fighting poses a dilemma for the fighters: the scuffling should be realistic enough to serve as a useful rehearsal for offense and defense, but each party wants the other to know the attack is a sham so the fight doesn't escalate and do real damage. Chimp laughter and other primate play faces have evolved as a signal that the aggression is, as we say, all in fun. So we have two candidates for precursors to laughter: a signal of collective aggression and a signal of mock aggression. They are not mutually exclusive, and both may shed light on humor in humans.

Humor is often a kind of aggression. Being laughed at is aversive and feels like an attack. Comedy often runs on slapstick and insult, and in less refined settings, including the foraging societies in which we evolved, humor can be overtly sadistic. Children often laugh hysterically when other children hurt themselves or suffer misfortune. Many reports in the literature on humor among foragers are similar. When the anthropologist Raymond Hames was living with the Ye'Kwana in the Amazon rainforest, he once smacked his head on the crossbar of the entrance to a hut and crumpled to the ground, bleeding profusely and writhing in pain. The onlookers were doubled over in laughter. Not that we are all that different. Executions in England used to be occasions for the whole family to turn out and laugh at the condemned man as he was led to the gallows and hanged. In *1984*, Orwell presents a satire of popular entertainment through Winston Smith's diary that comes uncomfortably close to a typical evening in today's cinemaplexes:

> Last night to the flicks. All war films. One very good one of a ship full of refugees being bombed somewhere in the Mediterranean. Audience much amused by shots of a great huge fat man trying to swim away with a helicopter after him. first you saw him wallowing along in the water like porpoise, then you saw him through the helicopters gunsights, then he was full of holes and the sea round him turned pink and he sank as suddenly as though the holes had let in the water. audience shouting with laughter when he sank. then you saw a lifeboat full of children with a helicopter hovering over it. there was a middleaged woman might have been a jewess sitting up in the bow with a little boy about three years old in her arms. little boy screaming with fright and hiding his head between her breasts as if he was trying to burrow right into her and the woman putting her arms around him and comforting him although she was blue with fright herself. all the time covering him up as much as possible as if she thought her arms could keep the bullets off him. then the helicopter planted a 20 kilo bomb in among them terrific flash and the boat went all to matchwood. then there was a wonderful shot of a childs arm going up up up right up into the air a helicopter with a camera in its nose must have followed it up and there was a lot of applause . . .

I can hardly bear to read it, but on the other hand I don't remember ever laughing so hard in the movies as when Indiana Jones pulled out his gun and shot the grinning, scimitar-twirling Egyptian.

The horror that Orwell elicits by his pathetic description of the victims' terror shows that cruelty alone is not the trigger for humor. The butt of a joke has to be seen as having some undeserved claim to dignity and respect, and the humorous incident must take him down a few pegs. Humor is the enemy of pomp and decorum, especially when they prop up the authority of an adversary or a superior. The most inviting targets of ridicule are teachers, preachers, kings, politicians, military officers, and other members of the high and mighty. (Even the *Schadenfreude* of the Ye'Kwana feels more familiar when we are told that they are a diminutive people and Hames is a strapping American.) Probably the funniest thing I have ever seen in real life was a military parade in Cali, Colombia. At the front of the parade was an officer strutting proudly, and in front of *him* was a ragamuffin of no more than seven or eight strutting even *more* proudly, his nose in the air and his arms swinging grandly. The officer tried to take swipes at the urchin without breaking his stride, but the boy always managed to skip a few steps ahead and stay just out of reach as he led the procession through the streets.

A descent in dignity also underlies the universal appeal of sexual and scatological humor. Most of the world's wit is more *Animal House* than Algonquin Round Table. When Chagnon began to gather genealogical data among the Yanomamö, he had to work around their taboo against mentioning the names of prominent people (a bit like the sensibility behind our own forms of address like *Sir* and *Your honor*). Chagnon asked his informants to whisper the names of a person and the person's relatives into his ear, and clumsily repeated it to make sure he had heard correctly. When the named one glowered at him and the onlookers giggled, Chagnon felt reassured that he had recorded the person's true name. After months of work he had assembled an elaborate genealogy, and during a visit to a neighboring village he tried to show off by dropping the name of the headman's wife.

> A stunned silence followed, and then a villagewide roar of uncontrollable laughter, choking, gasping, and howling. It seems that I thought the Bisaasi-teri headman was married to a woman named "hairy cunt." It also came out that I was calling the headman "long dong," his brother "eagle shit," one of his sons "asshole," and a daughter "fart breath." Blood welled in my temples as I realized that I had nothing but nonsense to show for my five months of dedicated genealogical effort.

Of course, *we* would never laugh at anything so puerile. *Our* humor is "salty," "earthy," "bawdy," "racy," "raunchy," "ribald," or "Rabelaisian." Sex and excretion are reminders that anyone's claim to round-the-clock dignity is tenuous. The so-called rational animal has a desperate drive to pair up and writhe and moan. And as Isak Dinesen wrote, "What is man, when you come to think upon him, but a minutely set, ingenious machine for turning, with infinite artfulness, the red wine of Shiraz into urine?"

But oddly enough, humor is also a prized tactic of rhetoric and intellectual argument. Wit can be a fearsome rapier in the hands of a skilled polemicist. Ronald Reagan's popularity and effectiveness as president owed much to his facility with one-liners that quashed debate and criticism, at least for the moment; for example, when deflecting questions about abortion rights he would say, "I notice that everyone in favor of abortion has already been born." Philosophers relish the true story of the theoretician who announced at a scholarly conference that while some languages use a double negative to convey an affirmative, no language uses a double affirmative to convey a negative. A philosopher standing at the back of the hall shouted in a singsong, "Yeah, yeah." Though it may be true, as Voltaire wrote, that "a witty saying proves nothing," Voltaire was famously not above using them himself. The perfect quip can give a speaker an instant victory, deserved or not, and leave opponents stammering. We often feel that a clever aphorism captures a truth that would require pages to defend in any other way.

And here we get to Koestler's attempt to reverse-engineer humor. Koestler was an early appreciator of cognitive science at a time when behaviorism ruled, and he called attention to the mind's inventory of rule systems, modes of construal, ways of thinking, or frames of reference. Humor, he said, begins with a train of thought in one frame of reference that bumps up against an anomaly: an event or statement that makes no sense in the context of what has come before. The anomaly can be resolved by shifting to a different frame of reference, one in which the event does makes sense. And within *that* frame, someone's dignity has been downgraded. He calls the shift "bisociation." Koestler's examples of humor have not aged well, so I'll illustrate the theory with a few that amuse me, at the cost of killing the jokes by explaining them.

Lady Astor said to Winston Churchill, "If you were my husband, I'd put poison in your tea." He replied, "If you were my wife, I'd drink it." The response is anomalous in the frame of reference of murder, because people resist being murdered. The anomaly is resolved by switching to the frame of reference of suicide, in which death is welcomed as an escape from misery. In that frame Lady Astor is the cause of marital misery, an ignominious role.

A mountain climber slips over a precipice and clings to a rope over a thousand-foot drop. In fear and despair, he looks to the heavens and cries, "Is there anyone up there who can help me?" A voice from above booms, "You will be saved if you show your faith by letting go of the rope." The man looks down, then up, and shouts, "Is there anyone *else* up there who can help me?" The response is incongruous in the frame of reference of religious stories, in which God grants miracles in return for signs of faith and people are grateful for the bargain. It is resolved by slipping into the frame of day-to-day life, in which people have a healthy respect for the laws of physics and are skeptical of anyone who claims to defy them. In that frame, God (and indirectly his propagandists in the religious establishment) may be a flimflam artist—though if he is not, the man's common sense is his undoing.

W. C. Fields was once asked, "Do you believe in clubs for young people?" He answered, "Only when kindness fails." The reply is not a sensible answer to a question about a recreational group, the usual meaning of *club*, but the anomaly may be resolved by switching to a second meaning, "weapon." Young people flip from being a target of beneficence to being a target of discipline.

Koestler's three ingredients of humor—incongruity, resolution, and indignity—have been verified in many experiments of what makes a joke funny. Slapstick humor runs off the clash between a psychological frame, in which a person is a locus of beliefs and desires, and a physical frame, in which a person is a hunk of matter obeying the laws of physics. Scatological humor runs off the clash between the psychological frame and a physiological frame, in which a person is a manufacturer of disgusting substances. Sexual humor also runs off a clash between the psychological frame and a biological one; this time the person is a mammal with all the instincts and organs necessary for internal fertilization. Verbal humor hinges on a clash between two meanings of one word, the second one unexpected, sensible, and insulting.

The rest of Koestler's theory suffered from two old-fashioned ideas: the hydraulic model of the mind, in which psychic pressure builds up and needs a safety valve, and a drive for aggression, which supplies the pressure. To complete the answer to the question "What, if anything, is humor for?" we need three new ideas.

First, dignity, stature, and the other balloons punctured by humor are part of the complex of dominance and status discussed in Chapter 7. Dominance and status benefit those who hold them at the expense of those who don't, so peons always have a motive to mount a challenge to the eminent. In humans, dominance is not just the spoil of victory in fighting but a nebulous aura earned by a recognition of effectiveness in any of the arenas in which humans interact: prowess, expertise, intelligence, skill, wisdom, diplomacy, alliances, beauty, or wealth. Many of these claims to stature are partly in the eye of the beholder and would disintegrate if the beholders changed their weightings of the strengths and weaknesses that sum to yield the person's worth. Humor, then, may be an anti-dominance weapon. A challenger calls attention to one of the many less-than-exalted qualities that any mortal, no matter how high and mighty, is saddled with.

Second, dominance is often enforceable one-on-one but impotent before a united mob. A man with a single bullet in his gun can hold a dozen hostages if they have no way to signal a single moment at which to overpower him. No government has the might to control an entire population, so when events happen quickly and people all lose confidence in a regime's authority at the same time, they can overthrow it. This may be the dynamic that brought laughter—that involuntary, disruptive, and contagious signal—into the service of humor. When scattered titters swell into a chorus of hilarity like a nuclear chain reaction, people are acknowledging that they have all noticed the same infirmity in an exalted target. A lone insulter would have risked the reprisals of the target, but a mob of them, unambiguously in cahoots in recognizing the target's foibles, is safe. Hans Christian Andersen's story of the emperor's new clothes is a nice parable of the subversive power of collective humor. Of course, in everyday life we don't have to overthrow tyrants or to humble kings, but we do have to undermine the pretensions of countless blowhards, blusterers, bullies, gasbags, goody-goodies, holier-than-thous, hotshots, know-it-alls, and prima donnas.

Third, the mind reflexively interprets other people's words and gestures by doing whatever it takes to make them sensible and true. If the words are sketchy or incongruous, the mind charitably fills in missing premises or shifts to a new frame of reference in which they make sense. Without this "principle of relevance," language itself would be impossible. The thoughts behind even the simplest sentence are so labyrinthine that if we ever expressed them in full our speech would sound like the convoluted verbiage of a legal document. Say I were to tell you, "Jane heard the jingling ice cream truck. She ran to get her piggy bank from her dresser and started to shake it. Finally some money came out." Though I didn't say it in so many words, you know that Jane is a child (not an eighty-seven-year-old woman), that she shook the piggy bank (not the dresser), that coins (not bills) came out, and that she wanted the money to buy ice cream (not to eat the money, invest it, or bribe the driver to turn off the jingling).

The jester manipulates this mental machinery to get the audience to entertain a proposition—the one that resolves the incongruity—against their will. People appreciate the truth of the disparaging proposition because it was not baldly asserted as a piece of propaganda they might reject but was a conclusion they deduced for themselves. The proposition must possess at least a modicum of warrant or the audience could not have deduced it from other facts and could not have gotten the joke. This explains the feeling that a witty remark may capture a truth that is too complex to articulate, and that it is an effective weapon that forces people, at least for a moment, to agree to things they would otherwise deny. Reagan's wisecrack that abortion-rights advocates had already been born is so trivially true—*everyone* has been born—that on first hearing it makes no sense. But it does make sense on the assumption that there are two kinds of individuals, the born and the unborn. Those are the terms in which abortion opponents want the issue to be framed, and anyone who understands the quip has implicitly acknowledged that the framing is possible. And within that frame, the abortion-rights advocate possesses a privilege but wants to deny it to others and hence is a hypocrite. The argument is not necessarily sound, but a rebuttal would need many more words than the dozen that sufficed for Reagan. The "higher" forms of wit are cases where an audience's cognitive processes have been commandeered against them to deduce a disparaging proposition from premises they cannot deny.

Not all humor is malicious. Friends spend a good deal of time in playful badinage in which no one gets hurt; indeed, an evening spent laughing with friends is one of life's greatest pleasures. Of course, much of the pleasure comes from disparaging people outside the circle, which reinforces the friendship by the principle that the enemy of my enemy is my friend. But much of it is mild self-deprecation and gentle teasing that everyone seems to enjoy.

Not only is convivial humor not particularly aggressive; it's not particularly funny. Robert Provine did something that no one in the two-thousand-year history of pontificating about humor had ever thought to do: he went out to see what makes people laugh. He had his assistants hang out on the college campus near groups of people in conversation and surreptitiously note what triggered their laughter. What did he find? A typical laugh line was, "I'll see you guys later," or "What is that supposed to mean?!" As they say, you had to be there. Only about ten to twenty percent of the episodes could be classified as humorous, and then only by the most indulgent standards. The funniest lines in twelve hundred examples were, "You don't have to drink; just buy us drinks," "Do you date within your species?" and "Are you working here or just trying to look busy?" Provine notes, "The frequent laughter heard at crowded social gatherings is not due to a furious rate of joke telling by guests. Most pre-laugh dialogue is like that of an interminable television situation comedy scripted by an extremely ungifted writer."

How do we explain the appeal of the barely humorous banter that incites most of our laughter? If humor is an anti-dominance poison, a dignicide, it need not be used only for harmful purposes. The point of Chapter 7 was that when people interact with each other they have to choose from a menu of different social psychologies, each with a different logic. The logic of dominance and status is based on implicit threats and bribes, and it vanishes when the superior can no longer make good on them. The logic of friendship is based on a commitment to mutual unmeasured aid, come what may. People want status and dominance, but they also want friends, because status and dominance can fade but a friend will be there through thick and thin. The two are incompatible, and that raises a signaling problem. Given any two people, one will always be stronger, smarter, wealthier, better-looking, or better connected than the other. The triggers of a dominant-submissive or celebrity-fan relationship are always there, but neither party may want the relationship to go in that direction. By deprecating the qualities that you *could* have lorded over a friend or that a friend could have lorded over you, you are conveying that the basis of the relationship, as far as you are concerned, is not status or dominance. All the better if the signal is involuntary and hence hard to fake.

If this idea is correct, it would explain the homology between adult human laughter and the response to mock aggression and tickling in children and chimpanzees. The laughter says, It may look like I'm trying to hurt you, but I'm doing something that both of us want. The idea also explains why kidding is a precision instrument for assessing the kind of relationship one has with a person. You don't tease a superior or a stranger, though if one of you floats a trial tease that is well received, you know the ice is breaking and the relationship is shifting toward friendship. And if the tease elicits a mirthless chuckle or a freezing silence, you are being told that the grouch has no desire to become your friend (and may even have interpreted the joke as an aggressive challenge). The recurring giggles that envelop good friends are reavowals that the basis of the relationship is still friendship, despite the constant temptations for one party to have the upper hand.

THE INQUISITIVE IN PURSUIT
OF THE INCONCEIVABLE

"The most common of all follies," wrote H. L. Mencken, "is to believe passionately in the palpably not true. It is the chief occupation of mankind." In culture after culture, people believe that the soul lives on after death, that rituals can change the physical world and divine the truth, and that illness and misfortune are caused and alleviated by spirits, ghosts, saints, fairies, angels, demons, cherubim, djinns, devils, and gods. According to polls, more than a quarter of today's Americans believe in witches, almost half believe in ghosts, half believe in the devil, half believe that the book of Genesis is literally true, sixty-nine percent believe in angels, eighty-seven percent believe that Jesus was raised from the dead, and ninety-six percent believe in a God or universal spirit. How does religion fit into a mind that one might have thought was designed to reject the palpably not true? The common answer—that people take comfort in the thought of a benevolent shepherd, a universal plan, or an afterlife—is unsatisfying, because it only raises the question of *why* a mind would evolve to find comfort in beliefs it can plainly see are false. A freezing person finds no comfort in believing he is warm; a person face-to-face with a lion is not put at ease by the conviction that it is a rabbit.

What is religion? Like the psychology of the arts, the psychology of religion has been muddied by scholars' attempts to exalt it while understanding it. Religion cannot be equated with our higher, spiritual, humane, ethical yearnings (though it sometimes overlaps with them). The Bible contains instructions for genocide, rape, and the destruction of families, and even the Ten Commandments, read in context, prohibit murder, lying, and theft only within the tribe, not against outsiders. Religions have given us stonings, witch-burnings, crusades, inquisitions, jihads, fatwas, suicide bombers, abortion-clinic gunmen, and mothers who drown their sons so they can be happily reunited in heaven. As Blaise Pascal wrote, "Men never do evil so completely and cheerfully as when they do it from religious conviction."

Religion is not a single topic. What we call religion in the modern West is an alternative culture of laws and customs that survived alongside those of the nation-state because of accidents of European history. Religions, like other cultures, have produced great art, philosophy, and law, but their customs, like those of other cultures, often serve the interests of the people who promulgate them. Ancestor worship must be an appealing idea to people who are about to become ancestors. As one's days dwindle, life begins to shift from an iterative prisoner's dilemma, in which defection can be punished and cooperation rewarded, to a one-shot prisoner's dilemma, in which enforcement is impossible. If you can convince your children that your soul will live on and watch over their affairs, they are less emboldened to defect while you are alive. Food taboos keep members of the tribe from becoming intimate with outsiders. Rites of passage demarcate the people who are entitled to the privileges of social categories (fetus or family member, child or adult, single or married) so as to preempt endless haggling over gray areas. Painful initiations weed out anyone who wants the benefits of membership without being committed to paying the costs. Witches are often mothers-in-law and other inconvenient people. Shamans and priests are Wizards of Oz who use special effects, from sleight-of-hand and ventriloquism to sumptuous temples and cathedrals, to convince others that they are privy to forces of power and wonder.

Let's focus on the truly distinctive part of the psychology of religion. The anthropologist Ruth Benedict first pointed out the common thread of religious practice in all cultures: religion is a technique for success. Ambrose Bierce defined *to pray* as "to ask that the laws of the universe be annulled on behalf of a single petitioner confessedly unworthy." People everywhere beseech gods and spirits for recovery from illness, for success in love or on the battlefield, and for good weather. Religion is a desperate measure that people resort to when the stakes are high and they have exhausted the usual techniques for the causation of success—medicines, strategies, courtship, and, in the case of the weather, nothing.

What kind of mind would do something as useless as inventing ghosts and bribing them for good weather? How does that fit into the idea that reasoning comes from a system of modules designed to figure out how the world works? The anthropologists Pascal Boyer and Dan Sperber have shown that it fits rather well. First, nonliterate peoples are not psychotic hallucinators who are unable to distinguish fantasy from reality. They know there is a humdrum world of people and objects driven by the usual laws, and find the ghosts and spirits of their belief system to be terrifying and fascinating precisely *because* they violate their own ordinary intuitions about the world.

Second, the spirits, talismans, seers, and other sacred entities are never invented out of whole cloth. People take a construct from one of the cognitive modules of Chapter 5—an object, person, animal, natural substance, or artifact—and cross out a property or write in a new one, letting the construct keep the rest of its standard-issue traits. A tool or weapon or substance will be granted some extra causal power but otherwise is expected to behave as it did before. It lives at one place at one time, is unable to pass through solid objects, and so on. A spirit is stipulated to be exempt from one or more of the laws of biology (growing, aging, dying), physics (solidity, visibility, causation by contact), or psychology (thoughts and desires are known only through behavior). But otherwise the spirit is recognizable as a kind of person or animal. Spirits see and hear, have a memory, have beliefs and desires, act on conditions that they believe will bring about a desired effect, make decisions, and issue threats and bargains. When the elders spread religious beliefs, they never bother to spell out these defaults. No one ever says, "If the spirits promise us good weather in exchange for a sacrifice, and they know we want good weather, they predict that we will make the sacrifice." They don't have to, because they know that the minds of the pupils will automatically supply these beliefs from their tacit knowledge of psychology. Believers also avoid working out the strange logical consequences of these piecemeal revisions of ordinary things. They don't pause to wonder why a God who knows our intentions has to listen to our prayers, or how a God can both see into the future and care about how we choose to act. Compared to the mind-bending ideas of modern science, religious beliefs are notable for their lack of imagination (God is a jealous man; heaven and hell are places; souls are people who have sprouted wings). That is because religious concepts are human concepts with a few emendations that make them wondrous and a longer list of standard traits that make them sensible to our ordinary ways of knowing.

But where do people get the emendations? Even when all else has failed, why would they waste time spinning ideas and practices that are useless, even harmful? Why don't they accept that human knowledge and power have limits and conserve their thoughts for domains in which they can do some good? I have alluded to one possibility: the demand for miracles creates a market that would-be priests compete in, and they can succeed by exploiting people's dependence on experts. I let the dentist drill my teeth and the surgeon cut into my body even though I cannot possibly verify for myself the assumptions they use to justify those mutilations. That same trust would have made me submit to medical quackery a century ago and to a witch doctor's charms millennia ago. Of course, witch doctors must have *some* track record or they would lose all credibility, and they do blend their hocus-pocus with genuine practical knowledge such as herbal remedies and predictions of events (for instance, the weather) that are more accurate than chance.

And beliefs about a world of spirits do not come from nowhere. They are hypotheses intended to explain certain data that stymie our everyday theories. Edward Tylor, an early anthropologist, noted that animistic beliefs are grounded in universal experiences. When people dream, their body stays in bed but some other part of them is up and about in the world. The soul and the body also part company in the trance brought on by an illness or a hallucinogen. Even when we are awake, we see shadows and reflections in still water that seem to carry the essence of a person without having mass, volume, or continuity in time and space. And in death the body has lost some invisible force that animates it in life. One theory that brings these facts together is that the soul wanders off when we sleep, lurks in the shadows, looks back at us from the surface of a pond, and leaves the body when we die. Modern science has come up with a better theory of shadows and reflections. But how well does it do at explaining the sentient self that dreams, imagines, and directs the body?

～

Some problems continue to baffle the modern mind. As the philosopher Colin McGinn put it in his summary of them, "The head spins in theoretical disarray; no explanatory model suggests itself; bizarre ontologies loom. There is a feeling of intense confusion, but no clear idea about where the confusion lies."

I discussed one of the problems in Chapter 2: consciousness in the sense of sentience or subjective experience (not in the sense of information access or self-reflection). How could an event of neural information-processing cause the feel of a toothache or the taste of lemon or the color purple? How could I know whether a worm, a robot, a brain slice in a dish, or *you* are sentient? Is your sensation of red the same as mine, or might it be like my sensation of green? What is it like to be dead?

Another imponderable is the self. What or where is the unified center of sentience that comes into and goes out of existence, that changes over time but remains the same entity, and that has a supreme moral worth? Why should the "I" of 1996 reap the rewards and suffer the punishments earned by the "I" of 1976? Say I let someone scan a blueprint of my brain into a computer, destroy my body, and reconstitute me in every detail, memories and all. Would I have taken a nap, or committed suicide? If two I's were reconstituted, would I have double the pleasure? How many selves are in the skull of a split-brain patient? What about in the partly fused brains of a pair of Siamese twins? When does a zygote acquire a self? How much of my brain tissue has to die before I die?

Free will is another enigma (see Chapter 1). How can my actions be a choice for which I am responsible if they are completely caused by my genes, my upbringing, and my brain state? Some events are determined, some are random; how can a choice be neither? When I hand my wallet to an armed man who threatens to kill me if I don't, is that a choice? What about if I shoot a child because an armed man threatens to kill me if I don't? If I choose to do something, I *could have* done otherwise—but what does that mean in a single universe unfolding in time according to laws, which I pass through only once? I am faced with a momentous decision, and an expert on human behavior with a ninety-nine percent success rate predicts that I will choose what at this point looks like the worse alternative. Should I continue to agonize, or should I save time and do what's inevitable?

A fourth puzzle is meaning. When I talk about *planets*, I can refer to all planets in the universe, past, present, and future. But how could I, right now, here in my house, be standing in some relationship to a planet that will be created in a distant galaxy in five million years? If I know what "natural number" means, my mind has commerce with an infinite set—but I am a finite being, who has tasted a tiny sample of the natural numbers.

Knowledge is just as perplexing. How could I have arrived at the certainty that the square of the hypotenuse is equal to the sum of the squares of the other two sides, everywhere and for all eternity, here in the comfort of my armchair with not a triangle or tape measure in sight? How do I know that I'm not a brain in a vat, or dreaming, or living a hallucination programmed by an evil neurologist, or that the universe was not created five minutes ago complete with fossils, memories, and historical records? If every emerald I have seen so far is green, why should I conclude "all emeralds are green" rather than "all emeralds are grue," where *grue* means "either observed before the year 2020 and green, or not so observed and blue"? All the emeralds I have seen are green, but then all the emeralds I've seen are grue. The two conclusions are equally warranted, but one predicts that the first emerald I see in 2020 will be the color of grass and the other predicts that it will be the color of the sky.

A final conundrum is morality. If I secretly hatchet the unhappy, despised pawnbroker, where is the evil nature of that act registered? What does it mean to say that I "shouldn't" do it? How did *ought* emerge from a universe of particles and planets, genes and bodies? If the aim of ethics is to maximize happiness, should we indulge a sicko who gets more pleasure from killing than his victims do from living? If it is to maximize lives, should we publicly execute a framed man if it would deter a thousand murderers? Or draft a few human guinea pigs for fatal experiments that would save millions?

People have thought about these problems for millennia but have made no progress in solving them. They give us a sense of bewilderment, of intellectual vertigo. McGinn shows how thinkers have cycled among four kinds of solutions over the ages, none satisfactory.

Philosophical problems have a feeling of the divine, and the favorite solution in most times and places is mysticism and religion. Consciousness is a divine spark in each of us. The self is the soul, an immaterial ghost that floats above physical events. Souls just exist, or they were created by God. God granted each soul a moral worth and the power of choice. He has stipulated what is good, and inscribes every soul's good and evil acts in the book of life and rewards or punishes it after it leaves the body. Knowledge is granted by God to the prophet or the seer, or guaranteed to all of us by God's honesty and omniscience. The solution is explained in the rejoinder to the limerick (p. 316) about why the tree continues to be when there's no one about in the quad:

> Dear Sir, Your astonishment's odd:
> *I* am always about in the quad.
> And that's why the tree
> will continue to be,
> Since observed by Yours Faithfully, God.

The problem with the religious solution was stated by Mencken when he wrote, "Theology is the effort to explain the unknowable in terms of the not worth knowing." For anyone with a persistent intellectual curiosity, religious explanations are not worth knowing because they pile equally baffling enigmas on top of the original ones. What gave *God* a mind, free will, knowledge, certainty about right and wrong? How does he infuse them into a universe that seems to run just fine according to physical laws? How does he get ghostly souls to interact with hard matter? And most perplexing of all, if the world unfolds according to a wise and merciful plan, why does it contain so much suffering? As the Yiddish expression says, If God lived on earth, people would break his windows.

Modern philosophers have tried three other solutions. One is to say that the mysterious entities are an irreducible part of the universe and to leave it at that. The universe, we would conclude, contains space, time, gravity, electromagnetism, nuclear forces, matter, energy, *and consciousness* (or will, or selves, or ethics, or meaning, or all of them). The answer to our curiosity about *why* the universe has consciousness is, "Get over it, it just does." We feel cheated because no insight has been offered, and because we know that the details of consciousness, will, and knowledge are minutely related to the physiology of the brain. The irreducibility theory leaves that a coincidence.

A second approach is to deny that there is a problem. We have been misled by fuzzy thinking or by beguiling but empty idioms of language, such as the pronoun *I*. Statements about consciousness, will, self, and ethics cannot be verified by mathematical proof or empirical test, so they are meaningless. But this answer leaves us incredulous, not enlightened. As Descartes observed, our own consciousness is the most indubitable thing there is. It is a datum to be explained; it cannot be defined out of existence by regulations about what we are allowed to call meaningful (to say nothing of ethical statements, such as that slavery and the Holocaust were wrong).

A third approach is to domesticate the problem by collapsing it with one we *can* solve. Consciousness is activity in layer 4 of the cortex, or the contents of short-term memory. Free will is in the anterior cingulate sulcus or the executive subroutine. Morality is kin selection and reciprocal altruism. Each suggestion of this kind, to the extent that it is correct, does solve *one* problem, but it just as surely leaves unsolved the main problem. *How* does activity in layer 4 of the cortex cause my private, pungent, tangy sensation of redness? I can imagine a creature whose layer 4 is active but who does not have the sensation of red or the sensation of anything; no law of biology rules the creature out. No account of the causal effects of the cingulate sulcus can explain how human choices are are *not caused at all*, hence something we can be held responsible for. Theories of the evolution of the moral sense can explain why we condemn evil acts against ourselves and our kith and kin, but cannot explain the conviction, as unshakable as our grasp of geometry, that some acts are inherently wrong even if their net effects are neutral or beneficial to our overall well-being.

I am partial to a different solution, defended by McGinn and based on speculations by Noam Chomsky, the biologist Gunther Stent, and before them David Hume. Maybe philosophical problems are hard not because they are divine or irreducible or meaningless or workaday science, but because the mind of *Homo sapiens* lacks the cognitive equipment to solve them. We are organisms, not angels, and our minds are organs, not pipelines to the truth. Our minds evolved by natural selection to solve problems that were life-and-death matters to our ancestors, not to commune with correctness or to answer any question we are capable of asking. We cannot hold ten thousand words in short-term memory. We cannot see in ultraviolet light. We cannot mentally rotate an object in the fourth dimension. And perhaps we cannot solve conundrums like free will and sentience.

We can well imagine creatures with *fewer* cognitive faculties than we have: dogs to whom our language sounds like "Blah-blah-blah-Ginger-blah-blah," rats that cannot learn a maze with food in the prime-numbered arms, autistics who cannot conceive of other minds, children who cannot understand what the fuss around sex is about, neurological patients who see every detail in a face except whose it is, stereoblind people who can understand a stereogram as a problem in geometry but cannot see it pop out in depth. If stereoblind people did not know better, they might call 3-D vision a miracle, or claim that it just *is* and needs no explanation, or write it off as some kind of trick.

So why should there not be creatures with *more* cognitive faculties than we have, or with *different* ones? They might readily grasp how free will and consciousness emerge from a brain and how meaning and morality fit into the universe, and would be amused by the religious and philosophical headstands we do to make up for our blankness when facing these problems. They could try to explain the solutions to us, but we would not understand the explanations.

The hypothesis is almost perversely unprovable, though it could be *dis*proved if anyone ever solved the age-old puzzles of philosophy. And there are indirect reasons to suspect it is true. One is that the species' best minds have flung themselves at the puzzles for millennia but have made no progress in solving them. Another is that they have a different character from even the most challenging problems of science. Problems such as how a child learns language or how a fertilized egg becomes an organism are horrendous in practice and may never be solved completely. But if they aren't, it will be for mundane practical reasons. The causal processes are too intertwined or chaotic, the phenomena are too messy to capture and dissect in the lab, the math is beyond the capacity of foreseeable computers. But scientists can imagine the kinds of theories that *might* be solutions, right or wrong, feasible to test or not. Sentience and will are different. Far from being too complicated, they are maddeningly simple—consciousness and choice inhere in a special dimension or coloring that is somehow pasted onto neural events without meshing with their causal machinery. The challenge is not to discover the correct explanation of how that happens, but to imagine a theory that *could* explain how it happens, a theory that would place the phenomenon as an effect of some cause, any cause.

It is easy to draw extravagant and unwarranted conclusions from the suggestion that our minds lack the equipment to solve the major problems of philosophy. It does not say that there is some paradox of self-reference or infinite regress in a mind's trying to understand itself. Psychologists and neuroscientists don't study their own minds; they study someone else's. Nor does it imply some principled limitation on the possibility of knowledge by any knower, like the Uncertainty Principle or Gödel's theorem. It is an observation about one organ of one species, equivalent to observing that cats are color-blind or that monkeys cannot learn long division. It does not justify religious or mystical beliefs but explains why they are futile. Philosophers would not be out of a job, because they clarify these problems, chip off chunks that *can* be solved, and solve them or hand them over to science to solve. The hypothesis does not imply that we have sighted the end of science or bumped into a barrier on how much we can ever learn about how the mind works. The computational aspect of consciousness (what information is available to which processes), the neurological aspect (what in the brain correlates with consciousness), and the evolutionary aspect (when and why did the neurocomputational aspects emerge) are perfectly tractable, and I see no reason that we should not have decades of progress and eventually a complete understanding—even if we never solve residual brain-teasers like whether your red is the same as my red or what it is like to be a bat.

In mathematics, one says that the integers are *closed* under addition: adding two integers produces another integer; it can never produce a fraction. But that does not mean that the set of integers is finite. Humanly thinkable thoughts are closed under the workings of our cognitive faculties, and may never embrace the solutions to the mysteries of philosophy. But the set of thinkable thoughts may be infinite nonetheless.

Is cognitive closure a pessimistic conclusion? Not at all! I find it exhilarating, a sign of great progress in our understanding of the mind. And it is my last opportunity to pursue the goal of this book: to get you to step outside your own mind for a moment and see your thoughts and feelings as magnificent contrivances of the natural world rather than as the only way that things could be.

First, if the mind is a system of organs designed by natural selection, why should we ever have expected it to comprehend all mysteries, to grasp all truths? We should be thankful that the problems of science are close enough in structure to the problems of our foraging ancestors that we have made the progress that we have. If there were *nothing* we were bad at understanding, we would have to question the scientific worldview that sees the mind as a product of nature. Cognitive closure *should* be true if we know what we are talking about. Still, one might have thought that the hypothesis was merely a daydream, a logical possibility that could go no further than late-night dorm-room bull sessions. McGinn's attempt to identify the humanly unsolvable problems is an advance.

Even better, we can glimpse *why* certain problems are beyond our ken. A recurring theme of this book is that the mind owes its power to its syntactic, compositional, combinatorial abilities (Chapter 2). Our complicated ideas are built out of simpler ones, and the meaning of the whole is determined by the meanings of the parts and the meanings of the relations that connect them: part-of-a-whole, example-of-a-category, thing-at-a-place, actor-exerting-force, cause-of-an-effect, mind-holding-a-belief. These logical and lawlike connections provide the meanings of sentences in everyday speech and, through analogies and metaphors, lend their structures to the esoteric contents of science and mathematics, where they are assembled into bigger and bigger theoretical edifices (see Chapter 5). We grasp matter as molecules, atoms, and quarks; life as DNA, genes, and a tree of organisms; change as position, momentum, and force; mathematics as symbols and operations. All are assemblies of elements composed according to laws, in which the properties of the whole are predictable from the properties of the parts and the way they are combined. Even when scientists grapple with seamless continua and dynamical processes, they couch their theories in words, equations, and computer simulations, combinatorial media that mesh with the workings of the mind. We are lucky that parts of the world behave as lawful interactions among simpler elements.

But there is something peculiarly holistic and everywhere-at-once and nowhere-at-all and all-at-the-same-time about the problems of philosophy. Sentience is not a combination of brain events or computational states: how a red-sensitive neuron gives rise to the subjective feel of redness is not a whit less mysterious than how the whole brain gives rise to the entire stream of consciousness. The "I" is not a combination of body parts or brain states or bits of information, but a unity of selfness over time, a single locus that is nowhere in particular. Free will is not a causal chain of events and states, by definition. Although the combinatorial aspect of meaning has been worked out (how words or ideas combine into the meanings of sentences or propositions), the *core* of meaning—the simple act of referring to something—remains a puzzle, because it stands strangely apart from any causal connection between the thing referred to and the person referring. Knowledge, too, throws up the paradox that knowers are acquainted with things that have never impinged upon them. Our thoroughgoing perplexity about the enigmas of consciousness, self, will, and knowledge may come from a mismatch between the very nature of these problems and the computational apparatus that natural selection has fitted us with.

If these conjectures are correct, our psyche would present us with the ultimate tease. The most undeniable thing there is, our own awareness, would be forever beyond our conceptual grasp. But if our minds are part of nature, that is to be expected, even welcomed. The natural world evokes our awe by the specialized designs of its creatures and their parts. We don't poke fun at the eagle for its clumsiness on the ground or fret that the eye is not very good at hearing, because we know that a design can excel at one challenge only by compromising at others. Our bafflement at the mysteries of the ages may have been the price we paid for a combinatorial mind that opened up a world of words and sentences, of theories and equations, of poems and melodies, of jokes and stories, the very things that make a mind worth having.

NOTES

1. Standard Equipment

Page

6 Robot's-eye view: Poggio, 1984.

6 Building a visual system: Marr, 1982; Poggio, 1984; Aloimonos & Rosenfeld, 1991; Wandell, 1995; Papathomas, et al., 1995.

7 Chicken-and-egg problems in vision: Adelson & Pentland, 1996; Sinha & Adelson, 1993a, b.

8 Multi-sized afterimage ("Emmert's Law"): Rock, 1983.

9 Template matching: Neisser, 1967; figure adapted from Lindsay & Norman, 1972, pp. 2–6.

10 Legged locomotion: Raibert & Sutherland, 1983; Raibert, 1990.

11 Walking disaster: French, 1994.

12 Arms and architect's lamps: Hollerbach, 1990; Bizzi & Mussa-Ivaldi, 1990.

12 Galen on the hand: Quoted in Williams, 1992, p. 192.

12 Grips: Trinkaus, 1992.

13 Bachelors: Winograd, 1976.

14 Not-so-common sense: Lenat & Guha, 1990.

14 Sensible inferences: Cherniak, 1983; Dennett, 1987.

14 Frame problem: Dennett, 1987; Pylyshyn, 1987.

15 Rules of robotics: Asimov, 1950.

17 Engineering aggression: Maynard Smith, 1982; Tooby & Cosmides, 1988.

18 Logic of love: Symons, 1979; Buss, 1994; Frank, 1988; Tooby & Cosmides, 1996; Fisher, 1992; Hatfield & Rapson, 1993.

19 Neglect of left visual space: Bisiach & Luzzatti, 1978. Achromatopsia (cortical color-blindness): Sacks & Wasserman, 1987. Motion-blindness: Hess, Baker, & Zihl, 1989.

20 Agnosia (difficulty recognizing objects): Farah, 1990. Prosopagnosia (difficulty rec-

ognizing faces): Etcoff, Freeman, & Cave, 1991. Capgras syndrome (lack of familiarity despite recognition): Alexander, Stuss, & Benson, 1979.

20 Multiple brain areas for vision: Van Essen & DeYoe, 1995.

21 Separated at birth: Lykken et al., 1992; Bouchard et al., 1990; Bouchard, 1994;
Plomin, 1989; Plomin, Owen, & McGuffin, 1994; L. Wright, 1995.

22 Reverse-engineering: Dennett, 1995. Psychology as reverse-engineering: Tooby &
Cosmides, 1992.

22 Biology as reverse-engineering: Williams, 1966, 1992; Mayr, 1983.

22 Psychology on a new foundation: Darwin, 1859/1964.

23 Evolutionary psychology: Symons, 1979, 1992; Tooby, 1985; Cosmides, 1985; Tooby
& Cosmides, 1992; Barkow, Cosmides, & Tooby, 1992; Cosmides & Tooby, 1994;
Wright, 1994a; Buss, 1995; Allman, 1994.

23 Cognitive revolution: Gardner, 1985; Jackendoff, 1987; Dennett, 1978a. Evolution
revolution: Williams, 1966; Hamilton, 1996; Dawkins, 1976/1989, 1986; Maynard
Smith, 1975/1993, 1982; Tooby, 1988; Wright, 1994a.

24 What is information?: Dretske, 1981.

24 Computational theory of mind: Turing, 1950; Putnam, 1960; Simon & Newell, 1964; Newell
& Simon, 1981; Haugeland, 1981a, b, c; Fodor, 1968a, 1975, 1994; Pylyshyn, 1984.

26 Humans that speak, ants that farm: Cosmides & Tooby, 1994.

28 Specialization all the way down: Gallistel, 1995.

28 Vision as inverse optics: Poggio, 1984.

29 Visual assumptions: Marr, 1982; Hoffman, 1983.

31 Modules according to Fodor: Fodor, 1983, 1985.

31 Chomsky on mental organs: Chomsky, 1988, 1991, 1993.

32 Specialization of artificial intelligence systems: Marr, 1982; Minsky, 1985; Minsky
& Papert, 1988b; Pinker & Prince, 1988.

32 Precocious children: Hirschfeld & Gelman, 1994a, b; Sperber, Premack, &
Premack, 1995. Human universals: Brown, 1991.

33 Mind not a mixture of biology and culture: Tooby & Cosmides, 1992. Learning
requires innate learning mechanism: Fodor, 1975, 1981; Chomsky, 1975; Pinker,
1984, 1994; Tooby & Cosmides, 1992.

36 Brain assembly: Stryker, 1994; Cramer & Sur, 1995; Rakic, 1995a, b.

37 Non-selectionist evolutionary forces: Williams, 1966; Gould & Lewontin, 1979.
Natural selection as engineer: Darwin, 1859/1964; Dawkins, 1983, 1986, 1995;
Williams, 1966, 1992; Dennett, 1995.

37 The eye as Cartesian bridge: Tooby & Cosmides, 1992.

38 Criteria for adaptation: Williams, 1966; Dawkins, 1986; Dennett, 1995.

39 Pregnancy sickness: Profet, 1992.

41 Evolution as innovator: Tooby & Cosmides, 1989.

41 Sociobiology versus evolutionary psychology: Symons, 1979, 1992; Tooby & Cosmides, 1990a.

42 Behavior is not now adaptive; the mind used to be adaptive: Symons, 1979, 1992;
Tooby & Cosmides, 1990a.

43 You can't take it with you: Gould, 1992. Gene's-eye view: Williams, 1966; Dawkins,
1976/1989, 1983, 1995; Sterelny & Kitcher, 1988; Kitcher, 1992; Cronin, 1992;
Dennett, 1995. Against the gene's-eye view: Gould, 1980b, 1983b.

44 The Standard Social Science Model: Tooby & Cosmides, 1992; Symons, 1979; Daly & Wilson, 1988.

45 Hysteria over sociobiology: Wright, 1988, 1994a; Wilson, 1994. Innuendo: Lewontin, Rose, & Kamin, 1984, p. 260. Not in his book: compare Dawkins, 1976/1989, p. 20, with Lewontin, Rose, & Kamin, 1984, p. 287, and with Levins & Lewontin, 1985, pp. 88, 128. Smears in *Scientific American*: Horgan, 1993, 1995a. Too dangerous to teach: Hrdy, 1994.

46 Freeman, Mead, and Samoa: Freeman, 1983, 1992.

46 Seville Statement: The Seville Statement on Violence, 1990.

48 Inauthentic preferences: Sommers, 1994.

49 Universal human nature: Tooby & Cosmides, 1990b.

50 Difference feminism: Sommers, 1994; Patai & Koertge, 1994.

51 Not so noble: Daly & Wilson, 1988; Chagnon, 1992; Keely, 1996.

51 Religion and modularity: Wright, 1994a.

52 The defining quality of being a woman: Gordon, 1996.

52 Blameless philandering mates: Rose, 1978.

53 The Abuse Excuse and other dubious extenuating factors: Dershowitz, 1994.

54 Creeping exculpation: Dennett, 1984; R. Wright, 1994a, 1995.

55 Moral responsibility compatible with neurophysiological and evolutionary causation: Dennett, 1984; Nozick, 1981, pp. 317–362.

56 Gay gene brouhaha: Hamer & Copeland, 1994.

57 Deconstructing gender: Lorber, 1994. Deconstructing binaries: Katz, 1995. Deconstructing deconstructionism: Carroll, 1995; Sommers, 1994; Paglia, 1992; Searle, 1983, 1993; Lehman, 1992.

2. Thinking Machines

59 Twilight Zone: Zicree, 1989.

60 Louis Armstrong on consciousness: Quoted in Block, 1978.

61 Good aliens: Interview by D. C. Denison, *Boston Globe Magazine*, June 18, 1995.

62 Idiotic filings versus intelligent lovers: James, 1890/1950.

62 What is intelligence?: Dennett, 1978b; Newell & Simon, 1972, 1981; Pollard, 1993.

63 Skinner skinned: Chomsky, 1959; Fodor, 1968a, 1986; Dennett, 1978c.

64 Beliefs and desires: Fodor, 1968a, b, 1975, 1986, 1994; Dennett, 1978d; Newell & Simon, 1981; Pylyshyn, 1980, 1984; Marr, 1982; Haugeland, 1981a, b, c; Johnson-Laird, 1988.

65 What is information?: Dretske, 1981.

67 Turing machines: Moore, 1964.

69 Production systems: Newell & Simon, 1972, 1981; Newell, 1990; Anderson, 1983, 1993.

77 Broad definition of computation: Fodor & Pylyshyn, 1988; Fodor, 1994.

77 The Ghost in the Machine: Ryle, 1949. Ghosts in the Mind's Machine: Kosslyn, 1983.

79 Stupid homunculi: Fodor, 1968b; Dennett, 1978d, pp. 123–124.

80 Meaning in mind: Loewer & Rey, 1991; McGinn, 1989a; Block, 1986; Fodor, 1994; Dietrich, 1994.

81 Biology of meaning: Millikan, 1984; Block, 1986; Pinker, 1995; Dennett, 1995; Field, 1977.

82 Everyday AI: Crevier, 1993; Hendler, 1994.

82 What computers can't do: Dreyfus, 1979; Weizenbaum, 1976; Crevier, 1993.

82 The experts speak: Cerf & Navasky, 1984.

83 Natural computation: Coined by Whitman Richards.

83 The computational brain: Churchland & Sejnowski, 1992.

86 Representation and generalization: Pylyshyn, 1984; Jackendoff, 1987; Fodor & Pylyshyn, 1988; Pinker, 1984a; Pinker & Prince, 1988.

88 Vastness of language: Pinker, 1994a; Miller, 1967.

88 Mill melancholy about melodies: Cited in Sowell, 1995.

89 Mental representations in the lab: Posner, 1978.

90 Multiple representations: Anderson, 1983. Visual images: Kosslyn, 1980, 1994; Pinker, 1984b, c. Short-term memory loops: Baddeley, 1986. Chunks: Miller, 1956; Newell & Simon, 1972. Grammar in the head: Chomsky, 1991; Jackendoff, 1987, 1994; Pinker, 1994.

90 Mentalese: Anderson & Bower, 1973; Fodor, 1975; Jackendoff, 1987, 1990, 1994; Pinker, 1989, 1994.

90 "Processed" inputs to the hippocampus: Churchland & Sejnowski, 1992, p. 286. "Processed" inputs to the frontal lobe: Crick & Koch, 1995.

90 Programming style: Kernighan & Plauger, 1978.

92 Architecture of complexity: Simon, 1969.

92 Hora and Tempus: Simon, 1969, p. 188.

93 The Chinese Room: Block, 1978; Searle, 1980.

94 Chinese Room commentary: Searle, 1980; Dietrich, 1994. Chinese Room update: Searle, 1992.

94 Chinese Room refutations: Churchland & Churchland, 1994; Chomsky, 1993; Dennett, 1995.

96 They're made out of meat: Bisson, 1991.

97 The emperor's new mind: Penrose, 1989, 1990. Update: Penrose, 1994.

97 The emperor's new book: Penrose, 1989; Wilczek, 1994; Putnam, 1994; Crick, 1994; Dennett, 1995.

98 Tortoise and Achilles: Carroll, 1895/1956.

99 Neuro-logical networks: McCulloch & Pitts, 1943.

101 Neural networks: Hinton & Anderson, 1981; Feldman & Ballard, 1982; Rumelhart, McClelland, & the PDP Research Group, 1986; Grossberg, 1988; Churchland & Sejnowski, 1992; Quinlan, 1992.

106 Necker network: Feldman & Ballard, 1982.

107 Pattern associators: Hinton, McClelland, & Rumelhart, 1986; Rumelhart & McClelland, 1986b.

109 Problems with perceptrons: Minsky & Papert, 1988a; Rumelhart, Hinton, & Williams, 1986.

111 Hidden-layer networks as function approximators: Poggio & Girosi, 1990.

112 Connectionism: Rumelhart, McClelland, & the PDP Research Group, 1986; McClelland, Rumelhart, & the PDP Research Group, 1986; Smolensky, 1988; Morris, 1989. Why humans are smarter than rats: Rumelhart & McClelland, 1986a, p. 143.

112 Past-tense debate: Rumelhart & McClelland, 1986b; Pinker & Prince, 1988, 1994; Prince & Pinker, 1988; Pinker, 1991; Prasada & Pinker, 1993; Marcus, Brinkmann, Clahsen, Wiese, & Pinker, 1995.

113 Problems with connectoplasm: Pinker & Mehler, 1988; Pinker & Prince, 1988; Prince & Pinker, 1988; Prasada & Pinker, 1993; Marcus, 1997a, b, in preparation; Fodor & Pylyshyn, 1988; Fodor & McClaughlin, 1990; Minsky & Papert, 1988b; Lachter & Bever, 1988; Anderson, 1990, 1993; Newell, 1990; Ling & Marinov, 1993; Hadley, 1994a, b.

113 Hume on contiguity and similarity: Hume, 1748/1955.

115 Vanishing cherry: Berkeley, 1713/1929, p. 324.

116 Identifying individuals: Bloom, 1996a.

117 Loving a twin: L. Wright, 1995.

117 Which Blick bit?: *Boston Globe*, 1990.

118 Wildebeests and zebras versus lions and hyenas: Personal communication from Daniel Dennett.

120 Systematicity of thoughts: Fodor & Pylyshyn, 1988.

121 Problems with representing propositions: Hinton, 1981.

122 Propositions in networks: Hinton, 1981; McClelland & Kawamoto, 1986; Shastri & Ajjanagadde, 1993; Smolensky, 1990, 1995; Pollack, 1990; Hadley & Hayward, 1994.

123 Amnesic networks: McCloskey & Cohen, 1989; Ratcliff, 1990. Bat-wielding bat: McClelland & Kawamoto, 1986.

124 Multiple memories: Sherry and Schacter, 1987. Multiple connectionist memories: McClelland, McNaughton, & O'Reilly, 1995.

125 Recursive transition networks for sentence understanding: Pinker, 1994, chap. 7.

125 Recurrent networks: Jordan, 1989; Elman, 1990; Giles et al., 1990. Failure of recurrent networks to handle propositions: Marcus, 1997a, in preparation. Connectionist proposition-crunchers: Pollack, 1990; Berg, 1991; Chalmers, 1990.

126 Fuzzy categories: Rosch, 1978; Smith & Medin, 1981. Fuzzy categories in connectoplasm: Whittlesea, 1989; McClelland & Rumelhart, 1985.

127 Problems with fuzzy categories: Armstrong, Gleitman, & Gleitman, 1983; Rey, 1983; Pinker & Prince, 1996; Marcus, 1997b; Medin, 1989; Smith, Langston, & Nisbett, 1992; Keil, 1989.

127 Gorillas and onions: Hinton, Rumelhart, & McClelland, 1986, p. 82.

128 Ape diets: Glander, 1992.

128 Explanation-based generalization: Pazzani, 1987, 1993; Pazzani & Dyer, 1987; Pazzani & Kibler, 1993; de Jong & Mooney, 1986.

129 Sorites: Fodor & Pylyshyn, 1988; Poundstone, 1988. Universality of long reasoning chains: Brown, 1991; Boyd & Silk, 1996.

130 Connectionist family tree: Rumelhart, Hinton, & Williams, 1986.

132 Johnson on mind and matter: Quoted in Minsky, 1985. Huxley on the Djin: Quoted in Humphrey, 1992. Water into wine: McGinn, 1989b.

133 Consciousness boom: Humphrey, 1992; Dennett, 1991; Crick, 1994; Penrose, 1994; Jackendoff, 1987; Searle, 1992, 1995; Marcel & Bisiach, 1988; Baars, 1988.

133 Gould on inventing consciousness: Gould, 1993, pp. 294–295.

133 Mirror, mirror: Gallup, 1991; Parker, Mitchell, & Boccia, 1994. Mirrors and mon-

keys revisited: Hauser et al., 1995. Unconscious ancients: Jaynes, 1976. Contagious consciousness: Dennett, 1991.

134 Sorting out the consciousness mess: Jackendoff, 1987; Block, 1995.

136 Consciousness among the neurons: Crick, 1994; Crick & Koch, 1995.

137 Bulletin board systems: Jagannathan, Dodhiawala, & Baum, 1989. Consciousness as bulletin board: Baars, 1988; Newell & Simon, 1972; Navon, 1989; Fehling, Baars, & Fisher, 1990.

137 Costs of computation: Minsky & Papert, 1988b; Ullman, 1984; Navon, 1985; Fehling, Baars, & Fisher, 1990; Anderson, 1990, 1991.

139 Intermediate-level consciousness: Jackendoff, 1987.

140 Visual attention: Treisman & Gelade, 1980; Treisman, 1988.

142 Floating letters: Mozer, 1991.

142 Memories of shocking news: Brown & Kulik, 1977; McCloskey, Wible, & Cohen, 1988; Schacter, 1996.

142 Optimality of memory: Anderson, 1990, 1991.

143 Function of emotional coloring: Tooby & Cosmides, 1990a, b.

144 Society of mind: Minsky, 1985. Multiple drafts: Dennett, 1991.

144 Will center discovered: Damasio, 1994; Crick, 1994.

144 Frontal lobes: Luria, 1966; Duncan, 1995.

145 Sentience versus access: Block, 1995.

146 Paradoxes of sentience: Nagel, 1974; Poundstone, 1988; Dennett, 1991; McGinn, 1989b, 1993; Block, 1995.

147 Debunking qualia: Dennett, 1991.

3. Revenge of the Nerds

149 Earth's greatest hits: Sullivan, 1993.

150 Little green men: Kerr, 1992. Evolutionary skeptics: Mayr, 1993.

150 Number of extraterrestrial civilizations: Sullivan, 1993.

151 We're only the first: Drake, 1993.

153 Human chauvinism: Gould, 1989, 1996.

153 Costs and benefits in evolution: Maynard Smith, 1984.

154 Costs and benefits of big brains: Tooby & DeVore, 1987.

155 Darwin and the universe: Dawkins, 1983, 1986; Williams, 1966, 1992; Maynard Smith, 1975/1993; Reeve & Sherman, 1993.

159 Photons don't wash an eye clear: Dawkins, 1986.

159 Macromutations cannot explain complex design: Dawkins, 1986. "Punctuated equilibria" are not the same as macromutations: Dawkins, 1986; Gould, 1987, p. 234.

160 "Adaptive mutation": Cairns, Overbaugh, & Miller, 1988; Shapiro, 1995. Problems with adaptive mutation: Lenski & Mittler, 1993; Lenski & Sniegowski; Shapiro, 1995.

160 Complexity theory: Kauffman, 1991; Gell-Mann, 1994.

161 Take a hike, Darwin: James Barham, *New York Times Book Review*, June 4, 1995; also Davies, 1995.

161 Limitations of complexity theory: Maynard Smith, 1995; Horgan, 1995b; Dennett, 1995.

162 Evidence for natural selection: Dawkins, 1986, 1995; Berra, 1990; Kitcher, 1982; Endler, 1986; Weiner, 1994.

163 Ascent of man: Bronowski, 1973, pp. 417–421.

164 Simulated evolving eye: Nilsson & Pelger, 1994; described in Dawkins, 1995.

165 Darwin-hating academics: Dawkins, 1982; Pinker & Bloom, 1990 (see commentaries and reply); Dennett, 1995.

165 Straw adaptationist: Lewontin, 1979.

166 Snagged seminal ducts: Williams, 1992.

166 Adaptationist advances: Mayr, 1983, p. 328.

167 Animal engineering excellence: Tooby & Cosmides, 1992; Dawkins, 1982, 1986; Williams, 1992; Griffin, 1974; Tributsch, 1982; French, 1994; Dennett, 1995; Cain, 1964.

168 The splendid camel: French, 1994, p. 239.

168 Howlers: Author's reply in Pinker & Bloom, 1990. Symmetry: Corballis & Beale, 1976. Sexy symmetry: Ridley, 1993.

170 Birds on the wing: Wilford, 1985.

170 Bugs on the wing: Kingsolver & Koehl, 1985.

171 Misunderstanding exaptation: Piattelli-Palmarini, 1989, p. 1.

171 Exaptation: Gould & Vrba, 1981. Problems with exaptation: Reeve & Sherman, 1993; Dennett, 1995. Housefly acrobatics: Wootton, 1990.

172 Debating design: Pinker & Bloom, 1990, including commentaries and reply; Williams, 1966, 1992; Mayr, 1983; Dennett, 1995; Reeve & Sherman, 1993; Dawkins, 1982, 1986; Tooby & Cosmides, 1990a, b, 1992; Tooby & DeVore, 1987; Sober, 1984a, b; Cummins, 1984; Lewontin, 1984.

172 Chomsky on natural selection: Personal communication, November 1989.

175 Value of information: Raiffa, 1968.

176 Tweaking the brain in evolution: Killackey, 1995; Rakic, 1995b; Stryker, 1994; Deacon, 1994.

177 Genetic algorithms: Mitchell, 1996.

177 Genetic algorithms and neural networks: Belew, 1990; Belew, McInerney, & Schraudolph, 1990; Nolfi, Elman, & Parisi, 1994; Miller & Todd, 1990.

178 Simultaneous evolution and learning: Hinton & Nowlan, 1987.

179 Baldwin effect: Dawkins, 1982; Maynard Smith, 1987.

179 Navigating ants: Wehner & Srinivasan, 1981. Dead reckoning: Gallistel, 1995, p. 1258.

180 Those amazing animals: Gallistel, 1990, 1995; J. Gould, 1982; Rozin, 1976; Hauser, 1996; Gaulin, 1995; Dawkins, 1986.

182 Conditioning as time-series analysis and other feats of animals: Gallistel, 1990, 1995.

183 Mammals' brains are not all the same: Preuss, 1993, 1995; Gaulin, 1995; Sherry & Schacter, 1987; Deacon, 1992a; Hauser, 1996.

183 Re-engineering the human brain: Deacon, 1992b; Holloway, 1995; Hauser, 1996; Killackey, 1995.

185 Broody hen: James, 1892/1920, pp. 393–394.

186 Zoologically unique or extreme human traits: Tooby & DeVore, 1987; Pilbeam, 1992.

188 Evolutionary arms race: Dawkins, 1982, 1986; Ridley, 1993. Cognitive niche: Tooby & DeVore, 1987.

189 Universal scientific and logical concepts: Brown, 1991.

189 Spoor analysis: Liebenberg, 1990, p. 80, quoted in Boyd & Silk, 1996.

189 High-tech hunter-gatherers: Brown, 1991; Kingdon, 1993.

190 Megafauna extinctions: Martin & Klein, 1984; Diamond, 1992.

190 Zoological uniqueness and the cognitive niche: Tooby & DeVore, 1987; Kingdon, 1993.

191 Primate vision: Deacon, 1992a; Van Essen & DeYoe, 1995; Preuss, 1995.

191 Vision co-opted by abstract concepts: Jackendoff, 1983, 1987, 1990; Lakoff, 1987; Talmy, 1988; Pinker, 1989.

192 Flatland: Gardner, 1991.

192 Madding crowd: Jones, Martin, & Pilbeam, 1992, part 4; Boyd & Silk, 1996.

193 Primate liars: Hauser, 1992; Lee, 1992; Boyd & Silk, 1996; Byrne & Whiten, 1988; Premack & Woodruff, 1978.

193 Primate yentas: Cheney & Seyfarth, 1990.

193 Cognitive arms race: Trivers, 1971; Humphrey, 1976; Alexander, 1987b, 1990; Rose, 1980; Miller, 1993. Problems with the cognitive arms race: Ridley, 1993.

194 Leisurely brain inflation: Williams, 1992.

194 Ape hands and posture: Jones, Martin, & Pilbeam, 1992, part 2; Boyd & Silk, 1996; Kingdon, 1993. Importance of hands: Tooby & DeVore, 1987.

195 Rehabilitating Man the Hunter: Tooby & DeVore, 1987; Boyd & Silk, 1996.

197 Carnal for carnal in apes and humans: Tooby & DeVore, 1987; Ridley, 1993; Symons, 1979; Harris, 1985; Shostak, 1981.

199 Hominid ancestors: Jones, Martin, & Pilbeam, 1992; Boyd & Silk, 1996; Kingdon, 1993; Klein, 1989; Leakey et al., 1995; Fischman, 1994; Swisher et al., 1996.

200 Fossils and the cognitive niche: Tooby & DeVore, 1987.

201 Australopithecine hands: L. Aiello, 1994. Australopithecine brains and tools: Holloway, 1995; Coppens, 1995. Vertically challenged habilines: Lewin, 1987.

202 African Eve refuses to die: Gibbons, 1994, 1995a.

202 Great leap forward: Diamond, 1992; Marschack, 1989; White, 1989; Boyd & Silk, 1996.

203 Anatomically not-so-modern humans: Boyd & Silk, 1996; Stringer, 1992.

204 Pontiac in Leonardo's attic: Shreeve, 1992; Yellen et al., 1995; Gutin, 1995.

204 Logic of Eve: Dawkins, 1995; Dennett, 1995; Ayala, 1995. Fantastic misunderstandings: Pinker, 1992.

204 Mixed-sex versus all-female line ancestors: Dawkins, 1995.

205 Recent bottlenecks: Gibbons, 1995b, c; Harpending, 1994; Cavalli-Sforza, Menozzi, & Piazza, 1993. Speed of evolution: Jones, 1992.

205 End of evolution: Jones, 1992; Cavalli-Sforza, Menozzi, & Piazza, 1993.

206 Darwinian social science: Turke & Betzig, 1985, p. 79; Alexander, 1987a; Betzig et al., 1988.

206 Functionalism: Bates & MacWhinney, 1990, p. 728; Bates & MacWhinney, 1982.

206 Lamarck on felt need: quoted in Mayr, 1982, p. 355.

207 Rats: Personal communication from B. F. Skinner, 1978. Chimps: Nagell, Olguin, & Tomasello, 1993.

207 Adaptation a thing of the past: Tooby & Cosmides, 1990a; Symons, 1979, 1992.

208 Cultural evolution: Dawkins, 1976/1989; Durham, 1982; Lumsden & Wilson, 1981; Diamond, 1992; Dennett, 1995. Problems with cultural evolution: Tooby & Cosmides, 1990a, 1992; Symons, 1992; Daly, 1982; Maynard Smith & Warren, 1988; Sperber, 1985.

208 Genes and memes: Dawkins, 1976/1989.

210 Culture as disease: Cavalli-Sforza & Feldman, 1981; Boyd & Richerson, 1985; Sperber, 1985.

4. The Mind's Eye

211 Autostereograms: N.E. Thing Enterprises, 1994; *Stereogram*, 1994; *Superstereogram*, 1994.

211 Birth of the autostereogram: Tyler, 1983.

212 Perception as an ill-posed problem; illusions as violations of assumptions: Gregory, 1970; Marr, 1982; Poggio, 1984; Hoffman, 1983.

213 Perception as description: Marr, 1982; Pinker, 1984c; Tarr & Black, 1994a, b.

215 Pictures, perspective, and perception: Gregory, 1970; Kubovy, 1986; Solso, 1994; Pirenne, 1970. Pictures in New Guinea: Ekman & Friesen, 1975.

215 Adelbert Ames: Ittelson, 1968.

218 Binocular parallax and stereo vision: Gregory, 1970; Julesz, 1971, 1995; Tyler, 1991, 1995; Marr, 1982; Hubel, 1988; Wandell, 1995.

219 Wheatstone: From Wandell, 1995, p. 367.

223 Stereoscopes: Gardner, 1989.

230 Random-dot stereograms: Julesz, 1960, 1971, 1995; Tyler, 1991, 1995.

233 Lemurs and leaf rooms: Tyler, 1991. Penetrating camouflage: Julesz, 1995.

234 Modeling the cyclopean eye: Marr, 1982; Tyler, 1995; Weinshall & Malik, 1995; Anderson & Nakayama, 1994.

235 Stereo networks that cooperate and relax: Marr & Poggio, 1976. Diagram adapted from Johnson-Laird, 1988.

236 Da Vinci stereo: Nakayama, He, & Shimojo, 1995; Anderson & Nakayama, 1994.

237 Stereoblindness and stereodeficiency: Richards, 1971. Binocular neurons: Poggio, 1995. Update on stereo pools: Cormack, Stevenson, & Schor, 1993.

238 Binocular babies: Shimojo, 1993; Birch, 1993; Held, 1993; Thorn et al., 1994.

238 Prewired stereo circuitry: Birch, 1993; Freeman & Ohzawa, 1992.

239 Monocular monkeys: Hubel, 1988; Stryker, 1993. Sharpening neurons: Stryker, 1994; Miller, Keller, & Stryker, 1989.

240 Crossed eyes, lazy eyes: Birch, 1993; Held, 1993; Thorn et al., 1994.

240 Neural sensitivity and growing skulls: Timney, 1990; Pettigrew, 1972, 1974.

242 Shading, shape, and lighting: Adelson & Pentland, 1996.

243 Perception as playing the odds: Knill & Richards, 1996. Nonaccidental properties: Lowe, 1987; Biederman, 1995.

244 Betting on a regular world: Attneave, 1982; Jepson, Richards, & Knill, 1996; Knill & Richards, 1996.

245 Straight lines in nature: Sanford, 1994; Montello, 1995.

245 Lightness, brightness, and illumination: Marr, 1982; Adelson & Pentland, 1996.

247 Retinex theory: Land & McCann, 1971; Marr, 1982; Brainard & Wandell, 1986. Newer models: Brainard & Wandell, 1991; Maloney & Wandell, 1986.

248 Shape from shading: Marr, 1982; Pentland, 1990; Ramachandran, 1988; Nayar & Oren, 1995.

249 Moonstruck: Nayar & Oren, 1995.

249 Seeing the simplest world: Adelson & Pentland, 1996; Attneave, 1972, 1981, 1982; Beck, 1982; Kubovy & Pomerantz, 1981; Jepson, Knill, & Richards, 1996.

255 Flipped shape, flipped light source: Ramachandran, 1988.

256 Sandbox in the head: Attneave, 1972. Problems with the sandbox: Pinker, 1979, 1980, 1984c, 1988; Pinker & Finke, 1980.

257 Eye movements: Rayner, 1992; Kowler, 1995; Marr, 1982.

258 Two-dimensionality of vision: French, 1987.

258 Objects versus surfaces: Marr, 1982, p. 270; Nakayama, He, & Shimojo, 1995.

260 2½-D sketch: Marr, 1982; Pinker, 1984c, 1988. Visible surface representation: Jackendoff, 1987; Nakayama, He, & Shimojo, 1995.

262 Compensating for eye movements: Rayner, 1992.

264 The visual field and the visual world: Gibson, 1950, 1952; Boring, 1952; Attneave, 1972, 1982; Hinton & Parsons, 1981; Pinker, 1979, 1988.

264 Gravity and vision: Rock, 1973, 1983; Shepard & Cooper, 1982; Pinker, 1984c.

264 Heave ho!: Mazel, 1992.

265 Space sickness: Oman, 1982; Oman et al., 1986; Young et al., 1984.

265 Motion sickness and neurotoxins: Treisman, 1977.

266 What's up in shape perception?: Rock, 1973; Shepard & Cooper, 1982; Corballis, 1988.

267 Dancing triangles: Attneave, 1968.

269 Shape recognition as matching object-centered descriptions: Marr & Nishihara, 1978; Marr, 1982; Corballis, 1988; Biederman, 1995; Pinker, 1984c; Hinton & Parsons, 1981; Dickinson, Pentland, & Rosenfeld, 1992.

270 Geons: Biederman, 1995.

271 Shapes in the left and right hemispheres: Kosslyn, 1994; Farah, 1990. Fragmented inner vision: Farah, 1990.

271 Finding parts in the 2½-D sketch: Hoffman & Richards, 1984; Lowe, 1987; Dickinson, Pentland, & Rosenfeld, 1992.

272 Psychology of clothing: Bell, 1992, pp. 50–51.

272 Faces: Etcoff, Freeman, & Cave, 1991; Landau, 1989; Young & Bruce, 1991; Bruce, 1988; Farah, 1995. Babies and faces: Morton & Johnson, 1991.

272 Man who could not recognize faces: Etcoff, Freeman, & Cave, 1990; Farah, 1995.

273 Man who could recognize only faces: Behrmann, Winocur, & Moscovitch, 1992; Moscovitch, Winocur, & Behrmann, in press.

274 Sphere with a toy shows we need all views: Thanks to Jacob Feldman.

274 Multiple views: Poggio & Edelman, 1991; Bülthoff & Edelman, 1992.

275 Recognizing shapes by mentally rotating them: Shepard & Cooper, 1982; Tarr & Pinker, 1989, 1990; Tarr, 1995; Ullman, 1989.

275 Mental rotation: Cooper & Shepard, 1973; Shepard & Cooper, 1982; Tarr & Pinker, 1989, 1990; Corballis, 1988.

276 Handedness and the universe: Gardner, 1990. The psychology of left and right: Corballis & Beale, 1976.

277 Obliviousness to left and right: Corballis & Beale, 1976; Corballis, 1988; Hinton & Parsons, 1981; Tarr & Pinker, 1989.

279 How people recognize shapes: Tarr & Pinker, 1989, 1990; Tarr, 1995; Tarr & Bülthoff, 1995; Biederman, 1995; Bülthoff & Edelman, 1992; Sinha, 1995.

284 Mental imagery: Kosslyn, 1980, 1983, 1994; Paivio, 1971; Finke, 1989; Block, 1981; Pinker, 1984c, 1988; Tye, 1991; Logie, 1995; Denis, Engelkamp, & Richardson, 1988; Hebb, 1968.

284 Yanomamö imagery: Chagnon, 1992.

285 Creativity and imagery: Finke, 1990; Shepard, 1978; Shepard & Cooper, 1982; Kosslyn, 1983.

285 Corrugator muscle: Buss, 1994, p. 128.

286 Pictures versus propositions: Pylyshyn, 1973, 1984; Block, 1981; Kosslyn, 1980, 1994; Tye, 1991; Pinker, 1984; Kosslyn, Pinker, Smith, & Shwartz, 1979. Imagery in computers: Funt, 1980; Glasgow & Papadias, 1992; Stenning & Oberlander, 1995; Ioerger, 1994.

287 Cortical maps: Van Essen & DeYoe, 1995.

288 Cloying hunger by imagining a feast: *Richard II*, act 1, scene 3.

288 Perky effect: Perky, 1910; Segal & Fusella, 1970; Craver-Lemley & Reeves, 1992; Farah, 1989.

288 Imagery and coordination: Brooks, 1968; Logie, 1995.

288 Imagery and illusions: Wallace, 1984. Imagery and alignment: Freyd & Finke, 1984.

288 Confusing images and reality: Johnson & Raye, 1981.

289 Neglect of imaginary space: Bisiach & Luzzatti, 1978.

289 Imagery lights up the visual cortex: Kosslyn et al., 1993; Kosslyn, 1994.

289 Images with and without both halves of visual cortex: Farah, Soso, & Dasheiff, 1992.

289 Dreams and images: Symons, 1993. Reality monitoring: Johnson & Raye, 1981.

290 Medium underlying imagery: Pinker, 1984c, 1988; Cave, Pinker, et al., 1994; Kosslyn, 1980, 1994.

291 Computing with imagery: Funt, 1980; Glasgow & Papadias, 1992; Stenning & Oberlander, 1995; Ioerger, 1994.

292 Mental animation: Ullman, 1984; Jolicoeur, Ullman, & MacKay, 1991.

292 Answering questions using imagery: Kosslyn, 1980.

293 Flipping duck-rabbits in imagery: Chambers & Reisberg, 1985; Finke, Pinker, & Farah, 1989; Peterson et al., 1992; Hyman & Neisser, 1991.

294 Piecemeal fading images: Kosslyn, 1980.

294 Imagery and vantage point: Pinker, 1980, 1984c, 1988.

294 Multiple perspective in paintings: Kubovy, 1986; Pirenne, 1970. Cro-Magnon perspective: Boyd & Silk, 1996.

294 Filing images: Pylyshyn, 1973; Kosslyn, 1980.

295 Visual memory in chess masters: Chase & Simon, 1973.

295 Memory for a penny: Nickerson & Adams, 1979.

295 Mental map distortions: Stevens & Coupe, 1978.

296 Images aren't concepts: Pylyshyn, 1973; Fodor, 1975; Kosslyn, 1980; Tye, 1991.

297 Imagery gone mad: Titchener, 1909, p. 22.

5. Good Ideas

299 Darwin vs. Wallace: Gould, 1980c; Wright, 1994a.

300 Brain as overkill: Davies, 1995, pp. 85–87.

301 Exapted computer: Gould, 1980e, p. 57.

301 Cerebral savages: Brown, 1991; Kingdon, 1993.

303 Cane juice syllogism: Cole et al., 1971, pp. 187–188; Neisser, 1976.

303 Logic and lame puppies: Carroll, 1896/1977.

304 Ecological rationality: Tooby & Cosmides, 1997. Dissimilarities between thinking and science: Harris, 1994; Tooby & Cosmides, 1997; Neisser, 1976.

305 Flimflam shamans: Harris, 1989, pp. 410–412.

305 Caste-society know-nothings: Brown, 1988.

307 Concepts as predictors: Rosch, 1978; Shepard, 1987; Bobick, 1987; Anderson, 1990, 1991; Pinker & Prince, 1996.

308 Fuzziness and similarity versus rules and theories: Armstrong, Gleitman, & Gleitman, 1983; Pinker & Prince, 1996; Murphy, 1993; Medin, 1989; Kelly, 1992; Smith, Langston, & Nisbett, 1992; Rey, 1983; Pazzani, 1987, 1993; Pazzani & Dyer, 1987; Pazzani & Kibler, 1993; Rips, 1989.

310 Species according to biologists: Mayr, 1982; Ruse, 1986.

311 Piss-poor reptile: Quoted in Konner, 1982. Fuzzy fish: Dawkins, 1986; Gould, 1983c; Ridley, 1986; Pennisi, 1996. Shoehorning extinct animals: Gould, 1989.

311 All is fuzzy: Lakoff, 1987.

312 Crisp idealizations: Pinker & Prince, 1996.

313 Nonsense stereotypes of outsiders: Brown, 1985.

313 Statistically accurate negative stereotypes: McCauley & Stitt, 1978; Brown, 1985.

314 Ways of explaining: Dennett, 1978b, 1995, 1990; Hirschfeld & Gelman, 1994a, b; Sperber, Premack, & Premack, 1995; Carey, 1985; Carey & Spelke, 1994; Baron-Cohen, 1995; Leslie, 1994; Schwartz, 1979; Keil, 1979.

314 Dead bird, live bird: Dawkins, 1986, pp. 10–11.

316 Innate AI systems: Lenat & Guha, 1990.

317 Babies as physicists: Spelke, 1995; Spelke et al., 1992; Spelke, Phillips, & Woodward, 1995; Spelke, Vishton, & Hofsten, 1995; Baillargeon, 1995; Baillargeon, Kotovsky, & Needham, 1995.

320 Intuitive impetus theory: McCloskey, Caramazza, & Green, 1980; McCloskey, 1983. Intuitive physics: Proffitt & Gilden, 1989.

320 College students' understanding of force: Redish, 1994.

322 Dot drama: Heider & Simmel, 1944; Michotte, 1963; Premack, 1990.

322 Infants and oomph: Premack, 1990; Leslie, 1994, 1995a; Mandler, 1992; Gelman, Durgin, & Kaufman, 1995; Gergely et al., 1995.

323 Universality of folk biology: Konner, 1982; Brown, 1991; Atran, 1990, 1995; Berlin, Breedlove, & Raven, 1973.

323 Lions, tigers, and other natural kinds: Quine, 1969; Schwartz, 1979; Putnam, 1975; Keil, 1989.

324 Darwin and natural kinds: Kelly, 1992; Dawkins, 1986.

325 Essentialism and resistance to evolution: Mayr, 1982.

326 Children as essentialists: Keil, 1989, 1994, 1995; Gelman, Coley, & Gottfried, 1994; Gelman & Markman, 1987. Skepticism on children as essentialists: Carey, 1995.

327 Children distinguishing psychology from biology: Hatano & Inagaki, 1995; Carey, 1995.

327 Babies and artifacts: Brown, 1990.

327 Artifacts and natural kinds stored separately in the brain: Hillis & Caramazza, 1991; Farah, 1990.

328 What is an artifact?: Keil, 1979, 1989; Dennett, 1990; Schwartz, 1979; Putnam, 1975; Chomsky, 1992, 1993; Bloom, 1996b.

329 Folk psychology and the intentional stance: Fodor, 1968a, 1986; Dennett, 1978b, c; Baron-Cohen, 1995.

330 Theory of mind module: Leslie, 1994, 1995a, b; Premack & Premack, 1995; Gopnik & Wellman, 1994; Hirschfeld & Gelman, 1994b; Wimmer & Perner, 1983; Baron-Cohen, Leslie, & Frith, 1985; Baron-Cohen, 1995.

330 Young children and false beliefs: Leslie, 1994, 1995b.

331 Noisy skin-bags: Gopnik, 1993.

331 Autism: Baron-Cohen, 1995; Baron-Cohen et al., 1985; Frith, 1995; Gopnik, 1993.

331 Iceboxes, toilets, and autism: Bettelheim, 1959.

332 False photographs: Zaitchik, 1990.

333 Brain creates world: Miller, 1981.

334 Illogical undergraduates: Johnson-Laird, 1988.

334 Logic and thought: Macnamara, 1986, 1994; Macnamara & Reyes, 1994.

334 Defending the mind's logic: Macnamara, 1986; Braine, 1994; Bonatti, 1995; Rips, 1994; Smith, Langston, & Nisbett, 1992.

336 Falsification by card selection: Wason, 1966; Manktelow & Over, 1987.

337 Reasoning and cheater-detection: Cosmides, 1985, 1989; Cosmides & Tooby, 1992. Employer/employee problem: Gigerenzer & Hug, 1992. Other effects and alternative interpretations: Cheng & Holyoak, 1985; Sperber, Cara, & Girotto, 1995.

338 Psychology of number: Geary, 1994, 1995; Gelman & Gallistel, 1978; Gallistel, 1990; Dehaene, 1992; Wynn, 1990. Counting by babies: Wynn, 1992. Counting by monkeys: Hauser, MacNeilage, & Ware, 1996.

339 Math and basic human activities: Mac Lane, 1981; Lakoff, 1987. Blind toddlers take shortcuts: Landau, Spelke, & Gleitman, 1984.

341 American dunces: Geary, 1994, 1995.

342 Why Johnny still can't add: Geary, 1995.

342 Why Johnny still can't read: Levine, 1994; McGuinness, 1997.

343 *Informavore*: Coined by George Miller.

343 *Innumeracy*: Coined by John Allen Paulos.

343 Probability-blindness: Tversky & Kahneman, 1974, 1983; Kahneman, Slovic, & Tversky, 1982; Kahneman & Tversky, 1982; Nisbett & Ross, 1980; Sutherland, 1992; Gilovich, 1991; Piattelli-Palmarini, 1994; Lewis, 1990.

345 People as intuitive statisticians: Gigerenzer & Murray, 1987; Gigerenzer, 1991, 1996a; Gigerenzer & Hoffrage, 1995; Cosmides & Tooby, 1996; Lopes & Oden, 1991; Koehler, 1996. Reply: Kahneman & Tversky, 1996. Bees as intuitive statisticians: Staddon, 1988.

347 History of probability and statistics: Gigerenzer et al., 1989. Probabilities gathered from experience: Gigerenzer & Hoffrage, 1995; Gigerenzer, 1997; Cosmides & Tooby, 1996; Kleiter, 1994.

348 People are good statisticians with frequency information: Tversky & Kahneman, 1983; Fiedler, 1988; Cosmides & Tooby, 1996; Gigerenzer, 1991, 1996b, 1997; Hertwig & Gigerenzer, 1997.

349 Von Mises and the probability of a single event: Example adapted by Cosmides & Tooby, 1996.

350 O.J., wife-battering, and murder: Good, 1995.

351 "Conjunction fallacy" (feminist bankteller) is not a fallacy: Hertwig & Gigerenzer, 1997.

352 Spatial metaphor: Gruber, 1965; Jackendoff, 1983, 1987, 1990, 1994; Pinker, 1989.

353 Communication as giving: Pinker, 1989.

354 Force dynamics in language and thought: Talmy, 1988; Pinker, 1989.

355 Space and force in language and thought: Jackendoff, 1983, 1987, 1990, 1994; Pinker, 1989; Levin & Pinker, 1992; Wierzbicka, 1994; Miller & Johnson-Laird, 1976; Schanck & Riesbeck, 1981; Pustejovsky, 1995. Universality of space and force: Talmy, 1985; Pinker, 1989.

355 Leibniz's remarkable thought: Leibniz, 1956.

356 Spatial metaphor as cognitive vestige: Pinker, 1989.

356 Chimps and causation: Premack, 1976.

356 Universality of space and force metaphors: Talmy, 1985; Pinker, 1989.

356 Children's spatial metaphors: Bowerman, 1983; Pinker, 1989.

357 Basic metaphors in language versus poetic metaphors: Jackendoff and Aaron, 1991.

357 Metaphors we live by: Lakoff & Johnson, 1980; Lakoff, 1987.

359 Graphs: Pinker, 1990.

359 Mathematization of physics intuitions: Carey & Spelke, 1994; Carey, 1986; Proffitt & Gilden, 1989.

361 But is it dental work?: Allen, 1983.

361 Genius and creativity: Weisberg, 1986; Perkins, 1981.

6. Hotheads

364 Running amok: B. B. Burton-Bradley, quoted in Daly & Wilson, 1988, p. 281.

365 Universality of emotions: Brown, 1991; Lazarus, 1991; Ekman & Davidson, 1994; Ekman, 1993, 1994; Ekman & Friesen, 1975; Etcoff, 1986. Controversies on universality: Ekman & Davidson, 1994; Russell, 1994.

365 Darwin and emotional expression: Darwin, 1872/1965, pp. 15–17.

366 Anthropological correctness: Ekman, 1987. Emotion in blind and deaf children: Lazarus, 1991.

367 Mad pain: Lewis, 1980, p. 216.

368 Cow urine: Shweder, 1994, p. 36.

368 Mellow Inuits: Lazarus, 1991, p. 193. Mellow Samoans: Freeman, 1983.

368 Ethnography and etiquette: Quoted in Asimov & Shulman, 1986.

370 Triune brain: MacLean, 1990. Refutation: Reiner, 1990.

371 The emotional brain: Damasio, 1994; LeDoux, 1991, 1996; Gazzaniga, 1992.

372 Indispensability of emotion: Tooby & Cosmides, 1990a; Nesse & Williams, 1994; Nesse, 1991; Minsky, 1985.

374 Emotional robots: Minsky, 1985; Pfeiffer, 1988; Picard, 1995; Crevier, 1993.

374 Fight or flight: Marks & Nesse, 1994.

375 Habitat selection and environmental aesthetics: Orians & Heerwagen, 1992; Kaplan, 1992; Cosmides, Tooby, & Barkow, 1992.

375 Lifelong camping trip: Cosmides, Tooby, & Barkow, 1992, p. 552.

376 Native Americans and ersatz savannas: Christopher, 1995. Australian aborigines and ersatz savannas: Harris, 1992.

377 Reference frames in large terrains: Subbiah et al., 1996.

378 Disgust: Rozin & Fallon, 1987; Rozin, 1996.

379 Eating insects: Harris, 1985, p. 159.

380 Grossing out the Yanomamö: Chagnon, 1992.

381 Learning what is good to eat: Cashdan, 1994.

381 Mom and Dad as food tasters: Cashdan, 1994.

383 Contamination by contact: Tooby & Cosmides, personal communication.

383 Animalitos and optimal foraging: Harris, 1985.

384 Ecology and food taboos: Harris, 1985.

386 *Phobophobia*: Coined by Richard Lederer.

386 Fears and phobias: Brown, 1991; Marks & Nesse, 1994; Nesse & Williams, 1994; Rachman, 1978; Seligman, 1971; Marks, 1987; Davey, 1995.

387 Lion phobia in Chicago: Maurer, 1965.

388 Relative rarity of screaming meemies: Rachman, 1978; Myers & Diener, 1995.

388 Monkeys learning snake phobias: Mineka & Cook, 1993.

389 Conquering fear: Rachman, 1978.

390 Happiness and social comparisons: Kahneman & Tversky, 1984; Brown, 1985. Violence and inequality: Daly & Wilson, 1988, p. 288.

391 Who is happy?: Myers & Diener, 1995. Heritability of happiness baseline: Lykken & Tellegen, 1996.

392 Gains versus losses: Kahneman & Tversky, 1984; Ketelaar, 1995, 1997.

393 Hedonic treadmill: Brickman & Campbell, 1971; Campbell, 1975.

394 Murray and Esther: From Arthur Naiman's *Every Goy's Guide to Yiddish*.

395 Crime and discounting the future: Wilson & Herrnstein, 1985; Daly & Wilson, 1994; Rogers, 1994.

395 Myopic discounting: Kirby & Herrnstein, 1995.

395 Self-control and rational consumers: Schelling, 1984, p. 59

396 Two selves: Schelling, 1984, p. 58.

397 The selfish replicator: Williams, 1966, 1992; Dawkins, 1976/1989, 1982; Dennett, 1995; Sterelny & Kitcher, 1988; Maynard Smith, 1982; Trivers, 1981, 1985; Cosmides & Tooby, 1981; Cronin, 1992.

397 Selection of replicators, groups, and branches: Gould, 1980b; Wilson & Sober, 1994; Dennett, 1995; Williams, 1992; Dawkins, 1976/1989, 1982.

400 Kin selection: Williams & Williams, 1957; Hamilton, 1963, 1964; Maynard Smith, 1964; Dawkins, 1976/1989; Trivers, 1985.

402 Reciprocal altruism: Williams, 1966; Trivers, 1971, 1985; Dawkins, 1976/1989; Cosmides & Tooby, 1992; Brown, 1985, p. 93.

404 Reciprocal altruism and the emotions: Trivers, 1971, 1985; Alexander, 1987a; Axelrod, 1984; Wright, 1994a. The moral sense: Wilson, 1993.

404 Reciprocal altruism and social psychology research: Trivers, 1971, 1981.

406 Within-group amity = between-group enmity: Dawkins, 1976/1989; Alexander, 1987.

408 Dr. Strangelove: from Peter George, *Dr. Strangelove*, Boston: G. K. Hall, 1963/1979, pp. 98–99.

409 Thinking the unthinkable: Poundstone, 1992.

409 Paradoxical tactics: Schelling, 1960.

412 The emotions as doomsday machines and other paradoxical tactics: Schelling, 1960; Trivers, 1971, 1985; Frank, 1988; Daly & Wilson, 1988; Hirshleifer, 1987.

413 Fairness and the Falklands: Frank, 1988. Vengeance: Daly & Wilson, 1988. Honor: Nisbett & Cohen, 1996.

414 Facial expressions: Darwin, 1872/1965; Ekman & Friesen, 1975; Fridlund, 1991, 1995. Darwin's anti-Darwinism: Fridlund, 1992.

415 Voluntary and involuntary facial expressions, method acting, and the brain: Damasio, 1994.

415 Honest signaling in animals: Dawkins, 1976/1989; Trivers, 1981; Cronin, 1992; Hauser, 1996; Hamilton, 1996.

416 Emotions and the body: Ekman & Davidson, 1994; Lazarus, 1991; Etcoff, 1986.

417 Theory of mad love: Frank, 1988.

417 Marriage market: Buss, 1994; Fisher, 1992; Hatfield & Rapson, 1993.

419 Tactics for controlling self and others: Schelling, 1984.

420 Grief as a deterrent: Tooby & Cosmides, 1990a.

421 Self-deception: Trivers, 1985; Alexander, 1987a; Wright, 1994a; Lockard & Paulhaus, 1988. Self-deception and Freudian defense mechanisms: Nesse & Lloyd, 1992.

422 Split brains: Gazzaniga, 1992.

422 Lake Wobegon effect: Gilovich, 1991.

422 Beneffectance: Greenwald, 1988; Brown, 1985. Cognitive dissonance: Festinger, 1957. Cognitive dissonance as self-presentation: Aronson, 1980; Baumeister & Tice, 1984. Beneffectance and cognitive dissonance as self-deception: Wright, 1994a.

424 Argument between husband and wife: Trivers, 1985, p. 420.

424 Explaining Hitler: Rosenbaum, 1995.

7. *Family Values*

426 *Greening of America* controversy: Nobile, 1971.

426 Nineteenth-century utopias: Klaw, 1993.

427 Human universals: Brown, 1991.

427 The thirty-six dramatic situations: Polti, 1921/1977.

427 Darwinian competitors: Williams, 1966; Dawkins, 1976/1989, 1995.

428 Homicide rates: Daly & Wilson, 1988. Universal conflict resolution: Brown, 1991.

430 Biology of kinship: Hamilton, 1964; Wilson, 1975; Dawkins, 1976/1989. Psychology of kinship: Daly & Wilson, 1988; Daly, Salmon, & Wilson, in press; Alexander, 1987a; Fox, 1984; van den Berghe, 1974; Wright, 1994a.

431 Frost's definition of "home": From "The Death of the Hired Man," in *North of Boston*.

432 Kinship nonsense: Daly, Salmon, & Wilson, in press; Mount, 1992; Shoumatoff, 1985; Fox, 1984.

433 Stepparents, stepchildren: Daly & Wilson, 1988, 1995.

434 Cinderella stories: Daly & Wilson, 1988, p. 85.

435 Homicide as conflict resolution: Daly & Wilson, 1988, p. ix.

435 Nepotism: Shoumatoff, 1985; Alexander, 1987a; Daly, Salmon, & Wilson, in press. Yanomamö kinship: Chagnon, 1988, 1992.

437 Cousin marriages: Thornhill, 1991.

437 The reality of romantic love: Symons, 1978; Fisher, 1992; Buss, 1994; Ridley, 1993; H. Harris, 1995.

438 Fictive kin: Daly, Salmon, & Wilson, in press.

439 The subversive family: Shoumatoff, 1985; Mount, 1992.

440 Royalty versus families: Thornhill, 1991. Church versus families: Betzig, 1992.

441 Parent-offspring conflict: Trivers, 1985; Dawkins, 1976/1989; Wright, 1994a; Daly & Wilson, 1988, 1995; Haig, 1992, 1993.

441 Sibling rivalry: Dawkins, 1976/1989; Trivers, 1985; Sulloway, 1996; Mock & Parker, in press.

443 Raised voices in the womb: Haig, 1993.

443 Infanticide: Daly & Wilson, 1988, 1995.

444 Postpartum depression: Daly & Wilson, 1988.

444 Bonding: Daly & Wilson, 1988.

444 Cuteness: Gould, 1980d; Eibl-Eibesfeldt, 1989; Konner, 1982; Daly & Wilson, 1988.

445 Children's psychological tactics: Trivers, 1985; Schelling, 1960.

446 Oedipus revisited: Daly & Wilson, 1988.

447 Controlling daughters: Wilson & Daly, 1992.

447 Socializing children against themselves: Trivers, 1985.

448 Nature, nurture, and none of the above in personality: Plomin, 1989; Plomin & Daniels, 1987; Bouchard, 1994; Bouchard et al., 1990; J. Harris, 1995; Sulloway, 1995, 1996.

449 Switching parents around: J. Harris, 1995.

449 Clique leaders first to date: Dunphy, 1963.

449 Socialization by peers: J. Harris, 1995.

451 Mothers' ambivalence: Interview with Shari Thurer by D. C. Denison, *The Boston Globe Magazine*, May 14, 1995; Eyer, 1996.

451 Sex ed: Whitehead, 1994.

452 Sibling rivalry: Trivers, 1985; Sulloway, 1995, 1996; Dawkins, 1976/1989; Wright, 1994a.

452 Expected grandchildren: Daly & Wilson, 1988; Sulloway, 1996; Wright, 1994a. Filicide: Daly & Wilson, 1988. Grief: Wright, 1994a.

453 Family dynamics: Sulloway, 1995, 1996.

455 The girl next door: Fisher, 1992; Hatfield & Rapson, 1993; Buss, 1994.

456 Incest avoidance and incest taboos: Tooby, 1976a, b; Brown, 1991; Daly & Wilson, 1988; Thornhill, 1991.

456 Costs of inbreeding in mammals: Ralls, Ballou, & Templeton, 1988.

457 Costing out incest: Tooby, 1976a, b.

458 Incest statistics: Buss, 1994; Brown, 1991; Daly & Wilson, 1988.

459 Incest between people who have not grown up together: Brown, 1991.

461 The battle between the sexes: Symons, 1979; Dawkins, 1976/1989; Trivers, 1985. The psychology of sexuality: Symons, 1979; Ridley, 1993; Wright, 1994a, b; Buss, 1994.

461 Reality of some gender stereotypes: Eagly, 1995.

461 Why sex?: Tooby, 1982, 1988; Tooby & Cosmides, 1990b; Hamilton, Axelrod, and Tanese, 1990; Ridley, 1993.

462 Why sexes?: Cosmides & Tooby, 1981; Hurst & Hamilton, 1992; Anderson, 1992.

463 Why so few animal hermaphrodites?: Cosmides & Tooby, 1981.

463 Sexual selection and differences in parental investment: Trivers, 1985; Cronin, 1992; Dawkins, 1976/1989; Symons, 1979; Ridley, 1993; Wright, 1994a, b.

465 Apes and sex: Trivers, 1985; Ridley, 1993; Boyd & Silk, 1996; Mace, 1992; Dunbar, 1992. Primate infanticide: Hrdy, 1981.

465 Sperm competition: Baker & Bellis, 1996.

467 Adulterous birds: Ridley, 1993.

468 Humans and sex: Ridley, 1993; Wright, 1994a; Mace, 1992; Dunbar, 1992; Boyd & Silk, 1996; Buss, 1994.

468 Environment in which the mind evolved: Symons, 1979.

469 Fatherless children in foraging societies: Hill & Kaplan, 1988.

469 Male desire for variety: Symons, 1979; Buss, 1994; Ridley, 1993; Wright, 1994a.

470 *Voulez-vous coucher avec moi ce soir?*: Clark & Hatfield, 1989.

470 Coolidge effect in roosters and men: Symons, 1979; Buss, 1994.

472 Pornography more popular than movies or sports: Anthony Flint in the *Boston Globe*, December 1, 1996.

472 Pornography and bodice-rippers: Symons, 1979; Ridley, 1993; Buss, 1994.

473 Homosexuality as a window on heterosexuality: Symons, 1979, p. 300. Number of homosexual partners: Symons, 1980.

474 Sexual economics: Symons, 1979. Dworkin: Quoted in Wright, 1994b.

475 Monogamy and the mensch: Symons, 1979, p. 250.

476 Men's sexual tastes modulated by their attractiveness: Waller, 1994.

476 Polygyny: Symons, 1979; Daly & Wilson, 1988; Shoumatoff, 1985; Altman & Ginat, 1996; Ridley, 1993; Chagnon, 1992.

476 Despots and harems: Betzig, 1986.

477 Polyandry: Symons, 1979; Ridley, 1993.

477 Co-wives: Shoumatoff, 1985. Betzig on Bozo: Cited in Ridley, 1993. Monogamy as a cartel: Landsburg, 1993, p. 170; Wright, 1994a.

478 Monogamy and male competition: Betzig, 1986; Wright, 1994a; Daly & Wilson, 1988; Ridley, 1993.

479 Adulteresses: Buss, 1994; Ridley, 1993; Baker & Bellis, 1996.

479 Meat for sex: Harris, 1985; Symons, 1979; Hill & Kaplan, 1988. Women's tastes in short-term lovers: Buss, 1994.

480 High-status lovers: Baker & Bellis, 1996; Buss, 1994; Symons, 1979.

480 Virgin birth in the Trobriand Islands: Symons, 1979.

480 Short-term versus long-term men: Buss, 1994; Ellis, 1992. Madonna-whore dichotomy: Wright, 1994a.

480 Tastes in husbands and wives: Buss, 1992a, 1994; Ellis, 1992.

482 Mate preferences: Buss, 1992a, 1994. Age preferences in mates: Kenrick & Keefe, 1992.

482 Personal ads, dating services, marriages: Ellis, 1992; Buss, 1992a, 1994.

482 *Moko dudei*: Chagnon, 1992; Symons, 1995.

482 Husband's wealth and wife's looks: Buss, 1994. Schroeder on animal magnetism: Quoted in Wright, 1995, p. 72.

482 Prestigious women want prestigious men: Buss, 1994. Feminist leaders want prestigious men: Ellis, 1992.

483 Lebowitz: Quoted in J. Winokur, 1987, *The portable curmudgeon*. New York: New American Library.

483 Decorating bodies for beauty versus other reasons: Etcoff, 1998. Universality of beauty: Brown, 1991; Etcoff, 1998; Symons, 1979, 1995; Ridley, 1993; Perrett, May, & Yoshikawa, 1994.

483 Ingredients of beauty: Etcoff, 1998; Symons, 1979, 1995.

484 Average faces are attractive: Symons, 1979; Langlois & Roggman, 1990.

485 Youth and beauty: Symons, 1979, 1995; Etcoff, 1998.

485 Waist-to-hip ratio: Singh, 1993, 1994, 1995. Hourglass figures in the Upper Paleolithic: Unpublished research by Singh & R. Kruszynski.

486 Size versus shape: Singh, 1993, 1994, 1995; Symons, 1995; Etcoff, 1998.

487 Beauty and power: Bell, 1992; Wilson & Daly, 1992; Ellis, 1992; Etcoff, 1998; Paglia, 1990, 1992, 1994.

487 Virtual beauty and real life: Buss, 1994.

488 Universality of sexual jealousy: Brown, 1991.

488 Sex differences in sexual jealousy: Symons, 1979; Buss, 1994; Buunk et al., 1996. Debate on the sex differences: Harris & Christenfeld, 1996; DeSteno & Salovey, 1996; Buss, Larson, & Westen, 1996; Buss et al., 1997.

489 Violence and male sexual jealousy: Daly & Wilson, 1988; Wilson & Daly, 1992; Symons, 1979. Myth of sexual symmetry in marital violence: Dobash et al., 1992.

490 Bride-wealth and dowries: Daly & Wilson, 1988.

491 Boswell, Johnson, and the double standard: Daly & Wilson, 1988, pp. 192–193.

492 Feminism without orthodox social science: Sommers, 1994; Patai & Koertge, 1994; Paglia, 1992; Eagly, 1995; Wright, 1994b; Ridley, 1993; Denfeld, 1995.

493 Status as a spiritual need: Veblen, 1899/1994. Sartorial morality: Bell, 1992.

493 Animal signals: Zahavi, 1975; Dawkins, 1976/1989, 1983; Hauser, 1996; Cronin, 1992.

494 Aggressive strategies and dominance hierarchies: Maynard Smith, 1982; Dawkins, 1976/1989; Trivers, 1985.

495 Dominance in humans: Ellis, 1992; Buss, 1994; Eibl-Eibesfeldt, 1989. Height and salary: Frieze, Olson, & Good, 1990. Height and presidential elections: Ellis, 1992; Mathews, 1996. Beards and Brezhnev: Kingdon, 1993. Height and dating: Kenrick & Keefe, 1992.

496 Killing over insults: Daly & Wilson, 1988; Nisbett & Cohen, 1996.

496 Men's reputations: Daly & Wilson, 1988, p. 128.

498 Reckless youth: Rogers, 1994.

498 Argumentation as coercion: Lakoff & Johnson, 1980; Nozick, 1981.

499 What is status?: Buss, 1992b; Tooby & Cosmides, 1996; Veblen, 1899/1994; Bell, 1992; Frank, 1985; Harris, 1989; Symons, 1979.

500 Potlatch: Harris, 1989.

500 Handicap principle: Zahavi, 1975; Dawkins, 1976/1989; Cronin, 1992; Hauser, 1996.

501 What is fashion?: Bell, 1992; Etcoff, 1998.

501 Mimicry in butterflies: Dawkins, 1976/1989; Cronin, 1992; Hauser, 1996.

502 Logic of reciprocation and exchange: Cosmides & Tooby, 1992; Axelrod, 1984. Reciprocal altruism: Trivers, 1985; Dawkins, 1976/1989; Axelrod, 1984; Axelrod & Hamilton, 1981.

503 Prisoner's Dilemma: Poundstone, 1992; Schelling, 1960; Rapoport, 1964.

503 Iterated Prisoner's Dilemma and tit-for-tat: Axelrod & Hamilton, 1981; Axelrod, 1984.

504 Reciprocation in everyday life: Cosmides & Tooby, 1992; Fiske, 1992.

504 Primitive communism within kin groups: Fiske, 1992.

505 Variance and food sharing among foragers: Cashdan, 1989; Kaplan, Hill, & Hurtado, 1990.

505 Luck versus laziness: Cosmides & Tooby, 1992.

506 Enforcing the sharing ethic through gossip: Eibl-Eibesfeldt, 1989, pp. 525–526. Selfish !Kung: Konner, 1982, pp. 375–376.

507 Friendship versus reciprocation: Fiske, 1992. Happy marriage versus reciprocation: Frank, 1988.

507 Logic of friendship and the Banker's Paradox: Tooby & Cosmides, 1996.

509 War among foragers and human evolution: Chagnon, 1988, 1992, 1996; Keeley, 1996; Diamond, 1992; Daly & Wilson, 1988; Alexander, 1987a, b.

510 Blood feuds: Daly & Wilson, 1988.

511 Fighting over diamonds, gold, meat, and sex: Chagnon, 1992, p. 115. Crowded or malnourished tribes not more warlike: Chagnon, 1992; Keeley, 1996.

511 Women as the spoils of war in the Bible: Hartung, 1992, 1995.

512 Hot and forcing violation: *Henry V*, act 2, scene 3.

512 Rape and war: Brownmiller, 1975.

513 Reproductive success of war leaders: Betzig, 1986.

513 Logic of war: Tooby & Cosmides, 1988.

513 The Kandinsky fans hate the Klee fans: Tajfel, 1981. Ethnocentrism from a coin flip: Locksley, Ortiz, & Hepburn, 1980. Boys wage war at summer camp: Sherif, 1966. Ethnic conflict: Brown, 1985.

515 Richer groups go to war more: Chagnon, 1992; Keeley, 1996.

516 Fighting under a veil of ignorance: Tooby & Cosmides, 1993. World War II example: Rapoport, 1964, pp. 88–89.

518 Declining homicide rates: Daly & Wilson, 1988.

519 The Dalai Lama: Interview by Claudia Dreifus in *New York Times Magazine*, November 28, 1993.

8. *The Meaning of Life*

521 Universality of art, literature, music, humor, religion, philosophy: Brown, 1991; Eibl-Eibesfeldt, 1989.

521 Living for music, selling blood to buy movie tickets: Tooby & Cosmides, 1990a.

522 The arts as status-seeking: Wolfe, 1975; Bell, 1992.

522 Art, science, and the elite: Brockman, 1994. Honorable futility: From Bell, 1992.

526 Art and illusion: Gombrich, 1960; Gregory, 1970; Kubovy, 1986. Adaptation and visual aesthetics: Shepard, 1990; Orians & Heerwagen, 1992; Kaplan, 1992.

526 Geometric patterning, evolution, and aesthetics: Shepard, 1990.

528 Music and the mind: Sloboda, 1985; Storr, 1992; R. Aiello, 1994.

529 Universal musical grammar: Bernstein, 1976; Jackendoff, 1977, 1987, 1992; Lerdahl & Jackendoff, 1983.

531 Overtones and scales: Bernstein, 1976; Cooke, 1959; Sloboda, 1985. Dissenters: Jackendoff, 1977; Storr, 1992.

531 Intervals and emotions: Bernstein, 1976; Cooke, 1959. Infant music appreciation: Zentner & Kagan, 1996; Schellenberg & Trehub, 1996.

534 Flow and ebb of grief: Cooke, 1959, pp. 137–138.

534 Emotional semantics of music: Cooke, 1959.

534 Music and language: Lerdahl & Jackendoff, 1983; Jackendoff, 1987.

535 Auditory scene analysis: Bregman & Pinker, 1978; Bregman, 1990; McAdams & Bigand, 1993.

536 The aesthetics of regular patterns in art and music: Shepard, 1990.

536 Music and auditory unsettledness: Bernstein, 1976; Cooke, 1959.

536 Darwin on music: Darwin, 1874. Melody of emotional calls: Fernald, 1992; Hauser, 1996.

537 Habitat selection: Orians & Heerwagen, 1992; Kaplan, 1992.

537 Music and movement: Jackendoff, 1992; Epstein, 1994; Clynes & Walker, 1982.

539 Horace: From Hobbs, 1990, p. 5. Dryden: From Carroll, 1995, p. 170.

539 Illusions of fiction and cinema: Hobbs, 1990; Tan, 1996.

540 The economics of happy endings: Landsburg, 1993.

540 Benign masochism: Rozin, 1996.

540 Evolution of the yenta: Barkow, 1992.

541 Fiction as experiment: Hobbs, 1990. Literature and cognition: Hobbs, 1990; Turner, 1991.

541 Plots as goal-seeking: Hobbs, 1990. The goals in fiction are the goals in natural selection: Carroll, 1995.

542 Tabloid headlines: *Native Son* by Richard Wright; *The Scarlet Letter* by Nathaniel Hawthorne; *Romeo and Juliet* by William Shakespeare; *Crime and Punishment* by Fyodor Dostoevsky; *The Great Gatsby* by F. Scott Fitzgerald; *Jane Eyre* by Charlotte Brontë; *A Streetcar Named Desire* by Tennessee Williams; *Eumenides* by Aeschylus. All from Lederer & Gilleland, 1994.

543 Case-based reasoning: Schanck, 1982.

543 Answers to life's conundrums: *Hamlet; The Godfather; Fatal Attraction; Madame Bovary; Shane*.

544 Repleteness of art: Goodman, 1976; Koestler, 1964.

545 Koestler on humor: Koestler, 1964, p. 31.

546 Evolution of humor: Provine, 1996; Eibl-Eibesfeldt, 1989; Weisfeld, 1993. Studies of humor: Provine, 1996; Chapman & Foot, 1977; McGhee, 1979; Weisfeld, 1993.

546 Laughter: Provine, 1991, 1993, 1996.

546 Laughter as a mobbing call: Eibl-Eibesfeldt, 1989. Laughter in chimpanzees: Provine, 1996; Weisfeld, 1993. Tickling and play: Eibl-Eibesfeldt, 1989; Weisfeld, 1993. Play as practice fighting: Symons, 1978; Boulton & Smith, 1992.

547 Humor in *1984*: Orwell, 1949/1983, p. 11.

548 The Rabelaisian Yanomamö: Chagnon, 1992, pp. 24–25.

550 Mountain climber joke: Thanks to Henry Gleitman. W. C. Fields: Thanks to Thomas Shultz.

550 Studies of incongruity resolution in humor: Shultz, 1977; Rothbart, 1977; McGhee, 1979.

551 Humor as puncturing dominance: Schutz, 1977.

552 Mental interpolation in conversation: Pinker, 1994, chap. 7; Sperber & Wilson, 1986. Psychology of conversation and humor: Attardo, 1994.

553 Banality of banter: Provine, 1993, p. 296.

553 Logic of friendship: Tooby & Cosmides, 1996.

554 Beliefs in the untrue: witches, ghosts, the devil: *New York Times*, July 26, 1992. Genesis: Dennett, 1995. Angels: *Time* poll cited by Diane White, *Boston Globe*, October 24, 1994. Jesus: cited by Kenneth Woodward, *Newsweek*, April 8, 1996. God or spirit: Harris, 1989.

556 Anthropology of religion: Harris, 1989.

556 Cognitive psychology of religion: Sperber, 1982; Boyer, 1994a, b; Atran, 1995.

557 Empirical grounds for religious beliefs: Harris, 1989.

558 Philosophical bafflement: McGinn, 1993. Paradoxes of consciousness, self, will, meaning, and knowledge: Poundstone, 1988.

559 Cycles of philosophy: McGinn, 1993.

561 Philosophical bewilderment as a limitation of human conceptual equipment: Chomsky, 1975, 1988; McGinn, 1993.

565 Mismatch between the combinatorial mind and the problems of philosophy: McGinn, 1993.

REFERENCES

Adelson, E. H., & Pentland, A. P. 1996. The perception of shading and reflectance. In Knill & Richards, 1996.

Aiello, L. C. 1994. Thumbs up for our early ancestors. *Science, 265,* 1540–1541.

Aiello, R. (Ed.) 1994. *Musical perceptions.* New York: Oxford University Press.

Alexander, M., Stuss, M. P., & Benson, D. F. 1979. Capgras syndrome: A reduplicative phenomenon. *Neurology, 29,* 334–339.

Alexander, R. D. 1987a. *The biology of moral systems.* Hawthorne, N.Y.: Aldine de Gruyter.

Alexander, R. D. 1987b. Paper presented at the conference "The origin and dispersal of modern humans," Corpus Christi College, Cambridge, England, March 22–26. Reported in *Science, 236,* 668–669.

Alexander, R. D. 1990. How did humans evolve? Reflections on the uniquely unique species. Special Publication No. 1, Museum of Zoology, University of Michigan.

Allen, W. 1983. *Without feathers.* New York: Ballantine.

Allman, W. 1994. *The stone-age present: How evolution has shaped modern life.* New York: Simon & Schuster.

Aloimonos, Y., & Rosenfeld, A. 1991. Computer vision. *Science, 13,* 1249–1254.

Altman, I., & Ginat, J. 1996. *Polygynous families in contemporary society.* New York: Cambridge University Press.

Anderson, A. 1992. The evolution of sexes. *Science, 257,* 324–326.

Anderson, B. L., & Nakayama, K. 1994. Toward a general theory of stereopsis: Binocular matching, occluding contours, and fusion. *Psychological Review, 101,* 414–445.

Anderson, J. R. 1983. *The architecture of cognition.* Cambridge, Mass.: Harvard University Press.

Anderson, J. R. 1990. *The adaptive character of thought.* Hillsdale, N.J.: Erlbaum.

Anderson, J. R., & commentators. 1991. Is human cognition adaptive? *Behavioral and Brain Sciences, 14,* 471–517.

Anderson, J. R. 1993. *Rules of the mind.* Hillsdale, N.J.: Erlbaum.

Anderson, J. R., & Bower, G. H. 1973. *Human associative memory*. New York: Wiley.

Armstrong, S. L., Gleitman, L. R., & Gleitman, H. 1983. What some concepts might not be. *Cognition, 13*, 263–308.

Aronson, E. 1980. *The social animal*. San Francisco: W. H. Freeman.

Asimov, I. 1950. *I, robot*. New York: Bantam Books.

Asimov, I., & Shulman, J. A. (Eds.) 1986. *Isaac Asimov's book of science and nature quotations*. New York: Weidenfeld & Nicolson.

Atran, S. 1990. *The cognitive foundations of natural history*. New York: Cambridge University Press.

Atran, S. 1995. Causal constraints on categories and categorical constraints on biological reasoning across cultures. In Sperber, Premack, & Premack, 1995.

Attardo, S. 1994. *Linguistic theories of humor*. New York: Mouton de Gruyter.

Attneave, F. 1968. Triangles as ambiguous figures. *American Journal of Psychology, 81*, 447–453.

Attneave, F. 1972. Representation of physical space. In A. W. Melton & E. J. Martin (Eds.), *Processes in human memory*. Washington, D.C.: V. H. Winston.

Attneave, F. 1981. Three approaches to perceptual organization: Comments on views of Hochberg, Shepard, & Shaw. In Kubovy & Pomerantz, 1981.

Attneave, F. 1982. Prägnanz and soap bubble systems: A theoretical exploration. In Beck, 1982.

Axelrod, R. 1984. *The evolution of cooperation*. New York: Basic Books.

Axelrod, R., & Hamilton, W. D. 1981. The evolution of cooperation. *Science, 211*, 1390–1396.

Ayala, F. J. 1995. The myth of Eve: Molecular biology and human origins. *Science, 270*, 1930–1936.

Baars, B. 1988. *A cognitive theory of consciousness*. New York: Cambridge University Press.

Baddeley, A. D. 1986. *Working memory*. New York: Oxford University Press.

Baillargeon, R. 1995. Physical reasoning in infancy. In Gazzaniga, 1995.

Baillargeon, R., Kotovsky, L., & Needham, A. 1995. The acquisition of physical knowledge in infancy. In Sperber, Premack, & Premack, 1995.

Baker, R. R., & Bellis, M. A. 1996. *Sperm competition: Copulation, masturbation, and infidelity*. London: Chapman & Hall.

Barkow, J. H. 1992. Beneath new culture is old psychology: Gossip and social stratification. In Barkow, Cosmides, & Tooby, 1992.

Barkow, J. H., Cosmides, L., & Tooby, J. (Eds.) 1992. *The adapted mind: Evolutionary psychology and the generation of culture*. New York: Oxford University Press.

Baron-Cohen, S. 1995. *Mindblindness: An essay on autism and theory of mind*. Cambridge, Mass.: MIT Press.

Baron-Cohen, S., Leslie, A. M., & Frith, U. 1985. Does the autistic child have a theory of mind? *Cognition, 21*, 37–46.

Bates, E., & MacWhinney, B. 1982. Functionalist approaches to grammar. In E. Wanner & L. R. Gleitman (Eds.), *Language acquisition: The state of the art*. New York: Cambridge University Press.

Bates, E., & MacWhinney, B. 1992. Welcome to functionalism. In Pinker & Bloom, 1990.

Baumeister, R. F., & Tice, D. M. 1984. Role of self-presentation and choice in cognitive dissonance under forced compliance: Necessary or sufficient causes? *Journal of Personality and Social Psychology, 46,* 5–13.

Beck, J. (Ed.) 1982. *Organization and representation in perception.* Hillsdale, N.J.: Erlbaum.

Behrmann, M., Winocur, G., & Moscovitch, M. 1992. Dissociation between mental imagery and object recognition in a brain-damaged patient. *Nature, 359,* 636–637.

Belew, R. K. 1990. Evolution, learning, and culture: Computational metaphors for adaptive algorithms. *Complex Systems, 4,* 11–49.

Belew, R. K., McInerney, J., & Schraudolph, N. N. 1990. Evolving networks: Using the genetic algorithm with connectionist learning. In *Proceedings of the Second Artificial Life Conference.* Reading, Mass.: Addison-Wesley.

Bell, Q. 1992. *On human finery.* London: Allison & Busby.

Berg, G. 1991. Learning recursive phrase structure: Combining the strengths of PDP and X-bar syntax. *Proceedings of the International Joint Conference on Artificial Intelligence Workshop on Natural Language Learning.*

Berkeley, G. 1713/1929. Three dialogues between Hylas and Philonous. In M. W. Calkins (Ed.), *Berkeley Selections.* New York: Scribner's.

Berlin, B., Breedlove, D., & Raven, P. 1973. General principles of classification and nomenclature in folk biology. *American Anthropologist, 87,* 298–315.

Bernstein, L. 1976. *The unanswered question: Six talks at Harvard.* Cambridge, Mass.: Harvard University Press.

Berra, T. M. 1990. *Evolution and the myth of creationism.* Stanford, Calif.: Stanford University Press.

Bettelheim, B. 1959. Joey: A mechanical boy. *Scientific American,* March. Reprinted in Atkinson, R. C. (Ed.), 1971, *Contemporary psychology.* San Francisco: Freeman.

Betzig, L. 1986. *Despotism and differential reproduction.* Hawthorne, N.Y.: Aldine de Gruyter.

Betzig, L. 1992. Medieval monogamy. In S. Mithen & H. Maschner (Eds.), *Darwinian approaches to the past.* New York: Plenum.

Betzig, L., Borgerhoff Mulder, M., & Turke, P. (Eds.) 1988. *Human reproductive behavior: A Darwinian perspective.* New York: Cambridge University Press.

Biederman, I. 1995. Visual object recognition. In Kosslyn & Osherson, 1995.

Birch, E. E. 1993. Stereopsis in infants and its developmental relation to visual acuity. In Simons, 1993.

Bisiach, E., & Luzzatti, C. 1978. Unilateral neglect of representational space. *Cortex, 14,* 129–133.

Bisson, T. 1991. They're made out of meat. From a series of stories entitled "Alien/Nation." *Omni,* April.

Bizzi, E., & Mussa-Ivaldi, F. A. 1990. Muscle properties and the control of arm movements. In Osherson, Kosslyn, & Hollerbach, 1990.

Block, N. 1978. Troubles with functionalism. In C. W. Savage (Ed.), *Perception and cognition: Issues in the foundations of psychology. Minnesota Studies in the Philosophy of Science,* Vol. 9. Minneapolis: University of Minnesota.

Block, N. (Ed.) 1981. *Imagery.* Cambridge, Mass.: MIT Press.

Block, N. 1986. Advertisement for a semantics for psychology. In P. Rench, T. Uehling,

Jr., & H. Wettstein (Eds.), *Midwest Studies in Philosophy*, Vol. 10. Minneapolis: University of Minnesota Press.

Block, N., & commentators. 1995. On a confusion about a function of consciousness. *Behavioral and Brain Sciences, 18*, 227–287.

Bloom, P. 1996a. Possible individuals in language and cognition. *Current Directions in Psychological Science, 5*, 90–94.

Bloom, P. 1996b. Intention, history, and artifact concepts. *Cognition, 60*, 1–29.

Bobick, A. 1987. *Natural object categorization*. MIT Artificial Intelligence Laboratory Technical Report 1001.

Bonatti, L. 1995. Why should we abandon the mental logic hypothesis? *Cognition, 50*, 109–131.

Boring, E. G. 1952. The Gibsonian visual field. *Psychological Review, 59*, 246–247.

Bouchard, T. J., Jr. 1994. Genes, environment, and personality. *Science, 264*, 1700–1701.

Bouchard, T. J., Jr., Lykken, D. T., McGue, M., Segal, N. L., & Tellegen, A. 1990. Sources of human psychological differences: The Minnesota Study of Twins Reared Apart. *Science, 250*, 223–228.

Boulton, M. J., & Smith, P. K. 1992. The social nature of play fighting and play chasing: Mechanisms and strategies underlying cooperation and compromise. In Barkow, Cosmides, & Tooby, 1992.

Bowerman, M. 1983. Hidden meanings: The role of covert conceptual structures in children's development of language. In D. R. Rogers and J. A. Sloboda, (Eds.), *The acquisition of symbolic skills*. New York: Plenum.

Boyd, R., & Richerson, P. 1985. *Culture and the evolutionary process*. Chicago: University of Chicago Press.

Boyd, R., & Silk, J. R. 1996. *How humans evolved*. New York: Norton.

Boyer, P. 1994a. *The naturalness of religious ideas*. Berkeley: University of California Press.

Boyer, P. 1994b. Cognitive constraints on cultural representations: Natural ontologies and religious ideas. In Hirschfeld & Gelman, 1994a.

Brainard, D. H., & Wandell, B. A. 1986. Analysis of the retinex theory of color vision. *Journal of the Optical Society of America (A), 3*, 1651–1661.

Brainard, D. H., & Wandell, B. A. 1991. A bilinear model of the illuminant's effect on color appearance. In J. A. Movshon & M. S. Landy (Eds.), *Computational models of visual processing*. Cambridge, Mass.: MIT Press.

Braine, M. D. S. 1994. Mental logic and how to discover it. In Macnamara & Reyes, 1994.

Bregman, A. S. 1990. *Auditory scene analysis: The perceptual organization of sound*. Cambridge, Mass.: MIT Press.

Bregman, A. S., & Pinker, S. 1978. Auditory streaming and the building of timbre. *Canadian Journal of Psychology, 32*, 19–31.

Brickman, P., & Campbell, D. T. 1971. Hedonic relativism and planning the good society. In M. H. Appley (Ed.), *Adaptation-level theory: A symposium*. New York: Academic Press.

Brockman, J. 1994. *The third culture: Beyond the scientific revolution*. New York: Simon & Schuster.

Bronowski, J. 1973. *The ascent of man*. Boston: Little, Brown.

Brooks, L. 1968. Spatial and verbal components in the act of recall. *Canadian Journal of Psychology, 22*, 349–368.

Brown, A. L. 1990. Domain-specific principles affect learning and transfer in children. *Cognitive Science, 14*, 107–133.

Brown, D. E. 1988. *Hierarchy, history, and human nature: The social origins of historical consciousness*. Tucson: University of Arizona Press.

Brown, D. E. 1991. *Human universals*. New York: McGraw-Hill.

Brown, R. 1985. *Social psychology: The second edition*. New York: Free Press.

Brown, R., & Kulik, J. 1977. Flashbulb memories. *Cognition, 5*, 73–99.

Brownmiller, S. 1975. *Against our will: Men, women, and rape*. New York: Fawcett Columbine.

Bruce, V. 1988. *Recognizing faces*. Hillsdale, N.J.: Erlbaum.

Bülthoff, H. H., & Edelman, S. 1992. Psychophysical support for a two-dimensional view interpolation theory of object recognition. *Proceedings of the National Academy of Sciences, 89*, 60–64.

Buss, D. M. 1992a. Mate preference mechanisms: Consequences for partner choice and intrasexual competition. In Barkow, Cosmides, & Tooby, 1992.

Buss, D. M. 1992b. Human prestige criteria. Unpublished manuscript, Department of Psychology, University of Texas, Austin.

Buss, D. M. 1994. *The evolution of desire*. New York: Basic Books.

Buss, D. M. 1995. Evolutionary psychology: A new paradigm for psychological science. *Psychological Inquiry, 6*, 1–30.

Buss, D. M., Larsen, R. J., & Westen, D. 1996. Sex differences in jealousy: Not gone, not forgotten, and not explained by alternative hypotheses. *Psychological Science, 7*, 373–375.

Buss, D. M., Shackelford, T. K., Kirkpatrick, L. A., Choe, J., Hasegawa, T., Hasegawa, M., & Bennett, K. 1997. Jealousy and the nature of beliefs about infidelity: Tests of competing hypotheses about sex differences in the United States, Korea, and Japan. Unpublished manuscript, University of Texas, Austin.

Buunk, B. P., Angleitner, A., Oubaid, V., & Buss, D. M. 1996. Sex differences in jealousy in evolutionary and cultural perspective: Tests from the Netherlands, Germany, and the United States. *Psychological Science, 7*, 359–363.

Byrne, R. W., & Whiten, A. 1988. *Machiavellian intelligence*. New York: Oxford University Press.

Cain, A. J. 1964. The perfection of animals. In J. D. McCarthy & C. L. Duddington (Eds.), *Viewpoints in Biology*, Vol. 3. London: Butterworth.

Cairns, J., Overbaugh, J., & Miller, S. 1988. The origin of mutants. *Nature, 335*, 142–146.

Campbell, D. T. 1975. On the conflicts between biological and social evolution and between psychology and moral tradition. *American Psychologist, 30*, 1103–1126.

Carey, S. 1985. *Conceptual change in childhood*. Cambridge, Mass.: MIT Press.

Carey, S. 1986. Cognitive science and science education. *American Psychologist, 41*, 1123–1130.

Carey, S. 1995. On the origin of causal understanding. In Sperber, Premack, & Premack, 1995.

Carey, S., & Spelke, E. 1994. Domain-specific knowledge and conceptual change. In Hirschfeld & Gelman, 1994a.

Carroll, J. 1995. *Evolution and literary theory*. Columbia: University of Missouri Press.

Carroll, L. 1895/1956. What the tortoise said to Achilles and other riddles. In J. R. Newman (Ed.), 1956, *The world of mathematics*, Vol. 4. New York: Simon & Schuster.

Carroll, L. 1896/1977. *Symbolic logic*. In W. W. Bartley (Ed.), 1977, *Lewis Carroll's Symbolic Logic*. New York: Clarkson Potter.

Cashdan, E. 1989. Hunters and gatherers: Economic behavior in bands. In S. Plattner (Ed.), *Economic anthropology*. Stanford, Calif.: Stanford University Press.

Cashdan, E. 1994. A sensitive period for learning about food. *Human Nature*, 5, 279–291.

Cavalli-Sforza, L. L., Menozzi, P., & Piazza, A. 1993. Demic expansions and human evolution. *Science, 259*, 639–646.

Cavalli-Sforza, L. L., & Feldman, M. W. 1981. *Cultural transmission and evolution: A quantitative approach*. Princeton, N.J.: Princeton University Press.

Cave, K. R., Pinker, S., Giorgi, L., Thomas, C., Heller, L., Wolfe, J. M., & Lin, H. 1994. The representation of location in visual images. *Cognitive Psychology, 26*, 1–32.

Cerf, C., & Navasky, V. 1984. *The experts speak*. New York: Pantheon.

Chagnon, N. A. 1988. Life histories, blood revenge, and warfare in a tribal population. *Science, 239*, 985–992.

Chagnon, N. A. 1992. *Yanomamö: The last days of Eden*. New York: Harcourt Brace.

Chagnon, N. A. 1996. Chronic problems in understanding tribal violence and warfare. In G. Bock & J. Goode (Eds.), *The genetics of criminal and antisocial behavior*. New York: Wiley.

Chalmers, D. J. 1990. Syntactic transformations on distributed representations. *Connection Science, 2*, 53–62.

Chambers, D., & Reisberg, D. 1985. Can mental images be ambiguous? *Journal of Experimental Psychology: Human Perception and Performance, 11*, 317–328.

Changeux, J.-P., & Chavaillon, J. (Eds.) 1995. *Origins of the human brain*. New York: Oxford University Press.

Chapman, A. J., & Foot, H. C. (Eds.) 1977. *It's a funny thing, humor*. New York: Pergamon Press.

Chase, W. G., & Simon, H. A. 1973. Perception in chess. *Cognitive Psychology, 4*, 55–81.

Cheney, D., & Seyfarth, R. M. 1990. *How monkeys see the world*. Chicago: University of Chicago Press.

Cheng, P., & Holyoak, K. 1985. Pragmatic reasoning schemas. *Cognitive Psychology, 17*, 391–416.

Cherniak, C. 1983. Rationality and the structure of memory. *Synthèse, 53*, 163–186.

Chomsky, N. 1959. A review of B. F. Skinner's "Verbal behavior." *Language, 35*, 26–58.

Chomsky, N. 1975. *Reflections on language*. New York: Pantheon.

Chomsky, N. 1988. *Language and problems of knowledge: The Managua lectures*. Cambridge, Mass.: MIT Press.

Chomsky, N. 1991. Linguistics and cognitive science: Problems and mysteries. In A. Kasher (Ed.), *The Chomskyan turn*. Cambridge, Mass.: Blackwell.

Chomsky, N. 1992. Explaining language use. *Philosophical Topics, 20*, 205–231.

Chomsky, N. 1993. *Language and thought*. Wakefield, R.I., and London: Moyer Bell.

Christopher, T. 1995. In defense of the embattled American lawn. *New York Times*, July 23, The Week in Review, p. 3.

Churchland, P., & Churchland, P. S. 1994. Could a machine think? In Dietrich, 1994.

Churchland, P. S., & Sejnowski, T. J. 1992. *The computational brain*. Cambridge, Mass.: MIT Press.

Clark, R. D., & Hatfield, E. 1989. Gender differences in receptivity to sexual offers. *Journal of Psychology and Human Sexuality, 2*, 39–55.

Clynes, M., & Walker, J. 1982. Neurobiological functions of rhythm, time, and pulse in music. In M. Clynes (Ed.), *Music, mind, and brain: The neuropsychology of music*. New York: Plenum.

Cole, M., Gay, J., Glick, J., & Sharp, D. W. 1971. *The cultural context of learning and thinking*. New York: Basic Books.

Cooke, D. 1959. *The language of music*. New York: Oxford University Press.

Cooper, L. A., & Shepard, R. N. 1973. Chronometric studies of the rotation of mental images. In W. G. Chase (Ed.), *Visual information processing*. New York: Academic Press.

Coppens, Y. 1995. Brain, locomotion, diet, and culture: How a primate, by chance, became a man. In Changeux & Chavaillon, 1995.

Corballis, M. C. 1988. Recognition of disoriented shapes. *Psychological Review, 95*, 115–123.

Corballis, M. C., & Beale, I. L. 1976. *The psychology of left and right*. Hillsdale, N.J.: Erlbaum.

Cormack, L. K., Stevenson, S. B., & Schor, C. M. 1993. Disparity-tuned channels of the human visual system. *Visual Neuroscience, 10*, 585–596.

Cosmides, L. 1985. Deduction or Darwinian algorithms? An explanation of the "elusive" content effect on the Wason selection task. Ph.D. dissertation, Department of Psychology, Harvard University.

Cosmides, L. 1989. The logic of social exchange: Has natural selection shaped how humans reason? Studies with the Wason selection task. *Cognition, 31*, 187–276.

Cosmides, L., & Tooby, J. 1981. Cytoplasmic inheritance and intragenomic conflict. *Journal of Theoretical Biology, 89*, 83–129.

Cosmides, L., & Tooby, J. 1992. Cognitive adaptations for social exchange. In Barkow, Cosmides, & Tooby, 1992.

Cosmides, L., & Tooby, J. 1994. Beyond intuition and instinct blindness: Toward an evolutionarily rigorous cognitive science. *Cognition, 50*, 41–77.

Cosmides, L., & Tooby, J. 1996. Are humans good intuitive statisticians after all? Rethinking some conclusions from the literature on judgment under uncertainty. *Cognition, 58*, 1–73.

Cosmides, L., Tooby, J., & Barkow, J. 1992. Environmental aesthetics. In Barkow, Cosmides, & Tooby, 1992.

Cramer, K. S., & Sur, M. 1995. Activity-dependent remodeling of connections in the mammalian visual system. *Current Opinion in Neurobiology, 5*, 106–111.

Craver-Lemley, C., & Reeves, A. 1992. How visual imagery interferes with vision. *Psychological Review, 98*, 633–649.

Crevier, D. 1993. *AI: The tumultuous history of the search for artificial intelligence*. New York: Basic Books.

Crick, F. 1994. *The astonishing hypothesis: The scientific search for the soul*. New York: Simon & Schuster.

Crick, F., & Koch, C. 1995. Are we aware of neural activity in primary visual cortex? *Nature, 375*, 121–123.

Cummins, R. 1984. Functional analysis. In Sober, 1984a.

Daly, M. 1982. Some caveats about cultural transmission models. *Human Ecology, 10*, 401–408.

Daly, M., & Wilson, M. 1988. *Homicide.* Hawthorne, N.Y.: Aldine de Gruyter.

Daly, M., & Wilson, M. 1994. Evolutionary psychology of male violence. In J. Archer (Ed.), *Male violence.* London: Routledge.

Daly, M., & Wilson, M. 1995. Discriminative parental solicitude and the relevance of evolutionary models to the analysis of motivational systems. In Gazzaniga, 1995.

Daly, M., Salmon, C., & Wilson, M. In press. Kinship: The conceptual hole in psychological studies of social cognition and close relationships. In D. Kenrick & J. Simpson (Eds.), *Evolutionary social psychology.* Hillsdale, N.J.: Erlbaum.

Damasio, A. R. 1994. *Descartes' error: Emotion, reason, and the human brain.* New York: Putnam.

Darwin, C. 1859/1964. *On the origin of species.* Cambridge, Mass.: Harvard University Press.

Darwin, C. 1872/1965. *The expression of the emotions in man and animals.* Chicago: University of Chicago Press.

Darwin, C. 1874. *The descent of man, and selection in relation to sex.* 2d ed. New York: Hurst & Company.

Davey, G. C. L., & commentators. 1995. Preparedness and phobias: Specific evolved associations or a generalized expectancy bias? *Behavioral and Brain Sciences, 18*, 289–325.

Davies, P. 1995. *Are we alone? Implications of the discovery of extraterrestrial life.* New York: Basic.

Dawkins, R. 1976/1989. *The selfish gene.* New edition. New York: Oxford University Press.

Dawkins, R. 1982. *The extended phenotype.* New York: Oxford University Press.

Dawkins, R. 1983. Universal Darwinism. In D. S. Bendall (Ed.), *Evolution from molecules to man.* New York: Cambridge University Press.

Dawkins, R. 1986. *The blind watchmaker: Why the evidence of evolution reveals a universe without design.* New York: Norton.

Dawkins, R. 1995. *River out of Eden: A Darwinian view of life.* New York: Basic Books.

de Jong, G. F., & Mooney, R. J. 1986. Explanation-based learning: An alternative view. *Machine Learning, 1*, 145–176.

Deacon, T. 1992a. Primate brains and senses. In Jones, Martin, & Pilbeam.

Deacon, T. 1992b. The human brain. In Jones, Martin, & Pilbeam.

Dehaene, S. (Ed.) 1992. *Numerical cognition.* Special issue of *Cognition, 44.* Reprinted, Cambridge, Mass.: Blackwell.

Denfeld, R. 1995. *The new Victorians: A young woman's challenge to the old feminist order.* New York: Warner Books.

Denis, M., Engelkamp, J., & Richardson, J. T. E. (Eds.) 1988. *Cognitive and neuropsychological approaches to mental imagery.* Amsterdam, Netherlands: Martinus Nijhoff.

Dennett, D. C. 1978a. *Brainstorms: Philosophical essays on mind and psychology.* Cambridge, Mass.: Bradford Books/MIT Press.

Dennett, D. C. 1978b. Intentional systems. In Dennett, 1978a.

Dennett, D. C. 1978c. Skinner skinned. In Dennett, 1978a.

Dennett, D. C. 1978d. Artificial intelligence as philosophy and as psychology. In Dennett, 1978a.

Dennett, D. C. 1984. *Elbow room: The varieties of free will worth wanting*. Cambridge, Mass.: MIT Press.

Dennett, D. C. 1987. Cognitive wheels: The frame problem of AI. In Pylyshyn, 1987.

Dennett, D. C. 1990. The interpretation of texts, people, and other artifacts. *Philosophy and Phenomenological Research, 50,* 177–194.

Dennett, D. C. 1991. *Consciousness explained*. Boston: Little, Brown.

Dennett, D. C. 1995. *Darwin's dangerous idea: Evolution and the meanings of life*. New York: Simon & Schuster.

Dershowitz, A. M. 1994. *The abuse excuse*. Boston: Little, Brown.

DeSteno, D. A., & Salovey, P. 1996. Evolutionary origins of sex differences in jealousy? Questioning the "fitness" of the model. *Psychological Science, 7,* 367–372, 376–377.

Diamond, J. 1992. *The third chimpanzee: The evolution and future of the human animal*. New York: HarperCollins.

Dickinson, S. J., Pentland, A. P., & Rosenfeld, A. 1992. 3-D shape recovery using distributed aspect matching. *IEEE Transactions on Pattern Analysis and Machine Intelligence, 14,* 174–198.

Dietrich, E. (Ed.) 1994. *Thinking computers and virtual persons: Essays on the intentionality of machines*. Boston: Academic Press.

Dobash, R. P., Dobash, R. E., Wilson, M., & Daly, M. 1992. The myth of sexual symmetry in marital violence. *Social Problems, 39,* 71–91.

Drake, F. 1993. Extraterrestrial intelligence (letter). *Science, 260,* 474–475.

Dretske, F. I. 1981. *Knowledge and the flow of information*. Cambridge, Mass.: MIT Press.

Dreyfus, H. 1979. *What computers can't do*. 2d ed. New York: Harper & Row.

Dunbar, R. I. M. 1992. Primate social organization: Mating and parental care. In Jones, Martin, & Pilbeam, 1992.

Duncan, J. 1995. Attention, intelligence, and the frontal lobes. In Gazzaniga, 1995.

Dunphy, D. 1963. The social structure of early adolescent peer groups. *Sociometry, 26,* 230–246.

Durham, W. H. 1982. Interactions of genetic and cultural evolution: Models and examples. *Human Ecology, 10,* 299–334.

Eagly, A. H. 1995. The science and politics of comparing women and men. *American Psychologist, 50,* 145–158.

Eibl-Eibesfeldt, I. 1989. *Human ethology*. Hawthorne, N.Y.: Aldine de Gruyter.

Ekman, P. 1987. A life's pursuit. In T. A. Sebeok & J. Umiker-Sebeok (Eds.), 1987, *The semiotic web 86: An international yearbook*. Berlin: Mouton de Gruyter.

Ekman, P. 1993. Facial expression and emotion. *American Psychologist, 48,* 384–392.

Ekman, P. 1994. Strong evidence for universals in facial expression: A reply to Russell's mistaken critique. *Psychological Bulletin, 115,* 268–287.

Ekman, P., & Davidson, R. J. (Eds.) 1994. *The nature of emotion*. New York: Oxford University Press.

Ekman, P., & Friesen, W. V. 1975. *Unmasking the face*. Englewood Cliffs, N.J.: Prentice-Hall.

Ellis, B. J. 1992. The evolution of sexual attraction: Evaluative mechanisms in women. In Barkow, Cosmides, & Tooby, 1992.

Elman, J. L. 1990. Finding structure in time. *Cognitive Science, 14,* 179–211.

Endler, J. A. 1986. *Natural selection in the wild.* Princeton, N.J.: Princeton University Press.

Epstein, D. 1994. *Shaping time: Music, the brain, and performance.* New York: Schirmer.

Etcoff, N. L. 1986. The neuropsychology of emotional expression. In G. Goldstein & R. E. Tarter (Eds.), *Advances in Clinical Neuropsychology,* Vol. 3. New York: Plenum.

Etcoff, N. L. 1998. *Beauty.* New York: Doubleday.

Etcoff, N. L., Freeman, R., & Cave, K. R. 1991. Can we lose memories of faces? Content specificity and awareness in a prosopagnosic. *Journal of Cognitive Neuroscience, 3,* 25–41.

Eyer, D. 1996. *Motherguilt: How our culture blames mothers for what's wrong with society.* New York: Times Books.

Farah, M. J. 1989. Mechanisms of imagery-perception interaction. *Journal of Experimental Psychology: Human Perception and Performance, 15,* 203–211.

Farah, M. J. 1990. *Visual agnosia.* Cambridge, Mass.: MIT Press.

Farah, M. J. 1995. Dissociable systems for recognition: A cognitive neuropsychology approach. In Kosslyn & Osherson, 1995.

Farah, M. J., Soso, M. J., & Dasheiff, R. M. 1992. Visual angle of the mind's eye before and after unilateral occipital lobectomy. *Journal of Experimental Psychology: Human Perception and Performance, 18,* 241–246.

Fehling, M. R., Baars, B. J., & Fisher, C. 1990. A functional role for repression in an autonomous, resource-constrained agent. *Proceedings of the Twelfth Annual Meeting of the Cognitive Science Society.* Hillsdale, N.J.: Erlbaum.

Feldman, J., & Ballard, D. 1982. Connectionist models and their properties. *Cognitive Science, 6,* 205–254.

Fernald, A. 1992. Human maternal vocalizations to infants as biologically relevant signals: An evolutionary perspective. In Barkow, Cosmides, & Tooby.

Festinger, L. 1957. *A theory of cognitive dissonance.* Stanford, Calif.: Stanford University Press.

Fiedler, K. 1988. The dependence of the conjunction fallacy on subtle linguistic factors. *Psychological Research, 50,* 123–129.

Field, H. 1977. Logic, meaning and conceptual role. *Journal of Philosophy, 69,* 379–408.

Finke, R. A. 1989. *Principles of mental imagery.* Cambridge, Mass.: MIT Press.

Finke, R. A. 1990. *Creative imagery: Discoveries and inventions in visualization.* Hillsdale, N.J.: Erlbaum.

Finke, R. A., Pinker, S., & Farah, M. J. 1989. Reinterpreting visual patterns in mental imagery. *Cognitive Science, 13,* 51–78.

Fischman, J. 1994. Putting our oldest ancestors in their proper place. *Science, 265,* 2011–2012.

Fisher, H. E. 1992. *Anatomy of love: The natural history of monogamy, adultery, and divorce.* New York: Norton.

Fiske, A. P. 1992. The four elementary forms of sociality: Framework for a unified theory of social relations. *Psychological Review, 99,* 689–723.

Fodor, J. A. 1968a. *Psychological explanation: An introduction to the philosophy of psychology*. New York: Random House, 1968.

Fodor, J. A. 1968b. The appeal to tacit knowledge in psychological explanation. *Journal of Philosophy, 65*, 627–640.

Fodor, J. A. 1975. *The language of thought*. New York: Crowell.

Fodor, J. A. 1981. The present status of the innateness controversy. In J. A. Fodor, *RePresentations*. Cambridge, Mass.: MIT Press.

Fodor, J. A. 1983. *The modularity of mind*. Cambridge, Mass.: MIT Press.

Fodor, J. A., & commentators. 1985. Précis and multiple book review of "The modularity of mind." *Behavioral and Brain Sciences, 8*, 1–42.

Fodor, J. A. 1986. Why paramecia don't have mental representations. In P. Rench, T. Uehling, Jr., & H. Wettstein (Eds.), *Midwest Studies in Philosophy*, Vol. 10. Minneapolis: University of Minnesota Press.

Fodor, J. A. 1994. *The elm and the expert: Mentalese and its semantics*. Cambridge, Mass.: MIT Press.

Fodor, J. A., & McClaughlin, B. 1990. Connectionism and the problem of systematicity: Why Smolensky's solution doesn't work. *Cognition, 35*, 183–204.

Fodor, J. A., & Pylyshyn, Z. 1988. Connectionism and cognitive architecture: a critical analysis. *Cognition, 28*, 3–71. Reprinted in Pinker & Mehler, 1988.

Fox, R. 1984. *Kinship and marriage: An anthropological perspective*. New York: Cambridge University Press.

Frank, R. H. 1985. *Choosing the right pond: Human behavior and the quest for status*. New York: Oxford University Press.

Frank, R. H. 1988. *Passions within reason: The strategic role of the emotions*. New York: Norton.

Freeman, D. 1983. *Margaret Mead and Samoa: The making and unmaking of an anthropological myth*. Cambridge, Mass.: Harvard University Press.

Freeman, D. 1992. Paradigms in collision. *Academic Questions, 5*, 23–33.

Freeman, R. D., & Ohzawa, I. 1992. Development of binocular vision in the kitten's striate cortex. *Journal of Neuroscience, 12*, 4721–4736.

French, M. 1994. *Invention and evolution: Design in nature and engineering*. 2d ed. New York: Cambridge University Press.

French, R. E. 1987. *The geometry of vision and the mind-body problem*. New York: Peter Lang.

Freyd, J. J., & Finke, R. A. 1984. Facilitation of length discrimination using real and imagined context frames. *American Journal of Psychology, 97*, 323–341.

Fridlund, A. 1991. Evolution and facial action in reflex, social motive, and paralanguage. *Biological Psychology, 32*, 3–100.

Fridlund, A. 1992. Darwin's anti-Darwinism in "The expression of the emotions in man and animals." In K. T. Strongman (Ed.), *International Review of Studies of Emotion*, Vol. 2. New York: Wiley.

Fridlund, A. 1995. *Human facial expression: An evolutionary view*. New York: Academic Press.

Frieze, I. H., Olson, J. E., & Good, D. C. 1990. Perceived and actual discrimination in the salaries of male and female managers. *Journal of Applied Social Psychology, 20*, 46–67.

Frith, U. 1995. Autism: Beyond "theory of mind." *Cognition, 50,* 13–30.

Funt, B. V. 1980. Problem-solving with diagrammatic representations. *Artificial Intelligence, 13,* 210–230.

Gallistel, C. R. 1990. *The organization of learning.* Cambridge, Mass.: MIT Press.

Gallistel, C. R. 1995. The replacement of general-purpose theories with adaptive specializations. In Gazzaniga, 1995.

Gallup, G. G., Jr. 1991. Toward a comparative psychology of self-awareness: Species limitations and cognitive consequences. In G. R. Goethals & J. Strauss (Eds.), *The self: An interdisciplinary approach.* New York: Springer-Verlag.

Gardner, H. 1985. *The mind's new science: A history of the cognitive revolution.* New York: Basic Books.

Gardner, M. 1989. Illusions of the third dimension. In M. Gardner, *Gardner's whys and wherefores.* Chicago: University of Chicago Press.

Gardner, M. 1990. *The new ambidextrous universe.* New York: W. H. Freeman.

Gardner, M. 1991. Flatlands. In M. Gardner, *The unexpected hanging and other mathematical diversions.* Chicago: University of Chicago Press.

Gaulin, S. J. C. 1995. Does evolutionary theory predict sex differences in the brain? In Gazzaniga, 1995.

Gazzaniga, M. S. 1992. *Nature's mind: The biological roots of thinking, emotion, sexuality, language, and intelligence.* New York: Basic Books.

Gazzaniga, M. S. (Ed.) 1995. *The cognitive neurosciences.* Cambridge, Mass.: MIT Press.

Geary, D. C. 1994. *Children's mathematical development.* Washington, D.C.: American Psychological Association.

Geary, D. C. 1995. Reflections on evolution and culture in children's cognition. *American Psychologist, 50,* 24–37.

Gell-Mann, M. 1994. *The quark and the jaguar: Adventures in the simple and the complex.* New York: W. H. Freeman.

Gelman, R., Durgin, F., & Kaufman, L. 1995. Distinguishing between animates and inanimates: Not by motion alone. In Sperber, Premack, & Premack, 1995.

Gelman, R., & Gallistel, C. R. 1978. *The child's understanding of number.* Cambridge, Mass.: Harvard University Press.

Gelman, S. A., Coley, J. D., & Gottfried, G. M. 1994. Essentialist beliefs in children: The acquisition of concepts and theories. In Hirschfeld & Gelman, 1994a.

Gelman, S. A., & Markman, E. 1987. Young children's inductions from natural kinds: The role of categories and appearances. *Child Development, 58,* 1532–1540.

Gergely, G., Nádasdy, Z., Csibra, G., & Bíró, S. 1995. Taking the intentional stance at 12 months of age. *Cognition, 56,* 165–193.

Gibbons, A. 1994. African origins theory goes nuclear. *Science, 264,* 350–351.

Gibbons, A. 1995a. Out of Africa—at last? *Science, 267,* 1272–1273.

Gibbons, A. 1995b. The mystery of humanity's missing mutations. *Science, 267,* 35–36.

Gibbons, A. 1995c. Pleistocene population explosions. *Science, 267,* 27–28.

Gibson, J. J. 1950. *The perception of the visual world.* Boston: Houghton Mifflin.

Gibson, J. J. 1952. The visual field and the visual world: A reply to Professor Boring. *Psychological Review, 59,* 149–151.

Gigerenzer, G. 1991. How to make cognitive illusions disappear: Beyond heuristics and biases. *European Review of Social Psychology, 2,* 83–115.

Gigerenzer, G. 1996a. On narrow norms and vague heuristics: A reply to Kahneman and Tversky 1996. *Psychological Review, 103,* 592–596.

Gigerenzer, G. 1996b. The psychology of good judgment: Frequency formats and simple algorithms. *Journal of Medical Decision Making, 16,* 273–280.

Gigerenzer, G. 1997. Ecological intelligence: An adaptation for frequencies. In D. Cummins & C. Allen (Eds.), *The evolution of mind.* New York: Oxford University Press.

Gigerenzer, G., & Hoffrage, U. 1995. How to improve Bayesian reasoning without instruction: Frequency formats. *Psychological Review, 102,* 684–704.

Gigerenzer, G., & Hug, K. 1992. Domain specific reasoning: Social contracts, cheating and perspective change. *Cognition, 43,* 127–171.

Gigerenzer, G., & Murray, D. J. 1987. *Cognition as intuitive statistics.* Hillsdale, N.J.: Erlbaum.

Gigerenzer, G., Swijtink, Z., Porter, T., Daston, L., Beatty, J., & Krüger, L. 1989. *The empire of chance: How probability changed science and everyday life.* New York: Cambridge University Press.

Giles, C. L., Sun, G. Z., Chen, H. H., Lee, Y. C., & Chen, D. 1990. Higher order recurrent networks and grammatical inference. In D. S. Touretzky (Ed.), *Advances in Neural Information Processing Systems,* 2. San Mateo, Calif.: Morgan Kaufmann.

Gilovich, T. 1991. *How we know what isn't so: The fallibility of human reason in everyday life.* New York: Free Press.

Glander, K. E. 1992. Selecting and processing food. In Jones, Martin, & Pilbeam, 1992.

Glasgow, J., & Papadias, D. 1992. Computational imagery. *Cognitive Science, 16,* 355–394.

Gombrich, E. 1960. *Art and illusion: A study in the psychology of pictorial representation.* Princeton, N.J.: Princeton University Press.

Good, I. J. 1995. When batterer turns murderer. *Nature, 375,* 541.

Goodman, N. 1976. *Languages of art: An approach to a theory of symbols.* Indianapolis: Hackett.

Gopnik, A. 1993. Mindblindness. Unpublished manuscript, University of California, Berkeley.

Gopnik, A., & Wellman, H. M. 1994. The theory theory. In Hirschfeld & Gelman, 1994a.

Gordon, M. 1996. What makes a woman a woman? (Review of E. Fox-Genovese's "Feminism is not the story of my life"). *New York Times Book Review,* January 14, p. 9.

Gould, J. L. 1982. *Ethology.* New York: Norton.

Gould, S. J., & Vrba, E. 1981. Exaptation: A missing term in the science of form. *Paleobiology, 8,* 4–15.

Gould, S. J. 1980a. *The panda's thumb.* New York: Norton.

Gould, S. J. 1980b. Caring groups and selfish genes. In Gould, 1980a.

Gould, S. J. 1980c. Natural selection and the human brain: Darwin *vs.* Wallace. In Gould, 1980a.

Gould, S. J. 1980d. A biological homage to Mickey Mouse. In Gould, 1980a.

Gould, S. J. 1983a. *Hens' teeth and horses' toes.* New York: Norton.

Gould, S. J. 1983b. What happens to bodies if genes act for themselves? In Gould, 1983a.

Gould, S. J. 1983c. What, if anything, is a zebra? In Gould, 1983a.

Gould, S. J. 1987. *An urchin in the storm: Essays about books and ideas*. New York: Norton.

Gould, S. J. 1989. *Wonderful life: The Burgess Shale and the nature of history*. New York: Norton.

Gould, S. J. 1992. The confusion over evolution. *New York Review of Books*, November 19.

Gould, S. J. 1993. *Eight little piggies*. New York: Norton.

Gould, S. J. 1996. *Full house: The spread of excellence from Plato to Darwin*. New York: Harmony Books.

Gould, S. J., & Lewontin, R. C. 1979. The spandrels of San Marco and the Panglossian program: A critique of the adaptationist programme. *Proceedings of the Royal Society of London, 205*, 281–288.

Greenwald, A. 1988. Self-knowledge and self-deception. In Lockard & Paulhaus, 1988.

Gregory, R. L. 1970. *The intelligent eye*. London: Weidenfeld & Nicolson.

Griffin, D. R. (Ed.) 1974. *Animal engineering*. San Francisco: W. H. Freeman.

Grossberg, S. (Ed.) 1988. *Neural networks and natural intelligence*. Cambridge, Mass.: MIT Press.

Gruber, J. 1965. Studies in lexical relations. Ph.D. dissertation, MIT. Reprinted, 1976, as *Lexical structures in syntax and semantics*. Amsterdam: North-Holland.

Gutin, J. 1995. Do Kenya tools root birth of modern thought in Africa? *Science, 270*, 1118–1119.

Hadley, R. F. 1994a. Systematicity in connectionist language learning. *Mind and Language, 9*, 247–272.

Hadley, R. F. 1994b. Systematicity revisited: Reply to Christiansen and Chater and Niklasson and Van Gelder. *Mind and Language, 9*, 431–444.

Hadley, R. F., & Hayward, M. 1994. Strong semantic systematicity from unsupervised connectionist learning. Technical Report CSS-IS TR94–02, School of Computing Science, Simon Fraser University, Burnaby, BC.

Haig, D. 1992. Genetic imprinting and the theory of parent-offspring conflict. *Developmental Biology, 3*, 153–160.

Haig, D. 1993. Genetic conflicts in human pregnancy. *Quarterly Review of Biology, 68*, 495–532.

Hamer, D., & Copeland, P. 1994. *The science of desire: The search for the gay gene and the biology of behavior*. New York: Simon & Schuster.

Hamilton, W. D. 1963. The evolution of altruistic behavior. *American Naturalist, 97*, 354–356. Reprinted in Hamilton, 1996.

Hamilton, W. D. 1964. The genetical evolution of social behaviour (I and II). *Journal of Theoretical Biology, 7*, 1–16; 17–52. Reprinted in Hamilton, 1996.

Hamilton, W. D. 1996. *Narrow roads of gene land: The collected papers of W. D. Hamilton, Vol. 1: Evolution of social behavior*. New York: W. H. Freeman.

Hamilton, W. D., Axelrod, R., & Tanese, R. 1990. Sexual reproduction as an adaptation to resist parasites (a review). *Proceedings of the National Academy of Sciences, 87*, 3566–3573.

Harpending, H. 1994. Gene frequencies, DNA sequences, and human origins. *Perspectives in Biology and Medicine, 37*, 384–395.

Harris, C. R., & Christenfeld, N. 1996. Gender, jealousy, and reason. *Psychological Science*, 7, 364–366, 378–379.

Harris, D. R. 1992. Human diet and subsistence. In Jones, Martin, & Pilbeam, 1992.

Harris, H. Y. 1995. Human nature and the nature of romantic love. Ph.D. dissertation, Department of Anthropology, University of California, Santa Barbara.

Harris, J. R. 1995. Where is the child's environment? A group socialization theory of development. *Psychological Review*, 102, 458–489.

Harris, M. 1985. *Good to eat: Riddles of food and culture*. New York: Simon & Schuster.

Harris, M. 1989. *Our kind: The evolution of human life and culture*. New York: Harper-Collins.

Harris, P. L. 1994. Thinking by children and scientists: False analogies and neglected similarities. In Hirschfeld & Gelman, 1994a.

Hartung, J. 1992. Getting real about rape. *Behavioral and Brain Sciences*, 15, 390–392.

Hartung, J. 1995. Love thy neighbor: The evolution of in-group morality. *Skeptic*, 3, 86–100.

Hatano, G., & Inagaki, K. 1995. Young children's naive theory of biology. *Cognition, 50*, 153–170.

Hatfield, E., & Rapson, R. L. 1993. *Love, sex, and intimacy: Their psychology, biology, and history*. New York: HarperCollins.

Haugeland, J. (Ed.) 1981a. *Mind design: Philosophy, psychology, artificial intelligence*. Cambridge, Mass.: Bradford Books/MIT Press.

Haugeland, J. 1981b. Semantic engines: An introduction to mind design. In Haugeland, 1981a.

Haugeland, J. 1981c. The nature and plausibility of cognitivism. In Haugeland, 1981a.

Hauser, M. D. 1992. Costs of deception: Cheaters are punished in rhesus monkeys. *Proceedings of the National Academy of Sciences USA, 89*, 12137–12139.

Hauser, M. D. 1996. *The evolution of communication*. Cambridge, Mass.: MIT Press.

Hauser, M. D., Kralik, J., Botto-Mahan, C., Garrett, M., & Oser, J. 1995. Self-recognition in primates: Phylogeny and the salience of species-typical features. *Proceedings of the National Academy of Sciences USA, 92*, 10811–10814.

Hauser, M. D., MacNeilage, P., & Ware, M. 1996. Numerical representations in primates: Perceptual or arithmetic? *Proceedings of the National Academy of Sciences USA, 93*, 1514–1517.

Hebb, D. O. 1968. Concerning imagery. *Psychological Review, 75*, 466–477.

Heider, F., & Simmel, M. 1944. An experimental study of apparent behavior. *American Journal of Psychology, 57*, 243–259.

Held, R. 1993. Two stages in the development of binocular vision and eye alignment. In Simons, 1993.

Hendler, J. 1994. High-performance artificial intelligence. *Science, 265*, 891–892.

Hertwig, R., & Gigerenzer, G. 1997. The "conjunction fallacy" revisited: How intelligent inferences look like reasoning errors. Unpublished manuscript, Max Planck Institute for Psychological Research, Munich.

Hess, R. H., Baker, C. L., & Zihl, J. 1989. The "motion-blind" patient: Low level spatial and temporal filters. *Journal of Neuroscience, 9*, 1628–1640.

Hill, K., & Kaplan, H. 1988. Tradeoffs in male and female reproductive strategies among the Ache (parts 1 and 2). In Betzig, Borgerhoff Mulder, & Turke, 1988.

Hillis, A. E., & Caramazza, A. 1991. Category-specific naming and comprehension impairment: A double dissociation. *Brain, 114,* 2081–2094.

Hinton, G. E. 1981. Implementing semantic networks in parallel hardware. In Hinton and Anderson, 1981.

Hinton, G. E., & Anderson, J. A. 1981. *Parallel models of associative memory.* Hillsdale, N.J.: Erlbaum.

Hinton, G. E., & Nowlan, S. J. 1987. How learning can guide evolution. *Complex Systems, 1,* 495–502.

Hinton, G. E., McClelland, J. L., & Rumelhart, D. E. 1986. Distributed representations. In Rumelhart, McClelland, & the PDP Research Group, 1986.

Hinton, G. E., & Parsons, L. M. 1981. Frames of reference and mental imagery. In J. Long & A. Baddeley (Eds.), *Attention and Performance IX.* Hillsdale, N.J.: Erlbaum.

Hirschfeld, L. A., & Gelman, S. A. (Eds.) 1994a. *Mapping the mind: Domain specificity in cognition and culture.* New York: Cambridge University Press.

Hirschfeld, L. A., & Gelman, S. A. 1994b. Toward a topography of mind: An introduction to domain specificity. In Hirschfeld & Gelman, 1994a.

Hirshleifer, J. 1987. On the emotions as guarantors of threats and promises. In J. Dupré (Ed.), *The latest on the best: Essays on evolution and optimality.* Cambridge, Mass.: MIT Press.

Hobbs, J. R. 1990. *Literature and cognition.* Stanford, Calif.: Center for the Study of Language and Information.

Hoffman, D. D. 1983. The interpretation of visual illusions. *Scientific American,* December.

Hoffman, D. D., & Richards, W. A. 1984. Parts of recognition. *Cognition, 18,* 65–96. Reprinted in Pinker, 1984b.

Hollerbach, J. M. 1990. Planning of arm movements. In Osherson, Kosslyn, & Hollerbach, 1990.

Holloway, R. L. 1995. Toward a synthetic theory of human brain evolution. In Changeux & Chavaillon, 1995.

Horgan, J. 1993. Eugenics revisited. *Scientific American,* June.

Horgan, J. 1995a. The new Social Darwinists. *Scientific American,* October.

Horgan, J. 1995b. A theory of almost everything (Review of books by J. Holland, S. Kauffman, P. Davies, P. Coveney, and R. Highfield). *New York Times Book Review,* October 1, pp. 30–31.

Hrdy, S. B. 1981. *The woman that never evolved.* Cambridge, Mass.: Harvard University Press.

Hrdy, S. B. 1994. Interview. In T. A. Bass, *Reinventing the future: Conversations with the world's leading scientists.* Reading, Mass.: Addison Wesley.

Hubel, D. H. 1988. *Eye, brain, and vision.* New York: Scientific American.

Hume, D. 1748/1955. *Inquiry concerning human understanding.* Indianapolis: Bobbs-Merrill.

Humphrey, N. K. 1976. The social function of the intellect. In P. P. G. Bateson & R. A. Hinde (Eds.), *Growing points in ethology.* New York: Cambridge University Press.

Humphrey, N. K. 1992. *A history of the mind: Evolution and the birth of consciousness.* New York: Simon & Schuster.

Hurst, L., & Hamilton, W. D. 1992. Cytoplasmic fusion and the nature of the sexes. *Proceedings of the Royal Society of London B, 247,* 189–194.

Hyman, I. E., & Neisser, U. 1991. Reconstruing mental images: Problems of method. *Emory Cognition Project Technical Report Number 19*. Atlanta: Emory University.

Ioerger, T. R. 1994. The manipulation of images to handle indeterminacy in spatial reasoning. *Cognitive Science, 18,* 551–593.

Ittelson, W. H. 1968. *The Ames demonstrations in perception*. New York: Hafner.

Jackendoff, R. 1977. Review of Leonard Bernstein's "The unanswered question." *Language, 53,* 883–894.

Jackendoff, R. 1983. *Semantics and cognition*. Cambridge, Mass.: MIT Press.

Jackendoff, R. 1987. *Consciousness and the computational mind*. Cambridge, Mass.: MIT Press.

Jackendoff, R. 1990. *Semantic structures*. Cambridge, Mass.: MIT Press.

Jackendoff, R. 1992. Musical parsing and musical affect. In R. Jackendoff, *Languages of the mind: Essays on mental representation*. Cambridge, Mass.: MIT Press.

Jackendoff, R. 1994. *Patterns in the mind: Language and human nature*. New York: Basic Books.

Jackendoff, R., & Aaron, D. 1991. Review of Lakoff & Turner's "More than cool reason: A field guide to poetic metaphor." *Language, 67,* 320–339.

Jagannathan, V., Dodhiawala, R., & Baum, L. S. (Eds.) 1989. *Blackboard architectures and applications*. New York: Academic Press.

James, W. 1890/1950. *The principles of psychology*. New York: Dover.

James, W. 1892/1920. *Psychology: Briefer course*. New York: Henry Holt.

Jaynes, J. 1976. *The origin of consciousness in the breakdown of the bicameral mind*. Boston: Houghton Mifflin.

Jepson, A., Richards, W., & Knill, D. 1996. Modal structure and reliable inference. In Knill & Richards, 1996.

Johnson, M. K., & Raye, C. L. 1981. Reality monitoring. *Psychological Review, 88,* 67–85.

Johnson-Laird, P. 1988. *The computer and the mind*. Cambridge, Mass.: Harvard University Press.

Jolicoeur, P., Ullman, S., & MacKay, M. 1991. Visual curve tracing properties. *Journal of Experimental Psychology: Human Perception and Performance, 17,* 997–1022.

Jones, S. 1992. The evolutionary future of humankind. In Jones, Martin, & Pilbeam, 1992.

Jones, S., Martin, R., & Pilbeam, D. (Eds.) 1992. *The Cambridge encyclopedia of human evolution*. New York: Cambridge University Press.

Jordan, M. I. 1989. Serial order: A parallel distributed processing approach. In J. L. Elman & D. E. Rumelhart (Eds.), *Advances in connectionist theory*. Hillsdale, N.J.: Erlbaum.

Julesz, B. 1960. Binocular depth perception of computer-generated patterns. *Bell System Technical Journal, 39,* 1125–1162.

Julesz, B. 1971. *Foundations of cyclopean perception*. Chicago: University of Chicago Press.

Julesz, B. 1995. *Dialogues on perception*. Cambridge, Mass.: MIT Press.

Kahneman, D., & Tversky, A. 1982. On the study of statistical intuitions. *Cognition, 11,* 123–141.

Kahneman, D., & Tversky, A. 1984. Choices, values, and frames. *American Psychologist, 39,* 341–350.

Kahneman, D., & Tversky, A. 1996. On the reality of cognitive illusions: A reply to Gigerenzer's critique. *Psychological Review, 103*, 582–591.

Kahneman, D., Slovic, P., & Tversky, A. (Eds.) 1982. *Judgment under uncertainty: Heuristics and biases*. New York: Cambridge University Press.

Kaplan, H., Hill, K., & Hurtado, A. M. 1990. Risk, foraging, and food sharing among the Ache. In E. Cashdan (Ed.), *Risk and uncertainty in tribal and peasant economies*. Boulder, Colo.: Westview Press.

Kaplan, S. 1992. Environmental preference in a knowledge-seeking, knowledge-using organism. In Barkow, Cosmides, & Tooby, 1992.

Katz, J. N. 1995. *The invention of homosexuality*. New York: Dutton.

Kauffman, S. A. 1991. Antichaos and adaptation. *Scientific American*, August.

Keeley, L. H. 1996. *War before civilization: The myth of the peaceful savage*. New York: Oxford University Press.

Keil, F. C. 1979. *Semantic and conceptual development*. Cambridge, Mass.: Harvard University Press.

Keil, F. C. 1989. *Concepts, kinds, and cognitive development*. Cambridge, Mass.: MIT Press.

Keil, F. C. 1994. The birth and nurturance of concepts by domains: The origins of concepts of living things. In Hirschfeld & Gelman, 1994a.

Keil, F. C. 1995. The growth of causal understandings of natural kinds. In Sperber, Premack, & Premack, 1995.

Kelly, M. H. 1992. Darwin and psychological theories of classification. *Evolution and Cognition, 2*, 79–97.

Kenrick, D. T., Keefe, R. C., & commentators. 1992. Age preferences in mates reflect sex differences in human reproductive strategies. *Behavioral and Brain Sciences, 15*, 75–133.

Kernighan, B. W., & Plauger, P. J. 1978. *The elements of programming style*. 2d ed. New York: McGraw-Hill.

Kerr, R. A. 1992. SETI faces uncertainty on earth and in the stars. *Science, 258*, 27.

Ketelaar, P. 1995. Emotion as mental representations of fitness affordances I: Evidence supporting the claim that negative and positive emotions map onto fitness costs and benefits. Paper presented at the annual meeting of the Human Behavior and Evolution Society, Santa Barbara, June 28–July 2.

Ketelaar, P. 1997. Affect as mental representations of value: Translating the value function for gains and losses into positive and negative affect. Unpublished manuscript, Max Planck Institute, Munich.

Killackey, H. 1995. Evolution of the human brain: A neuroanatomical perspective. In Gazzaniga, 1995.

Kingdon, J. 1993. *Self-made man: Human evolution from Eden to extinction?* New York: Wiley.

Kingsolver, J. G., & Koehl, M. A. R. 1985. Aerodynamics, thermoregulation, and the evolution of insect wings: Differential scaling and evolutionary change. *Evolution, 39*, 488–504.

Kirby, K. N., & Herrnstein, R. J. 1995. Preference reversals due to myopic discounting of delayed reward. *Psychological Science, 6*, 83–89.

Kitcher, P. 1982. *Abusing science: The case against creationism*. Cambridge, Mass.: MIT Press.

Kitcher, P. 1992. Gene: current usages. In E. F. Keller & E. A. Lloyd (Eds.), *Keywords in evolutionary biology*. Cambridge, Mass.: Harvard University Press.

Klaw, S. 1993. *Without sin: The life and death of the Oneida community*. New York: Penguin.

Klein, R. G. 1989. *The human career: Human biological and cultural origins*. Chicago: University of Chicago Press.

Kleiter, G. 1994. Natural sampling: Rationality without base rates. In G. H. Fischer & D. Laming (Eds.), *Contributions to mathematical psychology, psychometrics, and methodology*. New York: Springer-Verlag.

Knill, D., & Richards, W. (Eds.) 1996. *Perception as Bayesian inference*. New York: Cambridge University Press.

Koehler, J. J., & commentators. 1996. The base rate fallacy reconsidered: Descriptive, normative, and methodological challenges. *Behavioral and Brain Sciences, 19*, 1–53.

Koestler, A. 1964. *The act of creation*. New York: Dell.

Konner, M. 1982. *The tangled wing: Biological constraints on the human spirit*. New York: Harper and Row.

Kosslyn, S. M. 1980. *Image and mind*. Cambridge, Mass.: Harvard University Press.

Kosslyn, S. M. 1983. *Ghosts in the mind's machine: Creating and using images in the brain*. New York: Norton.

Kosslyn, S. M. 1994. *Image and brain: The resolution of the imagery debate*. Cambridge, Mass.: MIT Press.

Kosslyn, S. M., Alpert, N. M., Thompson, W. L., Maljkovic, V., Weise, S. B., Chabris, C. F., Hamilton, S. E., Rauch, S. L., & Buonanno, F. S. 1993. Visual mental imagery activates topographically organized visual cortex: PET investigations. *Journal of Cognitive Neuroscience, 5*, 263–287.

Kosslyn, S. M., & Osherson, D. N. (Eds.) 1995. *An invitation to cognitive science*, Vol. 2: *Visual cognition*. 2d ed. Cambridge, Mass.: MIT Press.

Kosslyn, S. M., Pinker, S., Smith, G. E., Schwartz, S. P., & commentators. 1979. On the demystification of mental imagery. *Behavioral and Brain Sciences, 2*, 535–581. Reprinted in Block, 1981.

Kowler, E. 1995. Eye movements. In Kosslyn & Osherson, 1995.

Kubovy, M. 1986. *The psychology of perspective and Renaissance art*. New York: Cambridge University Press.

Kubovy, M., & Pomerantz, J. R. (Eds.) 1981. *Perceptual organization*. Hillsdale, N.J.: Erlbaum.

Lachter, J., & Bever, T. G. 1988. The relation between linguistic structure and associative theories of language learning—A constructive critique of some connectionist learning models. *Cognition, 28*, 195–247. Reprinted in Pinker & Mehler, 1988.

Lakoff, G. 1987. *Women, fire, and dangerous things: What categories reveal about the mind*. Chicago: University of Chicago Press.

Lakoff, G., & Johnson, M. 1980. *Metaphors we live by*. Chicago: University of Chicago Press.

Land, E. H., & McCann, J. J. 1971. Lightness and retinex theory. *Journal of the Optical Society of America, 61*, 1–11.

Landau, B., Spelke, E. S., & Gleitman, H. 1984. Spatial knowledge in a young blind child. *Cognition, 16*, 225–260.

Landau, T. 1989. *About faces: The evolution of the human face*. New York: Anchor.

Landsburg, S. E. 1993. *The armchair economist: Economics and everyday life*. New York: Free Press.

Langlois, J. H., & Roggman, L. A. 1990. Attractive faces are only average. *Psychological Science, 1*, 115–121.

Langlois, J. H., Roggman, L. A., Casey, R. J., & Ritter, J. M. 1987. Infant preferences for attractive faces: Rudiments of a stereotype? *Developmental Psychology, 23*, 363–369.

Lazarus, R. S. 1991. *Emotion and adaptation*. New York: Oxford University Press.

Leakey, M. G., Feibel, C. S., McDougall, I., & Walker, A. 1995. New four-million-year-old hominid species from Kanapoi and Allia Bay, Kenya. *Nature, 376*, 565–572.

Lederer, R., & Gilleland, M. 1994. *Literary trivia: Fun and games for book lovers*. New York: Vintage.

LeDoux, J. E. 1991. Emotion and the limbic system concept. *Concepts in Neuroscience, 2*, 169–199.

LeDoux, J. E. 1996. *The emotional brain: The mysterious underpinnings of emotional life*. New York: Simon & Schuster.

Lee, P. C. 1992. Testing the intelligence of apes. In Jones, Martin, & Pilbeam, 1992.

Lehman, D. 1992. *Signs of the times: Deconstructionism and the fall of Paul de Man*. New York: Simon & Schuster.

Leibniz, G. W. 1956. *Philosophical papers and letters*. Chicago: University of Chicago Press.

Lenat, D. B., & Guha, D. V. 1990. *Building large knowledge-based systems*. Reading, Mass.: Addison-Wesley.

Lenski, R. E., & Mittler, J. E. 1993. The directed mutation controversy and neo-Darwinism. *Science, 259*, 188–194.

Lenski, R. E., Sniegowski, P. D., & Shapiro, J. A. 1995. "Adaptive mutation": The debate goes on (letters). *Science, 269*, 285–287.

Lerdahl, F., & Jackendoff, R. 1983. *A generative theory of tonal music*. Cambridge, Mass.: MIT Press.

Leslie, A. M. 1994. ToMM, ToBY, and agency: Core architecture and domain specificity. In Hirschfeld & Gelman, 1994a.

Leslie, A. M. 1995a. A theory of agency. In Sperber, Premack, & Premack, 1995.

Leslie, A. M. 1995b. Pretending and believing: Issues in the theory of ToMM. *Cognition, 50*, 193–220.

Levin, B., & Pinker, S. (Eds.) 1992. *Lexical and conceptual semantics*. Cambridge, Mass.: Blackwell.

Levine, A. 1994. Education: The great debate revisited. *Atlantic Monthly*, December.

Levins, R., & Lewontin, R. C. 1985. *The dialectical biologist*. Cambridge, Mass.: Harvard University Press.

Lewin, R. 1987. The earliest "humans" were more like apes. *Science, 236*, 1061–1063.

Lewis, D. 1980. Mad pain and Martian pain. In N. Block (Ed.), *Readings in philosophy of psychology*, Vol. 1. Cambridge, Mass.: Harvard University Press.

Lewis, H. W. 1990. *Technological risk*. New York: Norton.

Lewontin, R. C. 1979. Sociobiology as an adaptationist program. *Behavioral Science, 24*, 5–14.

Lewontin, R. C. 1984. Adaptation. In Sober, 1984a.

Lewontin, R. C., Rose, S., & Kamin, L. J. 1984. *Not in our genes*. New York: Pantheon.

Liebenberg, L. 1990. *The art of tracking*. Cape Town: David Philip.

Lindsay, P. H., & Norman, D. A. 1972. *Human information processing*. New York: Academic Press.

Ling, C., & Marinov, M. 1993. Answering the connectionist challenge: A symbolic model of learning the past tenses of English verbs. *Cognition, 49*, 235–290.

Lockard, J. S., & Paulhaus, D. L. (Eds.) 1988. *Self-deception: An adaptive mechanism*. Englewood Cliffs, N.J.: Prentice Hall.

Locksley, A., Ortiz, V., & Hepburn, C. 1980. Social categorization and discriminatory behavior: Extinguishing the minimal group discrimination effect. *Journal of Personality and Social Psychology, 39*, 773–783.

Loewer, B., & Rey, B. (Eds.) 1991. *Meaning in mind: Fodor and his critics*. Cambridge, Mass.: Blackwell.

Logie, R. H. 1995. *Visuo-spatial working memory*. Hillsdale, N.J.: Erlbaum.

Lopes, L. L., & Oden, G. C. 1991. The rationality of intelligence. In E. Eells & T. Maruszewski (Eds.), *Rationality and reasoning*. Amsterdam: Rodopi.

Lorber, J. 1994. *Paradoxes of gender*. New Haven: Yale University Press.

Lowe, D. 1987. The viewpoint consistency constraint. *International Journal of Computer Vision, 1*, 57–72.

Lumsden, C., & Wilson, E. O. 1981. *Genes, mind, and culture*. Cambridge, Mass.: Harvard University Press.

Luria, A. R. 1966. *Higher cortical functions in man*. London: Tavistock.

Lykken, D. T., & Tellegen, A. 1996. Happiness is a stochastic phenomenon. *Psychological Science, 7*, 186–189.

Lykken, D. T., McGue, M., Tellegen, A., & Bouchard, T. J., Jr. 1992. Emergenesis: Genetic traits that may not run in families. *American Psychologist, 47*, 1565–1577.

Mac Lane, S. 1981. Mathematical models: A sketch for the philosophy of mathematics. *American Mathematical Monthly, 88*, 462–472.

Mace, G. 1992. The life of primates: Differences between the sexes. In Jones, Martin, & Pilbeam, 1992.

MacLean, P. D. 1990. *The triune brain in evolution*. New York: Plenum.

Macnamara, J. 1986. *A border dispute: The place of logic in psychology*. Cambridge, Mass.: MIT Press.

Macnamara, J. 1994. Logic and cognition. In Macnamara & Reyes.

Macnamara, J., & Reyes, G. E. (Eds.) 1994. *The logical foundations of cognition*. New York: Oxford University Press.

Maloney, L. T., & Wandell, B. 1986. Color constancy: A method for recovering surface spectral reflectance. *Journal of the Optical Society of America (A), 1*, 29–33.

Mandler, J. 1992. How to build a baby, II: Conceptual primitives. *Psychological Review, 99*, 587–604.

Manktelow, K. I., & Over, D. E. 1987. Reasoning and rationality. *Mind and Language, 2*, 199–219.

Marcel, A., & Bisiach, E. (Eds.) 1988. *Consciousness in contemporary science*. New York: Oxford University Press.

Marcus, G. F. 1997a. Rethinking eliminative connectionism. Unpublished manuscript, University of Massachusetts, Amherst.

Marcus, G. F. 1997b. Concepts, features, and variables. Unpublished manuscript, University of Massachusetts, Amherst.

Marcus, G. F. In preparation. *The algebraic mind*. Cambridge, Mass.: MIT Press.

Marcus, G. F., Brinkmann, U., Clahsen, H., Wiese, R., & Pinker, S. 1995. German inflection: The exception that proves the rule. *Cognitive Psychology, 29*, 189–256.

Marks, I. M. 1987. *Fears, phobias, and rituals*. New York: Oxford University Press.

Marks, I. M., & Nesse, R. M. 1994. Fear and fitness: An evolutionary analysis of anxiety disorders. *Ethology and Sociobiology, 15*, 247–261.

Marr, D. 1982. *Vision*. San Francisco: W. H. Freeman.

Marr, D., & Nishihara, H. K. 1978. Representation and recognition of the spatial organization of three-dimensional shapes. *Proceedings of the Royal Society of London, B, 200*, 269–294.

Marr, D., & Poggio, T. 1976. Cooperative computation of stereo disparity. *Science, 194*, 283–287.

Marshack, A. 1989. Evolution of the human capacity: The symbolic evidence. *Yearbook of Physical Anthropology, 32*, 1–34.

Martin, P., & Klein, R. 1984. *Quaternary extinctions*. Tucson: University of Arizona Press.

Masson, J. M., & McCarthy, S. 1995. *When elephants weep: The emotional lives of animals*. New York: Delacorte Press.

Mathews, J. 1996. A tall order for president: Picking a candidate of towering stature. *Washington Post*, May 10, D01.

Maurer, A. 1965. What children fear. *Journal of Genetic Psychology, 106*, 265–277.

Maynard Smith, J., 1964. Group selection and kin selection. *Nature, 201*, 1145–1147.

Maynard Smith, J. 1975/1993. *The theory of evolution*. New York: Cambridge University Press.

Maynard Smith, J. 1982. *Evolution and the theory of games*. New York: Cambridge University Press.

Maynard Smith, J. 1984. Optimization theory in evolution. In Sober, 1984a.

Maynard Smith, J. 1987. When learning guides evolution. *Nature, 329*, 762.

Maynard Smith, J. 1995. Life at the edge of chaos? (Review of D. Depew's & B. H. Weber's "Darwinism evolving"). *New York Review of Books*, March 2, pp. 28–30.

Maynard Smith, J., & Warren, N. 1988. Models of cultural and genetic change. In J. Maynard Smith, *Games, sex, and evolution*. New York: Harvester-Wheatsheaf.

Mayr, E. 1982. *The growth of biological thought*. Cambridge, Mass.: Harvard University Press.

Mayr, E. 1983. How to carry out the adaptationist program. *The American Naturalist, 121*, 324–334.

Mayr, E. 1993. The search for intelligence (letter). *Science, 259*, 1522–1523.

Mazel, C. 1992. *Heave ho! My little green book of seasickness*. Camden, Maine: International Marine.

McAdams, S., & Bigand, E. (Eds.) 1993. *Thinking in sound: The cognitive psychology of human audition*. New York: Oxford University Press.

McCauley, C., & Stitt, C. L. 1978. An individual and quantitative measure of stereotypes. *Journal of Personality and Social Psychology, 36*, 929–940.

McClelland, J. L., & Kawamoto, A. H. 1986. Mechanisms of sentence processing:

Assigning roles to constituents of sentences. In McClelland, Rumelhart, & the PDP Research Group.

McClelland, J. L., McNaughton, B. L., & O'Reilly, R. C. 1995. Why there are complementary learning systems in the hippocampus and neocortex: Insights from the successes and failures of connectionist models of learning and memory. *Psychological Review, 102,* 419–457.

McClelland, J. L., & Rumelhart, D. E. 1985. Distributed memory and the representation of general and specific information. *Journal of Experimental Psychology: General, 114,* 159–188.

McClelland, J. L., Rumelhart, D. E., & the PDP Research Group. 1986. *Parallel distributed processing: Explorations in the microstructure of cognition,* Vol. 2: *Psychological and biological models.* Cambridge, Mass.: MIT Press.

McCloskey, M. 1983. Intuitive physics. *Scientific American, 248,* 122–130.

McCloskey, M., Caramazza, A., & Green, B. 1980. Curvilinear motion in the absence of external forces: Naive beliefs about the motion of objects. *Science, 210,* 1139–1141.

McCloskey, M., & Cohen, N. J. 1989. Catastrophic interference in connectionist networks: The sequential learning problem. In G. H. Bower (Ed.), *The psychology of learning and motivation,* Vol. 23. New York: Academic Press.

McCloskey, M., Wible, C. G., & Cohen, N. J. 1988. Is there a special flashbulb-memory mechanism? *Journal of Experimental Psychology: General, 117,* 171–181.

McCulloch, W. S., & Pitts, W. 1943. A logical calculus of the ideas immanent in nervous activity. *Bulletin of Mathematical Biophysics, 5,* 115–133.

McGhee, P. E. 1979. *Humor: Its origins and development.* San Francisco: W. H. Freeman.

McGinn, C. 1989a. *Mental content.* Cambridge, Mass.: Blackwell.

McGinn, C. 1989b. Can we solve the mind-body problem? *Mind, 98,* 349–366.

McGinn, C. 1993. *Problems in philosophy: The limits of inquiry.* Cambridge, Mass.: Blackwell.

McGuinness, D. 1997. *Why our children can't read and what we can do about it.* New York: Free Press.

Medin, D. L. 1989. Concepts and conceptual structure. *American Psychologist, 44,* 1469–1481.

Michotte, A. 1963. *The perception of causality.* London: Methuen.

Miller, G. 1967. *The psychology of communication.* London: Penguin.

Miller, G. A. 1956. The magical number seven, plus or minus two: Some limits on our capacity for processing information. *Psychological Review, 63,* 81–96.

Miller, G. A. 1981. Trends and debates in cognitive psychology. *Cognition, 10,* 215–226.

Miller, G. A., & Johnson-Laird, P. N. 1976. *Language and perception.* Cambridge, Mass.: Harvard University Press.

Miller, G. F. 1993. Evolution of the human brain through runaway sexual selection: The mind as a protean courtship device. Ph.D. dissertation, Department of Psychology, Stanford University.

Miller, G. F., & Todd, P. M. 1990. Exploring adaptive agency. I: Theory and methods for simulating the evolution of learning. In D. S. Touretzky, J. L. Elman, T. Sejnowski, & G. E. Hinton (Eds.), *Proceedings of the 1990 Connectionist Models Summer School.* San Mateo, Calif.: Morgan Kaufmann.

Miller, K. D., Keller, J. B., & Stryker, M. P. 1989. Ocular dominance column development: Analysis and simulation. *Science, 245*, 605–615.

Millikan, R. 1984. *Language, thought, and other biological categories.* Cambridge, Mass.: MIT Press.

Mineka, S., & Cook, M. 1993. Mechanisms involved in the observational conditioning of fear. *Journal of Experimental Psychology: General, 122*, 23–38.

Minsky, M. 1985. *The society of mind.* New York: Simon & Schuster.

Minsky, M., & Papert, S. 1988a. *Perceptrons: Expanded edition.* Cambridge, Mass.: MIT Press.

Minsky, M., & Papert, S. 1988b. Epilogue: The new connectionism. In Minsky & Papert, 1988a.

Mitchell, M. 1996. *An introduction to genetic algorithms.* Cambridge, Mass.: MIT Press.

Mock, D. W., & Parker, G. A. In press. *The evolution of sibling rivalry.* New York: Oxford University Press.

Montello, D. R. 1995. How significant are cultural differences in spatial cognition? In A. U. Frank & W. Kuhn (Eds.), *Spatial information theory: A theoretical basis for GIS.* Berlin: Springer-Verlag.

Moore, E. F. (Ed.) 1964. *Sequential machines: Selected papers.* Reading, Mass.: Addison-Wesley.

Morris, R. G. M. (Ed.) 1989. *Parallel distributed processing: Implications for psychology and neurobiology.* New York: Oxford University Press.

Morton, J., & Johnson, M. H. 1991. CONSPEC and CONLERN: A two-process theory of infant face recognition. *Psychological Review, 98*, 164–181.

Moscovitch, M., Winocur, G., & Behrmann, M. In press. Two mechanisms of face recognition: Evidence from a patient with visual object agnosia. *Journal of Cognitive Neuroscience.*

Mount, F. 1992. *The subversive family: An alternative history of love and marriage.* New York: Free Press.

Mozer, M. 1991. *The perception of multiple objects: A connectionist approach.* Cambridge, Mass.: MIT Press.

Murphy, G. L. 1993. A rational theory of concepts. In G. H. Bower (Ed.), *The psychology of learning and motivation,* Vol. 29. New York: Academic Press.

Myers, D. G., & Diener, E. 1995. Who is happy? *Psychological Science, 6*, 10–19.

N. E. Thing Enterprises. 1994. *Magic Eye III: Visions: A new dimension in art.* Kansas City: Andrews and McMeel.

Nagel, T. 1974. What is it like to be a bat? *Philosophical Review, 83*, 435–450.

Nagell, K., Olguin, R., & Tomasello, M. 1993. Processes of social learning in the tool use of chimpanzees (*Pan troglodytes*) and human children (*Homo sapiens*). *Journal of Comparative Psychology, 107*, 174–186.

Nakayama, K., He, Z. J., & Shimojo, S. 1995. Visual surface representation: A critical link between lower-level and higher-level vision. In Kosslyn & Osherson, 1995.

Navon, D. 1985. Attention division or attention sharing? In M. I. Posner & O. Marin (Eds.), *Attention and performance XI.* Hillsdale, N.J.: Erlbaum.

Navon, D. 1989. The importance of being visible: On the role of attention in a mind viewed as an anarchic intelligence system. I: Basic tenets. *European Journal of Cognitive Psychology, 1*, 191–213.

Nayar, S. K., & Oren, M. 1995. Visual appearance of matte surfaces. *Science, 267*, 1153–1156.

Neisser, U. 1967. *Cognitive psychology*. Engelwood Cliffs, N.J.: Prentice-Hall.

Neisser, U. 1976. General, academic, and artificial intelligence: Comments on the papers by Simon and by Klahr. In L. Resnick (Ed.), *The nature of intelligence*. Hillsdale, N.J.: Erlbaum.

Nesse, R. M. 1991. What good is feeling bad? *The Sciences*, November/December, pp. 30–37.

Nesse, R. M., & Lloyd, A. T. 1992. The evolution of psychodynamic mechanisms. In Barkow, Cosmides, & Tooby, 1992.

Nesse, R. M., & Williams, G. C. 1994. *Why we get sick: The new science of Darwinian medicine*. New York: Times Books.

Newell, A. 1990. *Unified theories of cognition*. Cambridge, Mass.: Harvard University Press.

Newell, A., & Simon, H. A. 1972. *Human problem solving*. Englewood Cliffs, N.J.: Prentice-Hall.

Newell, A., & Simon, H. A. 1981. Computer science as empirical inquiry: Symbols and search. In Haugeland, 1981a.

Nickerson, R. A., & Adams, M. J. 1979. Long-term memory for a common object. *Cognitive Psychology, 11*, 287–307.

Nilsson, D. E., & Pelger, S. 1994. A pessimistic estimate of the time required for an eye to evolve. *Proceedings of the Royal Society of London, B, 256*, 53–58.

Nisbett, R. E., & Cohen, D. 1996. *Culture of honor: The psychology of violence in the South*. New York: HarperCollins.

Nisbett, R. E., & Ross, L. R. 1980. *Human inference: Strategies and shortcomings of social judgment*. Englewood Cliffs, N.J.: Prentice-Hall.

Nobile, P. (Ed.) 1971. *The Con III controversy: The critics look at "The greening of America."* New York: Pocket Books.

Nolfi, S., Elman, J. L., & Parisi, D. 1994. Learning and evolution in neural networks. *Adaptive Behavior, 3*, 5–28.

Nozick, R. 1981. *Philosophical explanations*. Cambridge, Mass.: Harvard University Press.

Oman, C. M. 1982. Space motion sickness and vestibular experiments in Spacelab. *Society of Automotive Engineers Technical Paper Series 820833*. Warrendale, Penn.: SAE.

Oman, C. M., Lichtenberg, B. K., Money, K. E., & McCoy, R. K. 1986. M.I.T./Canadian vestibular experiments on the Spacelab-1 mission: 4. Space motion sickness: Symptoms, stimuli, predictability. *Experimental Brain Research, 64*, 316–334.

Orians, G. H., & Heerwagen, J. H. 1992. Evolved responses to landscapes. In Barkow, Cosmides, & Tooby, 1992.

Orwell, G. 1949/1983. *1984*. New York: Harcourt Brace Jovanovich.

Osherson, D. I., Kosslyn, S. M., & Hollerbach, J. M. (Eds.) 1990. *An invitation to cognitive science*, Vol. 2: *Visual cognition and action*. Cambridge, Mass.: MIT Press.

Paglia, C. 1990. *Sexual personae: Art and decadence from Nefertiti to Emily Dickinson*. New Haven: Yale University Press.

Paglia, C. 1992. *Sex, art, and American culture*. New York: Vintage.

Paglia, C. 1994. *Vamps and tramps*. New York: Vintage.

Paivio, A. 1971. *Imagery and verbal processes*. Hillsdale, N.J.: Erlbaum.

Papathomas, T. V., Chubb, C., Gorea, A., & Kowler, E. (Eds.) 1995. *Early vision and beyond*. Cambridge, Mass.: MIT Press.

Parker, S. T., Mitchell, R. W. & Boccia, M. L. (Eds.) 1994. *Self-awareness in animals and humans*. New York: Cambridge University Press.

Patai, D., & Koertge, N. 1994. *Professing feminism: Cautionary tales from the strange world of women's studies*. New York: Basic Books.

Pazzani, M. 1987. Explanation-based learning for knowledge-based systems. *International Journal of Man-Machine Studies, 26*, 413–433.

Pazzani, M. 1993. Learning causal patterns: Making a transition for data-driven to theory-driven learning. *Machine Learning, 11*, 173–194.

Pazzani, M., & Dyer, M. 1987. A comparison of concept identification in human learning and network learning with the Generalized Delta Rule. In *Proceedings of the 10th International Joint Conference on Artificial Intelligence (IJCAI-87)*. Los Altos, Calif.: Morgan Kaufmann.

Pazzani, M., & Kibler, D. 1993. The utility of knowledge in inductive learning. *Machine Learning, 9*, 57–94.

Pennisi, E. 1996. Biologists urged to retire Linnaeus. *Science, 273*, 181.

Penrose, R. 1989. *The emperor's new mind: Concerning computers, minds, and the laws of physics*. New York: Oxford University Press.

Penrose, R., & commentators. 1990. Précis and multiple book review of "The emperor's new mind." *Behavioral and Brain Sciences, 13*, 643–705.

Penrose, R. 1994. *Shadows of the mind: A search for the missing science of consciousness*. New York: Oxford University Press.

Pentland, A. P. 1990. Linear shape from shading. *International Journal of Computer Vision, 4*, 153–162.

Perkins, D. N. 1981. *The mind's best work*. Cambridge, Mass.: Harvard University Press.

Perky, C. W. 1910. An experimental study of imagination. *American Journal of Psychology, 21*, 422–452.

Perrett, D. I., May, K. A., & Yoshikawa, S. 1994. Facial shape and judgments of female attractiveness: Preferences for non-average. *Nature, 368*, 239–242.

Peterson, M. A., Kihlstrom, J. F., Rose, P. M., & Klisky, M. L. 1992. Mental images can be ambiguous: Reconstruals and reference-frame reversals. *Memory and Cognition, 20*, 107–123.

Pettigrew, J. D. 1972. The neurophysiology of binocular vision. *Scientific American*, August. Reprinted in R. Held & W. Richards (Eds.), 1976, *Recent progress in perception*. San Francisco: W. H. Freeman.

Pettigrew, J. D. 1974. The effect of visual experience on the development of stimulus specificity by kitten cortical neurons. *Journal of Physiology, 237*, 49–74.

Pfeiffer, R. 1988. Artificial intelligence models of emotion. In V. Hamilton, G. H. Bower, & N. H. Frijda (Eds.), *Cognitive perspectives on emotion and motivation*. Netherlands: Kluwer.

Piattelli-Palmarini, M. 1989. Evolution, selection, and cognition: From "learning" to parameter setting in biology and the study of language, *Cognition, 31*, 1–44.

Piattelli-Palmarini, M. 1994. *Inevitable illusions: How mistakes of reason rule our minds*. New York: Wiley.

Picard, R. W. 1995. Affective computing. MIT Media Laboratory Perceptual Computing Section Technical Report #321.

Pilbeam, D. 1992. What makes us human? In Jones, Martin, & Pilbeam, 1992.

Pinker, S. 1979. The representation of three-dimensional space in mental images. Unpublished Ph.D. dissertation, Harvard University.

Pinker, S. 1980. Mental imagery and the third dimension. *Journal of Experimental Psychology: General, 109*, 254–371.

Pinker, S. 1984a. *Language learnability and language development*. Cambridge, Mass.: Harvard University Press.

Pinker, S. (Ed.) 1984b. *Visual cognition*. Cambridge, Mass.: MIT Press.

Pinker, S. 1984c. Visual cognition: an introduction. *Cognition, 18*, 1–63. Reprinted in Pinker.

Pinker, S. 1988. A computational theory of the mental imagery medium. In Denis, Engelkamp, & Richardson, 1988.

Pinker, S. 1989. *Learnability and cognition: The acquisition of argument structure*. Cambridge, Mass.: MIT Press.

Pinker, S. 1990. A theory of graph comprehension. In R. Friedle (Ed.), *Artificial intelligence and the future of testing*. Hillsdale, N.J.: Erlbaum.

Pinker, S. 1991. Rules of language. *Science, 253*, 530–535.

Pinker, S. 1992. Review of Bickerton's "Language and species." *Language, 68*, 375–382.

Pinker, S. 1994. *The language instinct*. New York: HarperCollins.

Pinker, S. 1995. Beyond folk psychology (Review of J. A. Fodor's "The elm and the expert"). *Nature, 373*, 205.

Pinker, S., Bloom, P., & commentators. 1990. Natural language and natural selection. *Behavioral and Brain Sciences, 13*, 707–784.

Pinker, S., & Finke, R. A. 1980. Emergent two-dimensional patterns in images rotated in depth. *Journal of Experimental Psychology: Human Perception and Performance, 6*, 244–264.

Pinker, S., & Mehler, J. (Eds.) 1988. *Connections and symbols*. Cambridge, Mass.: MIT Press.

Pinker, S., & Prince, A. 1988. On language and connectionism: Analysis of a parallel distributed processing model of language acquisition. *Cognition, 28*, 73–193. Reprinted in Pinker & Mehler, 1988.

Pinker, S., & Prince, A. 1994. Regular and irregular morphology and the psychological status of rules of grammar. In S. D. Lima, R. L. Corrigan, & G. K. Iverson (Eds.), *The reality of linguistic rules*. Philadelphia: John Benjamins.

Pinker, S., & Prince, A. 1996. The nature of human concept: Evidence from an unusual source. *Communication and Cognition 29*, 307–361.

Pirenne, M. H. 1970. *Optics, painting, and photography*. New York: Cambridge University Press.

Plomin, R. 1989. Environment and genes: Determinants of behavior. *American Psychologist, 44*, 105–111.

Plomin, R., Daniels, D., & commentators. 1987. Why are children in the same family so different from one another? *Behavioral and Brain Sciences, 10*, 1–60.

Plomin, R., Owen, M. J., & McGuffin, P. 1994. The genetic basis of complex human behaviors. *Science, 264*, 1733–1739.

Poggio, G. F. 1995. Stereoscopic processing in monkey visual cortex: A review. In Papathomas et al., 1995.

Poggio, T. 1984. Vision by man and machine. *Scientific American*, April.

Poggio, T., & Edelman, S. 1991. A network that learns to recognize three-dimensional objects. *Nature, 343*, 263–266.

Poggio, T., & Girosi, F. 1990. Regularization algorithms for learning that are equivalent to multilayer networks. *Science, 247*, 978–982.

Pollack, J. B. 1990. Recursive distributed representations. *Artificial Intelligence, 46*, 77–105.

Pollard, J. L. 1993. The phylogeny of rationality. *Cognitive Science, 17*, 563–588.

Polti, G. 1921/1977. *The thirty-six dramatic situations*. Boston: The Writer, Inc.

Posner, M. I. 1978. *Chronometric explorations of mind*. Hillsdale, N.J.: Erlbaum.

Poundstone, W. 1988. *Labyrinths of reason: John von Neumann, game theory, and the puzzle of the bomb*. New York: Anchor.

Poundstone, W. 1992. *Prisoner's dilemma: Paradox, puzzles, and the frailty of knowledge*. New York: Anchor.

Prasada, S., & Pinker, S. 1993. Generalizations of regular and irregular morphological patterns. *Language and Cognitive Processes, 8*, 1–56.

Premack, D. 1976. *Intelligence in ape and man*. Hillsdale, N.J.: Erlbaum.

Premack, D. 1990. Do infants have a theory of self-propelled objects? *Cognition, 36*, 1–16.

Premack, D., & Premack, A. J. 1995. Intention as psychological cause. In Sperber, Premack, & Premack, 1995.

Premack, D., & Woodruff, G. 1978. Does a chimpanzee have a theory of mind? *Behavioral and Brain Sciences, 1*, 512–526.

Preuss, T. 1993. The role of the neurosciences in primate evolutionary biology: Historical commentary and prospectus. In R. D. E. MacPhee (Ed.), *Primates and their relatives in phylogenetic perspective*. New York: Plenum.

Preuss, T. 1995. The argument from animals to humans in cognitive neuroscience. In Gazzaniga, 1995.

Prince, A., & Pinker, S. 1988. Rules and connections in human language. *Trends in Neurosciences, 11*, 195–202. Reprinted in Morris, 1989.

Profet, M. 1992. Pregnancy sickness as adaptation: A deterrent to maternal ingestion of teratogens. In Barkow, Cosmides, & Tooby, 1992.

Proffitt, D. L., & Gilden, D. L. 1989. Understanding natural dynamics. *Journal of Experimental Psychology: Human Perception & Performance, 15*, 384–393.

Provine, R. R. 1991. Laughter: A stereotyped human vocalization. *Ethology, 89*, 115–124.

Provine, R. R. 1993. Laughter punctuates speech: Linguistic, social, and gender contexts of laughter. *Ethology, 95*, 291–298.

Provine, R. R. 1996. Laughter. *American Scientist, 84* (January–February), 38–45.

Pustejovsky, J. 1995. *The generative lexicon*. Cambridge, Mass.: MIT Press.

Putnam, H. 1960. Minds and machines. In S. Hook (Ed.), *Dimensions of mind: A symposium*. New York: New York University Press.

Putnam, H. 1975. The meaning of 'meaning.' In K. Gunderson (Ed.), *Language, mind, and knowledge*. Minneapolis: University of Minnesota Press.

Putnam, H. 1994. The best of all possible brains? (Review of R. Penrose's "Shadows of the mind.") *New York Times Book Review*, November 20, p. 7.

Pylyshyn, Z. 1973. What the mind's eye tells the mind's brain: A critique of mental imagery. *Psychological Bulletin, 80*, 1–24.

Pylyshyn, Z. W., & commentators. 1980. Computation and cognition: Issues in the foundations of cognitive science. *Behavioral and Brain Sciences, 3*, 111–169.

Pylyshyn, Z. W. 1984. *Computation and cognition: Toward a foundation for cognitive science.* Cambridge, Mass.: MIT Press.

Pylyshyn, Z. W. (Ed.) 1987. *The robot's dilemma: The frame problem in artificial intelligence.* Norwood, N.J.: Ablex.

Quine, W. V. O. 1969. Natural kinds. In W. V. O. Quine, *Ontological relativity and other essays.* New York: Columbia University Press.

Quinlan, P. 1992. *An introduction to connectionist modeling.* Hillsdale, N.J.: Erlbaum.

Rachman, S. 1978. *Fear and courage.* San Francisco: W. H. Freeman.

Raibert, M. H. 1990. Legged robots. In P. H. Winston & S. A. Shellard (Eds.), *Artificial intelligence at MIT: Expanding frontiers*, Vol. 2. Cambridge, Mass.: MIT Press.

Raibert, M. H., & Sutherland, I. E. 1983. Machines that walk. *Scientific American*, January.

Raiffa, H. 1968. *Decision analysis.* Reading, Mass.: Addison-Wesley.

Rakic, P. 1995a. Corticogenesis in human and nonhuman primates. In Gazzaniga, 1995.

Rakic, P. 1995b. Evolution of neocortical parcellation: the perspective from experimental neuroembryology. In Changeux & Chavaillon, 1995.

Ralls, K., Ballou, J., & Templeton, A. 1988. Estimates of the cost of inbreeding in mammals. *Conservation Biology, 2*, 185–193.

Ramachandran, V. S. 1988. Perceiving shape from shading. *Scientific American*, August.

Rapoport, A. 1964. *Strategy and conscience.* New York: Harper & Row.

Ratcliff, R. 1990. Connectionist models of recognition memory: Constraints imposed by learning and forgetting functions. *Psychological Review, 97*, 285–308.

Rayner, K. (Ed.) 1992. *Eye movements and visual cognition.* New York: Springer-Verlag.

Redish, E. 1994. The implications of cognitive studies for teaching physics. *American Journal of Physics, 62*, 796–803.

Reeve, H. K., & Sherman, P. W. 1993. Adaptation and the goals of evolutionary research. *Quarterly Review of Biology, 68*, 1–32.

Reiner, A. 1990. An explanation of behavior (Review of MacLean, 1990). *Science, 250*, 303–305.

Rey, G. 1983. Concepts and stereotypes. *Cognition, 15*, 237–262.

Richards, W. 1971. Anomalous stereoscopic depth perception. *Journal of the Optical Society of America, 61*, 410–414.

Ridley, Mark 1986. *The problems of evolution.* New York: Oxford University Press.

Ridley, Matt 1993. *The Red Queen: Sex and the evolution of human nature.* New York: Macmillan.

Rips, L. J. 1989. Similarity, typicality, and categorization. In S. Vosniadou & A. Ortony (Eds.), *Similarity and analogical reasoning.* New York: Cambridge University Press.

Rips, L. J. 1994. *The psychology of proof.* Cambridge, Mass.: MIT Press.

Rock, I. 1973. *Orientation and form.* New York: Academic Press.

Rock, I. 1983. *The logic of perception.* Cambridge, Mass.: MIT Press.

Rogers, A. R. 1994. Evolution of time preference by natural selection. *American Economic Review,* 84, 460–481.

Rosch, E. 1978. Principles of categorization. In E. Rosch & B. B. Lloyd (Eds.), *Cognition and categorization.* Hillsdale, N.J.: Erlbaum.

Rose, M. 1980. The mental arms race amplifier. *Human Ecology,* 8, 285–293.

Rose, S. 1978. Pre-Copernican sociobiology? *New Scientist,* 80, 45–46.

Rosenbaum, R. 1995. Explaining Hitler. *New Yorker,* May 1, pp. 50–70.

Rothbart, M. K. 1977. Psychological approaches to the study of humor. In Chapman and Foot, 1977.

Rozin, P. 1976. The evolution of intelligence and access to the cognitive unconscious. In J. M. Sprague & A. N. Epstein (Eds.), *Progress in psychobiology and physiological psychology.* New York: Academic Press.

Rozin, P. 1996. Towards a psychology of food and eating: From motivation to module to model to marker, morality, meaning, and metaphor. *Current Directions in Psychological Science,* 5, 18–24.

Rozin, P., & Fallon, A. 1987. A perspective on disgust. *Psychological Review,* 94, 23–41.

Rumelhart, D. E., Hinton, G. E., & Williams, R. J. 1986. Learning representations by back-propagating errors. *Nature,* 323, 533–536.

Rumelhart, D. E., & McClelland, J. L. 1986a. PDP models and general issues in cognitive science. In Rumelhart, McClelland, & the PDP Research Group, 1986.

Rumelhart, D. E., & McClelland, J. L. 1986b. On learning the past tenses of English verbs. Implicit rules or parallel distributed processing? In Rumelhart, McClelland, & the PDP Research Group, 1986.

Rumelhart, D., McClelland, J., & the PDP Research Group. 1986. *Parallel distributed processing: Explorations in the microstructure of cognition,* Vol. 1: *Foundations.* Cambridge, Mass.: MIT Press.

Ruse, M. 1986. Biological species: Natural kinds, individuals, or what? *British Journal of the Philosophy of Science,* 38, 225–242.

Russell, J. A. 1994. Is there universal recognition of emotion from facial expression? A review of cross-cultural studies. *Psychological Bulletin,* 115, 102–141.

Ryle, G. 1949. *The concept of mind.* London: Penguin.

Sacks, O., & Wasserman, R. 1987. The case of the colorblind painter. *New York Review of Books,* 34, 25–34.

Sanford, G. J. 1994. Straight lines in nature. Visalia, California, *Valley Voice,* November 2. Reprinted as "Nature's straight lines," *Harper's,* 289 (February 1995), 25.

Schacter, D. L. 1996. *Searching for memory: The brain, the mind, and the past.* New York: Basic Books.

Schanck, R. C. 1982. *Dynamic memory.* New York: Cambridge University Press.

Schanck, R. C., & Riesbeck, C. K. 1981. *Inside computer understanding: Five programs plus miniatures.* Hillsdale, N.J.: Erlbaum.

Schellenberg, E. G., & Trehub, S. E. 1996. Natural musical intervals: Evidence from infant listeners. *Psychological Science,* 7, 272–277.

Schelling, T. C. 1960. *The strategy of conflict.* Cambridge, Mass.: Harvard University Press.

Schelling, T. C. 1984. The intimate contest for self-command. In T. C. Schelling, *Choice and consequence: Perspectives of an errant economist.* Cambridge, Mass.: Harvard University Press.

Schutz, C. E. 1977. The psycho-logic of political humor. In Chapman & Foot, 1977.

Schwartz, S. P. 1979. Natural kind terms. *Cognition* 7, 301–315.

Searle, J. R., & commentators. 1980. Minds, brains, and programs. *The Behavioral and Brain Sciences, 3,* 417–457.

Searle, J. R. 1983. The word turned upside down. *New York Review of Books,* October 27, pp. 74–79.

Searle, J. R., & commentators. 1992. Consciousness, explanatory inversion, and cognitive science. *Behavioral and Brain Sciences, 13,* 585–642.

Searle, J. R. 1993. Rationality and realism: What is at stake? *Daedalus, 122,* 55–83.

Searle, J. R. 1995. The mystery of consciousness. *New York Review of Books,* November 2, pp. 60–66; November 16, pp. 54–61.

Segal, S., & Fusella, V. 1970. Influence of imaged pictures and sounds on detection of visual and auditory signals. *Journal of Experimental Psychology, 83,* 458–464.

Seligman, M. E. P. 1971. Phobias and preparedness. *Behavior Therapy, 2,* 307–320.

The Seville Statement on Violence. 1990. *American Psychologist, 45,* 1167–1168.

Shapiro, J. A. 1995. Adaptive mutation: Who's really in the garden? *Science, 268,* 373–374.

Shastri, L., Ajjanagadde, V., & commentators. 1993. From simple associations to systematic reasoning: A connectionist representation of rules, variables, and dynamic bindings using temporal synchrony. *Behavioral and Brain Sciences, 16,* 417–494.

Shepard, R. N. 1978. The mental image. *American Psychologist, 33,* 125–137.

Shepard, R. N. 1987. Toward a universal law of generalization for psychological science. *Science, 237,* 1317–1323.

Shepard, R. N. 1990. *Mind sights: Original visual illusions, ambiguities, and other anomalies.* New York: W. H. Freeman.

Shepard, R. N., & Cooper, L. A. 1982. *Mental images and their transformations.* Cambridge, Mass.: MIT Press.

Sherif, M. 1966. *Group conflict and cooperation: Their social psychology* London: Routledge & Kegan Paul.

Sherry, D. F., & Schacter, D. L. 1987. The evolution of multiple memory systems. *Psychological Review, 94,* 439–454.

Shimojo, S. 1993. Development of interocular vision in infants. In Simons, 1993.

Shostak, M. 1981. *Nisa: The life and words of a !Kung woman.* New York: Vintage.

Shoumatoff, A. 1985. *The mountain of names: A history of the human family.* New York: Simon & Schuster.

Shreeve, J. 1992. The dating game. *Discover,* September.

Shultz, T. R. 1977. A cross-cultural study of the structure of humor. In Chapman & Foot.

Shweder, R. A. 1994. "You're not sick, you're just in love": Emotion as an interpretive system. In Ekman & Davidson, 1994.

Simon, H. A. 1969. The architecture of complexity. In H. A. Simon, *The sciences of the artificial.* Cambridge, Mass.: MIT Press.

Simon, H. A., & Newell, A. 1964. Information processing in computer and man. *American Scientist, 52,* 281–300.

Simons, K. (Ed.) 1993. *Early visual development: Normal and abnormal*. New York: Oxford University Press.

Singh, D. 1993. Adaptive significance of female physical attractiveness: Role of waist-to-hip ratio. *Journal of Personality and Social Psychology, 65*, 293–307.

Singh, D. 1994. Ideal female body shape: Role of body weight and waist-to-hip ratio. *International Journal of Eating Disorders, 16*, 283–288.

Singh, D. 1995. Ethnic and gender consensus for the effect of waist-to-hip ratio on judgment of women's attractiveness. *Human Nature, 6*, 51–65.

Sinha, P., 1995. Perceiving and recognizing three-dimensional forms. Ph.D. dissertation, Department of Electrical Engineering and Computer Science, MIT.

Sinha, P., & Adelson, E. H. 1993a. Verifying the 'consistency' of shading patterns and 3D structures. In *Proceedings of the IEEE Workshop on Qualitative Vision, New York*. Los Alamitos, Calif.: IEEE Computer Society Press.

Sinha, P., & Adelson, E. H. 1993b. Recovering reflectance and illumination in a world of painted polyhedra. In *Proceedings of the Fourth International Conference on Computer Vision, Berlin*. Los Alamitos, Calif.: IEEE Computer Society Press.

Sloboda, J. A. 1985. *The musical mind: The cognitive psychology of music*. New York: Oxford University Press.

Smith, E. E., & Medin, D. L. 1981. *Categories and concepts*. Cambridge, Mass.: Harvard University Press.

Smith, E. E., Langston, C., & Nisbett, R. 1992. The case for rules in reasoning. *Cognitive Science, 16*, 1–40.

Smolensky, P., & commentators. 1988. On the proper treatment of connectionism. *Behavioral and Brain Sciences, 11*, 1–74.

Smolensky, P. 1990. Tensor product variable binding and the representation of symbolic structures in connectionist systems. *Artificial Intelligence, 46*, 159–216.

Smolensky, P. 1995. Reply: Constituent structure and explanation in an integrated connectionist/symbolic cognitive architecture. In C. MacDonald & G. MacDonald (Eds.), *Connectionism: Debates on Psychological Explanations*, Vol. 2. Cambridge, Mass.: Blackwell.

Sober, E. (Ed.) 1984a. *Conceptual issues in evolutionary biology*. Cambridge, Mass.: MIT Press.

Sober, E. 1984b. *The nature of selection: Evolutionary theory in philosophical focus*. Cambridge, Mass.: MIT Press.

Solso, R. 1994. *Cognition and the visual arts*. Cambridge, Mass.: MIT Press.

Sommers, C. H. 1994. *Who stole feminism?* New York: Simon & Schuster.

Sowell, T. 1995. *The vision of the anointed: Self-congratulation as a basis for social policy*. New York: Basic Books.

Spelke, E. 1995. Initial knowledge: Six suggestions. *Cognition, 50*, 433–447.

Spelke, E. S., Breinlinger, K., Macomber, J., & Jacobson, K. 1992. Origins of knowledge. *Psychological Review, 99*, 605–632.

Spelke, E. S., Phillips, A., & Woodward, A. L. 1995. Infants' knowledge of object motion and human action. In Sperber, Premack, & Premack, 1995.

Spelke, E., Vishton, P., & von Hofsten, C. 1995. Object perception, object-directed action, and physical knowledge in infancy. In Gazzaniga, 1995.

Sperber, D. 1982. Apparently irrational beliefs. In M. Hollis & S. Lukes (Eds.), *Rationality and relativism*. Cambridge, Mass.: Blackwell.

Sperber, D. 1985. Anthropology and psychology: Towards an epidemiology of representations. *Man, 20,* 73–89.

Sperber, D., Cara, F., & Girotto, V. 1995. Relevance theory explains the selection task. *Cognition, 57,* 31–95.

Sperber, D., Premack, D., & Premack, A. J. (Eds.) 1995. *Causal cognition.* New York: Oxford University Press.

Sperber, D., & Wilson, D. 1986. *Relevance: Communication and cognition.* Cambridge, Mass.: Harvard University Press.

Staddon, J. E. R. 1988. Learning as inference. In R. C. Bolles & M. D. Beecher (Eds.), *Evolution and learning.* Hillsdale, N.J.: Erlbaum.

Stenning, K., & Oberlander, J. 1995. A cognitive theory of graphical and linguistic reasoning: Logic and implementation. *Cognitive Science, 19,* 97–140.

Sterelny, K., & Kitcher, P. 1988. The return of the gene. *Journal of Philosophy, 85,* 339–361.

Stereogram. 1994. San Francisco: Cadence Books.

Stevens, A., & Coupe, P. 1978. Distortions in judged spatial relations. *Cognitive Psychology, 10,* 422–437.

Storr, A. 1992. *Music and the mind.* New York: HarperCollins.

Stringer, C. 1992. Evolution of early humans. In Jones, Martin, & Pilbeam, 1992.

Stryker, M. P. 1993. Retinal cortical development: Introduction. In Simons, 1993.

Stryker, M. P. 1994. Precise development from imprecise rules. *Science, 263,* 1244–1245.

Subbiah, I., Veltri, L., Liu, A., & Pentland, A. 1996. Paths, landmarks, and edges as reference frames in mental maps of simulated environments. CBR Technical Report 96–4, Cambridge Basic Research, Nissan Research & Development, Inc.

Sullivan, W. 1993. *We are not alone: The continuing search for extraterrestrial intelligence.* Revised edition. New York: Penguin.

Sulloway, F. J. 1995. Birth order and evolutionary psychology: A meta-analytic overview. *Psychological Inquiry, 6,* 75–80.

Sulloway, F. J. 1996. *Born to rebel: Family conflict and radical genius.* New York: Pantheon.

Superstereogram. 1994. San Francisco: Cadence Books.

Sutherland, S. 1992. *Irrationality: The enemy within.* London: Penguin.

Swisher, C. C., III, Rink, W. J., Antón, S. C., Schwarcz, H. P., Curtis, G. H., Surpijo, A., & Widiasmoro. 1996. Latest *Homo erectus* of Java: Potential contemporaneity with *Homo sapiens* in Southeast Asia. *Science, 274,* 1870–1874.

Symons, D. 1978. *Play and aggression: A study of rhesus monkeys.* New York: Columbia University Press.

Symons, D. 1979. *The evolution of human sexuality.* New York: Oxford University Press.

Symons, D., & commentators. 1980. Précis of "The evolution of human sexuality." *Behavioral and Brain Sciences, 3,* 171–214.

Symons, D. 1992. On the use and misuse of Darwinism in the study of human behavior. In Barkow, Cosmides, & Tooby, 1992.

Symons, D. 1993. The stuff that dreams aren't made of: Why wake-state and dream-state sensory experiences differ. *Cognition, 47,* 181–217.

Symons, D. 1995. Beauty is in the adaptations of the beholder: The evolutionary psychology of human female sexual attractiveness. In P. R. Abramson & S. D. Pinkerton (Eds.), *Sexual nature, sexual culture*. Chicago: University of Chicago Press.

Tajfel, H. 1981. *Human groups and social categories*. New York: Cambridge University Press.

Talmy, L. 1985. Lexicalization patterns: Semantic structure in lexical forms. In T. Shopen (Ed.), *Language typology and syntactic description*. Vol. III: *Grammatical categories and the lexicon*. New York: Cambridge University Press.

Talmy, L. 1988. Force dynamics in language and cognition. *Cognitive Science, 12,* 49–100.

Tan, E. S. 1996. *Emotion and the structure of narrative film*. Hillsdale, N.J.: Erlbaum.

Tarr, M. J. 1995. Rotating objects to recognize them: A case study on the role of viewpoint dependency in the recognition of three-dimensional shapes. *Psychonomic Bulletin and Review, 2,* 55–82.

Tarr, M. J., & Black, M. J. 1994a. A computational and evolutionary perspective on the role of representation in vision. *Computer Vision, Graphics, and Image Processing: Image Understanding, 60,* 65–73.

Tarr, M. J., & Black, M. J. 1994b. Reconstruction and purpose. *Computer Vision, Graphics, and Image Processing: Image Understanding, 60,* 113–118.

Tarr, M. J., & Bülthoff, H. H. 1995. Is human object recognition better described by geon-structural-descriptions or by multiple views? *Journal of Experimental Psychology: Human Perception and Performance, 21,* 1494–1505.

Tarr, M. J., & Pinker, S. 1989. Mental rotation and orientation-dependence in shape recognition. *Cognitive Psychology, 21,* 233–282.

Tarr, M. J., & Pinker, S. 1990. When does human object recognition use a viewer-centered reference frame? *Psychological Science, 1,* 253–256.

Thorn, F., Gwiazda, J., Cruz, A. A. V., Bauer, J. A., & Held, R. 1994. The development of eye alignment, convergence, and sensory binocularity in young infants. *Investigative Ophthalmology and Visual Science, 35,* 544–553.

Thornhill, N., & commentators. 1991. An evolutionary analysis of rules regulating human inbreeding and marriage. *Behavioral and Brain Sciences, 14,* 247–293.

Timney, B. N. 1990. Effects of brief monocular occlusion on binocular depth perception in the cat: A sensitive period for the loss of stereopsis. *Visual Neuroscience, 5,* 273–280.

Titchener, E. B. 1909. *Lectures on the experimental psychology of the thought processes*. New York: Macmillan.

Tooby, J. 1976a. The evolutionary regulation of inbreeding. Institute for Evolutionary Studies Technical Report 76(1), 1–87. University of California, Santa Barbara.

Tooby, J. 1976b. The evolutionary psychology of incest avoidance. Institute for Evolutionary Studies Technical Report 76(2), 1–92. University of California, Santa Barbara.

Tooby, J. 1982. Pathogens, polymorphism, and the evolution of sex. *Journal of Theoretical Biology, 97,* 557–576.

Tooby, J. 1985. The emergence of evolutionary psychology. In D. Pines (Ed.), *Emerging syntheses in science*. Santa Fe, N.M.: Santa Fe Institute.

Tooby, J. 1988. The evolution of sex and its sequelae. Ph.D. dissertation, Harvard University.

Tooby, J., & Cosmides, L. 1988. The evolution of war and its cognitive foundations.

Paper presented at the annual meeting of the Human Behavior and Evolution Society, Ann Arbor, Mich. Institute for Evolutionary Studies Technical Report 88–1. University of California, Santa Barbara.

Tooby, J., & Cosmides, L. 1989. Adaptation versus phylogeny: The role of animal psychology in the study of human behavior. *International Journal of Comparative Psychology*, 2, 105–118.

Tooby, J., & Cosmides, L. 1990a. The past explains the present: Emotional adaptations and the structure of ancestral environments. *Ethology and Sociobiology, 11*, 375–424.

Tooby, J., & Cosmides, L. 1990b. On the universality of human nature and the uniqueness of the individual: The role of genetics and adaptation. *Journal of Personality, 58*, 17–67.

Tooby, J., & Cosmides, L. 1992. Psychological foundations of culture. In Barkow, Cosmides, & Tooby, 1992.

Tooby, J., & Cosmides, L. 1993. Cognitive adaptations for threat, cooperation, and war. Plenary address, Annual Meeting of the Human Behavior and Evolution Society, Binghamton, New York, August 6.

Tooby, J., & Cosmides, L. 1996. Friendship and the Banker's Paradox: Other pathways to the evolution of adaptations for altruism. In J. Maynard Smith (Ed.), *Proceedings of the British Academy: Evolution of social behavior patterns in primates and man*. London: British Academy.

Tooby, J., & Cosmides, L. 1997. Ecological rationality and the multimodular mind: Grounding normative theories in adaptive problems. Unpublished manuscript, University of California, Santa Barbara.

Tooby, J., & DeVore, I. 1987. The reconstruction of hominid evolution through strategic modeling. In W. G. Kinzey (Ed.), *The evolution of human behavior: Primate models*. Albany, N.Y.: SUNY Press.

Treisman, A. 1988. Features and objects. *Quarterly Journal of Experimental Psychology, 40A*, 201–237.

Treisman, A., & Gelade, G. 1980. A feature-integration theory of attention. *Cognitive Psychology, 12*, 97–136.

Treisman, M. 1977. Motion sickness: An evolutionary hypothesis. *Science, 197*, 493–495.

Tributsch, H. 1982. *How life learned to live: Adaptation in nature*. Cambridge, Mass.: MIT Press.

Trinkaus, E. 1992. Evolution of human manipulation. In Jones, Martin, & Pilbeam, 1992.

Trivers, R. 1971. The evolution of reciprocal altruism. *Quarterly Review of Biology, 46*, 35–57.

Trivers, R. 1981. Sociobiology and politics. In E. White (Ed.), *Sociobiology and human politics*. Lexington, Mass.: D. C. Heath.

Trivers, R. 1985. *Social evolution*. Reading, Mass.: Benjamin/Cummings.

Turing, A. M. 1950. Computing machinery and intelligence. *Mind, 59*, 433–460.

Turke, P. W., & Betzig, L. L. 1985. Those who can do: Wealth, status, and reproductive success on Ifaluk. *Ethology and Sociobiology, 6*, 79–87.

Turner, M. 1991. *Reading minds: The study of English in the age of cognitive science*. Princeton: Princeton University Press.

Tversky, A., & Kahneman, D. 1974. Judgment under uncertainty: Heuristics and biases. *Science, 185*, 1124–1131.

Tversky, A., & Kahneman, D. 1983. Extensions versus intuitive reasoning: The conjunction fallacy in probability judgment. *Psychological Review, 90*, 293–315.

Tye, M. 1991. *The imagery debate*. Cambridge, Mass.: MIT Press.

Tyler, C. W. 1983. Sensory processing of binocular disparity. In C. M. Schor & K. J. Ciuffreda (Eds.), *Vergence eye movements: Basic and clinical aspects*. London: Butterworths.

Tyler, C. W. 1991. Cyclopean vision. In D. Regan (Ed.), *Vision and visual dysfunction*, Vol. 9: *Binocular vision*. New York: Macmillan.

Tyler, C. W. 1995. Cyclopean riches: Cooperativity, neurontropy, hysteresis, stereoattention, hyperglobality, and hypercyclopean processes in random-dot stereopsis. In Papathomas et al., 1995.

Ullman, S. 1984. Visual routines. *Cognition, 18*, 97–159. Reprinted in Pinker, 1984b.

Ullman, S. 1989. Aligning pictorial descriptions: An approach to object recognition. *Cognition, 32*, 193–254.

van den Berghe, P. F. 1974. *Human family systems: An evolutionary view*. Amsterdam: Elsevier.

Van Essen, D. C., & DeYoe, E. A. 1995. Concurrent processing in the primate visual cortex. In Gazzaniga, 1995.

Veblen, T. 1899/1994. *The theory of the leisure class*. New York: Penguin.

Wallace, B. 1984. Apparent equivalence between perception and imagery in the production of various visual illusions. *Memory and Cognition, 12*, 156–162.

Waller, N. G. 1994. Individual differences in age preferences in mates. *Behavioral and Brain Sciences, 17*, 578–581.

Wandell, B. A. 1995. *Foundations of vision*. Sunderland, Mass.: Sinauer.

Wason, P. 1966. Reasoning. In B. M. Foss (Ed.), *New horizons in psychology*. London: Penguin.

Wehner, R., & Srinivasan, M. V. 1981. Searching behavior of desert ants, genus *Cataglyphis* (*Formicidae*, Hymenoptera). *Journal of Comparative Physiology, 142*, 315–338.

Weiner, J. 1994. *The beak of the finch*. New York: Vintage.

Weinshall, D., & Malik, J. 1995. Review of computational models of stereopsis. In Papathomas et al., 1995.

Weisberg, R. 1986. *Creativity: Genius and other myths*. New York: Freeman.

Weisfeld, G. E. 1993. The adaptive value of humor and laughter. *Ethology and Sociobiology, 14*, 141–169.

Weizenbaum, J. 1976. *Computer power and human reason*. San Francisco: W. H. Freeman.

White, R. 1989. Visual thinking in the Ice Age. *Scientific American*, July.

Whitehead, B. D. 1994. The failure of sex education. *Atlantic Monthly, 274*, 55–61.

Whittlesea, B. W. A. 1989. Selective attention, variable processing, and distributed representation: Preserving particular experiences of general structures. In Morris, 1989.

Wierzbicka, A. 1994. Cognitive domains and the structure of the lexicon: The case of the emotions. In Hirschfeld & Gelman, 1994a.

Wilczek, F. 1994. A call for a new physics (Review of R. Penrose's "The emperor's new mind"). *Science, 266*, 1737–1738.

Wilford, J. N. 1985. *The riddle of the dinosaur*. New York: Random House.

Williams, G. C. 1966. *Adaptation and natural selection: A critique of some current evolutionary thought*. Princeton, N.J.: Princeton University Press.

Williams, G. C. 1992. *Natural selection: Domains, levels, and challenges*. New York: Oxford University Press.

Williams, G. C., & Williams, D. C. 1957. Natural selection of individually harmful social adaptations among sibs with special reference to social insects. *Evolution, 11*, 32–39.

Wilson, D. S., Sober, E., & commentators. 1994. Re-introducing group selection to the human behavior sciences. *Behavioral and Brain Sciences, 17*, 585–608.

Wilson, E. O. 1975. *Sociobiology: The new synthesis*. Cambridge, Mass.: Harvard University Press.

Wilson, E. O. 1994. *Naturalist*. Washington, D.C.: Island Press.

Wilson, J. Q. 1993. *The moral sense*. New York: Free Press.

Wilson, J. Q., & Herrnstein, R. J. 1985. *Crime and human nature*. New York: Simon & Schuster.

Wilson, M., & Daly, M. 1992. The man who mistook his wife for a chattel. In Barkow, Cosmides, & Tooby, 1992.

Wimmer, H., & Perner, J. 1983. Beliefs about beliefs: Representation and constraining function of wrong beliefs in young children's understanding of deception. *Cognition, 13*, 103–128.

Winograd, T. 1976. Towards a procedural understanding of semantics. *Revue Internationale de Philosophie, 117–118*, 262–282.

Wolfe, T. 1975. *The painted word*. New York: Bantam Books.

Wootton, R. J. 1990. The mechanical design of insect wings. *Scientific American*, November.

Wright, L. 1995. Double mystery. *New Yorker*, August 7, pp. 45–62.

Wright, R. 1988. *Three scientists and their gods: Looking for meaning in an age of information*. New York: HarperCollins.

Wright, R. 1994a. *The moral animal: Evolutionary psychology and everyday life*. New York: Pantheon.

Wright, R. 1994b. Feminists, meet Mr. Darwin. *New Republic*, November 28.

Wright, R. 1995. The biology of violence. *New Yorker*, March 13, pp. 67–77.

Wynn, K. 1990. Children's understanding of counting. *Cognition, 36*, 155 193.

Wynn, K. 1992. Addition and subtraction in human infants. *Nature, 358*, 749–750.

Yellen, J. E., Brooks, A. S., Cornelissen, E., Mehlman, M. J., & Steward, K. 1995. A Middle Stone Age worked bone industry from Katanda, Upper Semliki Valley, Zaire. *Science, 268*, 553–556.

Young, A. W., & Bruce, V. 1991. Perceptual categories and the computation of 'grandmother.' *European Journal of Cognitive Psychology, 3*, 5–49.

Young, L. R., Oman, C. M., Watt, D. G. D., Money, K. E., & Lichtenberg, B. K. 1984. Spatial orientation in weightlessness and readaptation to earth's gravity. *Science, 225*, 205–208.

Zahavi, A. 1975. Mate selection—A selection for a handicap. *Journal of Theoretical Biology, 53*, 205–214.

Zaitchik, D. 1990. When representations conflict with reality: The preschooler's problem with false beliefs and "false" photographs. *Cognition, 35*, 41–68.

Zentner, M. R., & Kagan, J. 1996. Perception of music by infants. *Nature, 383*, 29.

Zicree, M. S. 1989. *The Twilight Zone companion*. 2d ed. Hollywood: Silman-James Press.

CREDITS

INDEX

Page ranges in **boldface** indicate a section devoted to the topic.